Materials Horizons: From Nature to Nanomaterials

Series Editor

Vijay Kumar Thakur, School of Aerospace, Transport and Manufacturing, Cranfield University, Cranfield, UK

Materials are an indispensable part of human civilization since the inception of life on earth. With the passage of time, innumerable new materials have been explored as well as developed and the search for new innovative materials continues briskly. Keeping in mind the immense perspectives of various classes of materials, this series aims at providing a comprehensive collection of works across the breadth of materials research at cutting-edge interface of materials science with physics, chemistry, biology and engineering.

This series covers a galaxy of materials ranging from natural materials to nanomaterials. Some of the topics include but not limited to: biological materials, biomimetic materials, ceramics, composites, coatings, functional materials, glasses, inorganic materials, inorganic-organic hybrids, metals, membranes, magnetic materials, manufacturing of materials, nanomaterials, organic materials and pigments to name a few. The series provides most timely and comprehensive information on advanced synthesis, processing, characterization, manufacturing and applications in a broad range of interdisciplinary fields in science, engineering and technology.

This series accepts both authored and edited works, including textbooks, monographs, reference works, and professional books. The books in this series will provide a deep insight into the state-of-art of Materials Horizons and serve students, academic, government and industrial scientists involved in all aspects of materials research.

More information about this series at https://link.springer.com/bookseries/16122

Sreerag Gopi · Preetha Balakrishnan ·
Nabisab Mujawar Mubarak
Editors

Nanotechnology for Biomedical Applications

Springer

Editors
Sreerag Gopi
ADSO Naturals Private Limited
Bengaluru, India

Curesupport B V
Deventer, The Netherlands

Preetha Balakrishnan
ADSO Naturals Private Limited
Bengaluru, India

Curesupport B V
Deventer, The Netherlands

Nabisab Mujawar Mubarak
Petroleum and Chemical Engineering
Faculty of Engineering
Universiti Teknologi Brunei
Bandar Seri Begawan, Brunei

ISSN 2524-5384 ISSN 2524-5392 (electronic)
Materials Horizons: From Nature to Nanomaterials
ISBN 978-981-16-7482-2 ISBN 978-981-16-7483-9 (eBook)
https://doi.org/10.1007/978-981-16-7483-9

This Springer imprint is published by the registered company Springer Nature Singapore Pte Ltd.
The registered company address is: 152 Beach Road, #21-01/04 Gateway East, Singapore 189721, Singapore

Contents

About the Editors

Dr. Sreerag Gopi is a Chief Scientific Officer at Centre for Innovations and Technologies (CIT), ADSO Naturals Private Limited company, India. He also serves as Vice President (Research) at Curesupport Holding B.V, Netherlands. In research, Dr. Sreerag has published more than 50 journal papers, several conference proceedings and authored 15 book chapters. He has more than 7 years of total experience in industrial and academic research in chemistry, material science including nanocomposites for drug delivery, environmental remediation and bioprinting. His area of interests are synthesis and functionalization of nanoparticles from biopolymers, application in drug delivery and liposomal encapsulations. He is a recipient of the prestigious Erasmus Mundus fellowship by European Union during his Ph.D. period. Dr. Sreerag is a member of Royal Society of Chemistry, Member of Royal Australian Chemical Institute and Chartered Chemist at Royal chemical institute, Australia. He is a editor for several books published by Elsevier such as Handbook of Chitin and Chitosan 3 volumes and several on going projects in his hand with the collaboration of RSC, Springer, Apple Academic Press, Wiley and ACS.

Dr. Preetha Balakrishnan is the principal scientist, QA, QC ADSO naturals India and Curesupport Netherlands. She did her graduation in Chemistry from Calicut University Kerala India and postgraduation from Mahatma Gandhi University Kerala with Gold medal and first rank. She is a recipient of prestigious INSPIRE Fellowship from Government of India. She completed her Ph.D. in Chemistry from Mahatma Gandhi University under the guidance of prof. Thomas, vice-chancellor, a renowned scientist in this area. She is an outstanding scientist with sustained international acclaims for his work in Polymer Science and Engineering, Polymer Nanocomposites, Elastomers, Polymer Blends, Interpenetrating Polymer Networks, Polymer Membranes, Green Composites and Nanocomposites, Nanomedicine and Green Nanotechnology. She visited many foreign universities as a part of her research activities and published around 15 research articles, 20+ book chapters in peer reviewed international journals. She edited 10 books with leading publishers like Elsevier, Springer, Wiley, RSC across the globe. Dr. Preetha has received a number of National, international presentation award. She worked as a post-doctoral researcher in the research group of Prof. Thomas and did enormous works in biomaterials for tissue engineering. She also worked as a guest lecturer in Chemistry, at Department of Chemistry, Morning star Home science College Angamaly Kerala, India.

Dr. Nabisab Mujawar Mubarak is presently working as an Associate Professor in the Petroleum and Chemical Engineering Department, Faculty of Engineering, UTB. He has 15 years of working experience in academics and industry. His research interests include advanced carbon nanomaterials synthesis via microwave technology, graphene/CNT buckypaper for strain sensor application, biodiesel, biofuels, magnetic buckypaper, immobilisation of enzymes, protein purification, magnetic biochar production using microwaves, and wastewater treatment using advanced materials. He also serves as a scientific reviewer for numerous chemical engineering and nanotechnology journals. He is an editorial board member of Nature's Scientific Report, the Journal of Environment and Biotechnology Research, Acta Materialia Turcica, and Materials.

He has secured many internal and external grants from the Ministry of Higher Education (MoHE) under its Fundamental Research Grant Scheme (FRGS) as Co-Principal Investigator (Co-PI) for several projects and British council the UK respectively and completed the projects meeting their objectives. He has published more than 210 journal papers, 30 conference proceedings, 30 book chapters, and 9 Malaysian patents to his credits. He has also attended numerous international conferences and has been invited to be a key speaker at many national and international conferences. He is the only faculty in Curtin University who is twice the recipient of the Curtin Malaysia Most Productive Researcher Award (2020 and 2021). He also received numerous accolades for his research, including the Outstanding Faculty of Chemical Engineering Award (2018), Best Scientific Research Award London (2018), and Outstanding Scientist in Publication and Citation awarded by i-Proclaim Malaysia (2017). He also has the distinction of being listed in the top two per cent of the world's most influential scientists in the area of chemical and energy. The List of the Top 2% Scientists in the World compiled and published by Stanford University is based on their international scientific publications, a number of scientific citations for research, and participation in the review and editing of scientific research. He is also a Fellow Member of The Institution of Engineers Australia, a Chartered Professional Engineer (CPEng) of The Institution of Engineers Australia, a Chartered Chemical Engineer of the Institute of Chemical Engineering (IChemE), and a Fellow of the Higher Education Academy (FHEA), UK. He published 3 edited books and is a co-editor of 5 ongoing Elsevier edited books: (1) Sustainable Nanotechnology for Environmental Remediation, (2) Nanomaterials for Carbon Capture and Conversion Technique, (3) Advanced nanomaterials and nanocomposites for Bioelectrochemical Systems, (4) Green Mediated Synthesis-based Nanomaterials for Photocatalysis. and (5) Hybrid Nanotechnology for Sustainable Applications.

Chapter 1
Introduction to Biomedical Applications in Nanotechnology

S. Archana, Devi Radhika, K. Yogesh Kumar, S. B. Benaka Prasad, and R. Deepak Kasai

1 Introduction

Nanomaterials have a unique advantage in medical applications because of their smaller structures and increased surface area [1–3]. Appropriate selection of substrate is essential to achieving the desired multifunctional properties that many applications need [4, 5]. One of the most common elements in the Earth's crust is carbon. In different forms, carbon atoms bond with each other to form various carbon allotropes, to produce a set of carbon-based Nanocomposites. These include nanodiamonds [6, 7] carbon dots [8, 9], carbon nanotubes [10, 11] graphene and its derivatives [12–14]. Metal nanocomposites are the types of materials that comprise metal or alloy as the substrate in which specific nanosized material is grafted. These composites include metal-ceramic characteristics, for example, ZnO, TiO_2, SiO_2, and CeO_2 [15, 16]. Polymer nanocomposites are generally used for their easy manufacturing, flexible and wear resistance properties. In contrast to ceramic materials, they have certain restrictions, such as limited strength and modulus [17, 18]. In a broad array of applications, nanomaterials are considered to be the efficient ones. This is a product of its remarkable electrical, optical, photocatalytic, and biochemical properties, large specific surface area, effective bandgap, including more significant biochemical activity.

S. Archana · D. Radhika (✉) · K. Yogesh Kumar · S. B. Benaka Prasad · R. Deepak Kasai
Department of Chemistry, Faculty of Engineering and Technology, Jain Deemed-To-Be University, Bangalore, Karnataka, India

S. Gopi et al. (eds.), *Nanotechnology for Biomedical Applications*,
Materials Horizons: From Nature to Nanomaterials,
https://doi.org/10.1007/978-981-16-7483-9_1

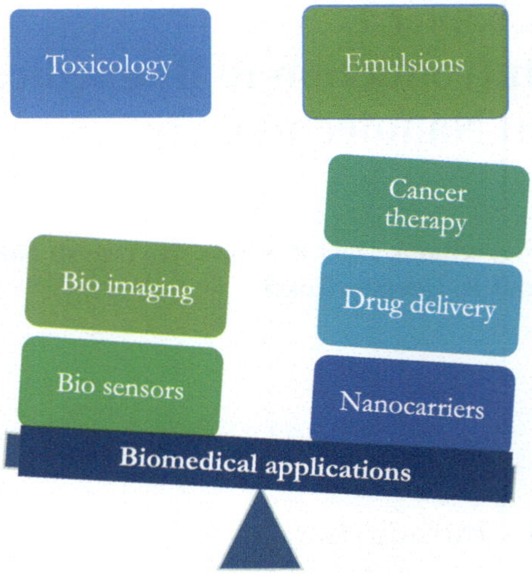

Fig. 1 An illustrafftion of some biomedical applications involving nanoparticles

2 Nanotechnology in Biomedical Applications

It is noted that many changes are needed for the transformation of a discipline from life science to technology and its applications. This development must involve Innovative design and simulation implementation, the potential to assess and evaluate, and extensive initiative in the advancement of technology. In life science, the functions and actions of a body, including cells, RNA, and DNA or Proteins, are on the nanoscale level. Thus the use of nanotechnology is the gateway to biotechnological progress [19]. Ultimately, nanotechnology can make it possible to transport and manipulate biomaterials and integrate them [20]. In biotechnology, nanotechnology has special applications, particularly in diagnosis and therapeutics [21]. It is possible to combine various biological compounds with nanomaterials using physicochemical processes and through particular biochemical reactions, such as Protein-protein interactions, antibody-antigen interactions [22]. The focus on using nanomaterials in biomedical applications, such as drug delivery, hypoxia, therapy, biosensors, and bioimaging, is increasingly gaining popularity [23] (Fig. 1).

3 Properties Involved in Biomedical Applications

3.1 Magnetic Property

The most frequently studied and widely applied material for biomedical applications are magnetic nanomaterials. Its effectiveness is attributed to specific structural, chemical, and magnetic properties like stability, non-poisonous, bioactivity, high magnetic flux [24]. These Magnetic nanoparticles are mostly comprised of Fe_3O_4, Fe_2O_4, Co-doped Fe_2O_4, and Mn-doped Fe_2O_4. These nanomaterials are most studied since they all have unique characteristics which are crucial for use in various medical applications, like selective delivery of drugs, bioimaging, magnetic hyperthermia, therapy, biosensors, and photoablation [25].

3.2 Optical Property

The optical properties of metal oxide nanoparticles are mainly focused on biomedical applications. Doped materials are excellent frequency converters covering the spectrum from ultraviolet (UV) through visible to near-infrared because of the distinctive electronic structure of transition metals. The probability of biomedical application is a further benefit of the optical approach, gaining the benefit of different absorption spectra. Metal oxide-based NPs attract growing attention as optical sensor indicators and therapeutic and diagnostic agents from the biotechnology, chemistry, optics, and biomedical community due to their optical properties [26, 27].

3.3 Surface Morphology

The selective behavior is encouraged by this surface arrangement through increasing the active spots, leading to biomolecules being spread across the surface. By reducing steric interference and improving accessibility to the binding sites, biomedical applications are carried out on this sort of surface. It relies on the fact that the thermodynamic and kinetic mechanisms of the surface active sites and analytes have the same order, allowing for more efficient biomedical behavior. Ultimately, when the surfaces are structured geometrically, this reflects unique optical and electric properties which are used in improving various applications like high functional bio-implants, efficient biosensors, biochips in neuronal computing, Medical diagnostics with accuracy, molecular separations, and biosynthesis [28].

4 Nanoparticles in Biomedical Application

4.1 Drug Delivery Systems

Lately, major developments have occurred in this area of drug carriers systems to deliver drugs to their specified location for treating the different health conditions. A number of new drug delivery technologies have been widely implemented. However, there are some issues to be resolved. Therefore the nano-based drug carriers which

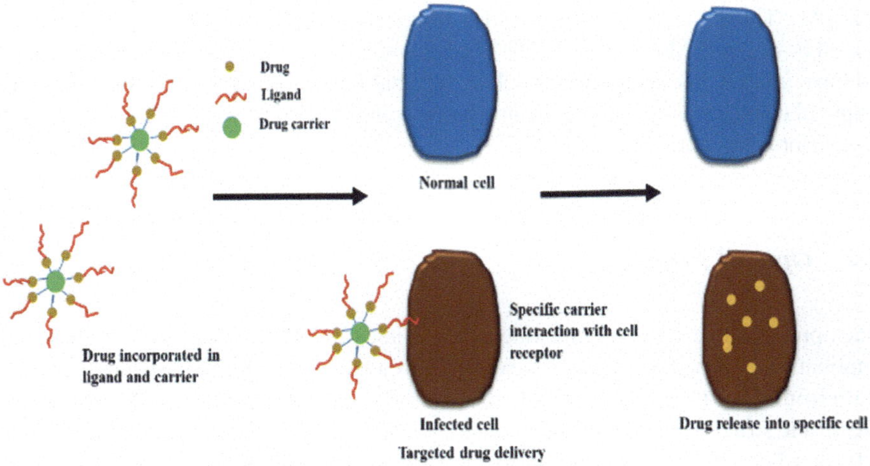

Fig. 2 Schematic representation of targeted drug delivery

will facilitate the highly developed drug delivery system are studied effectively [29]. This is due to its positive effects, like the capacity to change characteristics like solubility, patterns of drug release, diffusability, bioavailability, Suitable paths of administration, reduced toxic effects, lesser side effects, enhanced cellular uptake, and prolonged life cycle of medications and tolerability [30] (Fig. 2).

4.1.1 Polymeric Nanoparticles

Polymeric nanomaterials have more excellent biocompatibility with effective functional groups [31]. It's used in the binding or coating of nanomaterials of different sorts. Therefore, multiple nanoparticles of various functions are formed for effective use in the identification and treating the multiple forms of diseases. Xiaoping et al. [32] showed that a sequence of nanosized amphipathic cetirizine-chitosan polymer was efficiently used as a mucosal drug delivery system. In the presence of lysozyme, Cetirizine dihydrochloride (CedH): chitosan NPs demonstrated burst and persistent levels of drug release, with no major negative impacts on the body fluids [32]. Talitha et al. developed a chitosan film carrying PLGA nanoparticles packed with enhanced flavonoid fraction of *Cecropia glaziovii*. The result showed that the efficient chitosan nanocomposites were synthesized with an efficient capacity to overcome the less availability issue of EFF-Cg and proved as the potent delivery system in treating herpes infection [33]. Zhao et al. [34] discussed that the Glucose-sensitive polymer nanoparticles coupled with glucose oxidase, concanavalin A, and phenylboronic acid for self-controlled delivery of drugs which can give better control of blood level, and also delivers a precise dose of the medicine (e.g., insulin), Copying the pancreas' physiological control. GOD, Con A, or PBA [35]. Similarly, there are numerous polymeric nano drug carriers whose efficiency and the target application are listed in Table 1.

Table 1 Polymeric nano drug carriers with the target applications

Sl. No.	Polymer nanocomposite	Drug carried	Outcome	References
1	Tetraphenylethylene immobilized zirconium-based nanoscale coordination polymers	Curcumin	Promising medium for efficient delivery of drugs and continuous image analysis of fluorescence	[33]
2	Poly(lactic-co-glycolic acid)	Dox-HCl Dox-base	Enhanced hydrophilic drug miscibility in a hydrophobic PLA polymer will decrease the rate of discharge	[35]
3	NCPs, which consist of manganese ions (Mn^{2+}), as the metal connecting points, and dithiodiglycolic acid, as the organic bridging ligands	Doxorubicin	Enhanced in vivo inhibitory effects of tumor growth compared to free DOX	[34]
4	SiO2-PMAA-b-PNIPAM	Doxorubicin	DOX-loaded SiO2-PMAA-b-PNIPAM nanoparticles are Extremely effective towards Hela cells	[36]
5	Cellulose nanocrystals-HPG-HEBA	Epirubicin	Successfully accepted by cells, EPI nevertheless retains its biological activity for Attack of cancer cells	[37]
6	Oligo(ethylene glycol) methacrylate	Doxorubicin	Promoted drug release at pH 5.0, greater cellular uptake and cytotoxicity of Dox-loaded pH-sensitive micelles of PCL21-b-P(a-OEGMA)11 relative to the pH-insensitive analogs of PCL21-b-P(OEGMA)18	[38]
7	Polymer coated silica nanoparticles modified with guanidine containing co-polymers—γ-Fe_2O_3	Molsidomine	High capacity for drug loading because of the efficient electrostatic interactions of guanidine and molsidomine Which consists co-polymers	[39]

4.1.2 Metallic Nanoparticles

Metal-organic composites are flexible classes of hybrid materials made by metal comprising structures bound by organic linkers in three dimensions. A good number of different metallic nanoparticles with organic framework provides a wide range of properties that allow them to be useful in numerous applications including drug delivery. Due to several properties, like high pore size distribution and volume, they are becoming ideal for delivery of drugs. The capacity to modify organic linkers for stealth-tracking or biocompatibility, besides the higher carrying ability, indicates that changes can be made to metallic nanoparticles which are well developed to deliver drugs [40, 41].

Siamak et al. [42] synthesized carboxymethylcellulose/Zinc-based metal-organic framework/graphene oxide bio-nanocomposite to carry doxorubicin. The DOX release rate was considerably greater in the tumor cell pH 5 than in physiological conditions at pH 7.4. Also, the analysis suggests that DOX@CMC/MOF-5/GO exhibited substantial K562 cell cytotoxicity [42]. Wang et al. [43] showed that the mesoporous $FeSe_2$ hedgehogs can be tailored and used for tumor therapy using doxorubicin. Because of FeSe2 hedgehogs' powerful NIR-II photothermal activity, 1120 nm light irradiation into tumor cells leads to gelatin melting, regains the spiky structure, and thus promotes internalization of cells, this results in a particular aggregation in the tumor cells [43]. Zied et al. [44] Synthesized magnetic nanoparticles composed of iron oxide, 2-(2-methoxy)ethyl methacrylate (MEO2MA) and oligo(ethyleneglycol)methacrylate (OEGMA) for enhanced delivery of 100% doxorubicin after 52 h at 42 °C [44]. Milad et al. [45] stated that the prepared gold-iron oxide nanocomposites can be it will be used as a viable transport for Lipoic acid-curcumin (LA-CUR) a novel anticancer drug. Being a negatively charged carrier, studies showed a substantially increased cytotoxicity toward cancerous U87MGG in contrast to curcumin [45]. Carbon/calcium phosphate/Fe3O4 composite nanoparticles synthesized by Mingyu et al. can be rendered as a transverse relaxation (T2) contrast agent for MRI and when the cells are treated with carbon/CaP/Fe_3O_4, cell viability is as great as 95.6% demonstrating the composite NPs showed superior cytocompatibility [46]. There are diverse metallic nano drug carriers whose efficacy and target output are described in Table 2.

4.2 Biosensors

Nanobiotechnology implies methodologies that integrate nanomaterials or nanoparticles to build tools for biological processes as given in Fig. 3. As the active elements laid the groundwork for a major advance in the area, resulting in stable sensor devices, nanomaterials are integrated into the sensor applications. With their flexible surface chemistry, optoelectronic merits, the manufacturing processes, coupled with morphological characteristics, nanomaterials are by far the most frequently used in biomedical research [53, 54]. Usually, an electroanalytical biosensor comprises two main

Table 2 Metallic drug carriers with the target applications

Sl. No.	Metallic nanocomposite	Drug carried	Outcome	References
1	Gadolinium oxide-gold nanoclusters hybrid	Indocyanine green (ICG)	High loading capacity for the drug of 1.74 g/g	[47]
2	MIL-88A NPs composed of iron(III) and fumaric acid	Suberoyl bishydroxamic acid	strong therapeutic capacity without any early leakage when coated using exosome	[48]
3	UiO-66, a zirconium-based Metal–organic frame work	model cargo, RhB, and a corticosteroid, dex	UiO-66 NPs are a modern aerosol platform for a vast array of lung diseases, which include COPD, lung cancers and COVID-19, with possible targeted delivery	[49]
4	Multifunctional magnetite mesoporous silica nanoparticles	Tamoxifen	Research indicates that the highest biocompatibility of nanogels after 72 h is well above 80% viable cells	[50]
5	Zinc(II) metal–organic frameworks (Zn-MOFs)	5-FU and DOX	22.5% and 26.72% of DOX were released from the NPs after 12 and 24 h at pH 7.4, while 47.92% and 55.1% of the drug were released in the same time at pH 5.5, respectively	[51]
6	ZnO quantum dots	Doxorubicin	Could be fully biodegraded in the acidic environment, with almost 72% of DOX discharged after 80 h	[52]

sections. The analyte-recognizing biological factor in the sample. The segment of the detector that transforms the signal produced into a signal from biological activity, that can be calculated more effectively [55].

Qingzhou et al. synthesized In_2O_3 nanoribbon modified with the enzyme glucose oxidase, chitosan, and carbon nanotubes (SWCNTs) for glucose detection in various body fluids, such as sweat and saliva. This showed a mobility of $\sim$22 cm^2 V $-$ 1 s $-$ 1 in 0.1 $\times$ saline buffered using phosphate. It's been affixed on different surfaces, including watches and synthetic arms [56]. Hamed et al. [57] prepared a hydrogel by copolymerizing PEG linker and polyethylene glycol and to activate covalent cross-linking and gel formation, Eosin Y is taken as the photoinitiator. In order to efficiently facilitate the enzymatic reaction causing penicillin tracking down to 0.2 mM, the hydrogel-mediated activation of penicillinase was explained. To accomplish extremely accurate sensing, multiplexed surface modification was shown with penicillinase and acetylcholinesterase [57]. Montmorillonite clay was binded using PAMAM G2 dendrimers by Betul et al. and electrospinned using poly(vinyl) alcohol and pyranose oxidases. The identification limit was 0.7 μM glucose [58].

Samira et al. showed that iron (III) in the presence of 1, 10-phenanthroline detects hydroquinone and Catechol in the limits of 0.05 and 0.07 mg L-1. The linear dynamic range was 0.5–3.0 mg L−1 for both analytes [59].

4.3 Antibacterial Agents

Nanoparticles are usually able to interact with microbes as an effective antifungal and antibacterial agent. In recent years, the progress of nanotechnology has facilitated the discovery of new antibacterial drugs. In relation to traditional materials, as the size of materials reduces from micrometer to the nanometer scale, nanomaterials exhibit higher efficiency, like improved diffusivity, excessive material strength and chemical reactivity, and improved biological activities. Usually, through various forms of gram-negative and positive strains of bacteria, the antibacterial efficiency of nanoparticles is achieved [60]. This may be due to the occurrence of Reactive oxygen species generated, protein damage, DNA damage, Mutagenesis, Enzyme disruption, membrane damage, or destruction of electron transport [20] .

In order to battle pathogens, metals have been around since earlier times. Because of its wide inhibitory range towards microbes and pathogens, metal nanoparticles have gained increased curiosity as antimicrobial agents [61]. Qing et al. [62] proved that the synthesized copper nanoparticles damage Escherichia coli as high as 86.3 ± 0.2% within 12 h at the dosage of 100 μg/mL. The main explanation for the behavior is the production of oxygen radicals that destroys the constituents of the cell membrane and cytoplasm and inactivates lipid peroxidation and DNA damage [62]. By using a green synthesis technique, Tu Uyen et al. synthesized ZnO NPs using orange-peel extract as the reducing agent. The antimicrobial rate in the direction of *E. coli* was over 99.9%, while the bactericidal rate against Staphylococcus aureus in the relatively large range of 89–98% [63]. Dongdong et al. [64] synthesized remarkably effective antibacterial towards drug-resistant Escherichia coli (*E. coli*) and Staphylococcus aureus (*S. aureus*) are displayed by silver-decorated quercetin. Disruption of Nucleic acid assay presumed that the expression levels of DNA from both species steadily reduces with the concentrations of QA NP. Gene expression screening like RNA Seq is used to assess the sensing of toxicity pathways [64]. Shamkumar et al. [65] synthesized Ag NPs−PANI/MWCNT resulted in bacterial inactivation because of

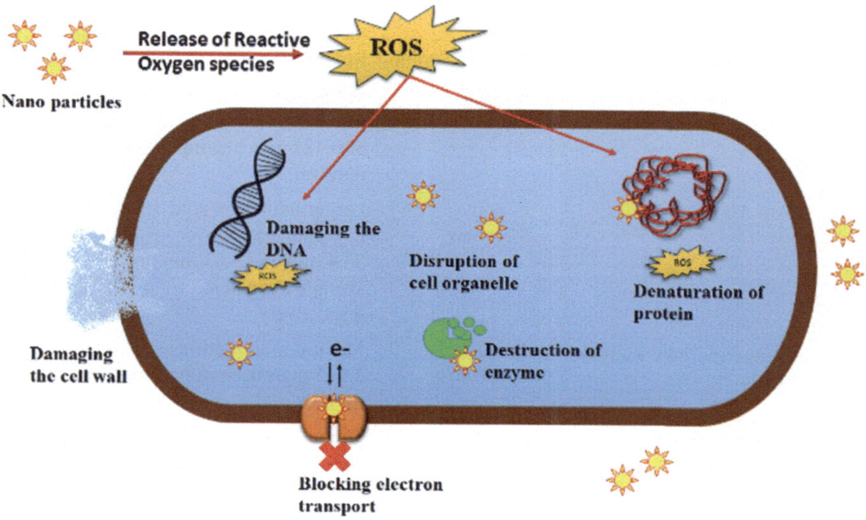

Fig. 3 Schematic diagram showing the mode of antibacterial activity

higher surface area of Ag NPs, and 1D MWCNT and acidic functional group of PANI [65].

Sl. No.	Nanomaterials	Bacterial strains	Report	References
1	Povidone-iodine nanoparticles	*E. coli* *S. aureus* *P. aeruginosa*	Iodine was mounted on P(NVP-MMA) NPs, with a contact period of 30 min displaying 100% elimination of E coli and S aureus	[66]
2	ZnO and CuO capped with polyvinyl alcohol, polyethylene glycol, and polyethylenimine	*E. coli* *S. aureus*	After 120 min of exposure, 99.9% bacterial destruction was exhibited by CuO-PEG and ZnO-PVA	[67]
3	Chitosan/Pd nanocomposites	*S. aureus* *B. anthracis* *B. subtilis* *B. cereus* *P. aeruginosa* *K. pneumoniae* *E. coli* *Proteus sp.*	Mic was recorded for CS/Pd-15%, i.e., 0.78, 1.56, 6.25, 0.78, 25, 50, 25, 0.78 μg/ml respectively	[68]
4	Ru(II) polypyridine complexes	*E. coli* *S. aureus* *Enterococcus*	16, 8, 16 μg/ml Mic were recorded respectively	[69]

(continued)

(continued)

Sl. No.	Nanomaterials	Bacterial strains	Report	References
5	Silver nanoparticles on mesoporous graphene	*E. coli*	When exposed for 2 h and showed an inhibition zone of 0.42 cm achieving 100% removal	[70]
6	Au–Ag NPs	*E. coli* *S. aureus*	Larger inhibition zone for *E. coli* and *S. aureus* 36.4 mm and 35.3 mm in average diameters, respectively	[71]
7	Tungsten oxide-graphene oxide	*E. coli* *B. subtilis*	Maximum inhibition at 2.5–5 mg/mL at irradiation for 6 h	[29]
8	Titanium dioxide	*E. coli* *S. aureus*	The minimum inhibitory concentration of 25 mg/mL^{-1} and 50 mg/mL^{-1} respectively	[72]
9	Au@Ag NPs	*E. coli* *S. aureus*	Minimum inhibitory concentration are 5 mg/mL^{-1} for *E. coli* and 7.5 mg/mL^{-1} for *S. aureus*	[73]

References

1. Abreu E (2006) Malsch NH, (editor): biomedical nanotechnology. Biomed Eng Online 5(1):1–2. https://doi.org/10.1186/1475-925x-5-20
2. Balcioglu M, Buyukbekar BZ, Yavuz MS, Yigit MV (2015) 73 Smart polymer functionalized graphene nano-devices for thermo-switch controlled biodetection. J Biomol Struct Dyn 33(sup1):47–48. https://doi.org/10.1080/07391102.2015.1032690
3. Zheng W, Huang P, Tu D, Ma E, Zhu H, Chen X (2015) Lanthanide-doped upconversion nano-bioprobes: electronic structures, optical properties, and biodetection. Chem Soc Rev 44(6):1379–1415. https://doi.org/10.1039/c4cs00178h
4. Manocha LM, Valand J, Patel N, Warrier A, Manocha S (2006) Nanocomposites for structural applications. Indian J Pure Appl Phys 44(2):135–142
5. Krasno S, Swathi K (2018) A review on types of nanocomposites and their applications. Int J Adv Res Ideas Innov Technol 4(6):235–236
6. Aversa R, Petrescu RVV, Apicella A, Petrescu FIT (2017) Nano-diamond hybrid materials for structural biomedical application. Am J Biochem Biotechnol 13(1):34–41. https://doi.org/10.3844/ajbbsp.2017.34.41
7. Simpson DA, Morrisroe E, McCoey JM, Lombard AH, Mendis DC, Treussart F, Hall LT, Petrou S, Hollenberg LCL (2018) Non-neurotoxic nanodiamond probes for intraneuronal temperature mapping. ArXiv 1–27
8. Anwar S, Ding H, Xu M, Hu X, Li Z, Wang J, Liu L, Jiang L, Wang D, Dong C, Yan M, Wang Q, Bi H (2019) Recent advances in synthesis, optical properties, and biomedical applications of carbon dots [Review-article]. ACS Appl Bio Mater 2(6):2317–2338. https://doi.org/10.1021/acsabm.9b00112

9. Su W, Wu H, Xu H, Zhang Y, Li Y, Li X, Fan L (2020) Carbon dots: a booming material for biomedical applications. Mater Chem Front 4(3):821–836. https://doi.org/10.1039/c9qm00658c
10. Alshehri R, Ilyas AM, Hasan A, Arnaout A, Ahmed F, Memic A (2016) Carbon nanotubes in biomedical applications: factors, mechanisms, and remedies of toxicity. J Med Chem 59(18):8149–8167. https://doi.org/10.1021/acs.jmedchem.5b01770
11. Sajid MI, Jamshaid U, Jamshaid T, Zafar N, Fessi H, Elaissari A (2016) Carbon nanotubes from synthesis to in vivo biomedical applications. Int J Pharm 501(1–2):278–299. https://doi.org/10.1016/j.ijpharm.2016.01.064
12. Banerjee AN (2018) Graphene and its derivatives as biomedical materials: future prospects and challenges. Interface Focus 8(3). https://doi.org/10.1098/rsfs.2017.0056
13. Shin YC, Song SJ, Hong SW, Jeong SJ, Chrzanowski W, Lee JC, Han DW (2017) Multifaceted biomedical applications of functional graphene nanomaterials to coated substrates, patterned arrays and hybrid scaffolds. Nanomaterials 7(11). https://doi.org/10.3390/nano7110369
14. Wang W, Su H, Wu Y, Zhou T, Li T (2019) Review—biosensing and biomedical applications of graphene: a review of current progress and future prospect. J Electrochem Soc 166(6):B505–B520. https://doi.org/10.1149/2.1231906jes
15. Jiang, J., Pi, J., & Cai, J. (2018). The advancing of zinc oxide nanoparticles for biomedical applications. Bioinorg Chem Appl 2018(2018), 18
16. McNamara K, Tofail SAM (2017) Nanoparticles in biomedical applications. Adv Phys X 2(1):54–88. https://doi.org/10.1080/23746149.2016.1254570
17. Kumar S, Raj S, Jain S, Chatterjee K (2016) Multifunctional biodegradable polymer nanocomposite incorporating graphene-silver hybrid for biomedical applications. Mater Des 108(June):319–332. https://doi.org/10.1016/j.matdes.2016.06.107
18. Swider E, Koshkina O, Tel J, Cruz LJ, de Vries IJM, Srinivas M (2018) Customizing poly(lactic-co-glycolic acid) particles for biomedical applications. Acta Biomater 73:38–51. https://doi.org/10.1016/j.actbio.2018.04.006
19. Chen CJ, Haik Y, Chatterjee J (2005) Development of nanotechnology for biomedical applications. Emerging Information Technology Conference 2005(2005):9–12. https://doi.org/10.1109/EITC.2005.1544329
20. Cavalcanti A, Shirinzadeh B, Freitas RA, Hogg T (2008) Nanorobot architecture for medical target identification. Nanotechnology 19(1). https://doi.org/10.1088/0957-4484/19/01/015103
21. Faraji AH, Wipf P (2009) Nanoparticles in cellular drug delivery. Bioorg Med Chem 17(8):2950–2962. https://doi.org/10.1016/j.bmc.2009.02.043
22. Castner DG, Ratner BD (2002) Biomedical surface science: Foundations to frontiers. Surf Sci 500(1–3). https://doi.org/10.1016/S0039-6028(01)01587-4
23. McNamara K, Tofail SAM (2015) Nanosystems: the use of nanoalloys, metallic, bimetallic, and magnetic nanoparticles in biomedical applications. Phys Chem Chem Phys 17(42):27981–27995. https://doi.org/10.1039/c5cp00831j
24. Karimi Z, Karimi L, Shokrollahi H (2013) Nano-magnetic particles used in biomedicine: core and coating materials. Mater Sci Eng, C 33(5):2465–2475. https://doi.org/10.1016/j.msec.2013.01.045
25. Nochehdehi AR, Thomas S, Sadri M, Afghahi SSS, Hadavi SM (2017) Iron oxide biomagnetic nanoparticles (IO-BMNPs); Synthesis, Characterization and biomedical application–a review. J Nanomed Nanotechnol 08(01). https://doi.org/10.4172/2157-7439.1000423
26. Hemmer E, Acosta-Mora P, Méndez-Ramos J, Fischer S (2017) Optical nanoprobes for biomedical applications: Shining a light on upconverting and near-infrared emitting nanoparticles for imaging, thermal sensing, and photodynamic therapy. J Mater Chem B 5(23):4365–4392. https://doi.org/10.1039/c7tb00403f
27. Obaid G, Broekgaarden M, Bulin AL, Huang HC, Kuriakose J, Liu J, Hasan T (2016) Photonanomedicine: a convergence of photodynamic therapy and nanotechnology. Nanoscale 8(25):12471–12503. https://doi.org/10.1039/c5nr08691d
28. Lisboa P, Valsesia A, Colpo P, Rossi F, Mascini M (2010) Nanopatterned surfaces for biodetection. Anal Lett 43(10–11):1556–1571. https://doi.org/10.1080/00032711003653916

29. Jeevitha G, Abhinayaa R, Mangalaraj D, Ponpandian N (2018) Tungsten oxide-graphene oxide (WO3-GO) nanocomposite as an efficient photocatalyst, antibacterial and anticancer agent. J Phys Chem Solids 116(December 2017):137–147. https://doi.org/10.1016/j.jpcs.2018.01.021
30. Patra JK, Das G, Fraceto LF, Campos EVR, Rodriguez-Torres MDP, Acosta-Torres LS, Diaz-Torres LA, Grillo R, Swamy MK, Sharma S, Habtemariam S, Shin HS (2018) Nano based drug delivery systems: recent developments and future prospects 10 technology 1007 nanotechnology 03 chemical sciences 0306 physical chemistry (incl. Structural) 03 chemical sciences 0303 macromolecular and materials chemistry 11 medical and he. J Nanobiotechnol 16(1):1–33. https://doi.org/10.1186/s12951-018-0392-8
31. Swierczewska M, Han HS, Kim K, Park JH, Lee S (2016) Polysaccharide-based nanoparticles for theranostic nanomedicine. Adv Drug Deliv Rev 99:70–84. https://doi.org/10.1016/j.addr.2015.11.015
32. Yu X, Mu Y, Xu M, Xia G, Wang J, Liu Y, Chen X (2017) Preparation and characterization of mucosal adhesive and two-step drug releasing cetirizine-chitosan nanoparticle. Carbohyd Polym 173:600–609. https://doi.org/10.1016/j.carbpol.2017.05.067
33. Wang L, Wang W, Xie Z (2016) Tetraphenylethylene-based fluorescent coordination polymers for drug delivery. J Mater Chem B 4(24):4263–4266. https://doi.org/10.1039/c6tb00952b
34. Zhao J, Yang Y, Han X, Liang C, Liu J, Song X, Ge Z, Liu Z (2017) Redox-sensitive nanoscale coordination polymers for drug delivery and cancer theranostics. ACS Appl Mater Interfaces 9(28):23555–23563. https://doi.org/10.1021/acsami.7b07535
35. Yuan Y, Choi K, Choi SO, Kim J (2018) Early stage release control of an anticancer drug by drug-polymer miscibility in a hydrophobic fiber-based drug delivery system. RSC Adv 8(35):19791–19803. https://doi.org/10.1039/c8ra01467a
36. Zheng Y, Wang L, Lu L, Wang Q, Benicewicz BC (2017) PH and thermal dual-responsive nanoparticles for controlled drug delivery with high loading content. ACS Omega 2(7):3399–3405. https://doi.org/10.1021/acsomega.7b00367
37. Wan W, Ouyang H, Long W, Yan W, He M, Huang H, Yang S, Zhang X, Feng Y, Wei Y (2019) Direct surface functionalization of cellulose nanocrystals with hyperbranched polymers through the anionic polymerization for pH-responsive intracellular drug delivery. ACS Sustain Chem Eng 7(23):19202–19212. https://doi.org/10.1021/acssuschemeng.9b05231
38. Zheng L, Zhang X, Wang Y, Liu F, Peng J, Zhao X, Yang H, Ma L, Wang B, Chang C, Wei H (2018) Fabrication of acidic pH-cleavable polymer for anticancer drug delivery using a dual functional monomer [Research-article]. Biomacromol 19(9):3874–3882. https://doi.org/10.1021/acs.biomac.8b01001
39. Timin AS, Khashirova SY, Rumyantsev EV, Goncharenko AA (2016) Magnetic silica hybrids modified with guanidine containing co-polymers for drug delivery applications. Mater Sci Eng, C 64:20–28. https://doi.org/10.1016/j.msec.2016.03.057
40. Abánades Lázaro I, Forgan RS (2019) Application of zirconium MOFs in drug delivery and biomedicine. Coord Chem Rev 380:230–259. https://doi.org/10.1016/j.ccr.2018.09.009
41. Simon-Yarza T, Mielcarek A, Couvreur P, Serre C (2018) Nanoparticles of metal-organic frameworks: on the road to in vivo efficacy in biomedicine. Adv Mater 30(37):1–15. https://doi.org/10.1002/adma.201707365
42. Javanbakht S, Pooresmaeil M, Namazi H (2018) Green one-pot synthesis of carboxymethylcellulose/Zn-based metal-organic framework/graphene oxide bio-nanocomposite as a nanocarrier for drug delivery system.
43. Wang J, Zhou J, Xu D, Li J, Deng D (2020) Tailoring viruslike mesoporous $FeSe_2$ hedgehogs for controlled drug delivery and synergistic tumor suppression. ACS Appl Mater Interfaces 12(42):47197–47207. https://doi.org/10.1021/acsami.0c10888
44. Ferjaoui Z, Jamal Al Dine E, Kulmukhamedova A, Bezdetnaya L, Soon Chang C., Schneider R, Mutelet F, Mertz D, Begin-Colin S, Quilès F, Gaffet E, Alem H (2019) Doxorubicin-loaded thermoresponsive superparamagnetic nanocarriers for controlled drug delivery and magnetic hyperthermia applications. ACS Appl Mater Interfaces 11(34):30610–30620. https://doi.org/10.1021/acsami.9b10444

45. Ghorbani M, Bigdeli B, Jalili-baleh L, Baharifar H., Akrami M, Dehghani S, Goliaei B, Amani A, Lotfabadi A, Rashedi H, Haririan I, Alam NR, Hamedani MP, Khoobi M (2018) Curcumin-lipoic acid conjugate as a promising anticancer agent on the surface of gold-iron oxide nanocomposites: a pH-sensitive targeted drug delivery system for brain cancer theranostics. Eur J Pharm Sci 114(May 2020):175–188. https://doi.org/10.1016/j.ejps.2017.12.008
46. Gou M, Li S, Zhang L, Li L, Wang C, Su Z (2016) Facile one-pot synthesis of carbon/calcium phosphate/Fe3O4 composite nanoparticles for simultaneous imaging and pH/NIR-responsive drug delivery. Chem Commun 52(74):11068–11071. https://doi.org/10.1039/c6cc05515j
47. Han L, Xia JM, Hai X, Shu Y, Chen XW, Wang JH (2017) Protein-stabilized gadolinium oxide-gold nanoclusters hybrid for multimodal imaging and drug delivery. ACS Appl Mater Interfaces 9(8):6941–6949. https://doi.org/10.1021/acsami.7b00246
48. Illes B, Hirschle P, Barnert S, Cauda V, Wuttke S, Engelke H (2017) Exosome-coated metal-organic framework nanoparticles: an efficient drug delivery platform. Chem Mater 29(19):8042–8046. https://doi.org/10.1021/acs.chemmater.7b02358
49. Jarai BM, Stillman Z, Attia L, Decker GE, Bloch ED, Fromen CA (2020) Evaluating UiO-66 metal-organic framework nanoparticles as acid-sensitive carriers for pulmonary drug delivery applications. ACS Appl Mater Interfaces 12(35):38989–39004. https://doi.org/10.1021/acsami.0c10900
50. Keshavarz H, Khavandi A, Alamolhoda S, Naimi-Jamal MR (2020) PH-Sensitive magnetite mesoporous silica nanocomposites for controlled drug delivery and hyperthermia. RSC Adv 10(64):39008–39016. https://doi.org/10.1039/d0ra06916g
51. Liu W, Pan Y, Xiao W, Xu H, Liu D, Ren F, Peng X, Liu J (2019) Recent developments on zinc(ii) metal–organic framework nanocarriers for physiological pH-responsive drug delivery. MedChemComm 10(12):2038–2051. https://doi.org/10.1039/c9md00400a
52. Cai X, Luo Y, Zhang W, Du D, Lin Y (2016) PH-Sensitive ZnO quantum dots-doxorubicin nanoparticles for lung cancer targeted drug delivery. ACS Appl Mater Interfaces 8(34):22442–22450. https://doi.org/10.1021/acsami.6b04933
53. Li J, Zhu Z, Zhu B, Ma Y, Lin B, Liu R, Song Y, Lin H, Tu S, Yang C (2016) Surface-enhanced raman scattering active plasmonic nanoparticles with ultrasmall interior nanogap for multiplex quantitative detection and cancer cell imaging. Anal Chem 88(15):7828–7836. https://doi.org/10.1021/acs.analchem.6b01867
54. Zhang Y, Wei Q (2016) The role of nanomaterials in electroanalytical biosensors: a mini review. J Electroanal Chem 781:401–409. https://doi.org/10.1016/j.jelechem.2016.09.011
55. Wang J (2012) Electrochemical biosensing based on noble metal nanoparticles. Microchim Acta 177(3–4):245–270. https://doi.org/10.1007/s00604-011-0758-1
56. Liu Q, Liu Y, Wu F, Cao X, Li Z, Alharbi M, Abbas AN, Amer MR, Zhou C (2018) Highly sensitive and wearable In2O3 nanoribbon transistor biosensors with integrated on-chip gate for glucose monitoring in body fluids. ACS Nano 12(2):1170–1178. https://doi.org/10.1021/acsnano.7b06823
57. Bay HH, Vo R, Dai X, Hsu HH, Mo Z, Cao S, Li W, Omenetto FG, Jiang X (2019) Hydrogel gate graphene field-effect transistors as multiplexed biosensors [Rapid-communication]. Nano Lett 19(4):2620–2626. https://doi.org/10.1021/acs.nanolett.9b00431
58. Unal B, Yalcinkaya EE, Demirkol DO, Timur S (2018) An electrospun nanofiber matrix based on organo-clay for biosensors: PVA/PAMAM-Montmorillonite. Appl Surf Sci 444:542–551. https://doi.org/10.1016/j.apsusc.2018.03.109
59. Boroumand S, Arab Chamjangali M, Bagherian G (2019) An asymmetric flow injection determination of hydroquinone and catechol: an analytic hierarchy and artificial neural network approach. Meas J Int Meas Confederation 139:454–466. https://doi.org/10.1016/j.measurement.2019.03.025
60. Shi LE, Li ZH, Zheng W, Zhao YF, Jin YF, Tang ZX (2014) Synthesis, antibacterial activity, antibacterial mechanism and food applications of ZnO nanoparticles: a review. Food Addit Contam Part A Chem Anal Control Exposure Risk Assess 31(2):173–186. https://doi.org/10.1080/19440049.2013.865147

61. Prabhu S, Poulose EK (2012) Silver nanoparticles: mechanism of antimicrobial action, synthesis, medical applications, and toxicity effects. Int Nano Lett 2(1):1–10. https://doi.org/10.1186/2228-5326-2-32
62. Lv Q, Zhang B, Xing X, Zhao Y, Cai R, Wang W, Gu Q (2018) Biosynthesis of copper nanoparticles using Shewanella loihica PV-4 with antibacterial activity: Novel approach and mechanisms investigation. J Hazard Mater 347(2010):141–149. https://doi.org/10.1016/j.jhazmat.2017.12.070
63. Doan Thi TU, Nguyen TT, Thi YD, Ta Thi KH, Phan BT, Pham KN (2020) Green synthesis of ZnO nanoparticles using orange fruit peel extract for antibacterial activities. RSC Adv 10(40):23899–23907. https://doi.org/10.1039/d0ra04926c
64. Sun D, Zhang W, Mou Z, Chen Y, Guo F, Yang E, Wang W (2017) Transcriptome analysis reveals silver nanoparticle-decorated quercetin antibacterial molecular mechanism. ACS Appl Mater Interfaces 9(11):10047–10060. https://doi.org/10.1021/acsami.7b02380
65. Deshmukh SP, Dhodamani AG, Patil SM, Mullani SB, More KV, Delekar SD (2020) Interfacially interactive ternary silver-supported polyaniline/multiwalled carbon nanotube nanocomposites for catalytic and antibacterial activity. ACS Omega 5(1):219–227. https://doi.org/10.1021/acsomega.9b02526
66. Gao T, Fan H, Wang X, Gao Y, Liu W, Chen W, Dong A, Wang YJ (2017) Povidone-iodine-based polymeric nanoparticles for antibacterial applications. ACS Appl Mater Interfaces 9(31):25738–25746. https://doi.org/10.1021/acsami.7b05622
67. Nagvenkar AP, Perelshtein I, Piunno Y, Mantecca P, Gedanken A (2019) sonochemical one-step synthesis of polymer-capped metal oxide nanocolloids: antibacterial activity and cytotoxicity. ACS Omega 4(9):13631–13639. https://doi.org/10.1021/acsomega.9b00181
68. Dhanavel S, Manivannan N, Mathivanan N, Gupta VK, Narayanan V, Stephen A (2018) Preparation and characterization of cross-linked chitosan/palladium nanocomposites for catalytic and antibacterial activity. J Mol Liq 257(2017):32–41. https://doi.org/10.1016/j.molliq.2018.02.076
69. Sun W, Boerhan R, Tian N, Feng Y, Lu J, Wang X, Zhou Q (2020) Fluorination in enhancing photoactivated antibacterial activity of Ru(ii) complexes with photo-labile ligands. RSC Adv 10(42):25364–25369. https://doi.org/10.1039/d0ra01806f
70. Yang L, Meng F, Qu X, Xia L, Huang F, Qin S, Zhang M, Xu F, Sun L, Liu H (2019) Multiple-twinned silver nanoparticles supported on mesoporous graphene with enhanced antibacterial activity. Carbon 155:397–402. https://doi.org/10.1016/j.carbon.2019.09.002
71. Li Q, Lu F, Ye H, Yu K, Lu B, Bao R, Xiao Y, Dai F, Lan G (2018) Silver inlaid with gold nanoparticles: enhanced antibacterial ability coupled with the ability to visualize antibacterial efficacy. ACS Sustain Chem Eng 6(8):9813–9821. https://doi.org/10.1021/acssuschemeng.8b00931
72. Hunagund SM, Desai VR, Kadadevarmath JS, Barretto DA, Vootla S, Sidarai AH (2016) Biogenic and chemogenic synthesis of TiO_2 NPs: via hydrothermal route and their antibacterial activities. RSC Adv 6(99):97438–97444. https://doi.org/10.1039/c6ra22163g
73. Yang L, Yan W, Wang H, Zhuang H, Zhang J (2017) Shell thickness-dependent antibacterial activity and biocompatibility of gold@silver core-shell nanoparticles. RSC Adv 7(19):11355–11361. https://doi.org/10.1039/c7ra00485k

Chapter 2
Lipid Nanocarriers: Applications in Biomedical Research and in Drug Delivery

Sujata Maurya, Manish Kumar Mishra, Brijesh Rathi, and Dhruv Kumar

1 Introduction

Lipid nanocarriers are the most advanced non-viral drug delivery systems. They are called nanocarriers, because of their size which is about few nanometres only. Nowadays, it is no doubt that nanoformulations are of extreme advantage in the arena of pharmaceutics. Lipid nanocarriers have become indispensable for use as drug delivery systems because of their complete biocompatibility and nontoxic nature [1]. There are numerous studies proving the safety and high efficacy of lipid nanoparticles in the fields of pharmacology, diagnostics, nutraceuticals, etc. Such studies have been the impetus in further research and development into this arena of nanoscience. Solid lipid nanocarriers (SLN) were the first generation nanocarriers. There were many advantages of SLN, but since the SLN is formed of a crystalline solid so it has a capacity to form gel, low incorporation rate. SLNs could not deliver the drug efficiently to the target site [2]. Due to the inefficacy of SLNs, NSL (nanostructured lipid carriers) were formulated. To overcome the disadvantage of solid lipid nanoparticles these lipid nanocarriers are formed of solid and liquid lipids. This possibility of drug incorporation in the lipid nanocarriers is a new technique which is highly advantageous and bio risk free. Oral administration of lipid nanocarrier based drugs

S. Maurya · D. Kumar (✉)
Amity Institute of Molecular Medicine & Stem Cell Research (AIMMSCR), Amity University Uttar Pradesh, J3-112, Sec-125, Noida, Uttar Pradesh 201313, India
e-mail: dkumar13@amity.edu

M. K. Mishra
Environmental Monitoring & Assessment Division, Bhabha Atomic Research Centre, Mumbai, India

B. Rathi
Laboratory for Translational Chemistry and Drug Discovery, Department of Chemistry, Hansraj College, University of Delhi, New Delhi, India

S. Gopi et al. (eds.), *Nanotechnology for Biomedical Applications*,
Materials Horizons: From Nature to Nanomaterials,
https://doi.org/10.1007/978-981-16-7483-9_2

is also possible, other traditional forms of sachets, tablets, etc. are also available. The major application of the ultra-deformable nanovesicles is their ability to transfer the drugs across the natural mammalian skin barrier. Various unstable proteins, peptides, drugs, and vaccines are transferred efficiently [3] (Figs. 1 and 2).

The major challenge faced by these lipid nanocarriers is that they cannot be administered by the parenteral route of drug administration because these are then recognized as foreign by the cells of reticuloendothelial system. This challenge can be overcome only if the size of nanoparticle is even smaller than 200-micron meters because these size nanocarriers are not treated as non self by the cells of RES.

Solid lipid nanocarriers (SLN) and Nanostructured lipid carriers (NSL)

Solid Lipid Nanoparticles (SLN)	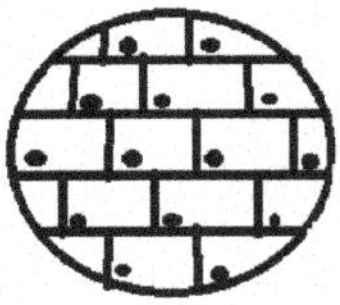	• SLN is a perfect crystal lattice structure. • There is less space for accommodation of drug inside the lipid core, resulting in the less drug loading and expulsion of drug out of the system.

Nanostructured Lipid Carriers (NSL)

NLC Type I		• It is an imperfect crystal core. • More space is available for drug accommodation inside the lipid core. • Hence, higher drug loading is possible and reduced/no possibility of drug expulsion from core.
NLC Type II	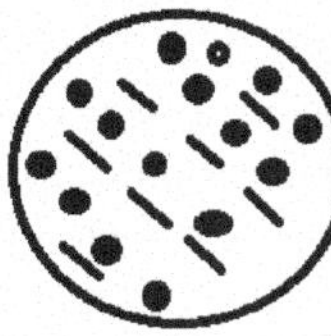	• This type is also known as structure less type. • Instead of conversion into a crystalline structure, solid lipids incorporated into this get converted into an amorphous form.
NLC Type III	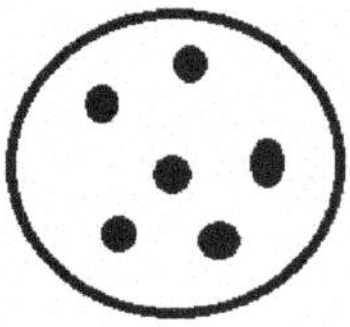	• This is multiple model known as O/F/W model. • Drugs having higher solubility in liquid lipids/ oils than solid lipid can be formulated into this type. • It can be prepared by phase separation method. • Drug is present in the dissolved state inside tiny oil droplets and uniformly distributed in the solid core.

Fig. 1 Shows the structures of SLNs and different types of NLCs [3]

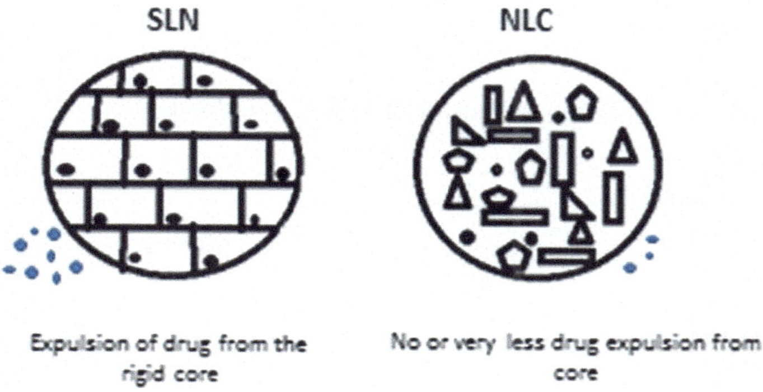

Fig. 2 Stability profile of SLNs and NLCs [5]

There remains a lot of research that must be carried out further in this field of nanotherapeutics [4].

The main advantages of lipid nanocarriers over the conventional carriers are as follows [6]:

1. Can store and extendedly release the target drug
2. Can store and release the target drug in an efficient and a stable manner
3. Both lipid soluble and water-soluble drug carriers
4. Most of the lipids used in formulation of lipid nanocarriers are biocompatible and non-allergens
5. Their production can be easily upscaled and are very easy to sterilize also.

Various lipid molecules interact with each other leading to formation of lipid-based nanostructures, which have no nonspecific interaction with other biomolecules, which in turn makes them a promising model for use in human body systems [7]. Lipid nanocarriers are one of the devices which have resulted out due to another revolution in the field of nanotechnology (Table 1).

Table 1 Differentiating parameters of SLNs and NLCs (Salvi and Pawar, 2019)

S. No.	Parameters	Solid lipid nanoparticles	Nanostructured lipid
1	Nature of lipids	Solid	Blend of solid and liquid lipids
2	Possible drug accommodation	Low	High
3	Degree of crystallinity	Higher (ordered matrix)	Lower (Amorphous/imperfect crystalline matrix)
4	Drug escape from matrix in dispersion media	Comparatively higher	Lower
5	Stability	Lower	Comparatively Higher

1.1 Classification

Different types of lipid nanocarriers are classified as follows:

1. Liposome
2. Transferosomes
3. Ethosomes

2 Liposome

2.1 Introduction

The first demonstration of liposome preparation was given by Prof. Alec Bangham [8], in Babrahm Institute, Cambridge, England in 1965. Since their discovery, the liposomes have been used as drug and pharmaceutical carriers (Fig 1). The liposomes consist of a central aqueous space (03–10 micrometer in diameter) surrounded by lipid bilayer comprising amphipathic lipids or phospholipids. So, basically, liposomes are nano sized lipid moieties of spherical shape [9]. There has been a lot of progress in the research on liposomes from conventional spherical liposomal bodies to second generation liposomes [10].

Second generation liposomes are those in which the size, charge, and composition of a lipid molecule are altered to some extent so as to make it a better delivery agent. Liposomes are nowadays widely used as drug delivery systems (for, e.g.,- doxorubicin, daunorubicin, cytarabine, etc.) for treatment of various infectious diseases and cancers also (Fig. 3). There are many advantages and disadvantages of using liposomes which are summarized in Table 2.

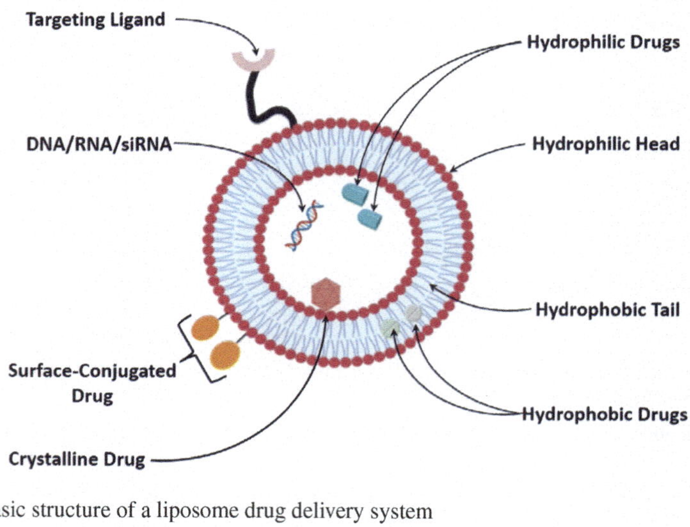

Fig. 3 Basic structure of a liposome drug delivery system

Table 2 Advantages and disadvantages of liposomes [11]

S. No.	Advantages of liposomes	Disadvantages of liposomes
1	Liposomes increased efficacy and therapeutic index of drug (actinomycin-D)	Low solubility
2	Liposome increased stability via encapsulation	Short half-life
3	Site avoidance effect	Fewer stables
4	Liposomes help reduce the exposure of sensitive tissues to toxic drugs	Production cost is high
5	Liposomes reduce the toxicity of the encapsulated agent (amphotericin B, Taxol)	Leakage and fusion of encapsulated drug/molecules

2.2 *Composition*

Liposomes formulation majorly consists of two types of phospholipids- glycerophospholipids and sphingomyelins. Glycerophospholipids (glycerol as backbone) and sphingomyelins are mainly the constituents of eukaryotic cells. The structure of a liposomal entity can be varied by altering the head groups of the glycerophospholipids. The different head groups can be phosphatidylcholine, phosphatidyl serine, phosphatidylethanolamine (Table 3), sphingomyelins have the property of efficient molecule entrapment, high stability in serum, also are readily released after delivery of the molecule to target organ [12].

2.3 *Methods for Preparation of Liposomes*

Three to four basic steps for liposome formation are as follows:

Step 1:- Lipid drying through organic solvent evaporation.
Step 2:- Dispersing the dried lipid in aqueous medium.
Step 3:- Involves the purification process of the obtained liposome
Step 4:- To structurally analyze and characterize the formed liposome.

Table 3 Liposome classification based on composition and mode of drug delivery [13]

S. No.	Type	Composition	Characteristics	References
1	Conventional liposome	Neutral and or negatively charged phospholipids + cholesterol	Subject to coated-pit endocytosis, contents ultimately liposomes delivered to lysosomes if they do not fuse from the endosomes, useful for RES targeting; rapid and saturable uptake by RES; short circulation half-life; dose dependent pharmacokinetics	[14]
2	PH sensitive	Phospholipids such as phosphatidyl ethanolamine, dioleoyl phosphatidyl ethanolamine with either CHEMS or OA	Subject to coated-pit endocytosis at low pH, fuse with liposomes cell or endosome membrane and release their contents in cytoplasm; suitable for intracellular delivery of weak base and macromolecules; biodistribution and pharmacokinetics similar to conventional liposomes	[15]
3	Cationic liposomes	Cationic lipids	Possibly fuse with cell or endosome membranes; suitable for delivery of negatively charged macromolecules (DNA, RNA); ease of formation; structurally unstable; transfection activity decreases with time; toxic at high dose, mainly restricted to local administration	[16]
4	Temperature (or) heat sensitive liposomes	Dipalmitoyl phophatidylcholine	Vesicles showed maximum release at 41°Y, the phase transition temperature of dipalmitoyl phophatidylcholine. Liposomes released the entrapped content at the target cell surface upon brief heating to the phase transition temperature of the liposome membrane	[17]

(continued)

Table 3 (continued)

S. No.	Type	Composition	Characteristics	References
5	Immuno-liposomes	Conventional or long circulating liposomes with attached Ab or recognition sequence	Subject to receptor-mediated endocytosis; cell specific binding (targeting); can release contents extracellularly near the target tissue and drugs diffuse through plasma membrane to produce their effects	[18]

3 Different Methods Used to Prepare Liposomes Are Given Below

3.1 Thin Film Hydration Method (Bangham Method)

This is one of the most used methods to prepare the liposomal entities. In this method, the lipids are dissolved in the organic solvents which can be either chloroform or chloroform and methanol mixture. Then these lipids are vacuum dried due to which a thin film of organic solvent is formed on the surface of the lipid moiety, which can be removed easily. Afterward, the lipid is dissolved in an aqueous solution. The lipid in aqueous buffer suspension swells to form liposomes of varied sizes (heterogenous liposomes) which are multilamellar, with a diameter measuring one micrometer. The resultant liposomes may be further processed by using different techniques like sonication, etc. [19]. But the major limitation of this technique is that it is not suitable for large scale production of liposomes [20].

3.2 Ethanol Injection Method

This interesting method of liposome (especially small unilamellar vesicles—SUVs) production without the need for sonication was introduced in 1973 by Batzri and Korn [21]. The lipids are dispersed in ethanol, this suspension is injected into the aqueous solution via a needle, thereby causing the dispersion of lipids into the aqueous medium and, hence forming the liposomes. This method is easy to perform and to scale up the formation of liposomal systems. This method of preparation of liposomal systems is of great advantage because it is not complicated, easily reproducible, does not cause oxidative damage or structural alterations to the resultant liposomal entities [22]. Also, this method has overcome the limitation of the previous thin film hydration method, which is used to prepare liposomes at laboratory scale only.

3.3 Ether Injection

In this method, either only ether or a mixture of ether and methanol is used to dissolve the lipids. This ethereal suspension of lipids is then very slowly injected into the aqueous suspension of the compound to be incorporated into the liposome cavity. The temperature at which this process is carried out is 55–65 °C and low pressure conditions are used. The major flaw of this method is the exposure of the compound to be incorporated to high temperatures, which can cause structural damage and hence alteration of the compound [23].

3.4 Sonication

This method involves the usage of acoustic energy (from bath or sonicator), which is used to break the large multilamellar lipid molecules into smaller uni- or multilamellar lipid particles. In this method, the size of the liposome generated depends upon the time for which sonication is done, with the minimum liposome size attained to 10.25 ± 0.55 nm. The chain length of the hydrocarbon used has no role to play in this process of liposome formation. This method is less time consuming as compared to extrusion and also the liposomes generated in this method do not exhibit homogeneity in size as exhibited by those generated by extrusion method [24].

3.5 Extrusion

Extrusion method involves the formation of liposomal systems by applying pressure. This is a very easy method and produces unilamellar and monodisperse liposomes. The desirable size of the liposome can be achieved by using this method and moreover the size once produced will always be reproducible. This is the major advantage of this method over the sonication method. This method basically involves the use of a polycarbonate membrane of a uniform pore size through which the lipid suspension is forced to pass under some pressure, thereby producing lipid particles of nearly the same size as the membrane pore size. The mean vesicle size of the lipid particle remains uniform from batch to batch, and there is no requirement of evaporation of the organic solvent [25].

3.6 Micro Emulsification Method

In this technique, a microfluidizer is used to form multi-lamellar vesicles from the concentrated lipid suspension [7]. The microfluidizer is used to force pass the lipid suspension under very high pressure of 10,000 psi through an orifice of 5 μm. From here, the suspension is channelized into two separate microchannels where two streams of lipid suspension are formed and collide at right angles with each other for an efficient energy transfer. Spherical lipid particles of 0.1–0.2 μm are formed as a result.

3.7 Applications

Liposomes are considered a promising technology for a variety of pharmaceutical and industrial applications since they are amphiphilic carriers open to modifications with different functional properties [26, 27].

1. The major use of liposomes in the food industry is the encapsulation of active food ingredients for example the essential oils, vitamin-C in the liposomal cavity, make them safer from degradation as natural food components are highly prone to degradation. The 50% natural activity of ascorbic acid is preserved when encapsulated in liposomes as compared with the encapsulated vitamin-C molecules.
2. In cosmetic industry: the liposomes also show a promising application in this industry also, because of their bio-properties like biocompatibility with the human skin, also skin adhesiveness of liposomes helps the skin to remain hydrated and healthy, the entrapped cosmetic components like oils, creams, dermal antibiotics are released to the cellular levels in the skin.
3. Liposomes also are used in gene delivery experiments, cancer therapy, and gene therapy also. Several gene delivery protocols have been developed for DNA delivery, for example, Haem agglutinating Virus of Japan (HVJ), in this the DNA-liposomal complex is used to deliver the viral DNA in the mammalian cell cytoplasm for further experiments.

4 Transferosomes

4.1 Introduction

The concept of ultra-deformable liposomes [28] or elastic nanovesicles [29] or transferosomes was first described by Gregor Cevc in 1991. These are the structures that are ultra-deformable, with an aqueous core and an outer lipid membrane. The transferosomes can transfer themselves through the pores even smaller than their own size. Although the liposomes are restricted only to the outermost layer of stratum corneum but transferosomes penetrate through the skin layers deeply because they are flexible liposomes. A transferosome can carry or transport or transfer the drug molecules from their sites of application to the active target site (Fig. 4). It is also shown that the transferosomes penetrate the intact human or mammalian skin [28].

4.2 Composition

Transferosomes are made up of the following constituents (Table 4):

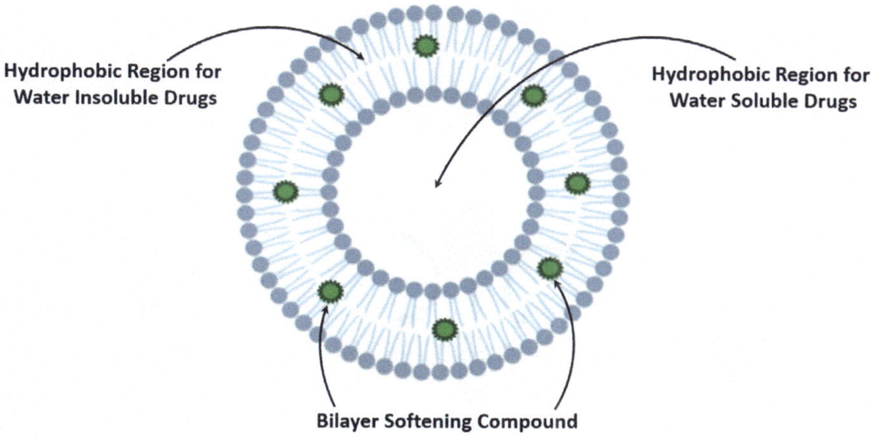

Fig. 4 The basic structure of transferosome. Composed on inner hydrophilic region and outer hydrophobic region loaded with water insoluble drugs

Table 4 Different Classes and their uses of transferosomes [30]

Class	Example	Uses
Phospholipids	Soya phosphatidyl choline, dipalmitoyl phosphatidyl choline, distearoyl phoshatidyl choline	Vesicles forming Component
Surfactant	Sodium cholate, sodium deoxycholate, tween-80, span-80	For providing flexibility
Alcohol	Ethanol, methanol	As a solvent
Buffering agent	Saline phosphate buffer (pH 6.4)	As a hydrating medium
Dye	Rhodamine-123, rhodamine-DHPE, fluorescein-DHPE nilered	For CSLM study

1. Phospholipids—phosphatidyl choline—folds to form a lipid bilayer in an aqueous environment.
2. Edge activator (EA)—this is usually a single chain surfactant, which destabilizes the transferosomes so that they become highly flexible and elastic.

4.3 Mechanism of Action of Transferosome

When the transferosome is applied on an unoccluded biological membrane, there is loss of water by evaporation, so two forces are responsible for the trans epidermal movement (skin permeation or penetration) of the transferosome—elasticity or deformability of transferosome and transepidermal water gradient. An unhampered flow of the transferosome occurs through very small skin pores (1/6 th of the size

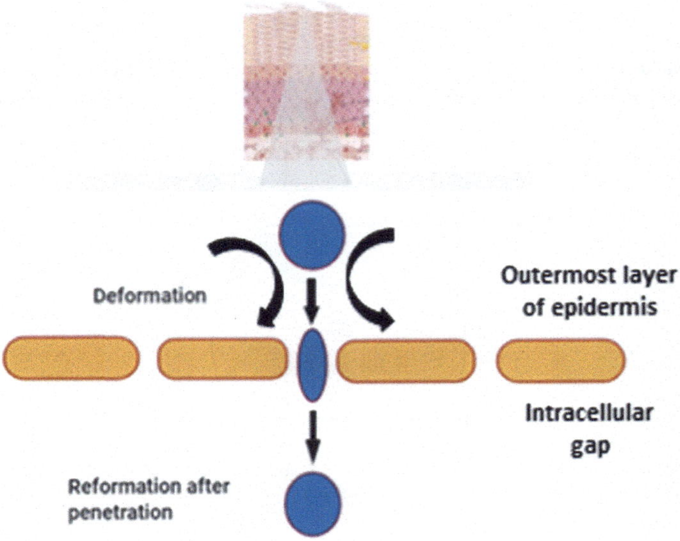

Fig. 5 Penetration of transferosome into the skin [13]

of transferosomes), due to the property of their high elasticity, and secondly, the flow is due to the water gradient (water content is lower in upper skin layers – 15% and is higher inside the deeper skin layers –75%) [31]. So, to hydrate its components, the transferosomes elastically deform and squeeze themselves deeper into the skin layers, carrying the drug of choice to target tissue from the area of application. The membrane lipid bilayer is least disturbed during the transferosome movement through the skin layers, because of its elastic nature it squeezes and moves through even the tightest pores. The major flaw of transferosome is that it is difficult to load hydrophobic drugs to the transferosome without altering its elasticity. The greatest advantage of using transferosome is that it enhances the skin permeation in comparison to other liposomes [32] (Fig. 5).

5 Methods of Preparation

5.1 Rotary Film Evaporation Method

This method was for the first time introduced by Bangham and is also known as the handshaking method. In this method, the phospholipids and the edge activator or the surfactant are dissolved in the organic solvent which can be either chloroform or chloroform and methanol both. Then this organic suspension is transferred to the round bottomed flask, which is shaked and rotated at a constantly maintained temperature and a low pressure. With time, a thin film of lipid and EA (edge activator) is formed on the inner walls of the flask, which is then hydrated using an aqueous solution which the drug of choice has dissolved in it. The lipids swell to form the vesicles, which are generally multilamellar in nature. These MLVs (multilamellar vesicles) can be later sonicated or can be made to undergo extrusion to fetch the desirably sized vesicles [33].

5.2 Reverse Phase Evaporation Method

In this method firstly the lipid organic solvent suspension is taken in a round bottomed flask. Then an aqueous solution of EA (edge activator) is added in the environment containing nitrogen gas. The drug of choice to be incorporated is dissolved either in the organic solvent or an aqueous solution depending upon its dissolving characteristics. The complete mixture suspension is sonicated till it becomes highly homogeneous. The mixture suspension is kept as such without any separation for at least thirty minutes. After this, the organic solvent is made to evaporate under the reduced pressure conditions. This converts the whole remaining mixture into a gel, and finally into vesicles [34].

5.3 Vortex/Sonication Method

In this method, the phospholipids and the surfactant chemicals are dissolved in the phosphate buffer by continuous shaking and agitation. The resultant suspension is then sonicated to get smaller vesicles. Further to fetch the vesicles of a desirable diameter, the same phosphate buffered suspension is extruded through the polycarbonate membrane of definite known pore size. This method gives the transferosome vesicles of desired diameters [26].

5.4 Ethanol Injection Method

This method involves the formation of two suspensions, solution A—Ethanolic suspension of the phospholipids and EAs (edge activators) formed by continuous heating and stirring, solution B—Aqueous solution of drug of choice to be incorporated into the transferosome. Then, solution A is added to solution B, drop wise, which leads to the formation of lipid vesicles. This method is highly reproducible and easy to perform [35].

5.5 *Freeze Thaw Method*

This method involves the use of pre-prepared multilamellar vesicle suspension. This suspension is transferred to alternate cycles of low temperature and high temperature. Finally, the suspension is poured into a tube which is then placed in liquid nitrogen at −30 °C for 30 s. Then again, this supercooled suspension is subjected to high temperatures. These steps are repeated 8 to 9 times.

5.6 *Applications*

The major application of the ultra-deformable nanovesicles is their ability to transfer the drugs across the natural mammalian skin barrier. Various unstable proteins, peptides, drugs, and vaccines are transferred efficiently. So, this property of transferosomes is enormously exploited nowadays clinically. Some of the drugs whose transferosomes have been prepared successfully and are being used by a clinician with promising results are mentioned below in Tables 5 and 6.

6 Ethosomes

6.1 *Introduction*

Discovered by Touitou et al. in 2020, these entities contain large amounts of ethyl alcohol (40–45%) so called by the name Ethosomes. This much high amount of ethanol causes the lipids to be more fluid in nature and tiny in size and enhances the lipid vesicle permeability into the mammalian skin. These ethanolic liposomes deliver the drug at a very controlled rate to the systemic circulation [44]. Like Transferosomes, Ethosomes also form an integral part of the transdermal delivery system (TDDS). Ethosomes are lipid vesicles comprising phospholipids, alcohol (ethanol or isopropyl alcohol) in maximum content, and water (Fig. 6). The major role of high amount of ethanol in these vesicles is that the alcohol provides an ultra-flexibility and softness to the vesicle, which in turn allows them to squeeze and pass through the gaps in stratum corneum easily without any hindrance and thereby delivering the drug to the target site [45].

6.2 *Composition*

Ethosomes consist of phospholipids, ethanol (up to 45%), glycerol, and water. Phospholipids like soya phosphatidylcholine, egg phosphatidylcholine, hydrogenated

Table 5 Drugs whose transferosomes are currently in use in clinical practice

S. No.	Drug	Drug category	Study conducted	Results achieved	References
1	5-fluorouracil	Anti-Cancer	Skin permeation and deposition	(a) Sustained drug release (b) Enhanced flux rate as compared to solution	[36]
2	Cyclosporine A	Immunosuppressive	Drug delivery through mice skin	Successful delivery of drug through and into the mice skin as compared to the conventional vesicles	[37]
3	Paclitaxel	Anti-neoplastic	Toxicity study	Maximum tolerable dose for elastic vesicular formulation was 120 mg/kg while 40 mg/kg for marketed formulation	[38]
4	Tetanus	Vaccine	(a) Elasticity (b) Immune response	(a) Increased elasticity as compared to liposome and noisome (b) Immune response achieved comparable to intramuscularly given tetanus toxoid	[39]
5	Oestradiol	Estrogens	Flux	17 times more flux compared to control	[40]

phosphatidylcholine, etc. are used in the formation of ethosomes. High amount of ethanol enhances the flexibility and permeability of the vesicle into the stratum corneum (Table 7).

Table 6 Transferosomes in clinical trials [18, 27, 37, 41–43]

S. No	Drug	Study Phase	Summary
1	Ketoprofen	Phase II	Increased safety and efficacy of ketoprofen (110 mg) in transfersome based gel is observed as compared to placebo and oral celecoxib (200 mg)
2	Ketoprofen	Phase III	Safety and efficacy of 100 mg ketoprofen in Directin®observed
3	Insulin	Phase I	Penetration of Transfersulin could result in a decrease in blood glucose level in alloxan induced diabetic mice with simultaneously increased blood insulin level
4	Triamcinolone acetonide (TAC)	Randomized Control trial	Ten fold lower dose of TAC in Transfersome (2.5 μg cm^2) was bioequivalent to 25 μg cm^2 dose of TAC in conventional formulations. Transfersome significantly improved the risk-benefit ratio of topically applied glucocorticosteroid

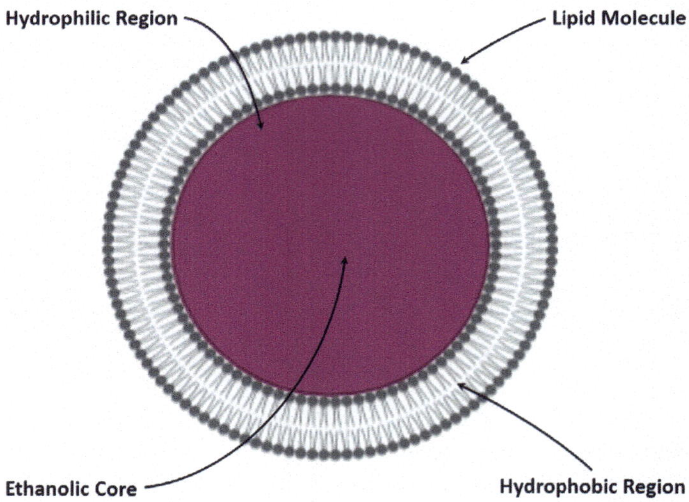

Fig. 6 Basic structure of ethosome. Consist of Central core ethanolic region and outer hydrophic region

Table 7 Different additives used in ethosome formulation [46]

S. No.	Class	Example	Uses
1	Phospholipid	Soya phosphatidyl choline, Egg phosphatidyl choline, Dipalmityl phosphatidyl choline, Distearyl phosphatidyl choline	Vesicles forming component
2	Alcohol	Ethanol, Isopropyl alcohol	For providing the softness for vesicle membrane as a penetration enhancer
3	Polyglycol	Propylene glycol, Transcutol RTM	As a skin penetration enhancer
4	Cholesterol	Cholesterol	For providing the stability to vesicle member
5	Dye	Rhodamine-123, Rhodamine red, Fluorescene Isothiocynate, (FITC)6-Carboxyfluorescence	For characterization study
6	Vehicle	Carbopol Đ934	As a gel former

6.3 Mechanism of Penetration

The mechanism of penetration is not clearly understood to date. But it is known that ethanol acts as the vesicle permeation enhancer for ethosomes. The mechanism occurs in two steps:

1. Ethanol effect
2. Effect by ethosomes.

Ethanol effect: It is clearly known that ethanol penetrates deeply into the skin cells and increases the flexibility and fluidity of their lipid bilayer. Thus, allowing the vesicles to move through easily and swiftly and not at the cost of the cell's membrane.

Effect by ethosomes: Major role is played by ethanol that is why the content of the latter is very high in these vesicles. As the vesicles slip down deeper into the skin layers, they initiate fusing themselves with the skin cells thereby delivering the drug of interest into the target skin tissues [47].

Some of the drugs which have been delivered successfully by ethosomes are [44]:

1. Acyclovir,
2. Minoxidil,
3. Trihexyphenidyl,
4. Testosterone,
5. Cannabidiol and
6. Zidovudine.

6.4 *Methods of Preparation*

Cold Method

This method involves the formation of a suspension containing phospholipids, the drug of choice, and other lipids in ethyl alcohol. The former suspension is then transferred into the covered vessel and is continuously stirred at room temperature. Propylene glycol is then added to this mixture and the complete mixture is heated to 30 °C. A separate plain aqueous solution at 30 °C containing only water is added into the lipid suspension which is already heated. Now both the suspensions were mixed and stirred for five minutes. The resultant vesicles formed are called ethosomes. The desirable size of these nanovesicles can be attained by sonication or extrusion. It is essential to store this formulation under refrigeration [47].

Hot Method

This method involves the heating of phospholipid in water to 40 °C to form a colloidal solution. Ethanol and propylene glycol are mixed and heated to 40 °C in a separate vessel. Both the mixtures are then mixed thoroughly once they attain the temperature of 40 °C. The drug can be mixed in aqueous or an organic solvent depending upon its solubility properties. The desirable vesicle size can be attained by sonication or extrusion [48].

6.5 *Applications*

Ethosomes and Pilosebaceous Targeting

Ethosomes entrapping finasteride have been prepared to the size extent of 100–200 nm, to treat alopecia in males. Ethosomes mediated delivery of finasteride is a highly effective treatment for alopecia or baldness in males because the former are able to cross the layers of stratum corneum and reach the pilosebaceous unit of the scalp for effective drug release to the target region [49].

Treatment of Herpetic Infection

Herpes is a viral infection causing mouth sores. To treat this infection, the topic acyclovir ethosome based gel (ACV) has been produced. This ethosomal based preparation is said to have deeper local skin penetration and effectively heal the sores [44].

Application for delivery of anti-arthritis drug [50]

The drug used for cure of arthritis is tetrandrine, an alkaloid, which has a plant origin, is extracted from the root system of Stephania tetrandra plant. Spherical liposome-based preparation of this drug has been lately used to cure this disease. The lipid

Table 8 Marketed or commercially available ethosomes products [42]

S. No.	Product name	Company name	Active Ingredient/Function
1	Cellutight EF	Hampden Health, USA	Topical cellulite cream contains a powerful combination of ingredients to increase metabolism and break down fat
2	Decorin cream	Sinere, Germany	Antiageing cream, treating, repairing, and delaying the visible aging signs of the skin including wrinkle lines, sagging, age spots, loss of elasticity, and hyperpigmentation
3	Supravir cream	Trima, Israel	For the treatment of herpes virus, formulation of acyclovir drug has a long shelf-life with no stability problems
4	Noicellex	Novel therapeutic Technologies, Israel	Topical anticellulite cream
5	Skin Genuity Physonics,	Nottingham, UK	Powerful cellulite buster, reduces orange pee

nanocarrier based tetrandrine medicine is able to penetrate in and across the skin for effective treatment (Table 8).

7 Conclusion

Lipid nanocarriers are much explored vesicles in the field of nanotechnology. Nowadays they have innumerable applications in various fields of medicine, pharmaceutics—as drug delivery systems through oral, topical, and parenteral route, food and cosmetic industry, and agricultural industry also. Solid lipid nanocarriers were the first generation nanocarriers but were inefficient in drug delivery mechanisms so were improved to form a new class of nanocarriers called nanostructured lipid carriers. Various lipid-based systems like ethosomes, transferosomes, and liposomes have shown promising results for drug delivery in various diseases like cancer, etc., vaccine delivery. These drug carriers are advantageous as they can carry and deliver both lipophilic and hydrophillic medicines to the target tissues or cells.

References

1. Montoto SS, Muraca G, Ruiz ME (2020) Solid lipid nanoparticles for drug delivery: pharmacological and biopharmaceutical aspects. Front Mol Biosci 7
2. Ghasemiyeh P, Mohammadi-Samani S (2018) Solid lipid nanoparticles and nanostructured lipid carriers as novel drug delivery systems: applications, advantages and disadvantages. Res Pharm Sci 13(4):288–303. https://doi.org/10.4103/1735-5362.235156
3. Naseri N, Valizadeh H, Zakeri-Milani P (2015) Solid lipid nanoparticles and nanostructured lipid carriers: structure, preparation and application. Adv Pharm Bull 5(3):305–313. https://doi.org/10.15171/apb.2015.043
4. Patra JK, Das G, Fraceto LF, Campos E, Rodriguez-Torres M, Acosta-Torres LS, Diaz-Torres LA, Grillo R, Swamy MK, Sharma S, Habtemariam S, Shin HS (2018) Nano based drug delivery systems: recent developments and future prospects. J Nanobiotechnol 16(1):71. https://doi.org/10.1186/s12951-018-0392-8
5. Das S, Ng WK, Tan RB (2012) Are nanostructured lipid carriers (NLCs) better than solid lipid nanoparticles (SLNs): development, characterizations and comparative evaluations of clotrimazole-loaded SLNs and NLCs? Eur J Pharm Sci Official J Eur Fed Pharm Sci 47(1):139–151. https://doi.org/10.1016/j.ejps.2012.05.010
6. Attama AA, Momoh MA, Builders PF (2012) Lipid nanoparticulate drug delivery systems: a revolution in dosage form design and development. Recent Adv Novel Drug Carrier Syst 5:107–140
7. Mashaghi S, Jadidi T, Koenderink G, Mashaghi A (2013) Lipid nanotechnology. Int J Mol Sci 14(2):4242–4282. https://doi.org/10.3390/ijms14024242
8. Puri A, Loomis K, Smith B, Lee JH, Yavlovich A, Heldman E, Blumenthal R (2009) Lipid-based nanoparticles as pharmaceutical drug carriers: from concepts to clinic. Crit Rev Ther Drug Carrier Syst 26(6):523–580. https://doi.org/10.1615/critrevtherdrugcarriersyst.v26.i6.10
9. Dwivedi C, Verma S (2013) Review on preparation and characterization of liposomes with application. Int J Sci Innov Res 2:486–508
10. Akbarzadeh A, Rezaei-Sadabady R, Davaran S, Joo SW, Zarghami N, Hanifehpour Y, Samiei M, Kouhi M, Nejati-Koshki K (2013) Liposome: classification, preparation, and applications. Nanoscale Res Lett 8(1):102. https://doi.org/10.1186/1556-276X-8-102
11. Kay JG, Grinstein S (2011) Sensing phosphatidylserine in cellular membranes. Sensors (Basel, Switzerland) 11(2):1744–1755. https://doi.org/10.3390/s110201744
12. Alberts B, Johnson A, Lewis J et al (2002) Molecular biology of the cell. Transport into the cell from the Plasma Membrane: Endocytosis. Garland Science, 4th edn. New York
13. Anwekar H, Patel S, Singhai AK (2011) Liposomes as drug carriers. IJPLS 2(7):945–951
14. Yingchoncharoen P, Kalinowski DS, Richardson DR (2016) Lipid-Based Drug Delivery Systems in Cancer Therapy: What Is Available and What Is Yet to Come. Pharmacol Rev 68(3):701–787. https://doi.org/10.1124/pr.115.012070
15. Bozzuto G, Molinari A (2015) Liposomes as nanomedical devices. Int J Nanomed 10:975–999. https://doi.org/10.2147/IJN.S68861
16. Chen W, Duša F, Witos J et al (2018) Determination of the main phase transition temperature of phospholipids by nanoplasmonic sensing. Sci Rep 8, 14815(2018). https://doi.org/10.1038/s41598-018-33107-5
17. Lodish H, Berk A, Zipursky SL et al (2000) Molecular cell biology. Section 17.9, Receptor-Mediated Endocytosis and the Sorting of Internalized Proteins, 4th edn. New York, W. H. Freeman
18. Biju SS, Talegaonkar S, Mishra PR, Khar RK (2006) Vesicular systems: an overview. Indian J Pharm Sci 68(2):141–153
19. Ahmed KS, Hussein SA, Ali AH, Korma SA, Lipeng Q, Jinghua C (2019) Liposome: composition, characterisation, preparation, and recent innovation in clinical applications. J Drug Target 27(7):742–761. https://doi.org/10.1080/1061186X.2018.1527337

20. Jaafar-Maalej C, Diab R, Andrieu V, Elaissari A, Fessi H (2010) Ethanol injection method for hydrophilic and lipophilic drug-loaded liposome preparation. J Liposome Res 20(3):228–243. https://doi.org/10.3109/08982100903347923
21. Charcosset C, Juban A, Valour JP, Urbaniak S, Fessi H (2015) Preparation of liposomes at large scale using the ethanol injection method: Effect of scale-up and injection devices. Chem Eng Res Des 94:508–515. https://doi.org/10.1016/j.cherd.2014.09.008
22. Melis Çağdaş M, Sezer AD, Bucak S (2014) Liposomes as potential drug carrier systems for drug delivery. IntechOpen
23. Huang X, Caddell R, Yu B, Xu S, Theobald B, Lee LJ, Lee RJ (2010) Ultrasound-enhanced microfluidic synthesis of liposomes. Anticancer Res 30(2):463–466
24. Lapinski MM, Castro-Forero A, Greiner AJ, Ofoli RY, Blanchard GJ (2007) Comparison of liposomes formed by sonication and extrusion: rotational and translational diffusion of an embedded chromophore. Langmuir ACS J Surf Colloids 23(23):11677–11683. https://doi.org/10.1021/la7020963
25. Lombardo D, Calandra P, Barreca D, Magazù S, Kiselev MA (2016) Soft interaction in liposome nanocarriers for therapeutic drug delivery. Nanomaterials 6(7):125. https://doi.org/10.3390/nano6070125
26. Rai S, Pandey V, Rai G (2017) Transfersomes as versatile and flexible nano-vesicular carriers in skin cancer therapy: the state of the art. Nano Rev Exp 8(1):1325708. https://doi.org/10.1080/20022727.2017.1325708
27. Sercombe L, Veerati T, Moheimani F, Wu SY, Sood AK, Hua S (2015) Advances and challenges of liposome assisted drug delivery. Front Pharmacol 6:286. https://doi.org/10.3389/fphar.2015.00286
28. Ascenso A, Raposo S, Batista C, Cardoso P, Mendes T, Praça FG, Bentley MV, Simões S (2015) Development, characterization, and skin delivery studies of related ultradeformable vesicles: transfersomes, ethosomes, and transethosomes. Int J Nanomed 10:5837–5851. https://doi.org/10.2147/IJN.S86186
29. Reddy YD, Sravani AB, Ravisankar V, Prakash PR, Reddy YSR, Bhaskar NV (2015) Transferosomes a novel vesicular carrier for transdermal drug delivery system. JIPBS. 2(2):193–208
30. Rajan R, Jose S, Mukund VP, Vasudevan DT (2011) Transferosomes—A vesicular transdermal delivery system for enhanced drug permeation. J Adv Pharm Technol Res 2(3):138–143. https://doi.org/10.4103/2231-4040.85524
31. Opatha S, Titapiwatanakun V, Chutoprapat R (2020) Transfersomes: a promising nanoencapsulation technique for transdermal drug delivery. Pharmaceutics 12(9):855. https://doi.org/10.3390/pharmaceutics12090855
32. Chaurasiya P, Ganju E, Upmanyu N, Ray SK, Jain P (2019) Transfersomes: a novel technique for transdermal drug delivery. J Drug Delivery Ther 9(1):279–285
33. Wu PS, Li YS, Kuo YC, Tsai SJ, Lin CC (2019) Preparation and Evaluation of Novel Transfersomes Combined with the Natural Antioxidant Resveratrol. Molecules (Basel, Switzerland) 24(3):600. https://doi.org/10.3390/molecules24030600
34. Sunitha Reddy M, Anusha M (2020) Transferosomes-novel drug delivery sysytem—a review. JCRT 8(9):2320–2882
35. Hussain A, Samad A, Ramzan M, Ahsan MN, Ur Rehman Z, Ahmad FJ (2016) Elastic liposome-based gel for topical delivery of 5-fluorouracil: in vitro and in vivo investigation. Drug Delivery 23(4):1115–1129. https://doi.org/10.3109/10717544.2014.976891
36. Guo J, Ping Q, Sun G, Jiao C (2000) Lecithin vesicular carriers for transdermal delivery of cyclosporin A. Int J Pharm 194(2):201–207. https://doi.org/10.1016/s0378-5173(99)00361-0
37. Utreja P, Jain S, Tiwary AK (2012) Evaluation of biosafety and intracellular uptake of Cremophor EL free paclitaxel elastic liposomal formulation. Drug Delivery 19(1):11–20. https://doi.org/10.3109/10717544.2011.621990
38. Gupta PN, Mishra V, Rawat A, Dubey P, Mahor S, Jain S, Chatterji DP, Vyas SP (2005) Non-invasive vaccine delivery in transfersomes, niosomes and liposomes: a comparative study. Int J Pharm 293(1–2):73–82. https://doi.org/10.1016/j.ijpharm.2004.12.022

39. El Maghraby GM, Williams AC, Barry BW (2000) Oestradiol skin delivery from ultradeformable liposomes: refinement of surfactant concentration. Int J Pharm 196(1):63–74. https://doi.org/10.1016/s0378-5173(99)00441-x
40. Rother M, Conaghan PG (2013) A randomized, double-blind, phase III trial in moderate osteoarthritis knee pain comparing topical ketoprofen gel with ketoprofen-free gel. J Rheumatol 40(10):1742–1748. https://doi.org/10.3899/jrheum.130192
41. Rother M, Lavins BJ, Kneer W, Lehnhardt K, Seidel EJ, Mazgareanu S (2007) Efficacy and safety of epicutaneous ketoprofen in Transfersome (IDEA-033) versus oral celecoxib and placebo in osteoarthritis of the knee: multicentre randomised controlled trial. Ann Rheum Dis 66(9):1178–1183. https://doi.org/10.1136/ard.2006.065128
42. Tiwari A, Mishra MK, Nayak K, Yadavand SK, Shukla A (2016) Ethosomes: a novel vesicular carrier system for therapeutic applications. IOSR J Pharm 6(9):25–33
43. Verma P, Pathak K (2010) Therapeutic and cosmeceutical potential of ethosomes: An overview. Journal of advanced pharmaceutical technology & research 1(3):274–282. https://doi.org/10.4103/0110-5558.72415
44. Abdulbaqi IM, Darwis Y, Khan NA, Assi RA, Khan AA (2016) Ethosomal nanocarriers: the impact of constituents and formulation techniques on ethosomal properties, in vivo studies, and clinical trials. Int J Nanomed 11:2279–2304. https://doi.org/10.2147/IJN.S105016
45. Aggarwal D, Nautiyal U (2016) Ethosomes: a review. Int J Pharm Med Res 4(4):354–363
46. Pilch E, Musiał W (2018) Liposomes with an ethanol fraction as an application for drug delivery. Int J Mol Sci 19(12):3806. https://doi.org/10.3390/ijms19123806
47. Hippalgaonkar K, Majumdar S, Kansara V (2010) Injectable lipid emulsions-advancements, opportunities and challenges. AAPS PharmSciTech 11(4):1526–1540. https://doi.org/10.1208/s12249-010-9526-5
48. Pandey V, Golhani D, Shukla R (2015) Ethosomes: versatile vesicular carriers for efficient transdermal delivery of therapeutic agents. Drug Delivery 22(8):988–1002
49. Mohammad Soleymani S, Salimi A (2019) Enhancement of dermal delivery of finasteride using microemulsion systems. Adv Pharm Bull 9(4):584–592. https://doi.org/10.15171/apb.2019.067
50. Fan C, Li X, Zhou Y, Zhao Y, Ma S, Li W, Liu Y, Li G (2013) Enhanced topical delivery of tetrandrine by ethosomes for treatment of arthritis. BioMed Res Int 161943.https://doi.org/10.1155/2013/161943
51. Fesq H, Lehmann J, Kontny A, Erdmann I, Theiling K, Rother M, Ring J, Cevc G, Abeck D (2003) Improved risk-benefit ratio for topical triamcinolone acetonide in transfersome in comparison with equipotent cream and ointment: a randomized controlled trial. Br J Dermatol 149(3):611–619. https://doi.org/10.1046/j.1365-2133.2003.05475.x

Chapter 3
Nanoemulsions

Shikha Indoria, Madhu Bala, and Vickramjeet Singh

1 Introduction

Emulsions are the colloidal systems having two immiscible liquids with one of the liquid phases dispersed as droplets and the second immiscible liquid act as the continuous phase [1, 2]. The dispersed phase forms droplets with the help of surface-active agents. The immiscible liquids forming emulsions are of different polarities [1, 2]. The dispersed droplets with size less than 100 nm are termed nano emulsions, however, such systems are called mini-emulsions, sub-micron-emulsions, translucent-emulsions, or ultrafine-emulsions [3, 4]. The upper size limits up to 500 nm are also referred to as nano emulsions [5]. These heterogeneous systems are kinetically stable but lack thermodynamic stability [6, 7]. The nanoemulsions (NEs) are distinct from microemulsions in terms of droplet size as well as stability [3, 4]. The microemulsions are thermodynamically stable systems [3, 4] and nanoemulsions can be stable against sedimentation and creaming [3, 4, 8]. Visually nanoemulsions are translucent or transparent [3, 4]. The nanoemulsions are being employed for a number of applications such as improving bioavailability as well as aqueous solubility of drugs, targeted drug delivery vehicles, oil clingage removal [9], in cosmetics, agrochemical, and also in food and beverage industries [8, 10–17].

The stability of nano emulsions is influenced by the composition, temperature/pressure, type of surfactant, and method of fabrication. The destabilization can

S. Indoria
Department of Chemistry, Kanya Maha Vidyalaya, Jalandhar, Punjab 144004, India

M. Bala · V. Singh (✉)
Department of Chemistry, Dr. B. R. Ambedkar National Institute of Technology, Jalandhar, Punjab 144011, India
e-mail: singhvj@nitj.ac.in

S. Gopi et al. (eds.), *Nanotechnology for Biomedical Applications*,
Materials Horizons: From Nature to Nanomaterials,
https://doi.org/10.1007/978-981-16-7483-9_3

occur due to coalescence, or Ostwald ripening, and flocculation [8]. The nanoemulsions can be oil-in-water or water-in-oil types [8, 17–28]. The O/W or W/O nanoemulsions can be synthesized by a number of techniques that require energy for reducing the size of dispersed droplets [29, 30]. High energy input or with fabrication process employing comparatively less amount of energy can form emulsions from heterogeneous polar-non-polar mixtures [29]. The former involves processes such as ultrasonication while the latter leads to spontaneous emulsification [29]. One of the benefits of droplet size shrinkage is related to improvement in biocompatibility of nanoformulations [29, 31]. These biofriendly systems incorporate non-polar molecules in polar continuous phase [29]. The main focus of this chapter is regarding the fabrication, characterization, and biomedical applications of nanoemulsions (NEs).

2 NE Fabrication

Nanoemulsions can be prepared by simply mixing oil and water but it will not be stable and drop coalescence will eventually occur [32, 33]. So, an emulsifier such as surface-active agent is needed to make sure the nanoemulsion is kinetically stable [32, 33]. These naturally occurring or synthetic emulsifiers can be used solely or can be used in combination with a cosurfactant [5, 32, 33]. The NE are thermodynamically unstable as the free energy of dispersed phase (droplet of one liquid in another immiscible continuous liquid phase) is higher than that of separated phases [34, 35]. Eventually, the thermodynamic instability will make nanoemulsions to be separated into individual components when sufficient time is provided [34]. The breakdown will depend upon the energy barrier and the mass transport process. The height of the energy barrier influences the kinetic stability [34] and for higher energy barriers, the breakdown of nanoemulsion into individual components (oil and water) will take a much longer time [34]. In other words, the NEs will persist for a longer period of time [34, 35]. The nanoemulsion system is of the following types: oil-in-water (O/W), water-in-oil (W/O), bi-continuous nanoemulsion, multiple nanoemulsion, Oil-in-water-in-oil (O/W/O), or Water-in-oil-in-water (W/O/W) [36].

Owing to smaller droplet sizes when in comparison to wavelength of light, the nanoemulsions are transparent or translucent (see Fig. 1). The added benefit of smaller size is that these systems are stable against aggregation and separation based on gravity [34]. The NE formation is not energetically favorable, so to achieve the

Fig. 1 Pictures of nanoemulsion on the left and microemulsion on the right with droplet size (diameter) of 35 nm and 1 Am, respectively. (Reproduced with permission from Ref. [3])

activation energy barrier for separated states, the input of energy externally is needed [34, 35]. Since this energy requirement requires droplet size reduction and hence can be of two types: high energy process or low-energy processes (spontaneous process) [34, 35].

2.1 Emulsifiers/Oil Phase

For stabilization of dispersed phase in the continuous phase, various emulsifiers have been used [11, 37–39]. Emulsifiers coat the droplet surface and stabilize emulsion by causing an increase in kinetic energy [40]. The emulsifier can be natural or synthetic chemicals and can be either in salt form or may be used as particle dispersion [37–40]. In case of NE used for food related applications, the edibility of formulation should be enhanced by using biopolymers or proteins as emulsifiers [40]. The commonly used emulsifier is surface-active agent, having solvent loving and also solvophobic parts. Surfactants can be cationic, anionic, amphoteric, or non-ionic. The concentration and chemical nature of surfactants influence the stability of NEs [41]. Lowering of surface tension to a low value helps in the dispersion process and the lipophilicity of surfactants deliver the correct curvature at the interface for the required NEs (W/O, O/W, or bi-continuous) [42–44]. W/O NEs are formed using surfactants like span which have a low HLB (hydrophilic-lipophilic) balance, whereas O/W NEs are formed by high HLB surfactants like Tween (HLB value of 8–18). However, according to Bancroft's rule, the O/W or W/O emulsion (micro/macro) formation depends upon the higher solubility of emulsifier in the continuous phase [40, 45, 46]. Table 1 shows some of the surfactant molecules used for the fabrication of NEs.

The surfactant can sometimes form NEs over a limited concentration range and this is due to the formation of a rather rigid film, but addition of a cosurfactant makes the interfacial film with adequate flexibility so as to achieve curvatures needed for NEs formation [42, 47]. Some of the cosurfactants used for NEs preparation are shown in Table 2 [42].

The selection of oil phase (either oil or lipid) depends upon the application of NEs, for instance, oil should be able to solubilize hydrophobic drug molecules when NEs are fabricated by considering the drug delivery applications [42]. Some of the oil/lipid used for NEs preparation is glycerol triacetate, propylene dicaprylate/dicaprate glycol, C8/C10 triglycerides, glyceryl tricaprylate/caprice, glyceryl tricaprylate, myristic acid isopropyl ester, isopropyl myristate, sunflower or corn oil (beta carotene), olive, sesame or soyabean oil, paraffin oil (alfa tocopherol) [42].

2.2 High Energy Emulsification Process (Top-Down)

The nanoemulsions are formed from a pre-emulsion state or more appropriately from macroemulsion. Thus, the dispersed droplets are made to shrink in size and

Table 1 List of surfactants/emulsifiers used for NE fabrication

Emulsifier/surfactant	Type	References
Alkanol-XC	Surfactant (anionic)	[48]
Brij-L4	Surfactant (non-anionic)	[49]
Cremophor-EL	Surfactant (non-anionic)	[49]
Decaglycerol monolaurate	Surfactant (non-anionic)	[50]
Dodecyl trimethyl ammonium bromide	Surfactant (cationic)	[51]
Gelatin	Protein (amphiphilic)	[52]
Lipoid S75-3	Phospholipid (amphiphilic)	[53]
Modified starch	Hydrocolloid (cationic)	[54]
Polyoxyethylene 6-lauryl ether	Surfactant (non-ionic)	[55]
Polyglycol ester of fatty acids	Surfactant (non-ionic)	[56]
Pectin	Hydrocolloid (anionic)	[57]
Ryotol	Surfactant (non-ionic)	[50]
Solutol–HS15	Surfactant (non-ionic)	[53]
Span-20	Surfactant (non-ionic)	[58]
Span-80	Surfactant (non-ionic)	[58]
Sodium caseinate	Protein (amphiphilic)	[50]
Sodium dodecyl sulfate	Surfactant (anionic)	[51]
Tween −20, −40, −60, and −80	Surfactant (non-ionic)	[58, 59]
Whey protein isolate	Protein (amphiphilic)	[60]
Whey protein concentrate	Protein (amphiphilic)	[61]
Whey protein hydrolysate	Protein (amphiphilic)	[61]

Table 2 Various cosurfactants employed for fabrication of NEs

Cosurfactant	Function during NE formation
Transcutol P	Nontoxic, used for negative interfacial tension (IFT) along with surface-active agent
Glycerin, ethylene glycol	
Propylene glycol	
Ethanol	
Propanol	
Methanol	Highly toxic not used for drug delivery systems, used for negative interfacial tension (IFT) along with surface-active agent

this droplet size reduction involves an enhancement of liquid–liquid interfacial area [62]. This will cause an increment in the Gibbs free energy and to make this process spontaneous, input of external energy is required [62]. Hakansson and Rayner [62], have explained the thermodynamics of oil-in-water (O/W) nanoemulsion formation by considering the low dispersed phase fraction (1%) and monodispersed droplet (size 10 μm.) [62]. The positive Gibbs free energy indicates that the process is non-spontaneous and nanoemulsions become destabilized either by separation into phases or form the larger drops. The suitable fabrication process, however, can provide long term kinetic stability [62].

For nanoemulsion preparation, energy can be provided mechanically using devices such as homogenizers, high stress mixing, ultrasonicators, vortexing, and microfluidic systems [63, 64]. The disruptive forces such as cavitation, compression, and collision are responsible for dispersion of nanodroplets of one liquid into another non-miscible continuous liquid phase. Nonetheless, the interface is stabilized by the presence of emulsifier [64, 65]. Since, the high energy input is associated with extreme conditions (temperature and pressure), which may not be suitable for biomolecules or some of the drug molecules [65].

The disruptive force required for droplet disruption has to overcome restoring forces that maintain the drops in spherical shape (drop shape) [16, 17, 65]. This restoring force or the Laplace pressure which is related to interfacial tension and droplet size is given as $\Delta P = \gamma/2r$, here γ is the interfacial tension and r is the drop radius. Size reduction can be achieved easier for larger drops and with low interfacial tension [16, 17, 65, 66]. The droplet size can be reduced with ease but to some extent, as size is reduced further breaking of droplets becomes challenging [16, 17]. Higher stress is required for much smaller dispersed droplets, thus high stress processes can form droplets with wide size distribution [65]. In addition, the type of surfactant, liquid viscosity (dispersed as well as continuous phase), and strength of mechanical agitation also influence the particle size distribution [16, 17, 65]. Thus, polydisperse droplets can be formed by careful manipulation of these factors [65, 67]. Some of the high energy processes are shown in Fig. 2 [16, 17]. These processes will be discussed briefly.

A. **Microfluidizer**

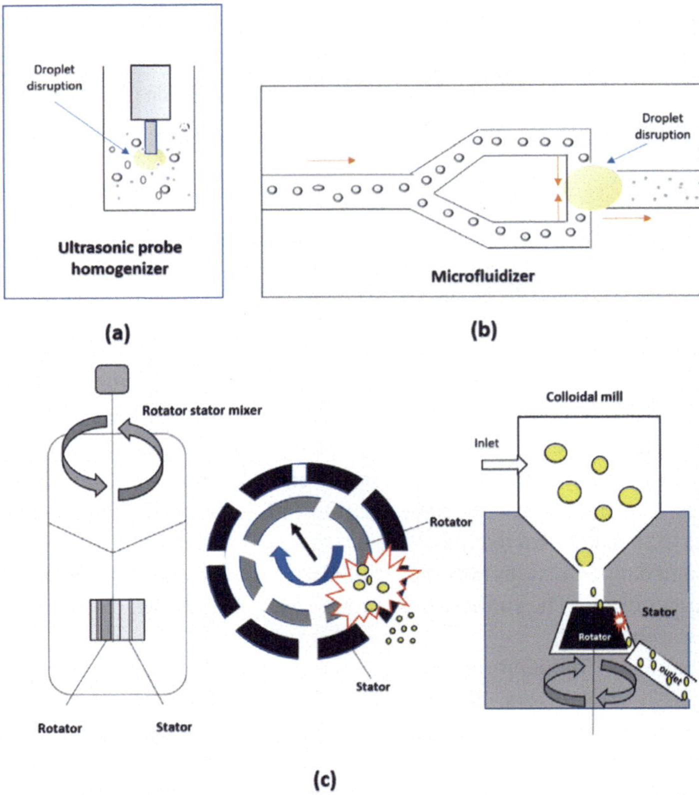

Fig. 2 The schematic diagrams of high energy processes showing **a** ultrasonic probe homogenizer, **b** microfluidizer, and **c** rotor–stator emulsification devices with rotor–stator mixer (on the left) and a colloidal mill (right)

This technique generally used in industries, involves synthesis of nanoemulsion owing to high-shear stress provided to the liquid mixture inside the microfluidic device [16, 17, 42, 65, 68]. This instrument used is known as microfuidizer having channels called microchannels and an interaction chamber as shown in Fig. 2b. Typically, microfluidizer can be Y-shaped or Z-shaped. Mechanical pump (5000–20,000 psi) can provide high pressure for the production of nanoemulsion [42, 65].

The microfluidizer separates the stream of emulsion flowing in a channel into two different streams flowing in separate channels and this is done by application of external pressure [16, 17]. The separated streams (jet) moving at high velocity are then made to interact in the interaction chamber [16, 17]. A high disruptive force gets generated when fast moving streams meet and collide inside the interaction chamber [68]. This interaction of fastmoving streams leads to reduction in the droplet size [16, 17, 68]. The methodology is used in pharmaceutical, food as well as beverage

Table 3 The NEs formed using microfluidizer

S. No.	Component	Emulsifier	Particle size	References
1	Vitamin C and A	soybean lecithin	Vit-A: 475.7 nm Vit-C: 225.4 nm	[69]
2	Antibacterial natural mixture (citral, trans-2-hexen-1-ol, and linalool, 1:1:1 w/w)	Tween 80	100–200 nm, depends upon pressure and number of cycles	[70]
3	α-Tocopherol	Decaglyceryl monooleate soyabean lecithin, Glycerol	80–400 nm	[71]
4	Tween20, SDS, beta-lactoglobulin		60–150 nm	[16, 17]
5	Non-ionic surfactant, glycerol, soybean lecithin		60 nm	[71]

industries [16, 17]. Some of the recent work reporting the fabrication of NEs using microfluidizers is shown in Table 3.

These studies indicate that smaller droplet sizes can be obtained by optimization of conditions for inhibition of droplet coalescence and to facilitate droplet disruption [16, 17]. Luo et al. [72] have fabricated various emulsions using natural surfactants (saponin and lecithin) by dual-channel microfluidizer [72]. The use of microfluidizer for nanoemulsion fabrication has also been reported by Komiako et al. [73] The emulsions with smaller droplet size with mean diameter less than 150 nm was achieved using microfluidic device. The particle size was optimized by the presence and concentration of natural emulsifier phospholipids. The presence of electrostatic interactions in stabilizing the nanoemulsion was studied by measuring the influence of ionic strength and pH on electrical charge and on aggregation behavior. Jafari, He, and Bhandari [54] have fabricated oil-in-water nanoemulsions using emulsifiers as whey protein concentrate, modified starch, maltodextrin as dissolved in the continuous phase with dispersed phase having d-limonene. Emulsions were made using microfluidizer and sonicator [54]. The nanoemulsions with particle size 150–170 nm were obtained and the comparison of the two techniques reveals that sonication produces particles with wider size distribution [54]. For both processes optimal conditions were required for particle size manipulation, otherwise, particle size can increase or may remain unaltered with processing conditions. The nanoemulsion fabrication process with droplet size distribution can be manipulated to form smaller particle size by increasing the number of microchannels in microfluidic devices or by increasing the pressure [63, 74, 75]. Obviously, the type and concentration of surfactants and the ratio of viscosities of continuous and dispersed phases are also important factors for size reduction [63, 75].

B. **High-pressure Homogenization**

This process employed for droplet size reduction has been used in food industries and also for cell disruption [76, 77]. In this technique, emulsion or suspension is made to flow with high velocity through a narrow orifice (passage) under high pressure (20 MPa to 100 MPa) [76, 78]. The pressure range (20–100 MPa) is employed for physical homogenization to form emulsion in cosmetic, beverage, and in pharmaceutical industries [76]. This methodology enhances emulsion stability against drop coalescence and creaming [76]. The solvent system with low viscosity is converted to emulsion using HPH (high-pressure homogenizer), whereas for high viscosity dispersions rotor–stator mixer (RSM) is used [79–81].

Homogenizer machines have a pump that forces the dispersion under pressure to pass through homogenizer valve [63, 78, 79]. The liquid mixture containing surfactant and cosurfactant is passed through the narrow orifice. The emulsion having large fraction of dispersed phase is formed first, which is diluted further [63]. The process can be manipulated by increasing the amount of surfactant, number of cycles for homogenization, or by modification in the equipment design to reduce the particle size [78, 79]. The equipment design (homogenizer chamber, valve, intensifier, and pressure-resistant material) modified to enhance the operating temperature up to 140–150 °C and pressure up to 400 MPa [78].

Basically, homogenization process involves pumping coarse emulsion through a narrow passage (with a gap of size in the range of few 10–100 μm). A high pressure is generated at the narrow passage (of the microstructured device) providing the flow conditions such that droplets are reduced in size [79, 82]. The downstream of the narrow passage a turbulent jet is formed and this jet encounters a comparatively stagnant fluid phase (see Fig. 3). This encounter occurs in the outlet chamber and the jet stream transfer kinetic energy leading to rise in temperature [79]. Such interactions between the turbulent jet and stagnant liquid lead to droplet break up [63, 78, 79].

C. **High-shear Mixer**

When highly viscous continuous phase is present or when fraction of dispersed phase volume is too high, then high-shear mixer or rotor–stator mixer (RSM) is employed for pumping the system through high-pressure homogenizer [79]. The detailed description of RSM can be checked in Ref. [79]. The diagrammatic representation of RSM is shown in Fig. 2c.

The powerful rotor causes fluid to be accelerated tangentially and the fluid is then made to pass through the slots present in the stator as shown in Fig. 2c. The powerful hydrodynamic stress exists due to velocity gradient and/or due to strong turbulence [79, 83]. This stress and the emulsion drop size depend upon the rotator speed which typically ranges from 10 to 30 ms^{-1} in industries [79, 84, 85].

D. **Membrane Based Process**

In this process, the emulsion preparation is carried out using a porous microporous membrane as shown in Fig. 4. This porous membrane not only acts as a separation between the continuous and dispersed phase but also allows the dispersed phase to pass through pores and mix with the continuous phase [79]. Basically, the continuous phase flows across the membrane as shown in Fig. 4 and the droplets are grown

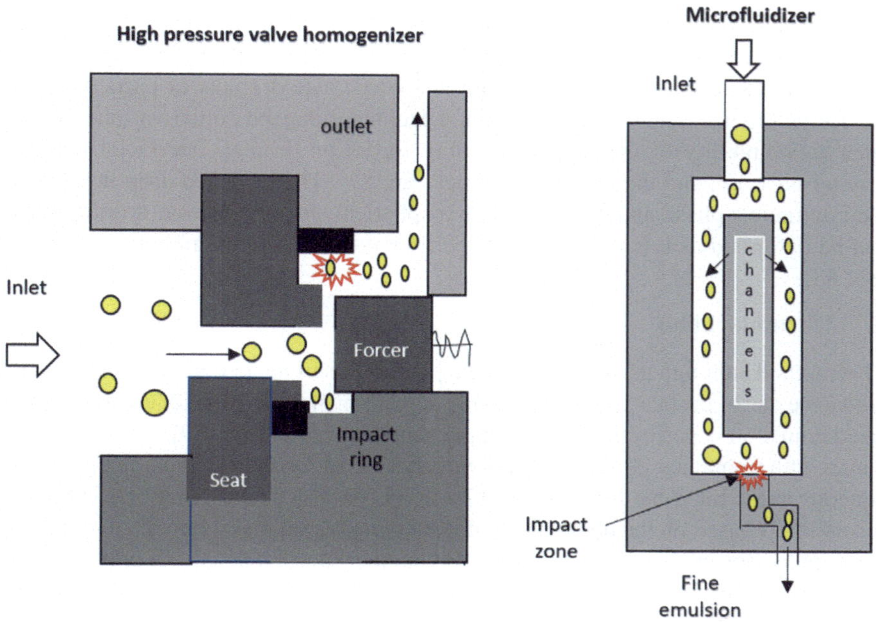

Fig. 3 The illustration of the principle of High-pressure homogenization (valve) shown on the left and micro-fluidization on the right

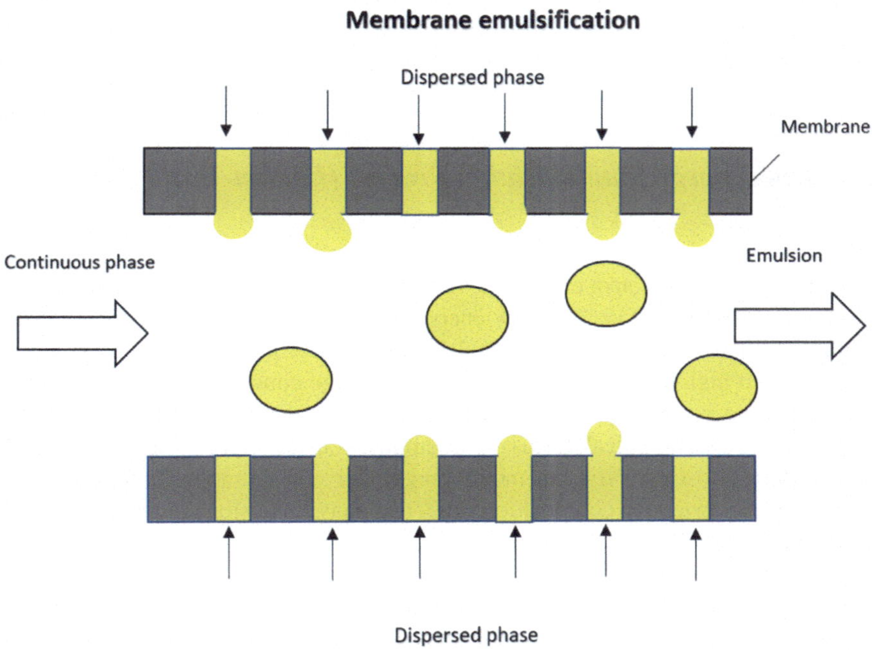

Fig. 4 Membrane emulsification method shown in schematic illustration

in the continuous phase from the porous outlets of the membrane [79, 86]. The droplets are formed as the dispersed phase is forced through the passages in the hole by application of suitable pressure [79, 86]. Once droplets of particular size are formed, detachment occurs and this detachment depends upon various factors such as wettability of the membrane, drag forces on droplet, interfacial tension, driving pressure, and the droplet buoyancy [86, 87]. The detached drop flows with the continuous phase and tends to attain a spherical shape; however, droplet shape can be deformed as it is influenced by the flow pressure and membrane wettability [86, 88].

E. **Ultrasonication**

This method although used for small scale preparation of nanoemulsions and considered to be one of the best techniques. In this, agitation of the system for creating nanosized emulsion is performed by ultrasonic sound waves [68, 79–89]. The ultrasonic waves with frequency 20–100 kHz provide disruptive forces for breaking up oil and aqueous phase into tiny droplets [16, 17, 63, 68, 79–89]. The nanoparticle (droplet) formation is based on the phenomena of dispersing the oil phase into the continuous water phase (or vice-versa) [68]. However, another mechanism also works, thereby ultrasonic waves creating the disruptive force (turbulence) making the oil drops be broken down into smaller fragments and these tiny fragments demonstrate enhanced stability [68, 90].

The ultrasonic homogenizer is shown in Fig. 2. For effectiveness of the process, premix should have the maximum exposure time at the region where disruption of droplet occurs [16, 17, 79, 91]. The process can be performed batchwise or can be adopted for continuous droplet disruption [16, 17, 79, 92].

2.3 Low-Energy Emulsification Process (Bottom-Up)

This low-energy process involves mixing the dispersed phase, continuous phase, and surfactants by modulation of environmental conditions (temperature) or composition of various components [40, 93]. Low-energy processes are phase inversion concentration or temperature, spontaneous emulsification [40, 93]. The low-energy process, as opposed to high energy process, may require large concentration of surfactants and reduce the possibility of NEs to be used for food related applications [93]. Low-energy process typical forms W/O emulsion using conventional mixer and then converted to O/W NEs but modulating either concentration or temperature, known as phase inversion composition (PIC) and phase inversion temperature (PIT), respectively [40, 94].

NE preparation caused due to rapid diffusion of solvent or surfactant molecules from dispersed to continuous phase by keeping the spontaneous curvature of surfactant unchanged is called self-emulsification [94]. As mentioned above, the modulation of surfactant spontaneous curvature during emulsification from positive to negative (forms W/O emulsion) or from negative to positive (O/W emulsion) [94] known as phase inversion methods [94].

A. **Spontaneous Emulsification**

When emulsions are formed without utilizing mechanical force or temperature variation are known as self-emulsification or spontaneous emulsification [40, 66]. The inversion of phases by addition of a phase (oil/water) to another phase [66]. This phase inversion can be in presence or absence of surfactants. When phase inversion occurs in absence of surfactant forming nanoemulsion is known as "Ouzo effect" [40, 66, 94]. For pharmaceutical applications, self-emulsifying drug delivery systems (SEDDS) or self-nanoemulsifying drug delivery systems (SNEDDS) are formulated using self-emulsification. The proposed mechanisms for NE formation are (a) interphase instability (Marangoni effect) by surface tension gradient causing individual droplet formation, (b) upon mixing of two phases, a new phase tend to condense in the supersaturation area or (c) dispersion by spontaneous increase in interfacial (surface) area when interfacial tension drops to zero [93]. In food related industries, the emulsions are prepared by spontaneous method, which involves mixing two phases, making one component become dispersed in the continuous phase causing increment in turbulent force, interfacial area (i.e., oil–water), and production of tiny droplets [93]. The spontaneous process in other words can be described as the one which makes use of chemical energy released during dilution [94]. The dilution causes formation of NEs as water soluble components diffuse from oil phase into the aqueous layer, increasing the interfacial area and forming a metastable state emulsion [94].

For the system containing water, SDS (sodium dodecyl sulfate), dodecane, and pentanol, the NEs were formed by self-emulsification upon dilution [95]. [Here, pentanol was used as a cosurfactant which causes the formation of NEs, as alcohol diffuses from oil to aqueous layer upon dilution [66, 94, 95]. The microemulsion becomes thermodynamically unstable, forming a NE, this diffusion process is depicted in Fig. 5.

B. **Phase Inversion Temperature Method**

In this method, firstly, the emulsion is formed simply by stirring oil–water mixture containing surfactant at room temperature. The emulsion thus formed is heated above phase inversion temperature. The NEs are formed and the system is diluted with cold water or rapidly cooled while stirring the mixture leading to the formation of O/W NEs [66, 94]. Thus, the solubility of surfactant in continuous phase is modulated by changes in temperature [93, 94]. This change in specific affinity of a surface-active agent (non-ionic) towards oil–water phase occurs by temperature variation, but at constant composition [93]. The surfactant (non-ionic) may be lipophilic at high temperatures and hydrophilic at low temperatures [66, 93]. The polyoxyethylene chains get dehydrated with enhancement of temperature responsible for distinct behavior [93, 94]. This temperature dependent relative surfactant solubility occurs at the cloud point [94]. The phase transfer occurs due to inversion of emulsifier film curvature from negative to positive or vice-versa. The surfactant monolayer which

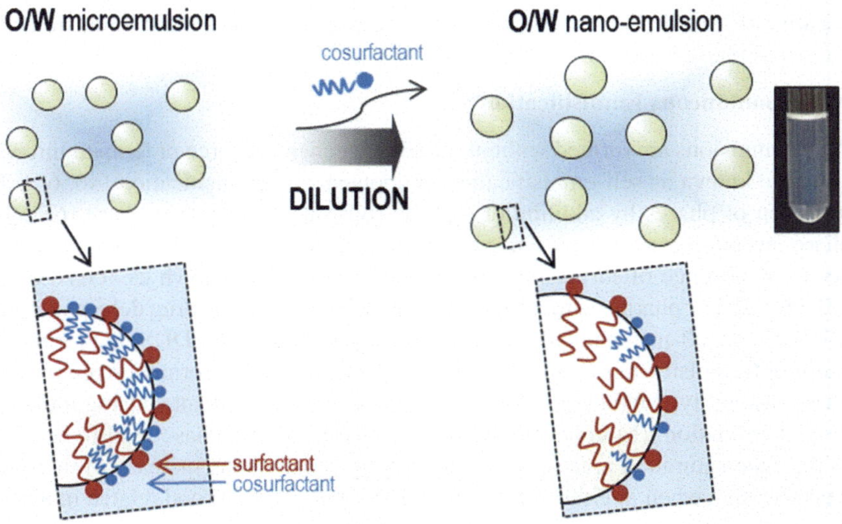

Fig. 5 The proposed mechanism for the self-emulsification process by dilution of O/W microemulsion. The dilution with water causing diffusion of cosurfactant from the oil–water interface to the aqueous phase, forming nanoemulsion. (Reproduced with permission from Ref. [94])

is positive at low temperature may coexist with an excess of oil (dispersed) phase (O/W) becomes negative at higher temperature coexisting with aqueous phase (W/O) [93, 94]. When the spontaneous curvature is zero (average), a bi-continuous phase with equal portions of oil and water phases coexists with both the phases [93]. Thus, transitions occurring from the surfactant film having zero curvature plays a crucial role in the formation of NEs [93, 94]. This inversion method is based on diverting the intrinsic feature of a rather thermodynamically stable microemulsion by dilution (with water/oil) forming kinetically stable inverse or direct NEs. Although the emulsification is spontaneous at the HLB temperature, still the NE system may lack stability [93]. The nature and type of surfactant (HLB) along with temperature are the factors influencing the stability of NEs.

C. Phase Inversion Composition

A method similar to PIT involves phase inversion triggered by modulating the composition instead of temperature [93]. The W/O emulsion system with high salt concentration can be changed to O/W by dilution with water, reducing the ionic strength. On the other hand, addition of salt causes phase inversion in an O/W stabilized by anionic surfactant to invert the phase forming W/O emulsion [93, 94].

3 Characterization

For the characterization of nanoemulsion, the droplet size (nanoscopic) and surface charge are evaluated using Dynamic Light Scattering along with Zeta-Potential Analyzer [96]. The particle size has also been evaluated using small angle X-ray scattering (SAXS), small angle neutron scattering (SANS), electron microscopic techniques scanning electron microscope (SEM), and transmission electron microscope (TEM). In addition, the morphology of nanoemulsion droplets has also been studied using atomic force microscope. The elastic behavior of nanoemulsions has also been studied by measuring the rheological properties [97–99]. Conductivity and optical turbidity are also measured for physicochemical characterization of nanoemulsion.

4 Biomedical Applications

The transparent nanoemulsions having unique properties such as tunable viscosity, enhanced stability, and small droplet size make them attractive candidates for various applications ranging from food to drug delivery systems [100]. This chapter discusses the biomedical applications of nano-system particularly in relation to drug delivery systems.

Properties such as lipophilic nature, tunable rheology, and surface charge enable water-insoluble drugs (or sparingly soluble) to be easily solubilized by nanoemulsion. Thus, hydrophobic drugs can be easily transported for drug delivery via nanocarriers. In addition, to drug delivery, nanoemulsions have also been evaluated as imaging agents and theranostic agents [31, 97, 101]. The nano-delivery system can be administered orally, internasal, or intravenously [100, 102].

A. **Transdermal Delivery**

Due to large available surface and ease of drug delivery, the drug molecules can be simply administered through the skin [96]. The skin barrier hindering the drug molecules entering the systemic circulation is one of the challenges for transdermal drug application [96, 103]. Otherwise, topical drug administration has various advantages as compared to conventional (e.g., oral) drug delivery, and these benefits are painless administration, no taste related issues, no gastric irritation, drug stability against first pass metabolism, and also no drug damage inside the gastrointestinal tract [102]. When O/W nanoemulsion based formulations are used, the lipophilic drug solubilization is enhanced by the dispersed oil droplets, while the hydrophilic continuous phase offers skin-friendly environment and is able to dissolve various biocompatible polymer (biopolymers) for improving drug formulations in terms of appearance, viscosity, and texture [100].

The transdermal drug delivery using nanoemulsion formulations has been reported [100, 104–107]. and some of these studies were conducted ex vivo on rat skin [108–113]. Lu et al. studied the fabrication and application of sorbitane trioleate, polyoxyethylene (20) oleyl ethe based nanoemulsions with drug D-limonene [110]. The results indicate that the HLB (hydrophilic-lipophilic) balance influences the dispersed droplet size as well as the encapsulation ability. The smaller dispersed droplet size demonstrated the better encapsulation ration. The drug delivery (permeation of D-limonene) using this nanoemulsion was studied on abdominal skin of male Sprague–Dawley rats. The light microscopy was used for histopathology and is shown in Fig. 6. The rat skin untreated and after nano-formulation treatment (skin permeation) was studied by Franz diffusion cell. The emulsion with the smallest droplet size of 54 nm has been shown to achieve the maximum permeation rate. The results indicate that nanoemulsions can be used safely for transdermal delivery of the drugs as the histopathology reveals no empty spaces or voids in the epidermal region of rat abdominal permeated skin [110].

Similarly, Shakeel and Ramdan [113], have prepared various W/O nanoemulsions using surfactants (Tween, 85, Tween 80, Labrasol, Cremophor, Plurol oleique, and Transcutol-HP) along with cosurfactants such as isopropyl alcohol, n-butanol, propylene glycol, and PEG-400). The nanoemulsions were characterized (morphology, viscosity, drop size, and refractive index) and used for in vitro skin permeation studies using the rat skin on Franz diffusion cell [113]. The nanoemulsion was optimized and the permeation results as shown in Fig. 7 were evaluated and compared with aqueous caffeine solutions. The effectiveness of the formulations was inferred on the basis of experimentally obtained parameters such as steady-state flux, permeability coefficient, and enhancement ration.

The nano-formulation can improve bioavailability of drug molecules as nanoemulsion can provide such a drug design allowing predetermined drug release and at a controlled rate [96, 114]. The topical route delivery of the drug can be self-administered with ease and without pain [16]. This methodology is adopted for the treatment of various diseases such as hypertension, diabetes, anticancer, arthritis (anti-inflammatories [96, 116] and also adopted for the administration of various active ingredients such as hepatoprotective drugs, antioxidants, herbs [96, 116,

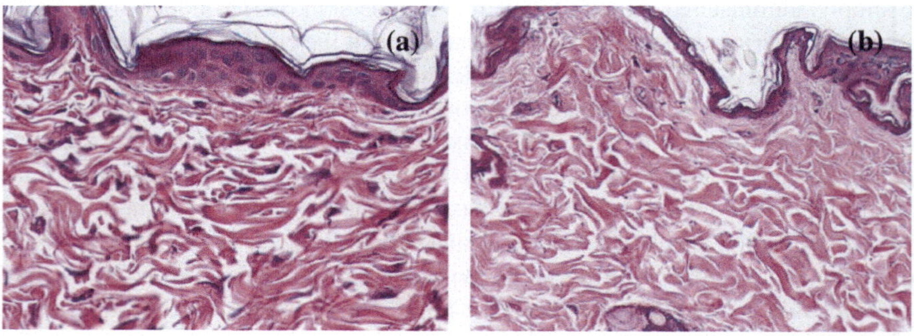

Fig. 6 Light microscopic images (×100) of rat skin **a** no treatment and **b** after 360 min of treatment with D-limonene NEs. (Reproduced with permission from Ref. [110])

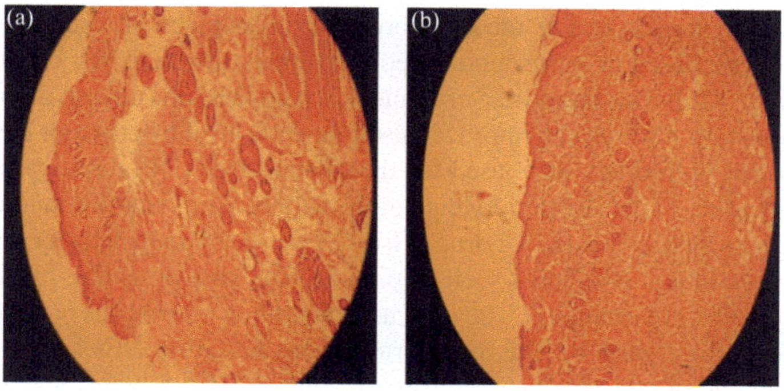

Fig. 7 Images (photomicrograph) of skin sample of **a** control animal group and **b** animal treated with C12 NE (40 × power) . (Reproduced with permission from Ref. [113])

117]. A few recent studies demonstrating the effectiveness of nanoemulsions for transdermal drug application have also been reported [96, 100, 115].

The studies suggest that nanoemulsion based formulations owing to low zeta-potential and smaller particle size can deliver hydrophobic drugs effectively when compared with drug suspension [100, 118, 119]. For an efficient delivery system with optimal therapeutic action, the factors which should be considered for nanoemulsions formation includes the nature and types of emulsifier, particle size, physicochemical properties of nanoemulsions, skin factors (e.g., thickness, nature of stratum corneum, anatomical location, surface charge on skin, etc. and obviously the choice of drug molecules [115]. Nanoemulsions are used to modify both the skin permeating conditions as well as drug distribution deep into the skin [100, 115]. In addition, the side effects related to skin toxicity and skin irritations should be minimized when the nanoemulsion transdermal delivery is being optimized for human use [115]. The nano-formulations have shown better therapeutic response in terms of transdermal drug permeation as compared to the drug delivery system based on gels or emulsion [103, 120–124].

B. **Intranasal Drug Delivery**

The non-invasive intranasal drug delivery is being employed for the treatment of disease/disorder related to brain by targeted drug delivery to brain cells [96]. For this, the drug is directly administered to brain by overcoming the blood–brain barrier (BBB). This cellular defensive blood–brain barrier also acts as interface controlling the central nervous system (CNS) micro-environment [125]. The BBB owing to its restrictive behavior restricts the drug transport and to overcome this challenge so as to achieve the required therapeutic amount of drug, various intranasal formulations have been designed [96, 125].

The BBB restrictions can hinder about 98–100% of the active drug molecules from passing the BBB, which depends upon the size and molecular weight of the drug molecules [96, 125], thus causing low bioavailability at the target [96, 125]. Methods for drug delivery to central nervous system (CNS) or brain included intracranial delivery, intra-parenchymal or intra-cerebroventricular injection, catheter infusion, electromagnetic based or ultrasound-based techniques, and requires surgical intervention [125, 126]. The nasal delivery thus is a pain-free non-invasive process for drug administration which can improve the therapeutic level by crossing BBB [125].

NEs loaded with drugs were used for the treatment of CNS diseases as well as for brain imaging [125]. Risperidone an antipsychotic drug bearing mucoadhesive nanoemulsions were evaluated for drug delivery via nose to brain delivery path [127]. The physicochemical properties such as particle size, zeta-potential, pH, and drug loading content were evaluated. The drug administration on live rats provides a way for studying drug transport. The drug formulations based on NEs have been used for the treatment of various diseases/disorders related to CNS/brain such as depression, Parkinson's diseases, schizophrenia, Meningitis, Alzheimer disease, cerebral ischemia, Epilepsy, Antiepileptic, and Tumor [128–131].

C. **Opthalmic Drug Delivery**

For treatment of various eye related diseases, the commonly preferred route for drug transport and administration is via ocular application [96, 132]. Greater than 90% ophthalmic formulations comprised of eye drops, however, the efficiency of eye drops is restricted owing to instant pre-corneal loss, low permeability/transport across corneal epithelium, high loss due to tears, and temporary residence time [132, 133]. The limitation, i.e., only about 5% or less amount of drug molecules reaches intraocular tissues after penetrating the corneal membrane [132, 134]. Thus, among various alternatives, NEs demonstrate promising results owing to several advantages such as better ocular retention time, better drug permeation across corneal membrane, lessen side effects, flexibility of incorporating lipophilic/lipophobic drugs, and sustained drug delivery [132]. Shah et al. fabricated antibacterial drug Moxifloxacin containing NEs by optimization of various physicochemical properties of NEs as well as components (oil, surfactant) required for NEs formation [132]. The drug administration through the use of NEs demonstrated the higher drug concentration achieved which is much greater than the minimum concentration required for the treatment of eye infections [132–136].

For treatment of eyes and ear related diseases/infections, a high concentration of drugs in eye/ear drops is used, as most of the drug is drained out due to nasolacrimal drainage or lacrimal secretion [103]. Thus, conventional drug drops suffer from low therapeutic response as well as poor bioavailability [103]. NEs are being explored to overcome these limitations normally encounter for the treatment of eyes/ear/nose related infections [96, 103]. Physicochemical properties of NEs can be studied to optimize the drug loading and sustain drug release ability of these formulations [96, 103]. Lim et al. [137] formulated self-emulsifying NE formulation for the solubilization and permeation of lutein. The NE with narrow size distribution (10–12 nm) was developed and evaluated for drug release ability. The O/W NEs showing potential for ophthalmic delivery have been commercialized [97]. The first NEs to be commercialized for the treatment of dry eye contains Cyclosporin A and cator oil, glycerine, carbomer copolymer type A, NaOH (pH adjuster), polysorbate 80, and purified water [97]. This medicine also known as artificial tears is used for treatment of dry eyes.

Among these artificial tears, the NE based eye drops are Soothe ® (Bausch & Lomb, Inc.), Novasorb® (Novigali Pharma), and Refresh Endura ® (Allergan, Inc.) [97, 138, 139]. The conjunctival cell membrane and corneal at physiological pH are negatively charged and to enhance the drug interactions, positively charged (cationic) NEs are being studied [97, 138, 139].

D. **Theranostic Agents (Also Include Cancer Treatment)**

Incorporation of imaging molecules, as well as therapeutic drug molecules in a single entity, is termed theranostic and these agents are known as theranostic agents [31]. Theranostic agents are not only used for disease detection but also for the targeted delivery of drugs, so as to simultaneous monitor the delivery system [31]. In addition, such modalities improve the limitations associated with conventional techniques [31]. Extensively, the theranostic agents (nanoparticle based) have been used for tumor detection and treatment [31]. The chemotherapy which is the destruction of rapidly growing cells by use of chemicals has been used for the treatment of cancer. The enhanced permeability and retention (EPR) effect of nanocarrier provides the opportunity for drug delivery [101]. This effect leads to longer circulation along with enhanced therapeutic response of loaded drugs towards cancer cells [31, 101].

Cancer biomedical imaging provides morphological, structural, metabolic, and functional information [101]. The hybrid technique of imaging can provide more information regarding the tumor cell such as staging, and this provides an opportunity for early detection and treatment planning [101, 140]. However, early detection of cancer is challenging and these nanoparticle-based formulations are being explored [31, 101]. The imaging can be done by use of hydrophobic dye molecules/contrast agent/imaging fluorescent/radioactive agents which are being administered as an encapsulated entity using nano-formulations including NEs [31, 101]. Hydrophobic drug molecules and/or contrast agents can be incorporated inside dispersed phase or can be attached to hydrophobic oil phase of NEs for simultaneous detection and diagnosis [101]. Formulations reduced in size causing miniaturization of the delivery system not only improves the EPR also favors the pharmacodynamics and -kinetics of entrapped active ingredients [101].

A schematic representation of a NE showing various modalities required for disease diagnosis and treatment is shown in Fig. 8. [101].

NEs with radiolabeled sonophores with zirconium-89 providing multi-spectral optoacoustic tomography of cancer cell was designed for non-invasive imaging [141]. The NEs were prepared by solvent exchange method. Similarly, gold nanosoheres, and ^{19}F based contrast agents were incorporated in NEs for cancer imaging by photoacoustic molecular imaging techniques [101, 142–144]. NEs based on perfluorocarbons having indocyanine used for biomedical bimodal imaging (MRI/near-IR fluorescence) [145]. The irradiation of NEs generates heat causing photothermal effect leading to cell death of glioblastoma cells (U87MG [145]. The photodynamic effect is based on fact that electromagnetic radiation (red light, 600–800 nm) excited photosensitizer, transferring its energy to oxygen forming reactive oxygen species [96, 146]. This process causes absorption of photosensitizer drug molecules by the diseased tissues [96]. The irreversible damage occurs to tumor cells [96, 147, 148].

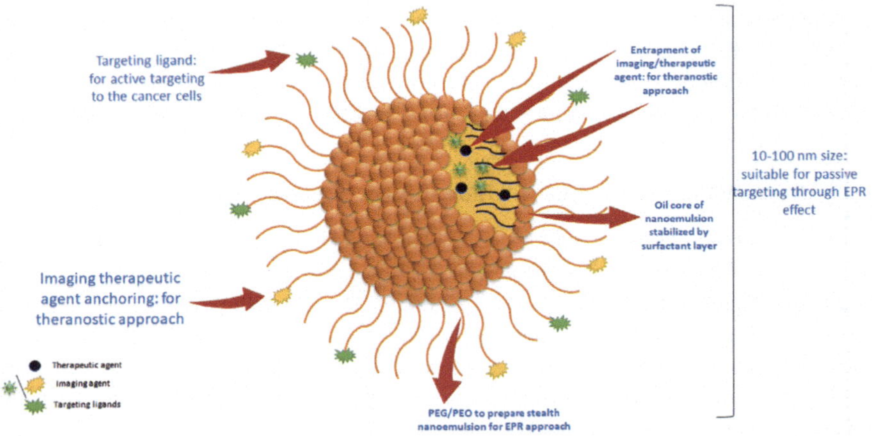

Fig. 8 The schematic representation of NE concept used as theranostic approach. PEG/PEO stands for poly(ethylene) glycol, polyethylene oxide, EPR for enhanced permeability and retention. (Reproduced with permission from Ref. [101])

This approach involving heating the photosensitive agent is not the only heat-based treatment, another method is based on hyperthermia [96]. Here radiofrequency waves are applied to NEs containing magnetic nanoparticles under the influence of alternating magnetic fields [96]. Using this methodology, photosensitizer Foscan® was delivered causing hyperthermia for cancer treatment [96, 149]. NEs based formulations have been used for the treatment of breast cancer, ovrian cancer [149–153]. Different kinds and types of surfactants are used for formulating enhanced NE systems for prolonged blood circulation, biocompatibility, and drug bioavailability [96, 97, 101, 103]. One such approach is the use of PEG (polyethylene glycol) based surfactants for NE formation. These PEG bearing molecules, not only provide opportunities for linking ligands targeting the cancer cell, but also improve the drug pharmacokinetics and therapeutic response [97, 154–157].

E. **Gene Therapy**

For the treatment of hereditary and acquired diseases, gene therapy by administration of nucleic acid is performed via NE formulations [96, 103]. It is an important therapeutic technique useful in the field of cancer therapy, tissue regeneration, and vaccination [97, 158]. Biocompatible PEG based NEs (cationic) have been used for ocular treatment/delivery of nucleic acids [97, 159, 160]. The non-viral vectors such as NEs, dendrimers, nanoparticles, and liposomes are used as a replacement of virus for alternatively delivering nucleic acids are being explored [96, 161]. The O/W cationic surfactants are mostly used delivery systems and are proposed as the transporting formulations for single as well as double-stranded nucleic acids [96, 162].

5 Conclusions

This chapter discusses the synthesis, characterization, and biomedical applications of NEs. The NEs are classified broadly having oil or water as dispersed phase, however, multiple NEs are also fabricated. The factors influencing the stability of NEs are also highlighted. For fabrication of NEs, high energy or low energy processes can be used. The production amount and the droplet size desired to govern the process used for NE preparation. From lab scale to industrial scale production of NE can be achieved.

Owing to its tunable physicochemical properties by choosing appropriate emulsifiers, dispersed and continuous phase, experimental fabrication parameters, NEs have found extensive applications in food industries as well as cosmetic industries. NEs can easily incorporate hydrophobic drugs, thereby enhancing drug solubilization. These NE formulations are being used for biomedical applications ranging from disease detection/imaging to treatment. The advantages of NE formulations are also highlighted.

Acknowledgements Authors (Bala, M & Singh, VJ) are highly thankful to Science and Engineering Research Board, New Delhi for the financial support. The authors are also thankful to Head, Department of Chemistry and Director, NIT J for the infrastructure and facilities. S. Indoria acknowledges the support and facilities from KMV Jalandhar.

References

1. Maali A, Mosavian MTH (2013) J Disp Sci Techn 34(1):92–105
2. Abdelrasoul MA, Eid AR, Badawy MEI (2020) Arch Phytopath Plant Prot 53:310–334
3. Solans C, Esquena J, Forgiarini A, Morales D, Usón N, Izquierdo P et al (2002) Surfactant science series. Marcel Dekker, New York, pp 525–554
4. Forgiarini A, Esquena J, González C, Solans C (2000) Langmuir 17:2076–2083
5. Dammak I, Sobral PJA, Aquino A, das Neves MA, Conte-Junior CA (2020) Compr Rev Food Sci Food Saf 19:2721–2746
6. Tayeb HH, Sainsbury F (2018) Nanomed 13. https://doi.org/10.2217/nnm-2018-0088
7. Wang L, Guan X, Zheng C, Wang N, Lu H, Huang Z (2020) Langmuir 36:10082–10090
8. Klaus A, Tidyy GJTY, Solans C, Harrar A, Touraud D, Kunz W (2–12) Langmuir 28:8318−8328
9. Meng R, Wang C, Shen Z, Wang R, Kuru E, Jin J (2020) J Mol Liq 314:113663
10. Wik J, Bansal KK, Assmuth T, Rosling A, Rosenholm JM (2020) Drug delivery. Transl Res 10:1228–1240
11. Sonneville-Aubrun O, Yukuyama M N, Pizzino A (2018) Application of nanoemulsions in cosmetics. In: Jafari SD, McClement DJ (eds) Nanoemulsions formulation, applications, and characterization. Academic Press: Cambridge, MA, USA, pp 435–475. ISBN 978-0-12-811838-2
12. Yukuyama MN, Ghisleni DDM, Pinto TJA, Bou-Chacra NA (2016) Int J Cosmet Sci 38(1):13–24
13. Kumari S, Kumaraswamy RV, Choudhary RC, Sharma SS, Pal A, Raliya R, Saharan V (2018) Sci Rep 8(1):6650
14. Lu W-C, Huang D-W, Wang C-CR, Yeh C-H, Tsai J-C, Huang Y-T, Li P-H (2018) J Food Drug Analy 26(1):82–9

15. Silva HD, Cerqueira MA, Souza BWS, Ribeiro C, Avides MC, Quintas MAC, Vicente AA (2011) J Food Eng 102(2):130–135
16. Rao J, McClements DJ (2011) J Agric Food Chem 59(9):5026–5035
17. McClements DJ, Rao J (2011) Food-grade. Crit Rev Food Sci Nutr 51(4):285–330
18. Graciaa A, Lachaise J, Cucuphat C, Bourrel M (1993) Salager J L. Langmuir 9:669–672
19. Graciaa A, Lachaise J, Cucuphat C, Bourrel M, Salager JL (1993) Langmuir 9:3371–3374
20. Miñna-Perez M, Graciaa A, Lachaise J, Salager JL (1995) Colloids Surf A 100:217–224
21. Velasquez J, Scorzza C, Vejar F, Forgiarini AM, Anton RE, Salager JL (2009) J Surfactants Deterg 13:69–73
22. Salager JL, Anton RE, Sabatini DA, Harwell JH, Acosta EJ, Tolosa LI (2005) J Surfactants Deterg 8:3–21
23. Witthayapanyanon A, Acosta EJ, Harwell JH, Sabatini DA (2006) J Surfactants Deterg 9:331–339
24. Witthayapanyanon A, Harwell JH, Sabatini DA (2008) J Colloid Interface Sci 325:259–266
25. Do LD, Witthayapanyanon A, Harwell JH (2009) Sabatini. J Surfactants Deterg 12:91–99
26. Witthayapanyanon A, Phan TT, Heitmann TC, Harwell JH, Sabatini DA (2009) J Surfactants Deterg 13:127–134
27. Klaus A, Tiddy GJT, Touraud D, Schramm A, Stühler G, Drechsler M, Kunz W (2010) Langmuir 26:5435–5443
28. Klaus A, Tiddy GJT, Touraud D, Schramm A, Stühler G, Kunz W (2010) Langmuir 26:16871–16883
29. Javanshir A, Karimi E, Maragheh AD, Tabrizi MH (2020) J Food Meas Charact 14:1356–1365
30. Jamoussi T, Zaiter T, Desrumaux C, Acar N, Pellequer Y, Beduneau A (2021) Colloids Surf B: Biointerfaces 197:111432
31. Indoria S, Singh V, Hsieh M-F (2020) Int J Pharm 582:119314
32. McClements DJ (2012) Soft Matter 8:1719–1729
33. McClements DJ (2011) Soft Matter 7:2297–2316
34. McClemens DJ, Jafari SM (2018). Ch. 1 General aspects of nanoemulsions and their formulation. In: Jafari SM, McClements, DJ, (eds) Nanoemulsions. Academic Press, pp 3–20. ISBN 9780128118382
35. Pavoni L, Perinelli DL, Ciacciarelli A, Quassinti L, Bramucci M, Miano A, Casettari L, Cespi M, Bonacucina G, Palmieri GF (2020) J Drug Delivery Sci Technol 58:101772
36. Basha SK, Muzammil MS, Dhandayuthabani R, Kumari VS, Kaviyarasu K (2020) Current Drug Res Rev 12:1–12
37. Marshall T, Marangoni AG, Lim L-T, Tchoukov P, Pensini E (2021) J Environ Chem Eng 9:104998
38. Taha A, Ahmed E, Ismaiel A, Muthupandian AK, Xu X, Pan S, Hu H (2020) Trends Food Sci Technol 105:363–377
39. Adjonu R, Doran G, Torley P, Agboola S (2014) J Food Eng 122:15–27
40. Espitia PJP, Fuenmayor CA, Otoni CG (2019) Comprehens Rev Food Sci Food Saf 18:264–285
41. Anton N, Vandamme TF (2016) Nano-Emulsions. In: Aliofkhazraei (ed) Handbook of nanoparticles. pp 93–116
42. Qadir A, Faiyazuddin MD, Hussain MDT, Alshammari TM, Shakeel F (2016) J Mol Liq 214:7–18
43. Grigoriev DO, Miller R (2009) Curr Opin Colloid Interf Sci 14:48–59
44. Huang M, Horwitz TS, Zweiben C, Singh SK (2011) J Pharm Sci 100:4617–4630
45. Singh V, Nguyen TP, Sheng Y-J, Tsao H-K (2018) J Phys Chem C 122:24750–24759
46. Ruckenstein E (1996) Langmuir 12(26):6351–6353
47. Warisnoicharoen W, Lansley AB, Lawrence MJ (2000) Int J Pharm 198:7–27
48. Rouessac F, Rouessac A (2007) Chemical analysis: modern instrumentation methods and techniques, 2nd edn. Wiley, France
49. Izquierdo P, Esquena J, Tadros TF, Dederen JC, Feng J, Garcia-Celma MJ, Azemar N, Solans C (2004) Langmuir 20:6594–6598

50. Silva HD, Cerqueira MA, Souza BWS, Ribeiro C, Avides MC, Quintas MAC, Coimbra JSR, Carneiro-da-Chunha MG, Vicente AA (2011) J Food Eng 102:130–135
51. D'Angelo M, Fioretto D, Onori G, Palmieri L, Santucci A (1996) Phys Rev E 54:993–996
52. McClements DJ (2000) J Agr Food Chem 48:5604–5611
53. Mulik RS, Mönkkönen J, Juvonen RO, Mahadik KR, Paradkar AR (2010) Int J Pharm 398:190–203
54. Anton N, Gayet P, Benoit JP, Saulnier P (2007) Int J Pharm 344:44–52
55. Jafari SM, He Y, Bhandari B (2007) J Food Eng 82:478–488
56. Izquierdo P, Feng J, Esquena J, Tadros TF, Dederen JC, Garcia MJ, Azemar N, Solans C (2005) J Colloid Interface Sci 285:388–394
57. Olla M, Monduzzi M, Ambrosone L (1999) Colloid Surf A 160:23–36
58. Porod G (1982) Small angle X-ray scattenng. In: Glatter O, Kratky O (eds) Academic Press, London, pp 17–51
59. Porras M, Solans C, González C, Gutiérrez JM (2008) Colloid Surf A 324:181–188
60. Connolly JR, Introduction to X-ray powder diffraction. Available at 2007. http://epswww.unm.edu/xrd/xrdclass/01-XRD-Intro
61. Lee SJ, McClements DJ (2010) Food Hydrocoll 24:560–569
62. Chu BS, Ichikawa S, Kanafusa S, Nakajima M (2007) J Am Chem Soc 84:1053–1062
63. Håkansson A, Rayner M (2018) General principles of nanoemulsion formation by high-energy mechanical methods. In: Jafari SM, McClements DJ (eds) Nanoemulsions. Academic Press, CH 5, pp 103–139, ISBN 9780128118382. https://doi.org/10.1016/B978-0-12-811838-2.00005-9
64. Jasmina H, Džana O, Alisa E, Edina V, Ognjenka R (2017) Preparation of nanoemulsions by high-energy and lowenergy emulsification methods. In: Badnjevic A (eds) CMBEBIH 2017. IFMBE proceedings, vol 62. Springer, Singapore. https://doi.org/10.1007/978-981-10-4166-2_48
65. Azmi NAN, Elgharbawy AAM, Motlagh SR, Samsudin N, Salleh HM (2019) Processes 7:617
66. Sheth T, Seshadri S, Prileszky T, Helgeson ME (2020) Nature Rev Materials 5:214–228
67. Saffarionpour S (2019) Food Eng Rev 11:259–289
68. Bilati U, Allémann E, Doelker E (2003) Pharm Dev Technol 8:1–9
69. Naseema A, Kovooru L, Behera AK, Kumar KPP, Srivastava P (2021) Adv Colloid Interface Sci 287:102–318
70. Kim T-I, Kim T-G, Lim D-H, Kim S-B, Park S-M, Hur T-Y, Ki K-S, Kwon E-G, Kumar MV, Kim Y-J (2019) Molecules 24:2566
71. Taghavi E, Mirhosseini H, Rukayadi Y (2018) Food Bioprocess Technol 11:645–659
72. Hatanaka J, Kimura Y, Lai-Fu Z, Onoue S, Yamada S (2008) Int J Pharm 363:112–117
73. Luo X, Zhou Y, Bai L, Liu F, Zhang R, Zhang Z, Zheng B, Deng Y, McClements DJ (2017) Food Res Int 96:103–112
74. Komaiko J, Sastrosubroto A, McClements DJ (2016) Food Chem 203:331–339
75. Jincy J, Krishnakumar K, Anish J, Dineshkumar B (2015) Current Res Drug Target 5:1–4
76. Kentish S, Wooster T, Ashokkumar M, Balachandran S, Mawson RL, Simons L (2008) Innovative Food Sci Emerging Technol 9:170–175
77. Patrignani F, Lanciotti R (2016) Front Microbiol 7:1132
78. Shirgaonkar IZS, Lothe RR, Pandit AB (1998) Biotechnol Prog 14:657–660
79. Levy R, Okun Z, Shpigelman A (2020) Food Eng Rev https://doi.org/10.1007/s12393-020-09239-8
80. Håkansson A (2019) Annu Rev Food Sci Technol 10:239–258
81. Schultz S, Wagner G, Urban K, Ulrich J (2004) Chem Eng Technol 27(4):361–368
82. Nguyen AV, Schulze HJ (2004) Colloid science of flotation. Marcel Dekker, New York, pp 27–34
83. Kelemen K, Gepperth S, Koch R, Bauer H-J, Schuchmann HP (2015) Microfluid Nanofluid 19(5):1139–1158
84. Hinze JO (1955) AIChE J 1(3):289–295
85. Davies JT (1985) Chem Eng Sci 40(5):839–842

86. Tcholakova S, Denkov ND, Danner T (2004) Langmuir 20:7444–7458
87. Joscelyne SM, Trägårdh G (2004) J Membr Sci 169:107–117
88. Schröder V, Behrend O, Schubert H (1998) J Colloid Interface Sci 202:334–340
89. Peng SJ, Williams RA (1998) Part Syst Charact 15:21–25
90. Jafari SM, He Y, Bhandari B (2006) Internat J Food Proper 9(3):475–485
91. Li M, Fogler H (1978) J Fluid Mech 88:513–528
92. Leong T, Wooster T, Kentish S, Ashokkumar M (2009) Ultrasonics Sonochem 16(6):721–727
93. De Cindio B, Grasso G, Cacace D (1991) Food Hydrocoll 4:339–353
94. Sugumar S, Ghosh V, Mukherjee A, Chandrasekaran N (2016) Flavor safety 93–100
95. Solans C, Solé I (2012) Curr Opin Colloid Interface Sci 17:246–254
96. Taylor P, Ottewill RH (1994) Colloid Surf A 88:303–316
97. Chircov C, Alexandru MG (2019) Nanoemulsion preparation, characterization, and application in the field of biomedicine. In: Grumezescu AM (eds) Nanoarchitectonics in Biomedicine. William Andrew Publishing, Ch-6, 169–188, ISBN 9780128162002, https://doi.org/10.1016/B978-0-12-816200-2.00019-0
98. Mazza M, Alonso-Sande M, Jones M-C, de la Fuente M (2008) Fundamentals of pharmaceutical nanoscience. Springer, pp 117–158
99. Howe AM, Pitt AR (2008) Adv Colloid Interface Sci 144:30–37
100. Pal R (2011) Curr Opin Colloid Interface Sci 16:41–60
101. Gupta A, Eral HB, Hatton TA, Doyle PS (2016) Nanoemulsions. Soft Matter 12:2826
102. Gorain B, Choudhury H, Nair AB, Dubey SK, Kesharwani P (2020) Drug Discov Today 25:1174–1188
103. Sutradhar KB, Amin ML (2013) Eur J Nanomed 5(2):97–110
104. Chime SA, Kenechukwu FC, Attama AA (July 25th 2014) Nanoemulsions—advances in formulation, characterization and applications in drug delivery, application of nanotechnology in drug delivery. Ali Demir Sezer, IntechOpen https://doi.org/10.5772/58673
105. Schwarz JS, Weisspapir MR, Friedman DI (1995) Pharm Res 12:687e92
106. Baboota S, Shakeel F, Ahuja A, Ali J, Shafiq S (2007) Acta Pharm 57:315e32
107. Shakeel F, Baboota S, Ahuja A, Ali J, Aqil M, Shafiq S (2007) AAPS PharmSciTech 8:191–199
108. Kong M, Chen XG, Kweon DK, Park HJ (2011) Carbohydr Polym 86:837–843
109. Kaur R, Ajitha M (2019) J Drug Delivery Sci Technol Technology 52:968–978
110. Ahmad N, Ahmad R, Buheazaha TM, Al-Homoud HS, Al-Nasif HA, Sarafroz MT (2020) Saudi J Boil Sci 27:1024–1040
111. Lu W-C, Chiang B-H, Huang D-W, Li P-H (2014) Ultrason Sonochem 21:826–832
112. Aziz ZAA, Nasir HM, Ahmad A, Setapar SHM, Ahmad H, Noor MHM et al (2019) Sci C Rep 9:13678
113. Mostafa DM, Kassem AA, Asfour MH, Okbi SYA, Mohamed DA, El-Sayed Hamed T (2015) J Mol Liq 212:6–15
114. Shakeel F, Ramadan W (2010) Colloids Surf B 75:356–362
115. Ahmad G, Sadda ER, Botchkina G, Ojima I, Egan J, Amiji M (2017) Cancer Lett 406:71–80
116. Rai VK, Mishra N, Yadav KS, Yadav NP (2018) J Controlled Release 270:203–225
117. Mostafa DM, Kassem AA, Asfour MH, Al Okbi SY, Mohamed DA (2015) Hamed TE-S. J Mol Liq 212:6–15
118. Nagi A, Iqbal B, Kumar S, Sharma S, Ali J, Baboota S (2017) J Drug Delivery Sci Technol 40:35–44
119. Tagne J-B, Kakumanu S, Nicolosi RJ (2008) Mol Pharmaceutics 5:1055–1063
120. Tagne J-B, Kakumanu S, Ortiz D, Shea T, Nicolosi RJ (2008) Mol Pharmaceutics 5:280–286
121. Ktistis G, Niopas I (1998) J Pharm Pharmacol 50:413–419
122. Gasco MR, Gallarate M, Pattarino F (1999) Int J Pharm 69:193–196
123. Kriwet K, Muller-Goymann CC (1995) Int J Pharm 125(2):231–242
124. Trotta M (1999) J Control Rel 60(2):399–405
125. Sheela AY, Sushil KP, Singh DK, Shaikh A (2012) Int J Pharm Pharm Sci 4(1):41–44
126. Bonferoni MC, Rossi S, Sandri G, Ferrari F, Gavini E, Rassu G, Giunchedi P (2019) Pharmaceutics 11:84

127. Dhuria SV, Hanson LR, Frey WH (2010) J Pharm Sci 99:1654–1673
128. Kumar M, Misra A, Babbar AK, Mishra AK, Mishra P, Pathak K (2008) Int J Pharm 358:285–291
129. Kaur A, Nigam K, Bhatnagar I, Sukhpal H, Awasthy S, Shankar S, Tyagi A, Dang S (2020) Drug Deliv Transl Res 10(6):1862–1875
130. Kaur A, Nigam K, Bhatnagar I et al (2020) Drug Deliv Transl Res 10:1862–1875
131. Kumar M, Nishad DK, Kumar A, Bhatnagar A, Karwasra R, Khanna K, Sharma N (2020) J Microencapsulation 1–21
132. Bahadur S, Pardh DM, Rautio J, Rosenholm JM, Pathak K (2020) Pharmaceutics 12:12–30
133. Shah J, Nair AB, Jacob S, Patel RK, Shah H, Shehata TM, Morsy MA (2019) Pharmaceutics 11:2–30
134. Agrahari V, Mandal A, Agrahari V, Trinh HM, Joseph M, Ray A, Hadji H, Mitra R, Pal D, Mitra AK (2016) Drug Deliv Transl Res 6:735–754
135. Patel A, Cholkar K, Agrahari V, Mitra AK (2013) World. J Pharmacol 2:47–64
136. Singh M, Bharadwaj S, Lee KE, Kang SG (2020) J Controlled Release 328:895–916
137. Qamar Z, Qizilbash FF, Iqubal MK, Ali A, Sahni JK, Ali J, Baboota S (2019) Recent Patents Drug Delivery Formulation 13:246–254
138. Sim T, Lim C, Hoang NH, Joo H, Lee JW, Kim D-W, Lee ES, Youn YS, Kim JO, Oh KT (2016) J Pharm Inv 46(4):351–362
139. Lallemand F, Daull P, Benita S, Buggage R, Garrigue JS (2012) J Drug Deliv 604204
140. Klang S, Abdulrazik M, Benita S (2000) Pharm Dev Technol 5:521–532
141. Fass L (2008) Mol Oncol 2:115–152
142. Roberts S, Andreou C, Choi C, Donabedian P, Jayaraman M, Pratt EC, Tang J, Carlos P-M, Jason de la CM, Mulder WJM, Grimm J, Kircher M, Reiner T (2018) Chem Sci 9:5646
143. Wei T-R, Tan G, Wu C-F, Chang C, Zhao LD, Li J-F, Snyder GJ, Kanatzidis MG (2017) Appl Phys Lett 110:053901
144. Kislukhin AA, Xu H, Adams SR, Narsinh KH, Tsien RY, Ahrens ET (2016) Nat Mater 15:662–668
145. Janjic JM, Ahrens ET (2009) WIREs Nanomed Nanobiotechnol 1:492–501
146. Wang Y-G et al (2013) Quant Imaging Med Surg 3:132–140
147. Furtado CM, de Faria FSEDV, Azevedo RB, Karen Py-D, Camara ALdS, da Silva JR, Oliveira Ede H, Rodriguez AFR, Degterev IA (2017) J Photochem Photobiol 167:242–248
148. Monge-Fuentes V, Muehlmann LA, Longo JPF, Silva JR, Fascineli ML, de Souza P, Azevedo RB (2017) J Photochem Photobiol 166:301–310
149. Castilho-Fernandes A, Lopes TG, Primo FL, Pinto MR, Tedesco AC (2017) Photodiagn Photodyn Ther 19:221–228
150. Sugumaran A, Ponnusamy C, Kandasamy P, Krishnaswami V, Palanichamy R, Kandasamy R, Natesan S (2018) Eur J Pharm Sci 116:15–25
151. Natesan S, Sugumaran A, Ponnusamy C, Thiagarajan V, Palanichamy R, Kandasamy R (2017) Int J Biol Macromol 104:1846–1852
152. Chen L, Chen B, Deng L, Gao B, Zhang Y, Wu C, Chen J (2017) Intern J Pharma 534:308–315
153. Kim J-E, Park Y-J (2017) Colloids Surf B150:362–372
154. Rajpoot P, Bali V, Pathak K (2012) Int J Pharm 426:219–230
155. Béduneau A, Saulnier P, Hindre F, Clavreul A, Leroux JC, Benoit JP (2007) Biomaterials 28:4978–4990
156. Allard E, Passirani C, Garcion E, Pigeon P, Vessières A, Jaouen G, Benoit JP (2008) J Control Release 130:146–153
157. Bourseau-Guilmain E, Béjaud J, Griveau A, Lautram N, Hindré F, Weyland M, Benoit JP, Garcion E (2012) Int J Pharm 423:93–101
158. Torrecilla D, Lozano MV, Lallana E, Neissa JI, Novoa-Carballal R, Vidal A, Fernández-Megia E, Torres D, Riguera R, Alonso MJ, Dominguez F (2013) Eur J Pharm Biopharm 83(3):330–337
159. Schatzlein AG (2001) Anticancer Drugs 12:275–304

160. Hagigit T, Abdulrazik M, Valamanesh F, Behar-Cohen F, Benita S (2012) J Control Release 160:225–231
161. Lozano MV, Lollo G, Alonso-Nocelo M, Brea J, Vidal A, Torres D, Alonso MJ (2013) J Nanopart Res 15(3):1–14
162. Silva AL, Marcelino HR, Veríssimo LM, Araujo IB, Agnez-Lima LF, do Egito EST (2016) J Nanosci Nanotechnol 16:1339–1345
163. Teixeira HF, Bruxel F, Fraga M, Schuh RS, Zorzi GK, Matte U, Fattal E (2017) Int J Pharma 534:356–367

Chapter 4
Quantum Dot Nanomaterials as the Aptasensing Platforms

Amir Khojastehnezhad, Zahra Khoshbin, Mohammad Ramezani, Mona Alibolandi, Khalil Abnous, and Seyed Mohammad Taghdisi

1 Introduction

Generally, the rapid, sensitive, and on-site detection of agents, pollutants, toxins, pesticides, and disease markers are beyond the scope of traditional detection methods; hence, biosensors have been introduced as the user-friendly diagnostic devices for low-cost, rapid analysis, and real-time detection of diverse target analytes [1–3]. In a conventional notion, a biosensor is an analytical tool comprises a bioreceptor component (e.g., enzymes, nucleic acids, antibodies, cells, peptides, and aptamers), which specifically links to the target and converts a biological response into a processable and quantifiable signal with the aid of a transducer component [4]. As one of the most efficient types of biosensors, aptamer-based biosensors (aptasensors) are novel sensing devices for rapid, real-time, and highly sensitive target detection, in which aptamers are intelligently utilized as the bioreceptor elements.

Amir Khojastehnezhad and Zahra Khoshbin contributed equally to the work.

A. Khojastehnezhad · Z. Khoshbin · M. Ramezani · M. Alibolandi · K. Abnous (✉)
Pharmaceutical Research Center, Pharmaceutical Technology Institute, Mashhad University of Medical Sciences, Mashhad, Iran
e-mail: abnouskh@mums.ac.ir

A. Khojastehnezhad · Z. Khoshbin · K. Abnous
Department of Medicinal Chemistry, Pharmaceutical Research Center, School of Pharmacy, Mashhad University of Medical Sciences, Mashhad, Iran

S. M. Taghdisi (✉)
Targeted Drug Delivery Research Center, Pharmaceutical Technology Institute, Mashhad University of Medical Sciences, Mashhad, Iran
e-mail: taghdisihm@mums.ac.ir

Department of Pharmaceutical Biotechnology, Targeted Drug Delivery Research Center, School of Pharmacy, Mashhad University of Medical Sciences, Mashhad, Iran

S. Gopi et al. (eds.), *Nanotechnology for Biomedical Applications*,
Materials Horizons: From Nature to Nanomaterials,
https://doi.org/10.1007/978-981-16-7483-9_4

Aptamers typically are short synthetic single-stranded DNA (ssDNA) isolated from large combinatorial oligonucleotide libraries through an artificial procedures and iterative selection processes, known as systematic evolution of ligands by exponential enrichment (SELEX) [5, 6]. The synthesis process of aptamers is generally in vitro that eliminates a need for laboratory animals and their fabrication. In spite of their small size (5–25 kDa), aptamers possess physicochemical stability and superior specificity in comparison with their antibody counterparts. Aptamers have the outstanding advantages including low susceptibility to denaturation, high chemical stability under a wide range of pH, temperature and buffer conditions, reversible thermal denaturation, versatility in labeling, resistance to treatments without loss of bioactivity, less immunogenicity, long storage life, biocompatibility, and ease of modification to facilitate their covalent binding to sensing substrates [7, 8]. Besides, the most significant superiority of aptamers compared to other biological molecules is the conformation change resulting from its specific binding to target, and consequently, a wide variety of aptamers can be constructed to develop various biological assays and diagnose of different targets [9]. In addition, reduction in detection process times, detection limits, test costs, and also their user friendliness convert aptasensors to the efficient analytical tools in the various scientific fields such as clinical diagnostics, healthcare, pharmaceuticals, food and water quality control, and environmental monitoring [10–12]. Thus, there is a pressing need to develop the methods based on aptasensing platforms to detect various biological targets.

The semiconductor nanocrystals with nanoscale dimensions (1–20 nm) that have unique electronic, optical, and magnetic properties and differ from larger particles due to quantum mechanics are famed as QDs [13]. These fluorescent-emitting crystals are able to emit the light and are used as fluorescent probes in biological assays [14, 15]. Generally, the smaller QDs (2–4 nm) emit shorter wavelengths with colors such as green and blue and the larger QDs (5–10 nm) emit longer wavelengths with colors such as red or orange. They have received considerable interests as the fundamental substances in the different fields of science such as solar cells, laser technology, light-emitting diodes as well as biological labeling, bioanalysis, and bioimaging [16, 17]. Moreover, different biological molecules like aptamers can be linked to the surface of QDs via physically and chemically binding and the QDs aptasensing platforms are formed based on interaction between QDs and aptamers [18].

In general, the QDs that are used as aptasensing platforms can be classified to two main groups; non-carbon-based (NC-based) and carbon-based (C-based) QDs. The NC-based QDs are consist of the elements of the groups of III_A–V_A, II_B–VI_A, and IV_A–VI_A which the QDs of the groups of II_B-VI_A are the most important ones and more studied in aptasensing platforms [19]. They are consisted of transition metals group II (Zn, Cd, Hg) and the main group VI (O, S, Se, Te, Po) in the periodic table [19, 20]. For example, the binary materials such as CdS, CdSe, CdTe, and ZnS and ternary materials like CdZnTe and CdSeTe have been intensively used for aptasensors [21]. In addition, there are the core-sell NC-based QDs that have been currently exploited for biosensing platforms. In fact, different elements are utilized for the core and shell sections. Basically, in these structures, the CdS, CdSe, or CdTe QDs can be utilized as either core and shell and the shell can be ZnS QDs, SiO_2, and TiO_2

nanomaterials or carbon QDs (CQDs). Thus, the optical and electronic properties of the semiconductor QDs can be exactly regulated with these core–shell materials which make them very attractive for a number of applications [22]. These QDs offer superior electrical, optical, and photophysical properties including high brightness, tunable emission maxima, narrow emission spectra, broad excitation spectra, high thermal stability, long fluorescence lifetime, high stability against chemical degradation and photobleaching, and great fluorescent quantum yield. However, they suffer from undesirable toxicity for living cells, tissues, and organisms. Hence, recently, the nontoxic and biocompatible QDs have been developed to overcome the limitation of the NC-based QDs such as C-based QDs [23–25]. C-based QDs including graphene QDs (GQDs), graphitic carbon nitride (g-C_3N_4), and carbon nanotube QDs (CNT-QDs) possess the outstanding properties such as low-cost synthesis, stable photoluminescence, good biocompatibility, great sensitivity, and high dispersibility that make them a prominent candidate for the electrochemical, fluorescence, luminescence, and chemiluminescence biosensing technology [26, 27].

2 Aptasensing Platforms

2.1 Optical Aptasensing Platforms

Recently, optical aptasensing platforms have been extended due to their simple operation, quick response, and high sensitivity and selectivity. Employing aptamer biomolecules as the recognition moiety and various optical methods as the signal transductions have been applied to assay different targets [28]. Indeed, there are various optical detection techniques including fluorescence, colorimetry, chemiluminescence, and surface-enhanced Raman scattering (SERS). However, in the case of optical QD nanomaterials, only the fluorescence-based aptasensing platforms have been studied. Thereby, in the next paragraphs, different aptasensing platforms are classified and reviewed based on fluorescence optical detection technique.

2.1.1 Fluorescence Aptasensing Platforms

Fluorescence aptasensing platforms can be categorized into two major groups: label-free and labeled fluorescent aptasensing platforms. Generally, fluorescence-based aptasensing platforms use fluorescence resonance energy transfer (FRET) in order to recognize different targets like proteins, peptides, nucleic acids, and biological molecules [29]. QDs are useful and versatile fluorophores that are extremely exploited as acceptor or donor fluorophores because, as we briefly explained in introduction, they have excellent fluorescent properties like sharp emission bands, resistance to photobleaching, and high spectral resolution [30]. Mainly, the fluorescent aptasensing platforms have been developed based on NC- and C-based QD platforms. Hence, in the following section, two fluorescence aptasensing platforms based on abovementioned QDs are reviewed and discussed.

NC-Based QD Fluorescent Aptasensing Platforms

As mentioned earlier, among NC-based QD fluorescent aptasensing platforms, cadmium and zinc metals have been studied more. Thus, in the following, we only reviewed the cadmium and zinc-based QD semiconductors for biosensing applications. In the first study, Chen et al. used the streptavidin-functionalized core–shell CdSe-ZnS QDs to detect adenosine triphosphate (ATP) [31]. ATP is an organic molecule that is the main energy currency of the cell and supplies energy to perform various processes in living cells. This aptasensing platform consisted of three modified DNA sections; 3′-biotin-DNA streptavidin-conjugated QD, 3′-Cy5-DNA, and a capture DNA with an ATP aptamer that was able to link to both abovementioned DNA. Without the target, the capture DNA conjugated to both 3′-biotin-DNA and 3′-Cy5-DNA and caused the close proximity of QDs and 3′-Cy5 and consequently, the FRET efficiency was increased. In the presence of the target, the 3′-Cy5-DNA was released from the system and because of the interruption of FRET, the fluorescence intensity of QD was increased while the fluorescence intensity of Cy5 was decreased. The limit of detection (LOD) of this method was 0.024 mM. After that study, the similar strategy was introduced by Li and co-workers. They employed CdTe QDs (525) for ultrasensitive aptasensing detection of ATP [32]. The differences of this study compared to the previous one were not only the QD structures, but also they used different modified DNA and its conjugation to CdTe QDs.

In the next research, the same core–shell QDs have been developed by Bala et al. for ultrasensitive detection of malathion (MLT) [33]. MLT is an organophosphate pesticide that is broadly utilized in public health pest control programs and agriculture [34]. They designed a core–shell CdTe-CdS QD aptasensing platform composed of malathion aptamer and poly(N-(3-guanidinopropyl) methacrylamide) homopolymer. In the attendance of MLT, the aptamer interacted with MLT and did not conjugate to cationic polymer and it was free. Thus, the fluorescence quenching of QDs happened. In the absence of target, the aptamer interacted with cationic polymer and as a consequence, the polymer was not free and was not able to turn off the fluorescence intensity of QDs (Fig. 1). This aptasensing platform showed a great sensitivity toward target with a LOD of 4 pM. Besides, in this method, the authors did not use any dyes and enzymes which they were the certain disadvantages of the previous approaches.

A binary CdTe QD modified MoS_2 nanosheets has been studied by Lu et al. for aptasensing detection of ochratoxin A (OTA) [35]. OTA is a naturally food-contaminating toxin which is produced by different fungi such as *Penicillium* and *Aspergillus* species. In this study, at first, the aptamer-conjugated QDs were prepared and the fluorescence quenching took place when the aptamers-conjugated QDs were immobilized on the surface of MoS_2 nanosheets. Conversely, when they used OTA as the target, the aptamers interacted with the OTA and did not bind to MoS_2

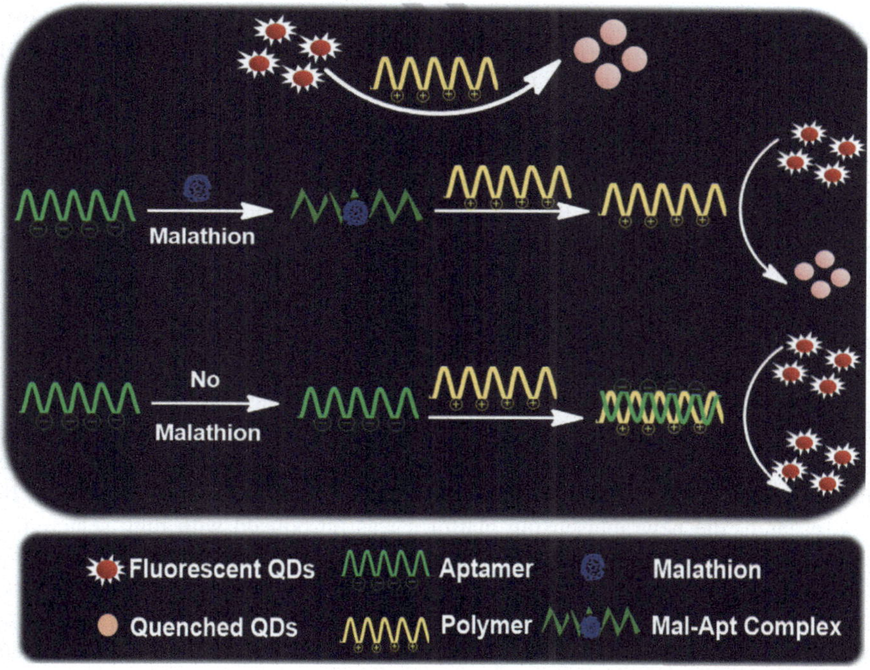

Fig. 1 Schematic representation of the NC-based aptasensor for the detection of MLT. Reprinted with permission from [33]

nanosheets and caused the fluorescence recovery of the complex. The observed LOD of this aptasensing platform was 1.0 ng mL^{-1}. The feasibility of the system was also investigated in some real red wine samples.

In another research reported by Miao et al., an "off–on" core–shell CdSe-ZnS QDs aptasensing platform was used for detection of chloramphenicol (CAP) in food samples based on the FRET [36]. CAP is an useful antibiotic for the treatment of various bacterial infections. Numerous aptasensing platforms have been presented for the detection of this antibiotic such as optical and electrochemical analyses [37]. In this study, firstly, the ssDNA binding protein (SSB) was immobilized on liposome-QDs and then, the final aptasensing platform was prepared by interaction of aptamer binding nano-gold (Apt-Au) and SSB/liposome-QDs. The system had great signal amplification values. In the absence of target, the system was in "off" state and could not emit florescence, because the florescence intensity was quenched by Apt-Au as an acceptor. In the presence of CAP, the aptamer linked to Au had the affinity to bind to the CAP and caused the "on" state. The observed LOD for this method was 0.3 pM with excellent linear response to target from 0.001 nM to 10 nM.

The ternary CdZnTe QD fluorescence aptasensor has been studied by Lu and co-workers for aflatoxin B1 (AFTB) determination [38]. AFTB is naturally generated by *Aspergillu parasiticus* and *Aspergillus flavus* that can be detected in different kinds of agricultural and natural products during the storage and maintenance processing [39]. To this purpose, the new ternary QD aptasensing platform has been designed based on FRET between QDs and gold nanoparticles (AuNPs). Hence, first of all, AFTB aptamers linked to silica coated QDs were considered as a donor (aptamer-QDs) and AuNPs linked to the modified complementary DNA (Au-DNA) was defined as an acceptor. When these two sections were brought into the close proximity via hybridization reaction, the FRET occurred from the donor to the acceptor. In the presence of the target, the specific affinity of the target to aptamer-QDs caused the dissociation of the Au-DNA section. The dissociation of the Au-DNA section led to the fluorescence recovery of aptamer-QDs. Moreover, upon the increasing concentration of target over a wide range of 50 pg mL^{-1} to 100 ng mL^{-1}, the fluorescence intensity dramatically increased.

In the last example of NC-based fluorescent aptasensing platform, Arvand et al. developed a strategy based on ZnS QDs and GOnanosheets for the edifenphos (EDI) detection [40]. EDI is a systemic fungicide with organic thiophosphate structure that controls various fungal diseases and inhibits phospholipid biosynthesis [41]. In this research, the new aptasensing platform was introduced with high selectivity, high sensitivity, and easy operation by conjugation of the aptamer on the cysteine capped QDs (aptamer-QDs) and subsequently, they were mixed with GO nanosheets. On the basis, the aptamer immobilized QDs interacted with the GO nanosheets via strong π–π stacking interactions to generate a GO-aptamer-QDs complex. Designed aptasensing platform worked based on FRET between QDs and GO nanosheets. Thus, the GO could quench the fluorescence intensity of the ZnS QDs. In the presence of target, due to the high affinity of aptamer to EDI, the aptamer-QDs were released from the GO and bound to the EDI. In this situation, the state of the complex was "on," because the fluorescence intensity of the QDs was recovered and there was not any FRET. This highly sensitive aptasensing platform showed the great performance for EDI detection with a satisfying LOD (1.3×10^{-4} mg L^{-1}).

C-based QD Fluorescent Aptasensing Platforms

Recently, C-based QDs have been attracted increasing attention due to the having fluorescence properties and use in optical biosensor technology. It is because C-based QDs show wonderful priority including facile synthesis, low toxicity, biocompatibility, super small particle size, and good stability in aqueous conditions [42]. There are different functional groups on the surface of C-based QDs like hydroxyl, epoxy, and carboxyl groups and because of these amazing functional groups, they can link easily to other biomaterials [43]. As discussed in introduction, different kinds of C-based QDs have been employed for construction of aptasensing platforms such as graphene, GO, g-C_3N_4, and CNTs that in the following, they have been discussed briefly.

GQDs are sheet-like graphene blocks with transverse size less than 100 nm with high thermal conductivity, great photoluminescence stability, chemical inertness, large surface area, high water solubility, and high mobility of charge carriers [44].

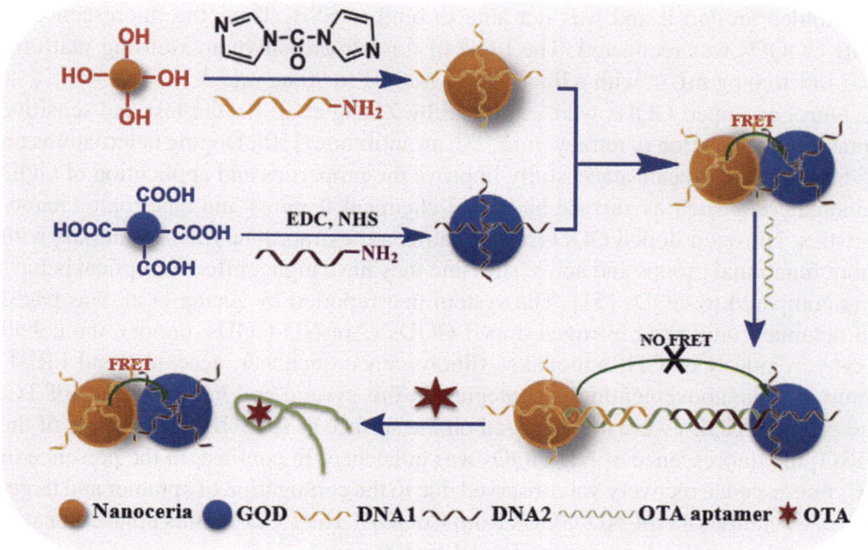

Fig. 2 Schematic process of FRET C-based aptasensor for OTA determination. Reprinted with permission from [45]

These unique properties are resulting from a combination of graphene nature and size-dependent quantum effects, making GQDs a better candidate for biosensing technology than graphene or GO [45–47]. Taking into account of these advantages of GQDs, Tian et al. reported a new FRET aptasensing platform based on GQDs for the OTA detection [45]. In this research, two kinds of DNA oligonucleotides have been selected to be hybridized with OTA aptamer and linked to colloidal cerium oxide nanoparticles (DNA1-nanoceria) and GQDs (DNA2-GQDs). In the absence of OTA, due to the electrostatic interactions between DNA1-nanoceria and DNA2-GQDs, the FRET has been performed effectively. In the presence of OTA aptamer, it linked to DNA sections of nanoceria and GQDs and the FRET was interrupted because of the large distance between nanoceria and GQDs. Conversely, by adding OTA to this specific aptasensing platform, the FRET has been recovered due to the high affinity of OTA to its aptamer (Fig. 2). This GQD aptasensing platform showed a low LOD (2.5 pg mL^{-1}) and a great linear range (0.01 to 20 ng mL^{-1}) for real samples.

In another study, He and co-workers used the GO quantum dots (GOQDs) for ultrasensitive aptasensing detection of kanamycin (KAN) [48]. KAN is a bactericidal antibiotic broadly utilized for treatment of serious bacterial infections [49]. In this respect, they used a new fluorescent probe based on SSB and GOQDs (SSB-GOQDs) and then, they combined this probe with aptamer labeled BHQ_1. In the absence of KAN, because of the high affinity of SSB to the aptamer-BHQ_1, it was conjugated with aptamer and the fluorescence of SSB-GOQDs was quenched by Apt-BHQ_1 via FRET. When they added the KAN to the system, the aptamer linked to the target and folded around it and was not able to bind to SSB. Thus, the fluorescence of SSB-GOQDs was recovered. The LOD of this ultrasensitive aptasensing platform was low to 6 pg mL^{-1} with a linear range of 0.01 to 90 ng mL^{-1}.

Nitrogen-doped GQDs were exploited by Zhang et al. for the fast and sensitive aptasensing detection of tetracycline (TC, an antibiotic) [50]. Doping heteroatoms on C–based materials can successfully improve the properties and application of GQD nanomaterials such as surface and local chemical features and electronic characteristics. Nitrogen-doped GQDs are versatile and extraordinary nanomaterials with more functional groups and active sites and they have higher effective optical behaviors compared to GQDs [51]. The system that reported by Zhang et al. was based on aptamer-conjugated nitrogen-doped GQDs (Apt-ND-GQDs, donor), the cobalt oxyhydroxide (CoOOH) nanoflakes (fluorescence quencher, acceptor) and FRET. Similar to the abovementioned strategies, in this system and in the absence of TC, the Apt-ND-GQDs were immobilized on the surface of CoOOH and because of the FRET, the fluorescence of ND-GQDs was quenched. In contrast, in the presence of TC, fluorescence recovery was observed due to the conjugation of aptamer and target and dissociation of Apt-ND-GQDs from CoOOH. The LOD for this aptasensor was 0.95 ng mL^{-1} with a linear range from 1 to 100 ng mL^{-1}.

G-C_3N_4 QDs are metal-free semiconductors with photoactive properties and have a direct band gap around 2.7 eV. In addition to graphene and GO, these CQDs have been extensively used for optical and electrochemical aptasensing platforms [52–55]. For example, very recently, Wang and co-workers reported an aptasensing method based on g-C_3N_4 for optical ultrasensitive detection of prostate specific antigen (PSA) [56]. Thereby, this aptasensing platform was designed based on the FRET from g-C_3N_4 QDs to Pd triangular surfaces. On this basis, Pd triangular surfaces have been conjugated with PSA aptamer (PSA-aptamer). Then, they were attached to g-C_3N_4 QDs, as a results, the high fluorescence quenching of g-C_3N_4 QDs was observed. By adding the PSA to the system, similar to the previous methods, the fluorescence intensity of g-C_3N_4 QDs was recovered due to the higher affinity of PSA aptamer to PSA. This proposed aptasensing platform detected PSA with ultra-low LOD (4.2 pg mL^{-1}). Also, this aptasensor can be used in biological detection and clinical assays.

2.2 *Electrochemical Aptasensing Platforms*

Electrochemistry is one the most important branches of physical and analytical chemistry that deals with the relationship between identifiable chemical changes and electricity. It is an analysis of electron and how it moves to chemical reactions. In this chemistry phenomenon, electricity can be produced by transfer of electrons between elements [57, 58]. Electrochemical aptasensing platforms with high selectivity and sensitivity are robust, portable, low-cost, easily miniaturized, and simple-to-operate and able to do rapid measurements and feasible application [59, 60]. Therefore, in recent years, they have been received much attention among researchers and scientists [61, 62]. In general, the electrochemical aptasensing platforms have been constructed based on electrochemiluminescence (ECL) and photoelectrochemical (PEC) phenomena. Therefore, in following, the QD aptasensors have been discussed based on ECL and PEC.

2.2.1 ECL Aptasensing Platforms

ECL is a blending of luminescence and electrochemical processes in which electron transfer reaction has been occurred at an electrode moiety to generate an excited photon emitting state and convert electrochemical energy into the light. Principally, ECL includes a wide variety of chemical reactions which take place on the surface of an electrode and in the presence of specified analytes, electrochemical energy converts to different light signals [63]. Basically, ECL is a hugely selective and sensitive detection strategy that have been received great attention in clinical diagnosis, pharmaceutical and biological analyzes, food contamination studies, and environmental pollutant determination [64–67]. QD materials-modified electrodes have been broadly employed as ECL aptasensing platforms. Similar to the previous section, QDs including NC-based and C-based materials have been used more for ECL aptasensing platforms. Hence, in the following sections, the ECL aptasensing platforms based on NC- and C-based QDs have been discussed.

NC-Based QD ECL Aptasensing Platforms

Similar to optical aptasensing platforms, the most common elements that used as ECL aptasensors are cadmium and zinc. Thereby, the first ECL aptasensing platform has been introduced by Wei et al. and it was based on CdS QDs, ECL resonance energy transfer (ECL-RET), and a nicking endonuclease-powered DNA walking machine for simple and sensitive OTA detection [68]. In this research, at first, the surface of glassy carbon electrode (GCE) was modified by QDs and then, two series of DNAs were immobilized on QDs covalently; a dsDNA including DNA walker and OTA aptamer and Cy5-labeled ssDNA including a seven-nucleotide nicking recognition sequence. In the absence of target, the ECL response has been decreased dramatically due to the ECL-RET between QDs and Cy5-DNA. On the other hand, by adding the target to the system, the higher affinity of the OTA aptamer to OTA caused the releasing of OTA aptamer from the dsDNA and it folded around OTA. Then, the free walker interacted with remained Cy5-DNA and the remained Cy5-DNA was cut under the action of Nb.BbvCI, releasing from electrode surface. As a consequence, the ECL response of QDs has been increased remarkably (Fig. 3). The LOD of this method was reported to be 0.012 nM with a linear range from 0.05 nM to 5 nM. Moreover, this aptasensing platform has been used for real beer and wine samples and showed the excellent results.

In the next research, Feng and co-workers employed the CdS QDs for aptasensing detection of both malachite green (MG) and CAP [69]. In this procedure, the complex

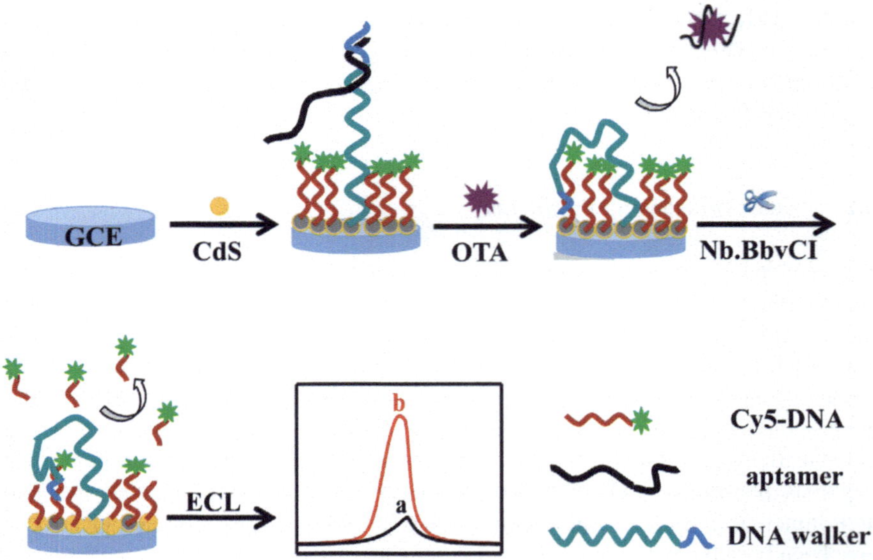

Fig. 3 Schematic illustration of the NC-based ECL detection of OTA. Reprinted with permission from [68]

aptasensing platform has been constructed on a homemade screen-printed carbon electrode (SPCE). The SPCE electrode composed of a carbon counter electrode, a Ag/AgCl reference electrode, and two carbon working electrodes (WE1 and WE2). WE1 included the CdS QDs and acted as a cathode ECL electrode and WE2 included the luminol-gold NPs and acted as an anode ECL electrode. The surface of QDs and luminol-gold NPs has been modified by the MG complementary aptamer (MG cDNA) and CAP complementary aptamer (CAP-cDNA), respectively. These two MG and CAP cDNA could interact with corresponding aptamers labeled with by Cy5 and chlorogenic acid, respectively. The Cy5 was a quencher of QDs, and chlorogenic acid was a quencher of luminol-gold NPs. In the presence of targets (MG and CAP), the labeled aptamers have been released from the surface of electrode due to the higher affinity of aptamers to targets. As a consequence, the ECL of QDs (cathode) and luminol-gold NPs (anode) has been recovered, simultaneously. This aptasensor had ability to detect the CAP and MG in the range of 0.2–150 nM and 0.1–100 nM with LOD of 0.07 nM and 0.03 nM, respectively. One of the worthwhile advantages of this study was the using of QDs with AuNPs and luminol that could help to improve the sensitivity and selectivity of biosensors. Moreover, this strategy was successfully used for detection of CAP and MG in real fish samples and showed satisfactory results.

The core–shell CdTe-CdS QDs were exploited by Wang and co-workers for aptasensing detection of thrombin (TB). [16]TB or coagulation factor 2 is a significant enzyme (protease) which has a physiological role in many life processes like maintaining blood coagulation, regulating hemostasis, and inflammation [70]. This turn-on near-infrared ECL platform worked based on ECL-RET from core–shell CdTe-CdS QDs to Au nanorods (AuNRs). For construction of system, the QDs have been immobilized on the surface of GCE and then, the chitosan organic ligand was used to link QDs to TB aptamer. Then, the probing DNA conjugated to AuNRs was hybridized as a quencher of QDs with aptamer. Briefly, in the absence of target, due to the ECL-RET of QDs to AuNRs, the ECL of QDs was decreased, while in the presence of TB, because of the big affinity of aptamer to its target, the AuNRs have been released from the system, and as a results, the ECL of QDs was recovered. The LOD of this system was 31 aM with the range of 100 aM to 10 fM. This aptasensing platform was successfully used for real serum samples and could detect the TB in very low concentrations (attomolar).

C-based QD ECL Aptasensing Platforms

The carbon structures that have been used for ECL aptasensing platforms are only CDs, graphene, and nitrogen-doped GQDs. As a consequence, You and co-workers reported the ECL aptasensing platform based on CDs capped gold nanoflowers for TB detection [71]. On this basis, a surface of GCE was modified with AuNPs and then, thiolated TB aptamer (Apt1) was immobilized on the surface AuNPs. Subsequently, the TB was linked to the aptamer due to the specific recognition affinity reaction between target and aptamer (GCE/AuNPs/Apt1/TB). Afterward, the aptamer 2 was linked to the TB through the affinity reaction (GCE/AuNPs/Apt1/TB/Apt2). Eventually, CDs capped gold nanoflowers were covalently bound to the aptamer 2 and the GCE/AuNPs/Apt1/TB/Apt2/CD-Au nanohybrids have been finalized that give the ECL response due to the TB presentation (Fig. 4). This CDs aptasensing platform showed a low LOD (0.08 nM) with a wide linear response range from 0.5 nM to 40 nM.

Lu and co-worker developed an ECL aptasensing platform based on silica coted GQDs for ATP detection [47]. In this project, the SiO_2 NPs were used as a signal carrier and a SiO_2@GQDs complex was considered as a signal amplification. In this regard, initially, the Au electrode was modified by thiolated-ATP-aptamer (ssDNA1) and the SiO_2@GQDs was linked to ssDNA2. In the presence of the target, a stable complex has been generated between the ssDNA1, ATP, and ssDNA2 which caused the interaction of SiO_2@GQDs to the electrode surface, and as a results, the ECL response has been occurred. This ECL aptasensing platform showed the great analytical performance for ATP detection, ranging from 5.0×10^{-12} to 5.0×10^{-9} mol L^{-1} with the LOD of 1.5×10^{-12} mol L^{-1}.

The ultrasensitive determination of AFTB has been described by Lu and co-workers based on GQDs and gold nanorods and flowers composites [72]. In this simple ECL aptasensor, at first, the surface of GCE electrode has been modified by poly(indole-6-carboxylic acid) gold flowers (PICA/F-Au). This modified electrode had a large specific surface area for aptamer immobilization and could increase the rate of electron transfer and conductivity. Then, GQDs, Au nanorods, and the thiolated AFTB aptamer were linked to the Au flowers surface, respectively. The

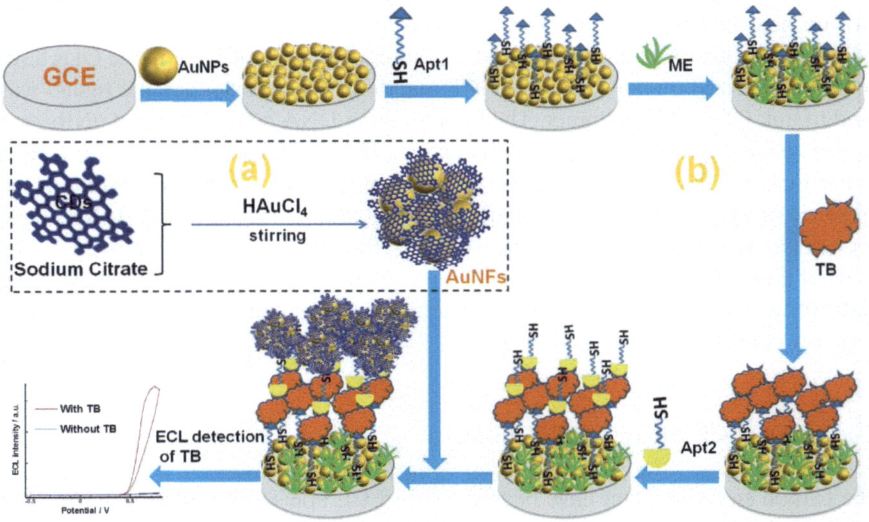

Fig. 4 Schematic representation of the C-based ECL detection of TB. Reprinted with permission from [71]

GQDs and Au nanorods were used to improve the ECL activity and sensitivity of the system. In presence of target, the ECL signal of the system has been changed that it was an indication of target existence. This system could detect the target in the range between 0.01 to 100 ng mL^{-1}, and the LOD of the method was 0.00375 ng mL^{-1}. Moreover, this aptasensing platform could detect the AFTB in real samples such as peanut, wheat, and maize successfully.

2.2.2 PEC Aptasensing Platforms

Photoelectrochemistry is a significant branches of physical chemistry that deals with to the interaction of light with electrochemical systems. This filed is a unique domain of investigation that offers a certain path for sensitive detection of biomolecules [73]. Since of the beginning of the present century, PEC detection has been actively attracted scientist attentions, because such detection platforms can provide great sensitivity due to the completely different forms of energy for excitation and also detection [74]. Owing to their noteworthy potential in future detection of biomolecules, recently, numerous attempts have been devoted to design and develop of new aptasensing platforms based on PEC [75]. Thus, in the next paragraphs, the PEC aptasensing platforms based on NC and CQDs have been discussed and reviewed.

NC-Based QD PEC Aptasensing Platforms

In this section, the cadmium hybrid semiconductors have been only investigated. In the first study that reported by Wang et al., Eu(III)-doped CdS QDs were used for PEC aptasensing detection of CAP [76]. On this basis, the surface of fluorine-doped tin oxide (FTO) electrode has been modified by nanorods TiO_2 and Eu(III)-doped CdS QDs. The introduction of nanorods TiO_2 as photoactive materials and Eu(III)-doped CdS QDs on the surface of electrode improved the photo-to-current conversion efficiency and promoted the charge transformations. After that, the CAP aptamer was linked on the QD materials covalently to capture the target. In the presence of the CAP, it was interacted with aptamer and because of the formation of aptamer-target complex, the PEC response of the system decreased due to the interruption of electron transfer. This system could detect the CAP in the linear range between 1.0 pM to 3.0 nM, and the LOD of the method was observed 0.36 pM. Also, the designed aptasensor was employed for the detection of CAP in real milk samples where it exhibited the satisfactory results.

A supersensitive PEC aptasensing platform has been developed by Zhang and co-workers based on TiO_2 and CdS QDs for carcinoembryonic antigen (CEA) derermination [77]. CEA is a set of highly-related glycoproteins that normally found in very low levels in the blood cells. It is a prominent tumor biomarker that can be used in the clinical diagnosis of different cancers [78]. In this project, similar to the previous studies, the GCE electrode has been modified by core–shell Br,N-codoped TiO_2@gold nanoparticles (TiO_2@AuNPs). Then, the DNA capture has been immobilized on the core–shell TiO_2@AuNPs through the Au–S bond. After that, in the presence of the target, a substantial single-chain DNA (T-DNA) was produced through the exonuclease III (Exo-III)-assisted cycle strategy that was able to hybridize with DNA capture on the electrode. Afterward, T-DNA was able to link to CdS QDs via CdS bond to generate the TiO_2@Au-QDs platform. In this platform, the QDs as a signal enhancer could amplify the photocurrent signal of TiO_2@AuNPs as a photoactive material. In the presence of the target, owing to the generation of the T-DNA and then linkage of QDs on the T-DNA, the photocurrent has been increased dramatically while in the absence of CEA, because, the complex was not formed completely, the photocurrent has been decreased (Fig. 5). This aptasensing platform showed a broad detection range from 1 fg mL^{-1} to 1 ng mL^{-1} and a low LOD (0.46 fg mL^{-1}). The similar strategy has been reported by Cong and co-workers for aptasensing detection of CEA [79]. The difference of this research with the previous study was not only the use of indium tin oxide (ITO) electrode, but also they utilized core-sell CdTe@CdS QDs for design of PEC aptasensing platform. The low LOD (0.12 pg mL^{-1}) and a wide linear range from 0.5 pg mL^{-1} to 10 ng mL^{-1} were the advantages of this platform for CEA determination.

In the next study of using NC-QDs for construction of PEC aptasensor, Liu et al. designed a "on–off-on" aptasensing platform based on RET from CdTe QDs to Au nanorods and using CNTs/GO as a signal amplifier for acetamiprid (ACP) detection [80]. ACP is a carboxamidine that acts as a neonicotinoid insecticide, an environmental contaminant, and a xenobiotic [81]. In this project, firstly, ITO electrode

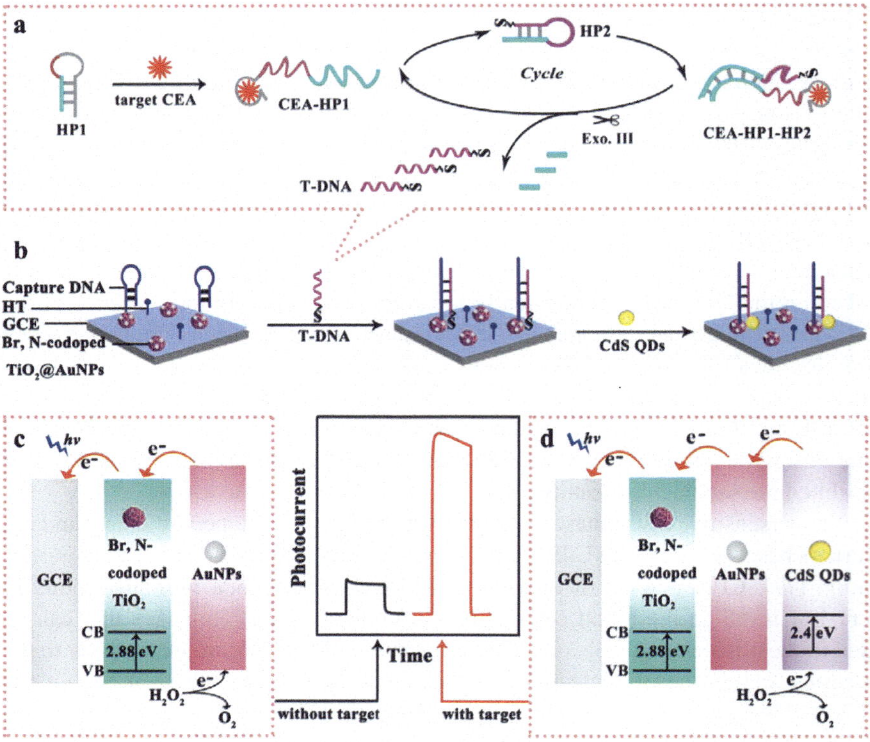

Fig. 5 Schematic representation of the NC-based PEC detection of CEA. Reprinted with permission from [77]

has been modified by QDs doped-CNTs/GO nanosheets and then, complementary DNA1 has been linked to the surface of QDs and subsequently, ACP aptamer has been hybridized with DNA1. Finally, the DNA2-Au nanorods have been conjugated to aptamer and the system. Before adding the DNA2-Au nanorods, because the RET did not occurre, the state of the system was "on," while by adding the DNA2-Au nanorods to the system, because of the RET from QDs to the Au nanorods, the state was "off" and eventually, in the presence of the target, due to the interaction of target with aptamer and release of the aptamer and DNA2-Au form the system, the state turned "on." The LOD of this aptasensor was 0.2 pM with linear range of 0.5 pM to 10 μM.

The last study was about the using of CdTe QDs and single walled carbon nanohorns (SWCNHs) for the sensitive detection of streptomycin (STR) based on "on-off-on" strategy [82]. STR is an aminoglycoside antibiotic which used to treat various bacterial infections and it is produced by the soil actinomycete *Streptomyces griseus* [83]. In this simple strategy, the CdTe-SWCNHs were immobilized on the surface of ITO electrode and the state of the system was "on" due to the strong cathodic photocurrent that could be obtained at the CdTe-SWCNs electrode, avoiding the intrinsic hole oxidation reactions happened at the photoanode/electrolyte interface. By adding the STR aptamer to the system, the photocurrent of the system decreased because of the steric effects of the aptamer to the system and thus the state of the system was "off." Finally, in the attendance of the target, owing to the high affinity of the target to its aptamer, the aptamer dissociated form the complex and photocurrent has been recovered and state turned to "on." This PEC aptasensing platform displayed a linear response ranging from 0.1 nM to 50 nM with a LOD of 0.033 nM.

C-based QD PEC Aptasensing Platforms

The carbon QD materials which were utilized for construction of PEC aptasensing platforms are only CDs, graphene, and nitrogen-doped GQDs. As a results, in the next paragraphs, three examples of using these C-based QDs have been discussed. For example, Qin et al. designed a PEC aptasensor based on GQDs-sensitized TiO_2 nanotube arrays (GQDs-TiO_2 NTs) for sensitive detection of CAP [84]. In this study, they used titanium foil as an electrochemical electrode. Then, its surface has been modified by GQDs-TiO_2 NTs to enhance visible light PEC activity of aptasensor. Next, the CAP aptamer has been conjugated to the GQDs-TiO_2 NTs via π–π stacking interactions between the nucleobases of the aptamer and GQDs basal plane. In the absence of CAP, the electron transfer resistance has been increased due to the electrostatic repulsion between the aptamer and redox species of electrode ($K_3[Fe(CN)_6]$ and therefore, the photocurrent of the system has been decreased. Conversely, in the presence of the CAP, it has been attached to the CAP aptamer and due to the formation of CAP-aptamer complex, the aptamer was dissociated from the electrode surface and thus, the photocurrent of the system has been raised (Fig. 6). This C-based aptasensing platform exhibited a wide linear range from 0.5 nM to 100 nM toward CAP determination with LOD of 57.9 pM. Moreover, the designed aptasensor was used to detect CAP in real honey samples where it showed the satisfactory results.

In another work, You and co-workers introduced the PEC aptasensing platform based on nitrogen-doped GQDs and MoS_2 nanoplates for ACP detection [85]. MoS_2 nanoplates and nitrogen-doped GQDs (NGQDs-MoS_2) have been introduced on the surface of ITO electrode. MoS_2 nanoplates have large optical absorption coefficients and relative long photoexcited carrier lifetimes. In this regard and in continuation, the ACP aptamer has been immobilized on the surface of NGQDs-MoS_2. Like to other abovementioned PEC aptasensors, with adding the target to the system, the complex of ACP aptamer has been formed on the surface of electrode which caused the interruption of electron transfer and as a results, the photocurrent of the system has been decreased. This C-based platform could detect the ACP in the linear range of 0.05 pM to 1.0 nM with LOD of 16.7 fM. Besides, this aptasensor had excellent ability to detect the ACP in real tomato and cucumber samples.

The last PEC aptasensing platform in this chapter has been reported by Cheng and co-workers for TB detection based on the use of CQDs-sensitized TiO_2 [86]. After the modification of ITO electrode with CQDs and TiO_2 NPs, the chitosan and

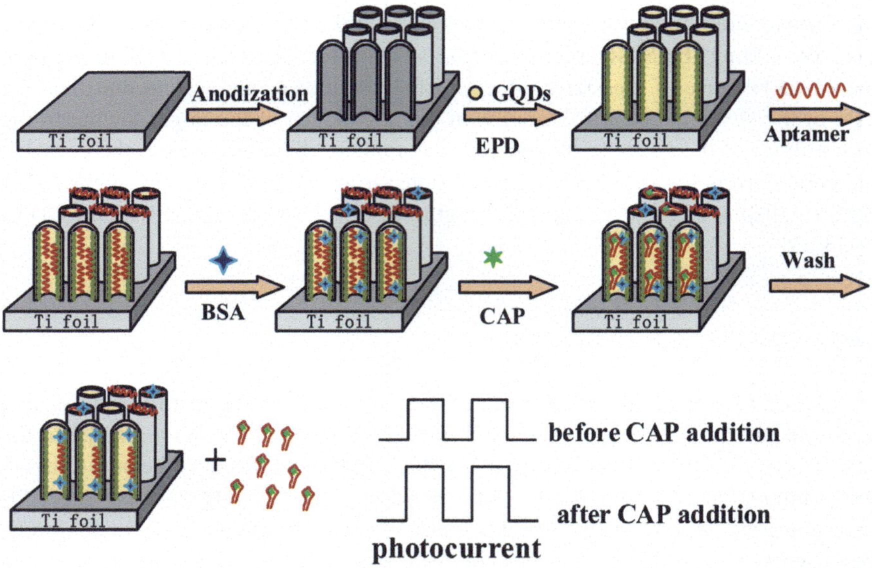

Fig. 6 Schematic illustration of the C-based PEC detection of CAP. Reprinted with permission from [84]

glutaraldehyde have been placed on the surface of CQDs to immobilize the carboxyl-functionalized TB aptamer. This system showed an enhanced and steady photocurrent response under irradiation by visible light. In the attendance of target, the complex of TB aptamer has been generated on the surface of modified ITO electrode which caused the interruption of electron transfer and as a consequence, the photocurrent of the system has been reduced. This PEC aptasensor showed a linear response in the range of 1.0 to 250 pM of target concentration, with LOD of 0.83 pM. Although, very high sensitivity, good stability, and excellent selectivity were the advantages of this aptasensing platform, but the proposed detection method was time-consuming which limited its applications.

3 Conclusion

The selective and sensitive detection of biomolecules are still an enormous challenge in biochemistry and chemistry. Optical and electrochemical platforms suggest an excellent strategy for this goal by employing a bioreceptor section against certain targets. As one of the most efficient types of biosensors, aptamer-based biosensors are novel sensing devices for rapid, real-time, and highly sensitive target detection, in which aptamers are intelligently utilized as the bioreceptor elements. They can interact well with a wide range of analytes from small molecules to the whole cells. On the other hand, the QDs as fluorescent-emitting crystals have received great attention, because they are able to emit the light and can be used as fluorescent probes in biological assays. Hence, in this chapter, the QDs aptasensing platforms were discussed and categorized to C- and NC-based materials and their optical and electrochemical properties were investigated in detection of different targets such as adenosine triphosphate, ochratoxin A, chloramphenicol, thrombin, acetamiprid, malathion, streptomycin, carcinoembryonic antigen, edifenphos, aflatoxin B1, kanamycin, tetracycline, prostate specific antigen, and malachite green. However, QDs aptasensing platforms promote the function of analytical methods including selectivity and sensitivity, they still require to be further optimized to ensure assurance of consumers.

References

1. Feng C, Dai S, Wang L (2014) Optical aptasensors for quantitative detection of small biomolecules: a review. Biosens Bioelectron 59:64–74
2. Khoshbin Z, Verdian A, Housaindokht MR, Izadyar M, Rouhbakhsh Z (2018) Aptasensors as the future of antibiotics test kits-a case study of the aptamer application in the chloramphenicol detection. Biosens Bioelectron 122:263–283
3. Meirinho SG, Dias LG, Peres AM, Rodrigues LR (2016) Voltammetric aptasensors for protein disease biomarkers detection: a review. Biotechnol Adv 34:941–953
4. Kaur H, Shorie M (2019) Nanomaterial based aptasensors for clinical and environmental diagnostic applications. Nanoscale Adv 1:2123–2138
5. Akki SU, Werth CJ (2018) Critical review: DNA aptasensors, are they ready for monitoring organic pollutants in natural and treated water sources? Environ Sci Technol 52:8989–9007
6. Yousefi M et al (2019) Aptasensors as a new sensing technology developed for the detection of MUC1 mucin: a review. Biosens Bioelectron 130:1–19
7. Hong P, Li W, Li J (2012) Applications of aptasensors in clinical diagnostics. Sensors 12:1181–1193
8. Torres-Chavolla E, Alocilja EC (2009) Aptasensors for detection of microbial and viral pathogens. Biosens Bioelectron 24:3175–3182
9. Hayat A, Yang C, Rhouati A, Marty JL (2013) Recent advances and achievements in nanomaterial-based, and structure switchable aptasensing platforms for ochratoxin A detection. Sensors 13:15187–15208
10. Ghorbani F, Abbaszadeh H, Dolatabadi JEN, Aghebati-Maleki L, Yousefi M (2019) Application of various optical and electrochemical aptasensors for detection of human prostate specific antigen: a review. Biosens Bioelectron 142:111484
11. Qi X, Yan X, Zhao Y, Li L, Wang S (2020) Highly sensitive and specific detection of small molecules using advanced aptasensors based on split aptamers: a review. TrAC Trends Anal Chem 116069
12. Khoshbin Z, Housaindokht MR, Izadyar M, Bozorgmehr MR, Verdian A (2020) Recent advances in computational methods for biosensor design. Biotechnol Bioeng
13. Jamieson T et al (2007) Biological applications of quantum dots. Biomaterials 28:4717–4732
14. Ghaderi S, Ramesh B, Seifalian AM (2011) Fluorescence nanoparticles "quantum dots" as drug delivery system and their toxicity: a review. J Drug Target 19:475–486
15. Wagner AM, Knipe JM, Orive G, Peppas NA (2019) Quantum dots in biomedical applications. Acta Biomater 94:44–63
16. Wang J, Jiang X, Han H (2016) Turn-on near-infrared electrochemiluminescence sensing of thrombin based on resonance energy transfer between CdTe/CdS coresmall/shellthick quantum dots and gold nanorods. Biosens Bioelectron 82:26–31

17. Petryayeva E, Algar WR, Medintz IL (2013) Quantum dots in bioanalysis: a review of applications across various platforms for fluorescence spectroscopy and imaging. Appl Spectrosc 67:215–252
18. Wen L, Qiu L, Wu Y, Hu X, Zhang X (2017) Aptamer-modified semiconductor quantum dots for biosensing applications. Sensors 17:1736
19. Bakirhan NK, Ozkan SA (2018) Handbook of nanomaterials for industrial applications. Elsevier, pp 520–529
20. Zhang Y, Liu Y, Li C, Chen X, Wang Q (2014) Controlled synthesis of Ag2S quantum dots and experimental determination of the exciton Bohr radius. J Phys Chem C 118:4918–4923
21. Girma WM, Fahmi MZ, Permadi A, Abate MA, Chang J-Y (2017) Synthetic strategies and biomedical applications of I-III–VI ternary quantum dots. J Mater Chem B 5:6193–6216
22. Cotta MA (2020) ACS Publications
23. Jacak L, Hawrylak P, Wojs A (2013) Quantum dots. Springer Science, Business Media
24. Kumar DS, Kumar BJ, Mahesh H (2018) Synthesis of inorganic nanomaterials. Elsevier, pp 59–88
25. Kargozar S, et al (2020) Quantum dots: a review from concept to clinic. Biotechnol J 2000117
26. Li M, Chen T, Gooding JJ, Liu J (2019) Review of carbon and graphene quantum dots for sensing. ACS sensors 4:1732–1748
27. Tajik S et al (2020) Carbon and graphene quantum dots: a review on syntheses, characterization, biological and sensing applications for neurotransmitter determination. RSC Adv 10:15406–15429
28. Robati RY et al (2016) Aptasensors for quantitative detection of kanamycin. Biosens Bioelectron 82:162–172
29. Shahdordizadeh M et al (2017) Aptamer based biosensors for detection of Staphylococcus aureus. Sens Actuators, B Chem 241:619–635
30. Dolatabadi JEN et al (2011) Optical and electrochemical DNA nanobiosensors. TrAC, Trends Anal Chem 30:459–472
31. Chen Z et al (2008) A new method for the detection of ATP using a quantum-dot-tagged aptamer. Anal Bioanal Chem 392:1185–1188
32. Li Z et al (2013) A novel aptasensor for the ultra-sensitive detection of adenosine triphosphate via aptamer/quantum dot based resonance energy transfer. Analyst 138:4732–4736
33. Bala R et al (2018) Ultra-sensitive detection of malathion using quantum dots-polymer based fluorescence aptasensor. Biosens Bioelectron 104:45–49
34. Bala R, Kumar M, Bansal K, Sharma RK, Wangoo N (2016) Ultrasensitive aptamer biosensor for malathion detection based on cationic polymer and gold nanoparticles. Biosens Bioelectron 85:445–449
35. Lu Z, Chen X, Hu W (2017) A fluorescence aptasensor based on semiconductor quantum dots and MoS2 nanosheets for ochratoxin A detection. Sens Actuators, B Chem 246:61–67
36. Miao Y-B et al (2016) A homogeneous and "off–on" fluorescence aptamer-based assay for chloramphenicol using vesicle quantum dot-gold colloid composite probes. Anal Chim Acta 929:49–55
37. Pilehvar S, Dierckx T, Blust R, Breugelmans T, De Wael K (2014) An electrochemical impedimetric aptasensing platform for sensitive and selective detection of small molecules such as chloramphenicol. Sensors 14:12059–12069
38. Lu X et al (2019) Target-driven switch-on fluorescence aptasensor for trace aflatoxin B1 determination based on highly fluorescent ternary CdZnTe quantum dots. Anal Chim Acta 1047:163–171
39. Soares RR et al (2018) Advances, challenges and opportunities for point-of-need screening of mycotoxins in foods and feeds. Analyst 143:1015–1035
40. Arvand M, Mirroshandel AA (2017) Highly-sensitive aptasensor based on fluorescence resonance energy transfer between l-cysteine capped ZnS quantum dots and graphene oxide sheets for the determination of edifenphos fungicide. Biosens Bioelectron 96:324–331
41. Marrs TC (1993) Organophosphate poisoning. Pharmacol Ther 58:51–66

42. Alavi M, Jabari E, Jabbari E (2020) Functionalized carbon-based nanomaterials and quantum dots with antibacterial activity: a review. Expert Rev Anti-infect Ther 1–10
43. Arkan E et al (2018) Green synthesis of carbon dots derived from walnut oil and an investigation of their cytotoxic and apoptogenic activities toward cancer cells. Adv Pharm Bull 8:149
44. Sun H, Wu L, Wei W, Qu X (2013) Recent advances in graphene quantum dots for sensing. Mater Today 16:433–442
45. Tian J et al (2018) Fluorescence resonance energy transfer aptasensor between nanoceria and graphene quantum dots for the determination of ochratoxin A. Anal Chim Acta 1000:265–272
46. Huang J-Y et al (2018) A high-sensitivity electrochemical aptasensor of carcinoembryonic antigen based on graphene quantum dots-ionic liquid-nafion nanomatrix and DNAzyme-assisted signal amplification strategy. Biosens Bioelectron 99:28–33
47. Lu J et al (2013) Electrochemiluminescence of blue-luminescent graphene quantum dots and its application in ultrasensitive aptasensor for adenosine triphosphate detection. Biosens Bioelectron 47:271–277
48. He Y, Wen X, Zhang B, Fan Z (2018) Novel aptasensor for the ultrasensitive detection of kanamycin based on grapheneoxide quantum-dot-linked single-stranded DNA-binding protein. Sens Actuators, B Chem 265:20–26
49. Wei Q et al (2012) Ultrasensitive detection of kanamycin in animal derived foods by label-free electrochemical immunosensor. Food Chem 134:1601–1606
50. Zhang L, Wang J, Deng J, Wang S (2020) A novel fluorescent "turn-on" aptasensor based on nitrogen-doped graphene quantum dots and hexagonal cobalt oxyhydroxide nanoflakes to detect tetracycline. Anal Bioanal Chem 412:1343–1351
51. Liu Q, Guo B, Rao Z, Zhang B, Gong JR (2013) Strong two-photon-induced fluorescence from photostable, biocompatible nitrogen-doped graphene quantum dots for cellular and deep-tissue imaging. Nano Lett 13:2436–2441
52. Dong Y et al (2016) Graphitic carbon nitride materials: sensing, imaging and therapy. Small 12:5376–5393
53. Chan M-H, Liu R-S (2016) Phosphors, up conversion nano particles, quantum dots and their applications. Springer, pp 485–502
54. Morozova S, Alikina M, Vinogradov A, Pagliaro M (2020) Silicon quantum dots: synthesis, encapsulation, and application in light-emitting diodes. Front Chem 8:191
55. Liu H et al (2020) Graphitic carbon nitride quantum dots embedded in carbon nanosheets for near-infrared imaging-guided combined photo-chemotherapy. ACS Nano 14:13304–13315
56. Wang H-M et al (2020) Construction of efficient "on-off-on" fluorescence aptasensor for ultra-sensitive detection of prostate specific antigen via covalent energy transfer between g-C3N4 quantum dots and palladium triangular plates. Anal Chim Acta 1104:53–59
57. Sassolas A, Blum LJ, Leca-Bouvier BD (2009) Electrochemical aptasensors. Electroanalysis: Int J Devoted Fundam Pract Aspects Electroanalysis 21:1237–1250
58. Hianik T, Wang J (2009) Electrochemical aptasensors–recent achievements and perspectives. Electroanalysis: Int J Devoted Fundam Pract Aspects Electroanalysis 21:1223–1235
59. Grieshaber D, MacKenzie R, Vörös J, Reimhult E (2008) Electrochemical biosensors-sensor principles and architectures. Sensors 8:1400–1458
60. Bahadır EB, Sezgintürk MK (2015) Electrochemical biosensors for hormone analyses. Biosens Bioelectron 68:62–71
61. Vasilescu A, Marty J-L (2016) Electrochemical aptasensors for the assessment of food quality and safety. TrAC, Trends Anal Chem 79:60–70
62. Rapini R, Marrazza G (2017) Electrochemical aptasensors for contaminants detection in food and environment: recent advances. Bioelectrochemistry 118:47–61
63. Khonsari YN, Sun S (2017) Recent trends in electrochemiluminescence aptasensors and their applications. Chem Commun 53:9042–9054
64. Zhang J-J et al (2014) Label-free and sensitive electrochemiluminescence aptasensor for the determination of 17β-estradiol based on a competitive assay with cDNA amplification. Anal Methods 6:6796–6801

65. Shi H-W, Wu M-S, Du Y, Xu J-J, Chen H-Y (2014) Electrochemiluminescence aptasensor based on bipolar electrode for detection of adenosine in cancer cells. Biosens Bioelectron 55:459–463
66. Luo L et al (2019) Monitoring zearalenone in corn flour utilizing novel self-enhanced electrochemiluminescence aptasensor based on NGQDs-NH_2-Ru@SiO_2 luminophore. Food Chem 292:98–105
67. Wang H, Gong W, Tan Z, Yin X, Wang L (2012) Label-free bifunctional electrochemiluminescence aptasensor for detection of adenosine and lysozyme. Electrochim Acta 76:416–423
68. Wei M et al (2019) A simple and sensitive electrochemiluminescence aptasensor for determination of ochratoxin A based on a nicking endonuclease-powered DNA walking machine. Food Chem 282:141–146
69. Feng X et al (2015) A novel "dual-potential" electrochemiluminescence aptasensor array using CdS quantum dots and luminol-gold nanoparticles as labels for simultaneous detection of malachite green and chloramphenicol. Biosens Bioelectron 74:587–593
70. Hwang K-K et al (2001) Identification of anti-thrombin antibodies in the antiphospholipid syndrome that interfere with the inactivation of thrombin by antithrombin. J Immunol 167:7192–7198
71. You X, Lin W, Wu H, Dong Y, Chi Y (2018) Carbon dot capped gold nanoflowers for electrochemiluminescent aptasensor of thrombin. Carbon 127:653–657
72. Lu Y et al (2020) An electrochemiluminescence aptasensor for the ultrasensitive detection of aflatoxin B1 based on gold nanorods/graphene quantum dots-modified poly (indole-6-carboxylic acid)/flower-gold nanocomposite. Microchem J 104959
73. Wang Z et al (2019) Nucleic acid-based ratiometric electrochemiluminescent, electrochemical and photoelectrochemical biosensors: a review. Microchim Acta 186:405
74. Zhao W-W, Xu J-J, Chen H-Y (2016) Photoelectrochemical aptasensing. TrAC, Trends Anal Chem 82:307–315
75. Peng B et al (2018) Current progress in aptasensors for heavy metal ions based on photoelectrochemical method: a review. Curr Anal Chem 14:4–12
76. Wang Y, Bian F, Qin X, Wang Q (2018) Visible light photoelectrochemical aptasensor for chloramphenicol by using a TiO_2 nanorod array sensitized with Eu(III)-doped CdS quantum dots. Microchim Acta 185:161
77. Zhang Y-H, Li M-J, Wang H-J, Yuan R, Wei S-P (2019) Supersensitive photoelectrochemical aptasensor based on Br, N-codoped TiO_2 sensitized by quantum dots. Anal Chem 91:10864–10869
78. Gan N, Jia L, Zheng L (2011) A sandwich electrochemical immunosensor using magnetic DNA nanoprobes for carcinoembryonic antigen. Int J Mol Sci 12:7410–7423
79. Cong X, Fan G-C, Wang X, Abdel-Halim E, Zhu J-J (2016) Enhanced photoelectrochemical aptasensing platform amplified through the sensitization effect of CdTe@CdS core–shell quantum dots coupled with exonuclease-I assisted target recycling. J Mater Chem B 4:6117–6124
80. Liu Q et al (2016) Resonance energy transfer from CdTe quantum dots to gold nanorods using MWCNTs/rGO nanoribbons as efficient signal amplifier for fabricating visible-light-driven "on-off-on" photoelectrochemical acetamiprid aptasensor. Sens Actuators, B Chem 235:647–654
81. Fan L, Zhao G, Shi H, Liu M, Li Z (2013) A highly selective electrochemical impedance spectroscopy-based aptasensor for sensitive detection of acetamiprid. Biosens Bioelectron 43:12–18
82. Xu X et al (2017) Photoelectrochemical aptasensor based on CdTe quantum dots-single walled carbon nanohorns for the sensitive detection of streptomycin. Sens Actuators, B Chem 251:564–571
83. Waksman SA (1953) Streptomycin: background, isolation, properties, and utilization. Science 118:259–266
84. Qin X, Wang Q, Geng L, Shu X, Wang Y (2019) A "signal-on" photoelectrochemical aptasensor based on graphene quantum dots-sensitized TiO_2 nanotube arrays for sensitive detection of chloramphenicol. Talanta 197:28–35

85. Jiang D, Du X, Zhou L, Li H, Wang K (2017) New insights toward efficient charge-separation mechanism for high-performance photoelectrochemical aptasensing: enhanced charge-carrier lifetime via coupling ultrathin MoS_2 nanoplates with nitrogen-doped graphene quantum dots. Anal Chem 89:4525–4531
86. Cheng W et al (2018) A photoelectrochemical aptasensor for thrombin based on the use of carbon quantum dot-sensitized TiO_2 and visible-light photoelectrochemical activity. Microchim Acta 185:263

Chapter 5
Carbon Dots: Fundamental Concepts and Biomedical Applications

Souravi Bardhan, Shubham Roy, and Sukhen Das

1 Introduction

Development of nanoscience and nanotechnology in recent times has paved the path for emergence of various new materials and prospects for betterment of the society. Nanoparticles have attracted tremendous attention compared to their bulk counterparts due to dramatic increment of various properties, mainly by virtue of increase of surface-to-volume ratio [1–3]. Although various new nano-engineered materials for targeted applications were synthesized, most of them suffered various issues like cost-effectiveness, lacking versatility in applications, requirement of high energy consumption, and use of toxic reagents or may cause secondary pollution or toxicity in environment [4, 5]. Hence, the search for alternative suitable eco-friendly nanomaterials with multiple advantages and application triggered extensive studies worldwide and led to the discovery of various new classes of carbon. Earlier, mainly three forms of carbon were well known, namely amorphous carbon, graphite, and diamond [6], but over the last decades, various carbonaceous nanoparticles such as carbon dots (CDs), carbon nanotubes (CNTs), graphene oxide, and quantum dots (GQDs) have gained immense popularity due to their unique and exceptional optoelectronic and physicochemical properties [7]. Among them, carbon dots, also known as "carbon quantum dots" or "carbon nanodots" [8], have attracted tremendous attention since discovery in 2004, due to their nontoxic, biocompatible nature, facile synthesis procedure, excellent photoluminescence, inherent ability for electron transfer, and magnificent fluorescence emission, having 0-dimensional, quasi-spherical appearance with

S. Bardhan · S. Roy · S. Das (✉)
Department of Physics, Jadavpur University, Kolkata 700032, India

S. Gopi et al. (eds.), *Nanotechnology for Biomedical Applications*,
Materials Horizons: From Nature to Nanomaterials,
https://doi.org/10.1007/978-981-16-7483-9_5

diameter less than 10 nm [9]. Having such advantages, carbon dots are widely considered as a potential material for pollutant sensing, biosening and bioimaging, biomonitoring, targeted drug delivery, photonic and electro-optical material synthesis, catalysis, and fabrication of fluorescent devices [10]. The tendency to exhibit tunable fluorescence emission, which is their capability to emit light of different colors formed the key factor for the versatile optical applications [11]. The size and fluorescence emission can be further modified or improved through surface passivation according to the requirement. Valcárcel's group had systematically classified carbon dots [12] based on their nature, surface groups, chemical and crystalline orientation, carbon core structure, and quantum confinement, which are heavily dependent on the different precursors and synthesis technique used. The spherical quantum dots having crystalline structure and quantum confinement are considered as quantum dots (CQDs); the amorphous nanodots lacking quantum confinement is referred as carbon nanodots (CNDs); and graphene quantum dots (GQDs) which are π-conjugated single sheets consisting of chemical groups in interlayer defect or on the edge and exhibit quantum confinement. Since carbon dots itself not always satisfy the photo-physical properties required for various optical and electrical applications, most researches focused on surface functionalization or passivation using various organic, inorganic, biological, or polymeric groups to enhance the fluorescence quality by manipulating their band gaps transitions, capping the voids on carbon dots surface or increasing their surface defects. This facilitates improvement in fluorescence properties, increases photostability, reactivity, solubility, and dispersing capacity [13], thus achieving much better results in various applications (Fig. 1).

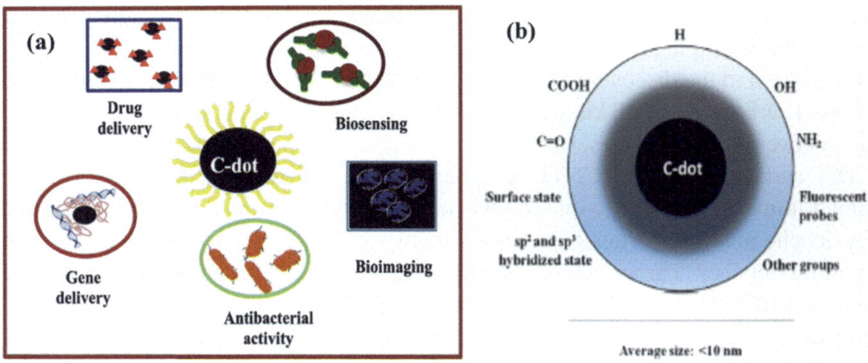

Fig. 1 **a** Schematic illustration of the various biomedical applications of carbon dots. Reproduced from [14] Copyright 2018, with permission from Elsevier, **b** a typical design of carbon dots depicting different surface functional groups. Reproduced from [15] Copyright 2018, with permission from Elsevier (Fig. 1b)

2 Development of Carbon Dots Over Past Decade

Although carbon dots are considerably quite new in the realm of carbon family, the extraordinary fluorescence coupled with the several benefits led to exponential growth of research on carbon dots, mainly in the domain of optical studies. Continual progress of research on carbon dots led to various developments to gain greater feasibility of the applications in the real world. Currently, many researchers focus on confining liquid or powder-form of carbon dots into polymer matrix such as PVDF [16] or PVP [17] for fabricating devices or for commercial purposes due to various advantages like reusability, durability, hydrophobicity, flexibility, and lacks any chances for agglomeration. Carbon dots are one of such inventions that originated accidently, but gained immense popularity within very small span of time due to its various attributes and applications, and carbon dots was considered as a boon in biomedical field since discovery.

2.1 Discovery of Carbon Dots: Newest Member of Carbon Family

Before the first report on carbon dots, carbon nanotubes were functionalized using oxidative acid treatment to obtain colorful photoluminescence emissions [18]. Origin of carbon dots occurred accidentally in 2004 [19] as a by-product during electrophoretic purification of single-walled nanotubes (SWNTs) which was derived from arc discharge soot. Xu et al. observed distinct, fast moving band of some fluorescent material, which exhibited different colors on exposure to 356 nm UV light during the course of investigation on separation of SWNTs in arc soot using preparative electrophoresis using agarose gel. Thus, initially, carbon dots or rather "carbon nanocrystals" were considered just a fluorescent side product of carbon nanotubes but this report received widespread interest for further study. Two years later, first stable and quantum-sized, colorful, and photoluminescent carbon nanoparticles were synthesized by Sun et al. [20] as a nontoxic alternative for fluorescent semiconductor quantum dots for in vitro and in vivo optical studies. Such benign analog for conventional but toxic semiconductor quantum dots that was designed for cell labeling and can exhibit bright fluorescence in both liquid and solid state was termed as "carbon dot" or "carbon quantum dot." Post-realization of the fact that origin and modulation of the fluorescence properties of the newly discovered nanoparticles is obtained through surface passivation, this report depicts the pathway for passivation of carbon dots using poly-propionylethylenimine-co-ethylenimine and suggested it for bioimaging purpose, even in presence of bioactive molecules. In order to get a better insight regarding obtaining such fluorescent carbon nanoparticle, Liu et al. in 2007 [21] conducted a detailed study on preparation of fluorescent carbon nanoparticles from candle soot and purification using polyacrylamide gel electrophoresis technique. Soon, after the optical properties of the new member of carbon family were reported, their application in the field of bioimaging was reported on 2007. Cao et al. [22] synthesized carbon dots as suggested by Sun et al. and explored photostability and luminescence on two-photon excitation and reported bright illumination of human breast cancer (MCF-7) cells. Thus, carbon dots were considered as highly potential candidate for biological studies since its discovery and a superior replacement for heavy-metal base quantum dots.

2.2 *Evolution of Diverse Routes for Carbon Dots Synthesis*

The discovery and successful application of carbon dots in bioimaging due to its brilliant tunable fluorescence, biocompatibility, cost-effectiveness, and ultrafine dimension inspired scientists worldwide to strive for more directions for synthesis and application of carbon dots. By 2010, significant breakthroughs took place and remarkable advancement was achieved since carbon dots became the center of attraction for desirable performances. Initially, carbon soot was considered as simple precursor for carbon dots which was obtained by refluxing in a strong acid (5 mol L^{-1} nitric acid), followed by size-dependent separation, but the quantum yield was very poor and lacked the required fluorescence for optical studies [23]. Then, laser ablation technique was considered as successful synthesis procedure [24], until diverse approaches for facile synthesis were adopted according to the application [25], which can be classified into two types: Top-down methods consisting of arc discharge, electrochemical oxidation, and chemical ablation techniques; and bottom-up approach, which consists of hydrothermal, solvothermal, pyrolysis or combustion, ultrasonic, microwave irradiation, plasma treatment, and acid oxidation. In most cases, carbon containing substrates like proteins, carbohydrates, and organic solvents were considered as building blocks for carbon dots. Significant advancement was also achieved on surface passivation of carbon dots using variety of molecular entities such as semiconductor salts like ZnO or ZnS for promoting surface defects to enhance fluorescence and quantum yield [26]. Liu et al. devised another technique involving "aqueous route" using silica sphere carrier to synthesize multi-colored photoluminescent carbon dots for bioimaging purpose [27]. Electrochemical synthesis [28] using 99% pure graphite rods and ultrapure water as electrolyte was also found to produce high yield of carbon dots in the electrolyte solution. Proliferation of carbon dots synthesis studies got oriented toward simpler schemes which allowed utilization of natural materials like wood or activated charcoal as precursor [29]. Since the first report on carbonization of organic molecules using microwave irradiation technique by Zhu et al. in 2009 [30] was reported, wide scale attention was focused into it for the convenience, eco-friendly, time-saving, and excellent results. After the first report on one-pot hydrothermal technique [31], this synthesis technique gained immense limelight due to the simplicity and eased to control the nucleation and functionality and proved to be a promising method for heteroatom or co-doped carbon dot synthesis (Fig. 2).

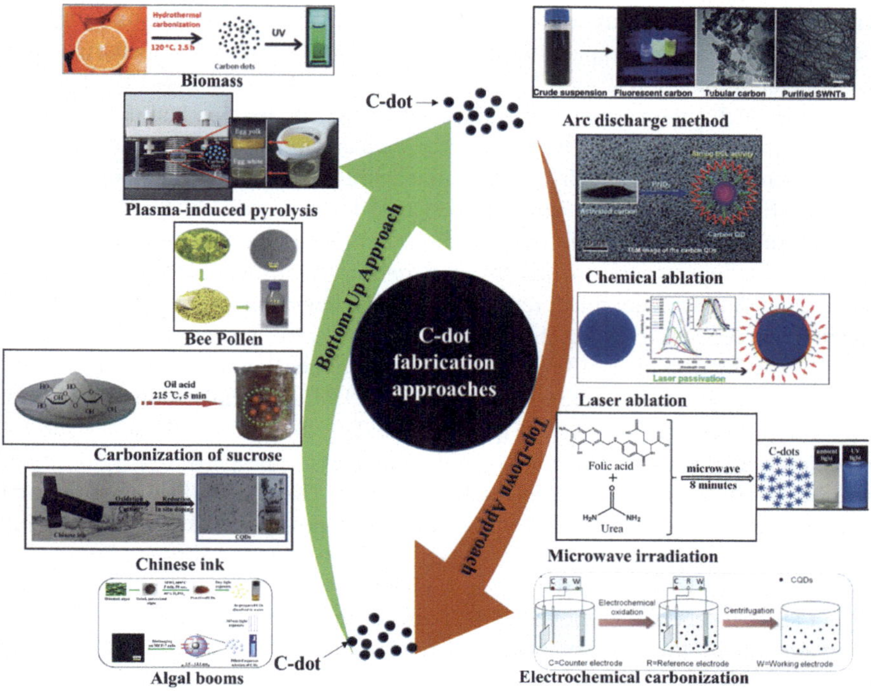

Fig. 2 Various routes for synthesis of carbon dots. Reproduced from [15] Copyright 2018, with permission from Elsevier

2.3 *Emergence of Heteroatom-Doping and Co-Doping*

In order to improve optical and electrical properties and to enhance quantum yield, various heteroatoms of comparable size were doped into the carbon dot structure singly or in combination (co-doping) over the course of synthesis. Generally, higher the content of heteroatom on carbon dots, greater will be the surface states formation and active sites, increasing the ability to trap a greater number of electrons, thus providing greater bright fluorescence and higher quantum yield value. On the quest for improved quantum yield, Liu et al. in 2011 [32] reported 12% enhancement of quantum yield on incorporation of amino group ($-NH_2$) over carbon dot surface by using 4,7,10-trioxo-1,13-tridecylenediamine in microwave irradiation. Since nitrogen and carbon has close similarity and electron can transfer from nitrogen atom into the carbon dot, thus modifying the electronic environment and improving photoluminescence properties for nitrogen-doped carbon dot [33]. Surface modulation using functional groups with high nitrogen content such as amines or amides resulted in distinct photo-physical properties like strong emission in shorter wavelengths (blue or green) or yellow emission (nearly 600 nm). Impact of variation of hydrothermal temperature and reaction time during carbon dots formation was widely studied, and it was found that there was an increase in quantum yield on increasing the reaction temperature from 150° to 200 °C, with subsequent reduction in diameter from 22 to 2 nm [34]. Study carried out Schneider and the team on nitrogen-doped carbon dots [35] displayed that ethylenediamine had the best fluorescence in blue region due to presence of greater amount of nitrogen compared to other amines like hexamethylenetetramine and triethanolamine, although similar emission-lifetime trend was found. Presence of C–N, C–N–C or C=N groups, or nitrogen in N–H coordination (pyrrolic nitrogen) [36] in the structure significantly amplifies the fluorescence quality. In comparison with nitrogen-doped carbon dots, sulfur atom doping is not much reported, although interest is given to the variety of precursors like waste cooking oil [37], sulfuric acid [30], or thiomalic acid (backbone for various pesticides) [38]. Sulfur doped fluorescent carbon dots owing to their high electronegativity have the tendency to bind to DNA or metal nanoparticles, with the ability to increase fluorescence quantum yield (11.8%) [38] without any surface passivation. In order to achieve great photoluminescence properties, several studies focused on co-doping nitrogen and sulfur. In 2013, Dong et al. [39] reported first co-doped carbon dots which were synthesized in one-pot hydrothermal method using mixture of citric acid (carbon source) and L-cysteine (sulfur and nitrogen source) and exhibited tremendously high fluorescence quantum yield of 73%. Three years later, Xue et al. [40] reported co-doped carbon dots of greater quantum yield of 74.15% using ammonium thiocyanate as the source for sulfur and nitrogen. Other than that, few reports also concentrate on boron-doping since it is the left-side neighbor of carbon in elemental periodic table, but the quantum yield obtained for the bright blue fluorescence was around 22% [41], 10–15% [42], or 14.8% [43]. Phosphorus was also used in few cases [44, 45] since phosphorus can act as n-type donor, which can be useful to alter or modify electrical or optical properties. In order to explore efficient and biocompatible heteroatom-doped carbon dots for bioimaging, Feng and co-workers [46] doped silicon on carbon dots. Subsequently, few reports also focused on introducing metals like copper, cobalt, zinc, etc., [26, 47] into the carbon dots matrix, such as a report proposed that doping magnesium into nitrogenous carbon dots can obtain quantum yield as high as 83% [33], but there is a concern regarding associated toxicity if metal releases from the matrix. As par Song et al. [48] formation of fluorescent co-doped carbon dots (N, S-CQDs) involves five stages, which include decomposition of sulfur and nitrogen source, their self-polymerization into small dots, carbon nuclei formation by aggregation, then carbonization at high temperature, followed by surface passivation, as illustrated in Fig. 3b.

Since the most striking feature of carbon dots is the brilliant, tunable, multicolored fluorescence emission whose origin was initially a debatable issue, but over a decade several theoretical progress and experimental evidence have been found. In 2015, Pang and colleagues [49] conducted a detailed study regarding different reaction conditions to obtain a series of emission wavelengths from 430 to 610 nm to demonstrate the prospects of carbon dots in multi-colored imaging applications. One of the major viewpoints behind origin of fluorescence is their surface state or surface defects, i.e., the amount of oxygen content or the surface functional groups or heteroatoms present on carbon dots. It was experimentally found that higher the

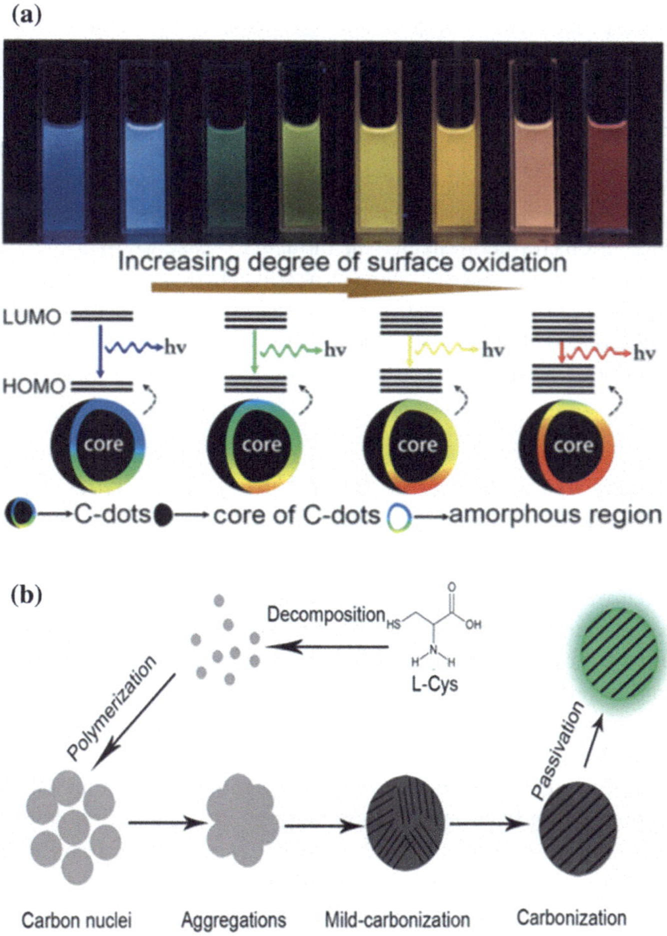

Fig. 3 **a** Schematic representation for the tunable photoluminescence of carbon dots by changing the surface state with different degrees of surface oxidation. Reproduced from [55] Copyright 2016, with permission from American Chemical Society; **b** schematic illustration of the possible mechanism for co-doped (N, S) carbon dots formation. Reproduced from [48] Copyright 2016, with permission from Elsevier

surface oxidation, greater is the number of surface defects, hence greater tendency to trap excitons, which ultimately results in reduction of band gap with subsequent red-shift emissions along with surface-state-related fluorescence [50]. Introducing various functional groups like C=N or C=O can also tune the emissions since they can influence the energy levels, producing new electronic transitions, or contribute as fluorophores, with enhanced trapping of electrons, thus facilitating better radiative recombination and higher and wider emission spectra [51]. Another reason for origin of fluorescence can be the quantum confinement effect and carbon core size-dependent properties. Experiments by Li et al. [52] revealed that on altering the carbon dots size from 1.2 to 3.8 nm yields different emission colors, as those near 1.2 nm size gives UV light emission (~350 nm), medium size of range 1.5–3 nm gives emission varying from bright blue to green, yellow and ultimately to red (400–700 nm), while larger carbon dots of 3.8 nm give near-infrared emission. Further, theoretical research using time-dependent density functional theory (TDDFT) and Gaussian calculations further indicated the influence of size sp^2 hybridized carbon on change of emission spectra since smaller size resulted in blue emission [53]. Recently, a third viewpoint regarding generation of fluorescence is impurities formed during bottom-up chemical synthesis or by-product induced molecular fluorescence [54] as confirmed using time-resolved electron paramagnetic resonance spectroscopy.

2.4 Synthesis of Carbon Dots from Natural Materials and Waste Products

Green synthesis of carbon dots using biodegradable waste, protein products, or direct plant sources is gaining immense prominence as an eco-friendly, economical, and safe alternative pathway to deal with the toxicity issues arising from use of harmful chemicals and in order to ensure better biocompatibility, without compromising on high quantum yield. Green synthesis also includes self-exothermic, base catalysis, or reduction techniques where carbon dots can be synthesized at room temperature or at temperature below that of chemical syntheses, without the aid of any costly instrument or application of external energy. It is a sustainable development strategy to utilize agro-industrial refuse, domestic waste, or excreta [56] as precursors, which additionally also combats pollution issues and facilitates zero-waste generation. Recycling or repurposing waste is always a beneficial way to combat pollution and using such waste can be a great alternative for utilization expensive or harsh chemicals in large scale synthesis of carbon dots. In the quest for development of fluorescent, sulfur-doped carbon dots in large scale with quantum yield as high as 47%, Mukherjee et al. [57] reported a facile synthesis method using bio-waste for real-time bioimaging purposes. Human waste likes urine can also be used as precursor of carbon dots (quantum yield of ~5.3%) via thermal treatment for cell imaging application [58]. Even algal bloom [59] can be made useful for in vitro imaging by synthesizing carbon dots through microwave route. A report by Zhang and Yu [60] showed various pathways for macroscale synthesis of fluorescent carbon dots for bioimaging and biosensing applications, mainly using Chinese ink, bee pollens, chicken eggs, coffee grounds, and sucrose as starting materials. Various reports focus on using peels of fruits and vegetables like onion [61], papaya [62], orange [63], or other kitchen waste [64] as precursors for green synthesis of nitrogenous or co-doped carbon dots. Apart from peels and fruit juices [65] for nitrogenous carbon dots preparation, various proteinaceous sources like milk [66], soy milk [67], meat [68] wool [69], or carbon-rich organic molecules like glucose, cellulose, lignin, chitosan, etc., are widely used [70, 71] to obtain fluorescent carbon dots. Plant sources are regularly used for carbon dots synthesis for various biomedical applications such as hydrothermal synthesis of carbon dots from onion [61], garlic [72], coriander leaves [73], carrot [74], kidney beans [75], and sweet potato [76] for cellular imaging purposes.

3 Biomedical Applications of Carbon Dots

The biocompatible or nontoxic nature of carbon dots along with their significant photo-physical attributes, nanoscale dimensionalities, and attractive functionalities made them a potential candidate for in vitro, in vivo studies, sensing or imaging, gene delivery, therapeutic applications, and other biologically-oriented applications. Several reports assessed the cytotoxicity of carbon dots, mainly against HeLa or MCF-7 cell model using various cell viability assays like MTT, WST-1, or CC8 [77–80]. Biosafety of utilization of carbon dots was also investigated using mice model [81]. Most of the in vitro and in vivo results indicated the low bio-toxic or the biocompatible nature of carbon dots, thus confirming safety for various biomedical applications. In 2014, Milosavljevic et al. [82] conducted a detailed study on interactions between carbon dots and single-stranded and double-stranded DNA using gel electrophoresis, electrochemical, and spectral methods and reported that there was no significant changes in single-stranded DNA, while fluorescence intensity increased in case of double-stranded DNA. It was assumed that the interaction of carbon dots with hydrogen bonding of double-stranded DNA resulted in such outcome, while the single-stranded DNA only interacts with π-π stacking of carbon dots. Furthermore, several biochemical analyzes like intracellular ROS generation, CCK, and lactate dehydrogenase (LDH) assays [83] were also conducted to determine the extent of interactions of carbon dots with cellular components like cellular membrane, phospholipids, cytoskeleton, or cell organelles, but no significant alteration of cellular membrane and structures or deformation of biomolecules was induced from administering carbon dots. Thus, such non-invasive, cost-effective, low cytotoxic nature with good water solubility, highly sensitive and stable photoluminescence, and excellent resistance to photobleaching made them one of the most suitable alternatives for organic dyes, metal ions, protein, or semiconductor quantum dots [84], which were earlier used as cellular biomarkers and other biomonitoring studies.

3.1 In Vitro Imaging

In the diverse field of biology, bioimaging has always proved to be an essential tool for identification of defective sites, monitoring drug delivery, or for measuring molecular interactions. For several decades, various quantum dots were used for this purpose, but toxicity issues and secondary reactions in biological system became the main concerns. Since Sun et al. [20] demonstrated the possibility of carbon dots as fluorescence tags for bioimaging along with their substantial merits and potentials over other conventional organic and inorganic fluorophores, worldwide attention was given to explore various avenues in bio-nanotechnology sectors. Confocal fluorescence microscopy images [85] revealed that carbon dots can be easily incorporated within the lipid vesicles of cells via endocytic pathways, owing to their amphiphilic nature, and can uniformly distribute within cellular cytosols and in nucleus, thus enabling effective multi-colored excitation/emission wavelengths dependent imaging. Moreover, cellular uptake, staining, or labeling have no adverse effect on shape or functionality of the respective cells [85]. Diverse set of cell lines like HEK293 [86], B16F11 [86], BGC823 [87], C6 glioma cells [88], human lung cancer (A549) [89], HepG2 [89], HeLa cell [90], etc., was utilized for bioimaging purposes. Carbon dots can also target folate receptors on the cancer cell surfaces or the cancer-expressing cells [91], thus revealing their significance in cancer cell studies. Since Zhu et al. [92] reported the first imaging of cancer cells using carbon dots, several studies were conducted and various modifications of carbon dots were reported such as conjugation of carbon dots with folic acid [93–95] or aptamers [96] to enable selective labeling of cancer cell, rather than normal cells. Carbon dots are also reported as effective imaging probe for determination of therapeutic mechanism stem cells [97, 98], which have immense significance in upcoming regenerative therapy. Presence of surface passivation molecules on carbon dots increases the chances of conjugation with specific bioactive species for required application [99]. Photostability is a crucial property for a bioimaging agent. Huo and team in 2010 [100] conducted a detailed analysis using COS-7 cells and reported successful penetration and labeling of cells and no reduction in fluorescence even upon 10 min continuous excitation, thus confirming excellent photostability (Fig. 4).

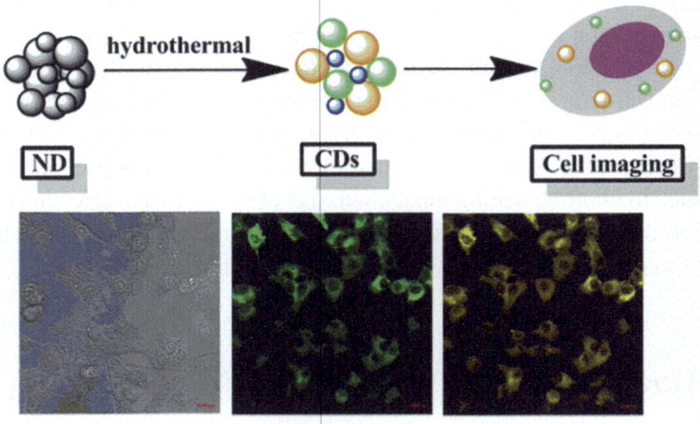

Fig. 4 Applications of carbon dots in bioimaging. Reproduced from [101] Copyright 2013, with permission from Elsevier

3.2 Microbial Detection Probe

Rapid detection of presence of microorganisms, diagnosis of infections by pathogenic bacteria, and development of effective bactericidal agents are vital clinical requirements to curb numerous diseases. Carbon dots have proved to be a promising sensitive probe for real-time monitoring of bacterial and fungal viability. Kasibabu et al. [102] synthesized carbon dots from pomegranate using facile hydrothermal route which acted as a probe for bacterial (*Pseudomonas aeruginosa*) and fungal (*Fusarium avenaceum*) cell imaging with rapid internalization of the carbon dots without any cytotoxic effects. Song et al. [103] demonstrated the pathway for bacterial viability evaluation using nitrogen, phosphorus, and sulfur co-doped carbon dots, which can selectively stain dead bacteria rather than live ones. A low-cost fluorescent probe synthesized by using sugarcane (*Saccharum officinarum*) juice as precursor for of bacteria (*E. coli* DH5α) and yeast (*Saccharomyces cerevisiae*) cells imaging was suggested by Mehta et al. [104] Singh and Chandra [105] developed pH sensitive, carbon dots-based microgels that can effectively detect even $\sim 10^4$ CFU of bacteria, as when bacteria gets encapsulated within microgels within 8 h and effectively changes the pH, resulting in enhanced green emission with increasing acidity. These microgels along with ampicillin can also serve as a platform for differentiation of resistant and non-resistant *E. coli* by observing changes in emission intensity within 4–6 h due to ampicillin inhibition. Carbon dots alongside bacterial imaging can also be used as bacterial growth inhibitor and antibacterial agent carriers like H_2O_2, Na_2CO_3, acetic acid, etc. [106]. Dong et al. [106] reported significant synergistic effect of carbon dots and H_2O_2 inhibiting growth both of Gram negative bacteria *E.coli* cells and Gram positive bacteria *Bacillus subtilis* cells. Carbon dots synthesized from *Lactobacillus plantarum* [107] can even prevent biofilm formation of *E. coli,* which otherwise is highly resilient and quite tough to prevent formation or removal due to highly structured nature and sturdy adherence. Hence, current advances in carbon dots development allow rapid, selective labeling, and growth inhibition [108] with can even induce bacteria apoptosis and death by DNA fragmentation, chromosomal condensation, and loss of structural integrity [109] on incubating with bacterial cells.

3.3 In Vivo Imaging

Although carbon dots were found to be efficient in vitro imaging agent, yet for better insight in biomedical studies, in vivo experiments are quite crucial. The first successful in vivo imaging using carbon dots was carried out in 2009 by Yang et al. [110] using mice model. On intravenous administration of carbon dots, which showed renal clearance within 24 h with minute accumulation. Histopathological analyzes of the organs post-exposure to carbon dots showed normal structure, without any sign of degeneration. No cytotoxicity or alteration was detected even after 28 days, while successful imaging of liver, spleen, and kidney could be done. This successful experimentation opened several other pathways like drug delivery, therapeutic, and nanomedicine administration. Tao et al. [111] synthesized carbon dots of 3–4 nm diameter and carried out a 3-month continuous experimentation on nude mice by injecting carbon dots (20 mg/kg body weight), but no mortality or weight loss was observed confirming nontoxic nature, even within animal model. Tao et al. also radiolabeled carbon dots with I^{125} to study the biodistribution of carbon dots on injecting into circulatory system. Similar to the findings of Yang et al., kidney intake was highest at early stage, indicating carbon dots can be excreted through urine along with accumulation in spleen and liver. Following such findings, several in vivo fluorescence imaging [112–114] was carried by injecting aqueous solution of carbon dots and collecting fluorescence images in red and green emission colors on using different excitation wavelengths. Most in vivo studies require emission in the near-infrared (NIR) spectral region on excitation with NIR light, whereas carbon dots have a drawback in this sector, with spatial resolution not much suitable for in vivo imaging. Hence, researchers are focusing to eliminating such disadvantages, causing emergence of techniques like photoacoustic and magnetic resonance imaging [115].

3.4 *Photoacoustic Imaging*

Photoacoustic imaging (optoacoustic imaging) is a new-age, emergent type of biomedical imaging modality based on the photoacoustic effect that operates by converting light into acoustic waves by absorbing electromagnetic waves and localized thermal excitation. This technique was found to be useful in in vivo bioimaging applications like methemoglobin measurement, skin melanoma detection, blood oxygenation mapping, or tumor angiogenesis [116]. Photoacoustic imaging overcomes optical scattering and uses ultrasonic detection-based effect, thus provides higher spatial resolution and deeper tissue penetration (few centimeters) and imaging with better spatial resolution [117]. Wu et al. [118] synthesized carbon dots from honey for developing high resolution, real-time intraoperative photoacoustic imaging probe that operated in conjunction with a near-infrared (NIR) for providing detailed information of progress of cancerous disease in sentinel lymph nodes (SLN). Drastic signal enhancement was observed due to presence of carbon dots, which have the perfect dimension needed for such studies. Parvin and Mandal [119] developed dual emissive co-doped (N, P) carbon dots for fluorescent and photoacoustic imaging on injecting in tumor bearing mice. Their findings confirmed the ability of carbon dots to distribute all over the body, especially in tumor area. Although limited studies were conducted using carbon dots [81, 120, 121], yet photoacoustic imaging has several prospects in future (Fig. 5).

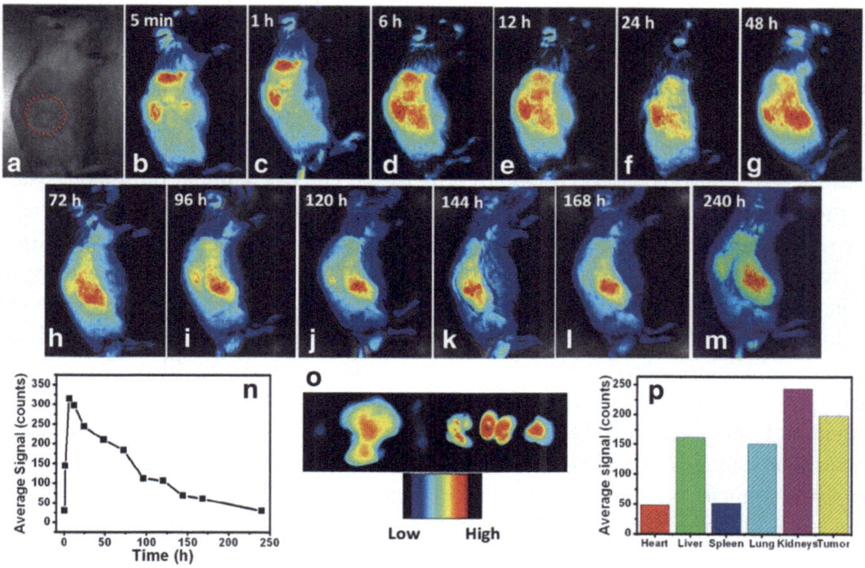

Fig. 5 Time-dependent in vivo red fluorescence images of BALB/c mouse bearing CT26 tumors after the intravenous injection of drug loaded carbon dots (the tumor is indicated with a red dotted circle). Here, **a** 0 min (taken under natural light; the red circle marks the position of tumor), **b** 5 min, **c** 1 h, **d** 6 h, **e** 12 h, **f** 24 h, **g** 48 h, **h** 72 h, **i** 96 h, **j** 120 h, **k** 144 h, **l** 168 h, and **m** 240 h. **n** average fluorescence intensity of the tumor area as a function of time. **o** ex vivo imaging of the tissues 48 h post-injection (from left to right: heart, liver, spleen, lung, kidneys, and tumor). **p** average fluorescent intensity of heart, liver, spleen, lung, kidneys, and tumor. Reproduced from [122] Copyright 2016, with permission from American Chemical Society

3.5 Biosensing

Since, early diagnosis of diseases is necessary for increasing chances of curing and reduction of treatment cost, rapid, sensitive, stable, and reliable biosensing are an essential tool for betterment of health and life. Due to excellent fluorescence, carbon dots are always considered as a remarkable biosensor for visual monitoring of cellular metal ions [123], glucose [124], nucleic acid [125], proteins, etc. Biosensing plays a significant role in early-stage disease detection, food safety, and biomonitoring. Specific surface passivation of carbon dots increased the fluorescence significantly and provided opportunities for interaction with the targeted analyte by π–π conjugation, electron transferring, or electrostatic interaction. Such interactions can ultimately result in "turn-on" or "turn-off" (quenching) of fluorescence. According to sensing strategies, carbon dots can be classified into various types such as electrochemical biosensors [126], optical biosensors [127], piezoelectric biosensors, thermoelectric biosensors [128], etc. Electrochemical biosensors are integrated devices that use bioreceptors (bio-recognition elements), which are in direct contact with electrochemical conduction elements, in order to collect information regarding target analytes. They directly convert biological information into detectable electrical signals [129]. Due to presence of high surface area, good conductivity and variety of functional groups which can be useful in modifying biological receptors along with excellent electron transport capabilities between the electrodes and sensing interface, they are considered as an efficient candidate for electrochemical biosensors fabrication [128]. Various carbon dots-based real-time electrochemical biosensing were studied, such as Jing et al. [130] designed nitrogen-doped carbon dots biosensor for rapid and direct detection of dopamine in human serum and urine, with limit of detection as low as detection limit of 1.2×10^{-9} mol/L. Optical biosensing is typically based on variation of fluorescence intensity or shift in emission wavelength on interaction with the target analytes. In case of "turn-off" sensing, target analytes concentration-dependent studies are followed and mechanisms like Förster resonance energy transfer (FRET) [131], photoinduced electron transfer (PET) [132], and inner filter effect (IFE) [133] mechanism are accordingly determined. One of the greatest advantages of optical biosensing is the rapid results without the need of any costly or sophisticated instruments, and in some cases (colorimetric sensing), detection is visible even under naked eyes. Since carbon dots are highly specific toward certain metal ions like Fe^{3+} [134] and Cu^{2+} [135], they are frequently used in such metal ion sensing in organisms. Bardhan et al. [136] experimentally and theoretically described the basis of selectivity of carbon dots toward specific metals in their recent report on dual sensing using a natural mineral-nitrogenous carbon dots-based sensor probe. Surface modification is done for selective detection of toxic heavy metals like Hg^{2+} [137], Pb^{2+} [138], Cr^{6+} [139] in living cells, which also has a pivotal role in toxicological and biosafety studies. Besides heavy metals, carbon dots can also be modified for fluorescence-based detection of various biological entities like cholesterol [140], glutathione [141], and enzymatic activities [142]. Carbon dots were also found useful in detecting and imaging cells based on pH and temperature variation. For better understanding of effect on pH in cellular functionality, Shangguan et al. [143] developed label-free carbon dots for intracellular ratiometric pH sensing at emission peaks of 475 and 545 nm. Similarly, intracellular ratiometric temperature sensing capability in living cells was analyzed by Wei et al. [144] at a temperature range of 5–50 °C. A recent report by Roy et al. [145] delineated the mechanism behind hexavalent chromium fluorometric detection and origin of fluorescence of the carbon dots probe using DFT studies. Additionally, Roy and group also developed a theoretical sensing model to detect presence of heavy metals like Cr^{6+} in living cells using molecular docking simulations. Recently, carbon dots-based nanoprobes are also used for detection of peroxynitrite production in mitochondria [146]. Peroxynitrite is a type of reactive oxygen/nitrogen species that is suspected to induce programmed cell death, hence sensitive and selective diagnosis of such intracellular reactive species are very important, and carbon dots have proved to be efficient for such purposes (Fig. 6).

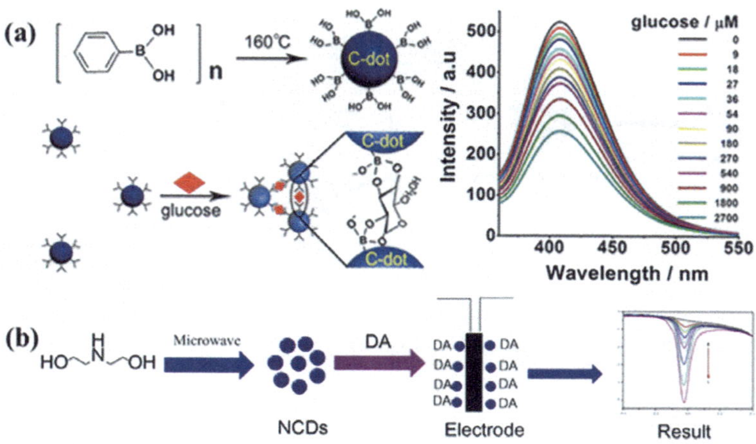

Fig. 6 **a** Detection of glucose by a carbon dots only biosensor through FRET. Reproduced from [147] Copyright 2014, with permission from American Chemical Society; **b** carbon dots electrode for detection of dopamine. Reproduced from [130] Copyright 2015, with permission from Elsevier

3.6 Targeted Drug/gene Delivery and Nanomedicine Applications

Drug or gene delivery systems require capability of transported them to specific target in the body and ensure proper absorption and distribution of drugs along with necessary interaction. Although, several conventional treatments like chemotherapy are available for treatment of cancer and other localized diseases, yet most of them lacks specificity, has multidrug resistance problems or causes toxicity. Biocompatible nature coupled with prominent fluorescence makes it suitable for dual applications, as therapeutic and bioimaging agent. Due to the nanoscale dimensions of carbon dots, they are quite appropriate as safe, potent, and good delivery vectors.

Currently, various gene carriers use positively charged polymeric materials for DNA fragment (negatively charged) delivery but most of them have shortcomings like tendency to accumulate within cells. Kim et al. [148] developed a polymeric carrier/plasmid DNA (pDNA) molecular complex by conjugating carbon dots gold nanoparticles with PEI-pDNA to monitor the association/dissociation of pDNA to the cells. The study revealed quenching of the fluorescence of the complex on entering the cytoplasm, due to release of pDNA into the cell nuclei, thus confirming successful transfection. In another report, Liu et al. [149] performed a comprehensive study on PEI functionalized carbon dots in order to compare the gene expression of pDNA in HepG2 and COS-7 cells with a control (PEI25k) along with the capacity to show tunable emission at different excitation wavelength. Carbon dots can serve as drug carrier moieties for various antibiotics and drugs like oxaliplatin [150], cisplatin (IV) [151], doxorubicin [152] due to their modified functional groups and maximum drug loading capacity. Tang et al. [152] developed FRET-based two-photon imaging

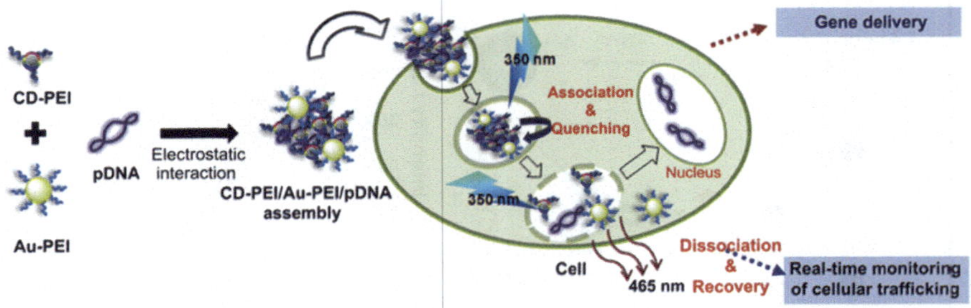

Fig. 7 Schematic representation for gene delivery using fluorescent carbon dots. Reproduced from [148] Copyright 2013, with permission from Elsevier

for real-time monitoring of drug delivery of one of the most common anticancer drug doxorubicin (DOX) using carbon dots. DOX was adsorbed onto carbon dots surface via π–π stacking, which during delivery undergone FRET mechanism as fluorescence intensity of carbon dots decreased while that of DOX increased as their interaction showed separation, thus revealing the real-time drug release. In order to avoid drug leakage and non-specific delivery, surface modification was done using PEG for functionalization of carbon dots and then conjugated with DOX via acid-labile Schiff base linkage. Such FRET-based drug delivery system was found to highly efficient for targeted drug delivery, especially in case of cancerous cells. For cancer cell targeted drug delivery, folic acid was combined with PEGylated carbon dots and DOX release from carbon dots surface was monitored at 498 nm as energy was transferred from carbon dots to DOX [152]. Another study was conducted using anti-psychotic drug haloperidol (HaLO)-grafted carbon dots, and cysteamine hydrochloride was used as linker to study the controlled release for more than 40 h [153]. The successful outcome thus confirmed the capability of carbon dots for successful drug delivery. Carbon dots other than being a nanocarrier can also act as medicine depending on its precursor. Li et al. [154] prepared carbon dots from ginger juice as precursor studied the uptake by HepG2 cells. The carbon dots proved to be selective and effective in killing cancer cells like hepatocellular carcinoma cells (HepG2), including human breast cancer cell line (MDA-MB-231) and human lung cancer cell line (A549), while showed no toxicity against normal cells (MCF-10A and FL83B). The π-plane in the aromatic rings of carbon dots helps in conjugating them with the drugs or therapeutic agents [155] which further increases drug loading capacity along with efficient targeting of tumors for their treatment (Fig. 7).

3.7 Photothermal and Photodynamic Therapy

Since most of the conventional therapeutic approaches like surgery, chemotherapy, or radiotherapy, especially for cancer treatment are quite aggressive, various new

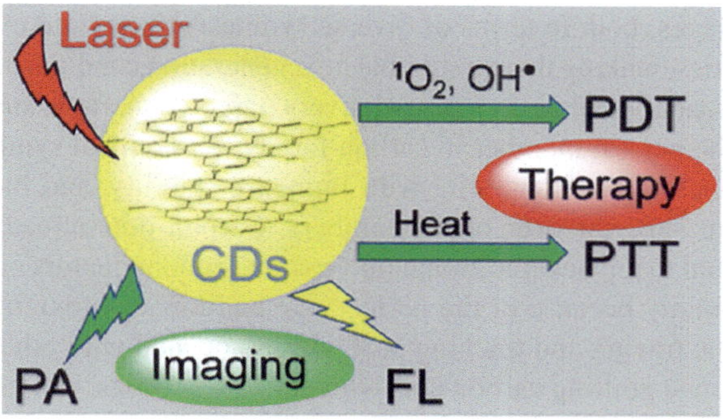

Fig. 8 Schematic illustration for application of carbon dots in photoacoustic and photothermal/photodynamic therapy. Reproduced from [173] Copyright 2020, with permission from Elsevier

classes of therapeutic strategy are emerging for better specificity, minimal invasiveness, high selectivity, and safety. Among various such new techniques, photothermal therapy (PTT) and photodynamic therapy (PDT) developed for various disease treatment using laser light [156]. PTT generally uses NIR photoabsorber to generate localized heat to destroy targeted cells, mainly the cancer cells, primary tumor site, or local metastatis in lymph node. Carbon dots can be useful as an economical alternative for traditionally used expensive photothermal agents like gold nanostructures [157]. Moreover, high conductivity of carbon dots facilitates weak electron–phonon interactions along with strong electron–electron interactions, thus having the ability to transform absorbed light into heat, thus becoming suitable for PTT applications [157]. Few reports [158–160] focused on using carbon dots for PTT and PDT purposes, but this field has already proved to have high future prospects for biomedical cancer research. Srivastava et al. [158] developed a novel cancer cell tracking probe coupled with PDT-PTT from waste leaves of Indian fig tree with quantum yield of 14.16%. They studied concentration-dependent photothermal response along with intracellular ROS generation using MDA-MB-231 breast cancer cells. The synthesized carbon dots great photostability despite continuous exposure to laser irradiation (30 min) proving the potentiality. In a recent report, Zhao et al. [160] elaborated the synergistic effect of carbon dots when coupled with PDT and PTT on generating ROS like singlet oxygen and hydroxyl radicals along with heat under laser irradiation (635 nm) into the tumor. Thus, this paved way toward tumor therapy along with their fluorescence imaging (Fig. 8).

4 Conclusion and Future Perspective

Advancement of nanoscience has already brought immense development in diagnosis and therapeutic purposes, including innovative treatment of complex diseases like cancer. Discovery of carbon dots was indeed revolutionary because of the of the multiple advantages, both in terms of diverse synthesis routes and excellent photophysical properties, making them a suitable next generation candidate for biosensing, cell imaging, bioanalytical assays, drug delivery, targeted therapy, and diagnosis. In spite of being the newest member in carbon family, economical synthesis methods, easy surface functionalization along with immense stability, and biocompatibility have made them superior over other members. Carbon dots already proved their immense potential to replace traditional fluorescent semiconductors or metal ions for optical study, mainly because of the perfect size and low cytotoxicity. Carbon dots have resolved the tracing and tracking problems associated with other conventional drug carriers, hence grafting carbon dots with anticancer drugs, and monitoring their controlled release has gained new directions in cancer treatment. Moreover, selectivity of carbon dots toward cancer cells and biocompatible nature toward normal and healthy cells has solved the issues associated with other cancer treatment techniques like chemotherapy. Recent applications in nanomedicine, PTT, PDT, and photoacoustic imaging opened new avenues for nanotheranostics since using carbon dots can lower their costs. Since most of the in vivo imaging generally operates at NIR range excitation-emission, many scientists are working toward improving this sector of carbon dots along with improvement of spatial resolution for better understanding. Although carbon dots synthesis is not always complex, but further purification issues often limit their utilization in healthcare centers. Many researchers are now developing pathways for large scale synthesis of carbon dots for industrial and commercial utilization.

Acknowledgements The authors would like to thank the Department of Physics, Jadavpur University, for extending their facilities.

Conflict of Interests The authors declare no conflict of interest.

Funding Details S.D. would like to acknowledge DST-SERB (Grant No. EEQ/2018/000747) for funding.

References

1. Nikolova M, Slavchov R, Nikolova G (2020) Nanotechnology in medicine. In: Hock F, Gralinski M (eds) Drug discovery and evaluation: methods in clinical pharmacology. Springer, Cham, pp 533–546
2. Li J (2015) Nanotechnology-based platform for early diagnosis of cancer. Sci Bull 60(4):488–490
3. Bardhan S, Pal K, Roy S, Das S, Chakraborty A, Karmakar P, Basu R, Das S (2019) Nanoparticle size-dependent antibacterial activities in natural minerals. J Nanosci Nanotechnol 19(11):7112–7122
4. Powell MC, Griffin MP, Tai S (2008) Bottom-up risk regulation? How nanotechnology risk knowledge gaps challenge federal and state environmental agencies. Environ Manage 42(3):426–443
5. Sotropa RMB (2018) The advantages and disadvantages of nanotechnology. J Oral Rehabil 10(2):113–119

6. Edwards IA, Marsh H, Menendez R (2013) Introduction to carbon science. Butterworth-Heinemann
7. Kang Z, Lee ST (2019) Carbon dots: advances in nanocarbon applications. Nanoscale 11(41):19214–19224
8. Lim SY, Shen W, Gao Z (2015) Carbon quantum dots and their applications. Chem Soc Rev 44(1):362–381
9. Himaja AL, Karthik PS, Singh SP (2015) Carbon dots: the newest member of the carbon nanomaterials family. Chem Rec 15(3):595–615
10. Zhao F, Wu J, Ying Y, She Y, Wang J, Ping J (2018) Carbon nanomaterial-enabled pesticide biosensors: design strategy, biosensing mechanism, and practical application. Trac-Trend Anal Chem 106:62–83
11. Wang L, Li W, Yin L, Liu Y, Guo H, Lai J, Han Y, Li G, Li M, Zhang J, Vajtai R (2020) Full-color fluorescent carbon quantum dots. Sci Adv 6(40):6772
12. Cayuela A, Soriano ML, Carrillo-Carrión C, Valcárcel M (2016) Semiconductor and carbon-based fluorescent nanodots: the need for consistency. Chem Commun 52(7):1311–1326
13. Li L, Dong T (2018) Photoluminescence tuning in carbon dots: Surface passivation or/and functionalization, heteroatom doping. J Mater Chem C 6(30):7944–7970
14. Jaleel JA, Pramod K (2018) Artful and multifaceted applications of carbon dot in biomedicine. J Control Release 269:302–321
15. Mishra V, Patil A, Thakur S, Kesharwani P (2018) Carbon dots: emerging theranostic nanoarchitectures. Drug Discov Today 23(6):1219–1232
16. Zhang D, Jiang W, Zhao Y, Dong Y, Feng X, Chen L (2019) Carbon dots rooted PVDF membrane for fluorescence detection of heavy metal ions. Appl Surf Sci 494:635–643
17. Lin Y, Zhang X, Shan X, Zeng T, Zhao X, Wang Z, Kang Z, Xu H, Liu Y (2020) Photo-tunable organic resistive random access memory based on PVP/N-doped carbon dot nanocomposites for encrypted image storage. J Mater Chem C 8(42):14789–14795
18. Riggs JE, Guo Z, Carroll DL, Sun YP (2000) Strong luminescence of solubilized carbon nanotubes. J Am Chem Soc 122:5879–5880
19. Xu X, Ray R, Gu Y, Ploehn HJ, Gearheart L, Raker K, Scrivens WA (2004) Electrophoretic analysis and purification of fluorescent single-walled carbon nanotube fragments. J Am Chem Soc 126(40):12736–12737
20. Sun YP, Zhou B, Lin Y, Wang W, Fernando KS, Pathak P, Meziani MJ, Harruff BA, Wang X, Wang H, Luo PG (2006) Quantum-sized carbon dots for bright and colorful photoluminescence. J Am Chem Soc 128(24):7756–7757
21. Liu H, Ye T, Mao C (2007) Fluorescent carbon nanoparticles derived from candle soot. Angew Chem Int 119(34):6593–6595
22. Cao L, Wang X, Meziani MJ, Lu F, Wang H, Luo PG, Lin Y, Harruff BA, Veca LM, Murray D, Xie SY (2007) Carbon dots for multiphoton bioimaging. J Am Chem Soc 129(37):11318–11319
23. Ray SC, Saha A, Jana NR, Sarkar R (2009) Fluorescent carbon nanoparticles: synthesis, characterization, and bioimaging application. J Phys Chem C 113(43):18546–18551
24. Hu SL, Niu KY, Sun J, Yang J, Zhao NQ, Du XW (2009) One-step synthesis of fluorescent carbon nanoparticles by laser irradiation. J Mater Chem 19(4):484–488
25. Baker SN, Baker GA (2010) Luminescent carbon nanodots: emergent nanolights. Angew Chem Int 49(38):6726–6744
26. Sun YP, Wang X, Lu F, Cao L, Meziani MJ, Luo PG, Gu L, Veca LM (2008) Doped carbon nanoparticles as a new platform for highly photoluminescent dots. J Phys Chem C 112(47):18295–18298
27. Liu R, Wu D, Liu S, Koynov K, Knoll W, Li Q (2009) An aqueous route to multi-color photoluminescent carbon dots using silica spheres as carriers. Angew Chem Int Ed 48(25):4598–4601
28. Ming H, Ma Z, Liu Y, Pan K, Yu H, Wang F, Kang Z (2012) Large scale electrochemical synthesis of high quality carbon nanodots and their photocatalytic property. Dalton Trans 41(31):9526–9531

29. Jelinek R (2017) Carbon-dot synthesis. In: Carbon quantum dots. Springer, Cham, pp 5–27
30. Zhu H, Wang X, Li Y, Wang Z, Yang F, Yang X (2009) Microwave synthesis of fluorescent carbon nanoparticles with electrochemiluminescence properties. Chem Commun 34:5118–5120
31. Zhang B, Liu CY, Liu Y (2010) A novel one-step approach to synthesize fluorescent carbon nanoparticles. Eur J Inorg 28:4411–4414
32. Liu C, Zhang P, Tian F, Li W, Li F, Liu W (2011) One-step synthesis of surface passivated carbon nanodots by microwave assisted pyrolysis for enhanced multicolor photoluminescence and bioimaging. J Mater Chem 21(35):13163–13167
33. Xu Q, Kuang T, Liu Y, Cai L, Peng X, Sreeprasad TS, Zhao P, Yu Z, Li N (2016) Heteroatom-doped carbon dots: synthesis, characterization, properties, photoluminescence mechanism and biological applications. J Mater Chem B 4(45):7204–7219
34. Jiang K, Sun S, Zhang L, Wang Y, Cai C, Lin H (2015) Bright-yellow-emissive N-doped carbon dots: preparation, cellular imaging, and bifunctional sensing. ACS Appl Mater Interfaces 7(41):23231–23238
35. Schneider J, Reckmeier CJ, Xiong Y, von Seckendorff M, Susha AS, Kasák P, Rogach AL (2017) Molecular fluorescence in citric acid-based carbon dots. J Phys Chem C 121(3):2014–2022
36. Sarkar S, Sudolska M, Dubecky M, Reckmeier CJ, Rogach AL, Zboril R, Otyepka M (2016) Graphitic nitrogen doping in carbon dots causes red-shifted absorption. J Phys Chem C 120(2):1303–1308
37. Hu Y, Yang J, Tian J, Jia L, Yu JS (2014) Waste frying oil as a precursor for one-step synthesis of sulfur-doped carbon dots with pH-sensitive photoluminescence. Carbon 77:775–782
38. Chandra S, Patra P, Pathan SH, Roy S, Mitra S, Layek A, Bhar R, Pramanik P, Goswami A (2013) Luminescent S-doped carbon dots: an emergent architecture for multimodal applications. J Mater Chem B 1(18):2375–2382
39. Dong Y, Pang H, Yang HB, Guo C, Shao J, Chi Y, Li CM, Yu T (2013) Carbon-based dots co-doped with nitrogen and sulfur for high quantum yield and excitation-independent emission. Angew Chem Int Ed Engl 52(30):7800–7804
40. Xue M, Zhang L, Zhan Z, Zou M, Huang Y, Zhao S (2016) Sulfur and nitrogen binary doped carbon dots derived from ammonium thiocyanate for selective probing doxycycline in living cells and multicolor cell imaging. Talanta 150:324–330
41. Shen C, Wang J, Cao Y, Lu Y (2015) Facile access to B-doped solid-state fluorescent carbon dots toward light emitting devices and cell imaging agents. J Mater Chem C 3(26):6668–6675
42. Bourlinos AB, Trivizas G, Karakassides MA, Baikousi M, Kouloumpis A, Gournis D, Bakandritsos A, Hola K, Kozak O, Zboril R, Papagiannouli I (2015) Green and simple route toward boron doped carbon dots with significantly enhanced non-linear optical properties. Carbon 83:173–179
43. Shan X, Chai L, Ma J, Qian Z, Chen J, Feng H (2014) B-doped carbon quantum dots as a sensitive fluorescence probe for hydrogen peroxide and glucose detection. Analyst 139(10):2322–2325
44. Han Y, Tang D, Yang Y, Li C, Kong W, Huang H, Liu Y, Kang Z (2015) Non-metal single/dual doped carbon quantum dots: a general flame synthetic method and electro-catalytic properties. Nanoscale 7(14):5955–5962
45. Zhou J, Shan X, Ma J, Gu Y, Qian Z, Chen J, Feng H (2014) Facile synthesis of P-doped carbon quantum dots with highly efficient photoluminescence. RSC Adv 4(11):5465–5468
46. Qian Z, Shan X, Chai L, Ma J, Chen J, Feng H (2014) Si-doped carbon quantum dots: a facile and general preparation strategy, bioimaging application, and multifunctional sensor. ACS Appl Mater Interfaces 6(9):6797–6805
47. Wu W, Zhan L, Fan W, Song J, Li X, Li Z, Wang R, Zhang J, Zheng J, Wu M, Zeng H (2015) Cu–N dopants boost electron transfer and photooxidation reactions of carbon dots. Angew Chem 127(22):6640–6644
48. Song Z, Quan F, Xu Y, Liu M, Cui L, Liu J (2016) Multifunctional N, S co-doped carbon quantum dots with pH-and thermo-dependent switchable fluorescent properties and highly selective detection of glutathione. Carbon 104:169–178

49. Bao L, Liu C, Zhang ZL, Pang DW (2015) Photoluminescence-tunable carbon nanodots: surface-state energy-gap tuning. Adv Mater 27(10):1663–1667
50. Ding H, Yu SB, Wei JS, Xiong HM (2016) Full-color light-emitting carbon dots with a surface-state-controlled luminescence mechanism. ACS Nano 10(1):484–491
51. Zhang Y, Yuan R, He M, Hu G, Jiang J, Xu T, Zhou L, Chen W, Xiang W, Liang X (2017) Multi-colour nitrogen-doped carbon dots: tunable photoluminescence and sandwich fluorescent glass-based light-emitting diodes. Nanoscale 9(45):17849–17858
52. Li H, He X, Kang Z, Huang H, Liu Y, Liu J, Lian S, Tsang CHA, Yang X, Lee ST (2010) Water-soluble fluorescent carbon quantum dots and photocatalyst design. Angew Chem 122(26):4532–4536
53. Eda G, Lin YY, Mattevi C, Yamaguchi H, Chen HA, Chen IS, Chen CW, Chhowalla M (2010) Blue photoluminescence from chemically derived graphene oxide. Adv Mater 22(4):505–509
54. Essner JB, Kist JA, Polo-Parada L, Baker GA (2018) Artifacts and errors associated with the ubiquitous presence of fluorescent impurities in carbon nanodots. Chem Mater 30(6):1878–1887
55. Do S, Kwon W, Kim YH, Kang SR, Lee T, Lee TW, Rhee SW (2016) N, S-induced electronic states of carbon nanodots toward white electroluminescence. Adv Opt Mater 4(2):276–284
56. Wei J, Zhang X, Sheng Y, Shen J, Huang P, Guo S, Pan J, Liu B, Feng B (2014) Simple one-step synthesis of water-soluble fluorescent carbon dots from waste paper. New J Chem 38(3):906–909
57. Sangam S, Gupta A, Shakeel A, Bhattacharya R, Sharma AK, Suhag D, Chakrabarti S, Garg SK, Chattopadhyay S, Basu B, Kumar V (2018) Sustainable synthesis of single crystalline sulphur-doped graphene quantum dots for bioimaging and beyond. Green Chem 20(18):4245–4259
58. Essner JB, Laber CH, Ravula S, Polo-Parada L, Baker GA (2016) Pee-dots: biocompatible fluorescent carbon dots derived from the upcycling of urine. Green Chem 18(1):243–250
59. Ramanan V, Thiyagarajan SK, Raji K, Suresh R, Sekar R, Ramamurthy P (2016) Outright green synthesis of fluorescent carbon dots from eutrophic algal blooms for in vitro imaging. ACS Sustain Chem Eng 4(9):4724–4731
60. Zhang J, Yu SH (2016) Carbon dots: large-scale synthesis, sensing and bioimaging. Mater Today 19(7):382–393
61. Bandi R, Gangapuram BR, Dadigala R, Eslavath R, Singh SS, Guttena V (2016) Facile and green synthesis of fluorescent carbon dots from onion waste and their potential applications as sensor and multicolour imaging agents. RSC Adv 6(34):28633–28639
62. Pooja D, Singh L, Thakur A, Kumar P (2019) Green synthesis of glowing carbon dots from Carica papaya waste pulp and their application as a label-freechemo probe for chromium detection in water. Sens Actuators B Chem 283:363–372
63. Prasannan A, Imae T (2013) One-pot synthesis of fluorescent carbon dots from orange waste peels. Ind Eng Chem Res 52(44):15673–15678
64. Himaja AL, Karthik PS, Sreedhar B, Singh SP (2014) Synthesis of carbon dots from kitchen waste: conversion of waste to value added product. J Fluoresc 24(6):1767–1773
65. Atchudan R, Edison TNJI, Chakradhar D, Perumal S, Shim JJ, Lee YR (2017) Facile green synthesis of nitrogen-doped carbon dots using *Chionanthus retusus* fruit extract and investigation of their suitability for metal ion sensing and biological applications. Sens Actuators B Chem 246:497–509
66. Wang L, Zhou HS (2014) Green synthesis of luminescent nitrogen-doped carbon dots from milk and its imaging application. Anal Chem 86(18):8902–8905
67. Zhu C, Zhai J, Dong S (2012) Bifunctional fluorescent carbon nanodots: green synthesis via soy milk and application as metal free electrocatalysts for oxygen reduction. Chem Commun 48:9367–9369
68. Kenneth N (2013) Versatility with carbon dots-from overcooked BBQ to brightly fluorescent agents and photocatalysts. RSC Adv 3:15604–15607
69. Wang L, Bi Y, Hou J, Li H, Xu Y, Wang B, Ding H, Ding L (2016) Facile, green and clean one-step synthesis of carbon dots from wool: application as a sensor for glyphosate detection based on the inner filter effect. Talanta 160:268–275

70. Ding Z, Li F, Wen J, Wang X, Sun R (2018) Gram-scale synthesis of single-crystalline graphene quantum dots derived from lignin biomass. Green Chem 20(6):1383–1390
71. Liang Z, Kang M, Payne GF, Wang X, Sun R (2016) Probing energy and electron transfer mechanisms in fluorescence quenching of biomass carbon quantum dots. ACS Appl Mater Interfaces 8(27):17478–17488
72. Zhao S, Lan M, Zhu X, Xue H, Ng TW, Meng X, Lee CS, Wang P, Zhang W (2015) Green synthesis of bifunctional fluorescent carbon dots from garlic for cellular imaging and free radical scavenging. ACS Appl Mater Interfaces 7(31):17054–17060
73. Sachdev A, Gopinath P (2015) Green synthesis of multifunctional carbon dots from coriander leaves and their potential application as antioxidants, sensors and bioimaging agents. Analyst 140(12):4260–4269
74. D'souza SL, Chettiar SS, Koduru JR, Kailasa SK (2018) Synthesis of fluorescent carbon dots using Daucus carota subsp. sativus roots for mitomycin drug delivery. Optik 158:893–900
75. Tripathi KM, Tran TS, Tung TT, Losic D, Kim T (2017) Water soluble fluorescent carbon nanodots from biosource for cells imaging. J Nanomater
76. Shen J, Shang S, Chen X, Wang D, Cai Y (2017) Facile synthesis of fluorescence carbon dots from sweet potato for Fe^{3+} sensing and cell imaging. Mater Sci Eng C 76:856–864
77. Zhang YP, Ma JM, Yang YS, Ru JX, Liu XY, Ma Y, Guo HC (2019) Synthesis of nitrogen-doped graphene quantum dots (N-GQDs) from marigold for detection of Fe^{3+} ion and bioimaging. Spectrochim Acta A 217:60–67
78. Tan X, Li Y, Li X, Zhou S, Fan L, Yang S (2015) Electrochemical synthesis of small-sized red fluorescent graphene quantum dots as a bioimaging platform. Chem Commun 51(13):2544–2546
79. Zhang M, Wang W, Cui Y, Zhou N, Shen J (2018) Near-infrared light-mediated photodynamic/photothermal therapy nanoplatform by the assembly of Fe_3O_4 carbon dots with graphitic black phosphorus quantum dots. ACS Biomater Sci Eng 13:2803
80. Yao H, Zhao W, Zhang S, Guo X, Li Y, Du B (2018) Dual-functional carbon dot-labeled heavy-chain ferritin for self-targeting bio-imaging and chemo-photodynamic therapy. J Mater Chem B 6(19):3107–3115
81. Jia Q, Zheng X, Ge J, Liu W, Ren H, Chen S, Wen Y, Zhang H, Wu J, Wang P (2018) Synthesis of carbon dots from Hypocrella bambusae for bimodel fluorescence/photoacoustic imaging-guided synergistic photodynamic/photothermal therapy of cancer. J Colloid Interface Sci 526:302–311
82. Milosavljevic V, Nguyen HV, Michalek P, Moulick A, Kopel P, Kizek R, Adam V (2015) Synthesis of carbon quantum dots for DNA labeling and its electrochemical, fluorescent and electrophoretic characterization. Chem Pap 69(1):192–201
83. Cui X, Xu S, Wang X, Chen C (2018) The nano-bio interaction and biomedical applications of carbon nanomaterials. Carbon 138:436–450
84. Pandit S, Behera P, Sahoo J, De M (2019) In situ synthesis of amino acid functionalized carbon dots with tunable properties and their biological applications. ACS Appl Bio Mater 2(8):3393–3403
85. Kong B, Zhu A, Ding C, Zhao X, Li B, Tian Y (2012) Carbon dot-based inorganic-organic nanosystem for two-photon imaging and biosensing of pH variation in living cells and tissues. Adv Mater 24(43):5844–5848
86. Goh EJ, Kim KS, Kim YR, Jung HS, Beack S, Kong WH, Scarcelli G, Yun SH, Hahn SK (2012) Bioimaging of hyaluronic acid derivatives using nanosized carbon dots. Biomacromol 13(8):2554–2561
87. Wang F, Xie Z, Zhang H, Liu CY, Zhang YG (2011) Highly luminescent organosilane-functionalized carbon dots. Adv Funct Mater 21(6):1027–1031
88. Ruan S, Qian J, Shen S, Chen J, Zhu J, Jiang X, He Q, Yang W, Gao H (2014) Fluorescent carbonaceous nanodots for noninvasive glioma imaging after angiopep-2 decoration. Bioconjug Chem 25(12):2252–2259
89. Luo TY, He X, Zhang J, Chen P, Liu YH, Wang HJ, Yu XQ (2018) Photoluminescent F-doped carbon dots prepared by ring-opening reaction for gene delivery and cell imaging. RSC Adv 8(11):6053–6062

90. Mehta VN, Jha S, Singhal RK, Kailasa SK (2014) Preparation of multicolor emitting carbon dots for HeLa cell imaging. New J Chem 38(12):6152–6160
91. Bhunia SK, Maity AR, Nandi S, Stepensky D, Jelinek R (2016) Imaging cancer cells expressing the folate receptor with carbon dots produced from folic acid. ChemBioChem 17(7):614–619
92. Zhu S, Zhang J, Qiao C, Tang S, Li Y, Yuan W, Li B, Tian L, Liu F, Hu R, Gao H (2011) Strongly green-photoluminescent graphene quantum dots for bioimaging applications. Chem Commun 47(24):6858–6860
93. Song Y, Shi W, Chen W, Li X, Ma H (2012) Fluorescent carbon nanodots conjugated with folic acid for distinguishing folate-receptor-positive cancer cells from normal cells. J Mater Chem 22(25):12568–12573
94. Zhang J, Zhao X, Xian M, Dong C, Shuang S (2018) Folic acid-conjugated green luminescent carbon dots as a nanoprobe for identifying folate receptor-positive cancer cells. Talanta 183:39–47
95. Zhao X, Zhang J, Shi L, Xian M, Dong C, Shuang S (2017) Folic acid-conjugated carbon dots as green fluorescent probes based on cellular targeting imaging for recognizing cancer cells. RSC Adv 7(67):42159–42167
96. Lee CH, Rajendran R, Jeong MS, Ko HY, Joo JY, Cho S, Chang YW, Kim S (2013) Bioimaging of targeting cancers using aptamer-conjugated carbon nanodots. Chem Commun 49(58):6543–6545
97. Malina T, Poláková K, Skopalík J, Milotová V, Holá K, Havrdová M, Tománková KB, Čmiel V, Šefc L, Zbořil R (2019) Carbon dots for in vivo fluorescence imaging of adipose tissue-derived mesenchymal stromal cells. Carbon 152:434–443
98. Han Y, Zhang F, Zhang J, Shao D, Wang Y, Li S, Lv S, Chi G, Zhang M, Chen L, Liu J (2019) Bioactive carbon dots direct the osteogenic differentiation of human bone marrow mesenchymal stem cells. Colloids Surf B 179:1–8
99. Li Q, Ohulchanskyy TY, Liu R, Koynov K, Wu D, Best A, Kumar R, Bonoiu A, Prasad PN (2010) Photoluminescent carbon dots as biocompatible nanoprobes for targeting cancer cells in vitro. J Phys Chem C 114(28):12062–12068
100. Qiao ZA, Wang Y, Gao Y, Li H, Dai T, Liu Y, Huo Q (2010) Commercially activated carbon as the source for producing multicolor photoluminescent carbon dots by chemical oxidation. Chem Commun 46(46):8812–8814
101. Zhang X, Wang S, Zhu C, Liu M, Ji Y, Feng L, Tao L, Wei Y (2013) Carbon-dots derived from nanodiamond: Photoluminescence tunable nanoparticles for cell imaging. J Colloid Interface Sci 397:39–44
102. Kasibabu BSB, D'souza SL, Jha S, Singhal RK, Basu H, Kailasa SK (2015) One-step synthesis of fluorescent carbon dots for imaging bacterial and fungal cells. Anal Methods 7(6):2373–2378
103. Song Y, Li H, Lu F, Wang H, Zhang M, Yang J, Huang J, Huang H, Liu Y, Kang Z (2017) Fluorescent carbon dots with highly negative charges as a sensitive probe for real-time monitoring of bacterial viability. J Mater Chem B 5(30):6008–6015
104. Mehta VN, Jha S, Kailasa SK (2014) One-pot green synthesis of carbon dots by using *Saccharum officinarum* juice for fluorescent imaging of bacteria (*Escherichia coli*) and yeast (*Saccharomyces cerevisiae*) cells. Mater Sci Eng C 38:20–27
105. Chandra A, Singh N (2018) Bacterial growth sensing in microgels using pH-dependent fluorescence emission. Chem Commun 54(13):1643–1646
106. Dong X, Awak MA, Tomlinson N, Tang Y, Sun YP, Yang L (2017) Antibacterial effects of carbon dots in combination with other antimicrobial reagents. PloS one 12(9):e0185324
107. Lin F, Li C, Chen Z (2018) Bacteria-derived carbon dots inhibit biofilm formation of *Escherichia coli* without affecting cell growth. Front microbiol 9:259
108. Otis G, Bhattacharya S, Malka O, Kolusheva S, Bolel P, Porgador A, Jelinek R (2018) Selective labeling and growth inhibition of *Pseudomonas aeruginosa* by aminoguanidine carbon dots. ACS Infect Dis 5(2):292–302

109. Bing W, Sun H, Yan Z, Ren J, Qu X (2016) Programmed bacteria death induced by carbon dots with different surface charge. Small 12(34):4713–4718
110. Yang ST, Cao L, Luo PG, Lu F, Wang X, Wang H, Meziani MJ, Liu Y, Qi G, Sun YP (2009) Carbon dots for optical imaging in vivo. J Am Chem Soc 131(32):11308–11309
111. Tao H, Yang K, Ma Z, Wan J, Zhang Y, Kang Z, Liu Z (2012) *In vivo* NIR fluorescence imaging, biodistribution, and toxicology of photoluminescent carbon dots produced from carbon nanotubes and graphite. Small 8(2):281–290
112. Li N, Liang X, Wang L, Li Z, Li P, Zhu Y, Song J (2012) Biodistribution study of carbogenic dots in cells and in vivo for optical imaging. J Nanopart Res 14(10):1177
113. Licciardello N, Hunoldt S, Bergmann R, Singh G, Mamat C, Faramus A, Ddungu JL, Silvestrini S, Maggini M, De Cola L, Stephan H (2018) Biodistribution studies of ultra-small silicon nanoparticles and carbon dots in experimental rats and tumor mice. Nanoscale 10(21):9880–9891
114. Zhang W, Wang R, Liu W, Wang X, Li P, Zhang W, Wang H, Tang B (2018) Te-containing carbon dots for fluorescence imaging of superoxide anion in mice during acute strenuous exercise or emotional changes. Chem Sci 9(3):721–727
115. Li Y, Lin TY, Luo Y, Liu Q, Xiao W, Guo W, Lac D, Zhang H, Feng C, Wachsmann-Hogiu S, Walton JH (2014) A smart and versatile theranostic nanomedicine platform based on nanoporphyrin. Nat Commun 5(1):1–15
116. Xu M, Wang LV (2006) Photoacoustic imaging in biomedicine. Rev Sci Instrum 77(4):041101
117. Wang LV, Hu S (2012) Photoacoustic tomography: in vivo imaging from organelles to organs. Science 335(6075):1458–1462
118. Wu L, Cai X, Nelson K, Xing W, Xia J, Zhang R, Stacy AJ, Luderer M, Lanza GM, Wang LV, Shen B (2013) A green synthesis of carbon nanoparticles from honey and their use in real-time photoacoustic imaging. Nano Res 6(5):312–325
119. Parvin N, Mandal TK (2017) Dually emissive P, N-co-doped carbon dots for fluorescent and photoacoustic tissue imaging in living mice. Microchim Acta 184(4):1117–1125
120. Jia Q, Ge J, Liu W, Liu S, Niu G, Guo L, Zhang H, Wang P (2016) Gold nanorod@silica-carbon dots as multifunctional phototheranostics for fluorescence and photoacoustic imaging-guided synergistic photodynamic/photothermal therapy. Nanoscale 8(26):13067–13077
121. Ge J, Jia Q, Liu W, Guo L, Liu Q, Lan M, Zhang H, Meng X, Wang P (2015) Red-emissive carbon dots for fluorescent, photoacoustic, and thermal theranostics in living mice. Adv Mater 27(28):4169–4177
122. Zheng M, Li Y, Liu S, Wang W, Xie Z, Jing X (2016) One-pot to synthesize multifunctional carbon dots for near infrared fluorescence imaging and photothermal cancer therapy. ACS Appl Mater Interfaces 8(36):23533–23541
123. Zhu A, Qu Q, Shao X, Kong B, Tian Y (2012) Carbon-dot-based dual-emission nanohybrid produces a ratiometric fluorescent sensor for in vivo imaging of cellular copper ions. Angew Chem 124(29):7297–7301
124. Shi W, Wang Q, Long Y, Cheng Z, Chen S, Zheng H, Huang Y (2011) Carbon nanodots as peroxidase mimetics and their applications to glucose detection. Chem Commun 47(23):6695–6697
125. Li H, Zhang Y, Wang L, Tian J, Sun X (2011) Nucleic acid detection using carbon nanoparticles as a fluorescent sensing platform. Chem Commun 47(3):961–963
126. Wang Z, Dai Z (2015) Carbon nanomaterial-based electrochemical biosensors: an overview. Nanoscale 7(15):6420–6431
127. Ng SM (2019) Carbon dots as optical nanoprobes for biosensors. In Nanobiosensors for Biomolecular Targeting. Elsevier, pp 269–300
128. Ji C, Zhou Y, Leblanc RM, Peng Z (2020) Recent developments of carbon dots in biosensing: A review. ACS Sens 5(9):2724–2741
129. Thevenot DR, Toth K, Durst RA, Wilson GS (1999) Electrochemical biosensors: recommended definitions and classification. Pure Appl Chem 71(12):2333–2348
130. Jiang Y, Wang B, Meng F, Cheng Y, Zhu C (2015) Microwave-assisted preparation of N-doped carbon dots as a biosensor for electrochemical dopamine detection. J Colloid Interface Sci 452:199–202

131. Cho MJ, Park SY (2019) Carbon-dot-based ratiometric fluorescence glucose biosensor. Sens Actuators B 282:719–729
132. Roy S, Pal K, Bardhan S, Maity S, Chanda DK, Ghosh S, Karmakar P, Das S (2019) Gd (III)-Doped boehmite nanoparticle: an emergent material for the fluorescent sensing of Cr (VI) in wastewater and live cells. Inorg Chem 58(13):8369–8378
133. Zu F, Yan F, Bai Z, Xu J, Wang Y, Huang Y, Zhou X (2017) The quenching of the fluorescence of carbon dots: a review on mechanisms and applications. Microchim Acta 184(7):1899–1914
134. Zhang H, Chen Y, Liang M, Xu L, Qi S, Chen H, Chen X (2014) Solid-phase synthesis of highly fluorescent nitrogen-doped carbon dots for sensitive and selective probing ferric ions in living cells. Anal Chem 86(19):9846–9852
135. Lu L, Feng C, Xu J, Wang F, Yu H, Xu Z, Zhang W (2017) Hydrophobic-carbon-dot-based dual-emission micelle for ratiometric fluorescence biosensing and imaging of Cu^{2+} in liver cells. Biosens Bioelectron 92:101–108
136. Bardhan S, Roy S, Chanda DK, Ghosh S, Mondal D, Das S, Das S (2020) Nitrogenous carbon dot decorated natural microcline: an ameliorative dual fluorometric probe for Fe^{3+} and Cr^{6+} detection. Dalton Trans 49(30):10554–10566
137. Mohapatra S, Sahu S, Sinha N, Bhutia SK (2015) Synthesis of a carbon-dot-based photoluminescent probe for selective and ultrasensitive detection of Hg^{2+} in water and living cells. Analyst 140(4):1221–1228
138. Kumar A, Chowdhuri AR, Laha D, Mahto TK, Karmakar P, Sahu SK (2017) Green synthesis of carbon dots from Ocimum sanctum for effective fluorescent sensing of Pb^{2+} ions and live cell imaging. Sens Actuators B 242:679–686
139. Yu L, Zhang L, Ren G, Li S, Zhu B, Chai F, Qu F, Wang C, Su Z (2018) Multicolorful fluorescent-nanoprobe composed of Au nanocluster and carbon dots for colorimetric and fluorescent sensing Hg^{2+} and Cr^{6+}. Sens Actuators B 262:678–686
140. Bui TT, Park SY (2016) A carbon dot–hemoglobin complex-based biosensor for cholesterol detection. Green Chem 18(15):4245–4253
141. Dong W, Wang R, Gong X, Dong C (2019) An efficient turn-on fluorescence biosensor for the detection of glutathione based on FRET between N, S dual-doped carbon dots and gold nanoparticles. Anal Bioanal Chem 411(25):6687–6695
142. Tong X, Shi S, Tong C, Iftikhar A, Long R, Zhu Y (2020) Quantum/carbon dots-based fluorescent assays for enzyme activity. Trac-Trend Anal Chem 131:116008
143. Shangguan J, He D, He X, Wang K, Xu F, Liu J, Tang J, Yang X, Huang J (2016) Label-free carbon-dots-based ratiometric fluorescence pH nanoprobes for intracellular pH sensing. Anal Chem 88(15):7837–7843
144. Wei L, Ma Y, Shi X, Wang Y, Su X, Yu C, Xiang S, Xiao L, Chen B (2017) Living cell intracellular temperature imaging with biocompatible dye-conjugated carbon dots. J Mater Chem B 5(18):3383–3390
145. Roy S, Bardhan S, Chanda DK, Roy J, Mondal D, Das S (2020) In Situ-grown cdot-wrapped boehmite nanoparticles for Cr (VI) sensing in wastewater and a theoretical probe for chromium-induced carcinogen detection. ACS Appl Mater Interfaces 12(39):43833–43843
146. Wu X, Sun S, Wang Y, Zhu J, Jiang K, Leng Y, Shu Q, Lin H (2017) A fluorescent carbon-dots-based mitochondria-targetable nanoprobe for peroxynitrite sensing in living cells. Biosens Bioelectron 90:501–507
147. Shen P, Xia Y (2014) Synthesis-modification integration: one-step fabrication of boronic acid functionalized carbon dots for fluorescent blood sugar sensing. Anal Chem 86(11):5323–5329
148. Kim J, Park J, Kim H, Singha K, Kim WJ (2013) Transfection and intracellular trafficking properties of carbon dot-gold nanoparticle molecular assembly conjugated with PEI-pDNA. Biomaterials 34(29):7168–7180
149. Liu C, Zhang P, Zhai X, Tian F, Li W, Yang J, Liu Y, Wang H, Wang W, Liu W (2012) Nanocarrier for gene delivery and bioimaging based on carbon dots with PEI-passivation enhanced fluorescence. Biomaterials 33(13):3604–3613
150. Zheng M, Liu S, Li J, Qu D, Zhao H, Guan X, Hu X, Xie Z, Jing X, Sun Z (2014) Integrating oxaliplatin with highly luminescent carbon dots: an unprecedented theranostic agent for personalized medicine. Adv Mater 26(21):3554–3560

151. Feng T, Ai X, Ong H, Zhao Y (2016) Dual-responsive carbon dots for tumor extracellular microenvironment triggered targeting and enhanced anticancer drug delivery. ACS Appl Mater Interfaces 8(29):18732–18740
152. Tang J, Kong B, Wu H, Xu M, Wang Y, Wang Y, Zhao D, Zheng G (2013) Carbon nanodots featuring efficient FRET for real-time monitoring of drug delivery and two-photon imaging. Adv Mater 25(45):6569–6574
153. Pandey S, Shah R, Mewada A, Thakur M, Oza G, Sharon M (2013) Gold nanorods mediated controlled release of doxorubicin: nano-needles for efficient drug delivery. J Mater Sci 24(7):1671–1681
154. Li CL, Ou CM, Huang CC, Wu WC, Chen YP, Lin TE, Ho LC, Wang CW, Shih CC, Zhou HC, Lee YC (2014) Carbon dots prepared from ginger exhibiting efficient inhibition of human hepatocellular carcinoma cells. J. Mat. Chem. B 2(28):4564–4571
155. Ehtesabi H, Amirfazli M, Massah F, Bagheri Z (2020) Application of functionalized carbon dots in detection, diagnostic, disease treatment, and desalination: a review. Adv Nat Sci-Nanosci 11(2):025017
156. Zou L, Wang H, He B, Zeng L, Tan T, Cao H, He X, Zhang Z, Guo S, Li Y (2016) Current approaches of photothermal therapy in treating cancer metastasis with nanotherapeutics. Theranostics 6(6):762
157. Ghosal K, Ghosh A (2019) Carbon dots: the next generation platform for biomedical applications. Mat Sci Eng C 96:887–903
158. Thakur M, Kumawat MK, Srivastava R (2017) Multifunctional graphene quantum dots for combined photothermal and photodynamic therapy coupled with cancer cell tracking applications. RSC Adv 7(9):5251–5261
159. Yang W, Wei B, Yang Z, Sheng L (2019) Facile synthesis of novel carbon-dots/hemin nanoplatforms for synergistic photo-thermal and photo-dynamic therapies. J Inorg Biochem 193:166–172
160. Zhao S, Wu S, Jia Q, Huang L, Lan M, Wang P, Zhang W (2020) Lysosome-targetable carbon dots for highly efficient photothermal/photodynamic synergistic cancer therapy and photoacoustic/two-photon excited fluorescence imaging. Chem Eng J 388:124212

Chapter 6
Liposomal Delivery System

Sarjana Raikwar, Pritish Kumar Panda, Pooja Das Bidla, Shivani Saraf, Ankit Jain, and Sanjay K. Jain

1 Introduction

Liposomes are vesicular carrier systems in which aqueous compartment is enclosed by phospholipid bilayer. From last few decades, liposomes were extensively used as carriers for the delivery of bioactive, as they are appropriate to entrap both hydrophobic and hydrophilic drugs [53]. Liposomes are associated with some advantages like biodegradability, biocompatibility, low-immunogenicity, increased efficacy and therapeutic index, reduction in toxicity of encapsulated agents, improved pharmacokinetic effects. Liposomes have displayed favorable outcomes in drug delivery. The liposomes formation occurs when phospholipids are hydrated. Phospholipids composed of hydrophilic polar head and hydrophobic non-polar tail. The hydrophobic tail (Fatty acid chains) and hydrophilic polar head (consist of phosphoric acid) are self-organized in the presence of aqueous phase to form vesicular structure [25, 30, 39, 89]. These amphiphiles are employed in the drug delivery, via soaps, detergents, polar lipids, which are used to design concentric bilayered structures. The hydration of lipid film leads to swelling of liquid crystalline bilayer. This hydrated lipid sheet gets detached during agitation and closes together to produce large, multilamellar vesicles (MLVs), which avoid interaction of water with hydrocarbon core of the bilayer at the edges (Vyas and Khar 2002). On the basis of sizes and characteristics of liposomes, they involve different methods of preparation. The thin-film hydration method is extensively used for the formation of MLVs. In this method, a thin-film (lipid film) is hydrated (using aqueous buffer) above the transition temperature (TT)

S. Raikwar · P. K. Panda · P. Das Bidla · S. Saraf · S. K. Jain (✉)
Pharmaceutics Research Project Laboratory, Department of Pharmaceutical Sciences, Dr. Harisingh Gour Vishwavidyalaya, Sagar, Madhya Pradesh 470 003, India

A. Jain
Department of Materials Engineering, Indian Institute of Science, Bangalore, Karnataka 560 012, India

S. Gopi et al. (eds.), *Nanotechnology for Biomedical Applications*,
Materials Horizons: From Nature to Nanomaterials,
https://doi.org/10.1007/978-981-16-7483-9_6

of lipids. Generally, this method is used to form MLVs which can be sonicated and passed through polycarbonate filters to form small unilamellar vesicles (SUVs). Major limitation associated with this method is its poor entrapment efficiency (5–15%) of hydrophilic drugs. However, reduction of liposomes size leads to reduce amount of entrapped drugs. Freeze-drying of SUV dispersion in an aqueous solution of the drug is another method to prepare MLV with high entrapment efficiency (up to 40%). The % EE of MLV can be increased by hydrating lipid in the presence of an organic solvent. Various methods have been described for the formation of large, unilamellar vesicles (LUV), including solvent (ether or ethanol) injection, detergent dialysis, calcium induced fusion, and reverse-phase evaporation (REV) method. The drug which showed lipophilic and hydrophilic properties (amphiphilic nature) on the basis of pH of solution may not be entrapped efficiently. It is due to diffusion of drugs in and out of lipid membrane. Therefore, it is challenging to entrap drugs inside the liposomes. However, these types of drugs can be entrapped into perforated liposomes with high efficiency (up to 90%) using the active loading technique [78]. Moreover, conventional liposomes exhibit some problems like lesser encapsulation efficiency and enhanced blood clearance by reticulo-endothelial system (RES). The recognition of liposomes (by opsonins) as foreign material leads to uptake by macrophages (mononuclear phagocyte system (MPS)) and RES. Several physicochemical properties like net surface charge, size, hydrophobicity, and hydrophilicity of liposomes have an essential role in the clearance of liposomes from circulation. However, the vesicles size (>200 nm) and hydrophobicity of liposomes help in opsonization and RES uptake. PEGylation is an essential method to avoid such problems of RES uptake. It is also used to develop long circulatory liposomes. Several other PEG alternatives (sialic acid, polyvinyl alcohol, and poly-N-vinylpyrrolidones) are used to form long circulatory liposomes. These agents are served to improve the blood circulation time but poor selectivity for tumors [74]. Thus, targeted liposomes were then designed for selective drug delivery to the targeted site. Several targeting moiety (ligands) like folate, transferrin, mannose, hyaluronic acid, asialoglycoprotein are being used in targeting liposomes. Recently, several characteristics of tumor microenvironment (TME) (such as acidic pH, raised in temperature, and hypoxia) have been utilized for the designing of stimuli-sensitive drug delivery systems (SDDS). Stimuli-sensitive materials are used to change their features in response to variations in the surroundings. SDDS specially deliver the drug in the presence of stimuli (intrinsic and extrinsic) [41]. Moreover, antibodies (Ab) and Ab fragments are extensively used targeting agents for liposomes (because of higher specificity for their target antigens). This led to developing novel nanocarriers known as immunoliposomes. They are formed by conjugation of Ab to the surface of liposomes and permit for active targeting via binding to cancer cell-specific receptors [62, 72]. Further, advanced techniques such as theranostic liposomes have been developed for diagnosis as well as treatment of cancer. The theranostic liposomes have both diagnostic agent and therapeutic agent in a single nanocarrier which can display the drug localization at a particular site, and improve therapeutic effects [41]. These theranostic liposomes may use diagnostic agents such as quantum dots (QDs) and graphene oxide along with therapeutic agent. The different types of liposomes are represented in Fig. 1.

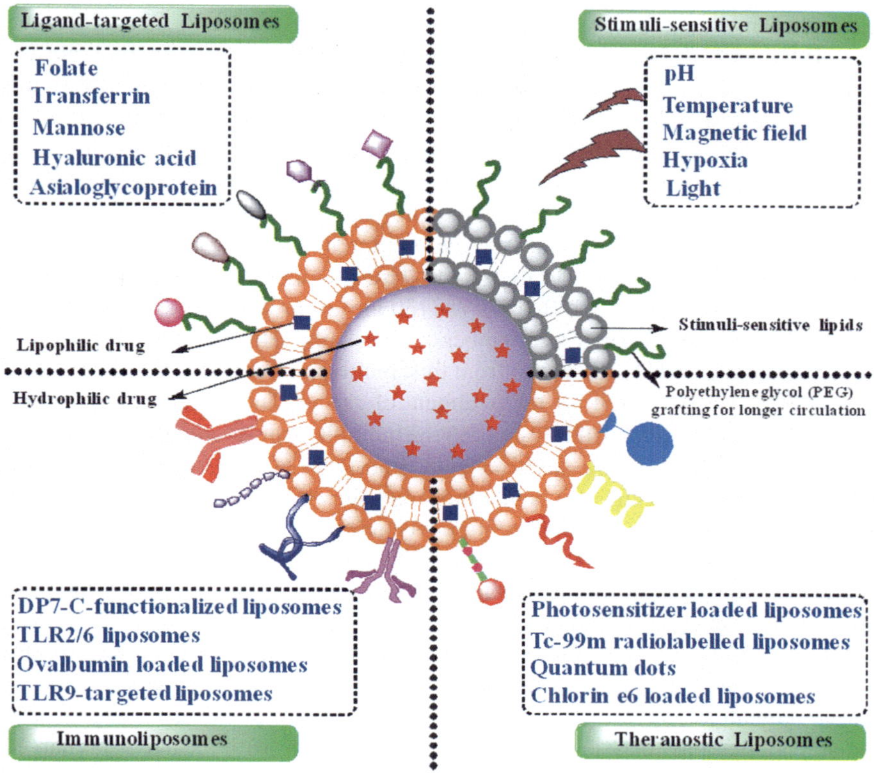

Fig. 1 Schematic representation of types of liposomes

Advances in nanotechnology have provided an opportunity to design site specific and individualized therapies for cancer [71, 90]. In this chapter, the application potential of several ligands such as folate, transferrin, mannose, hyaluronic acid, and asialoglyciprotein functionalized liposomes and SSDD (like pH, temperature, magnetic field, hypoxia, and photo-triggered liposomes) have been thoroughly discussed. This chapter also includes the advancement of liposomes and marketed formulation used for cancer treatment. Several liposomal formulations are approved for the treatment of cancer includes Onivyde™, Marqibo®, Visudyne®, Doxil®, and DepoCyt®.

2 Classification of Liposomes on the Basis of Size and Chemical Composition

Liposomes are vesicular carrier systems employed for targeting many solid tumors such as colon, prostate, breast, and brain [17, 31, 38, 60, 74]. Liposomes are classified on the basis of size (small, intermediate, or large) and lamellarity (uni-, oligo-, and multi-lamellar vesicles). These liposomes include MLVs, SUVs, and LUVs, etc. The MLVs are 500–5000 nm in size and comprised of different concentric bilayers while SUVs are about 100 nm in size and formed by a single bilayer and LUVs are 200–800 nm in size [25, 27, 29]. The development of unilamellar vesicles (ULVs) or multilamellar vesicles (MLVs) depends upon the method of preparation. Since ULVs contain a huge aqueous core and are appropriate for the entrapment of water soluble drugs while the MLVs favorably entrap the lipid-soluble drugs. They have the capability to entrap numerous drugs with different solubility features. MLVs are noticed to form more easily at larger hydrodynamic diameters and display higher entrapped volume than ULVs. But, ULVs (130 nm) show a much faster release rate than MLVs (250 nm) due to lesser number of phospholipid bilayer that it has to cross before being released [1, 78]. Apart from this, multivesicular liposomes (MVLs) are developed and widely used for the treatment of cancer. These multivesicular liposomes (DepoFoam) contain the vesicle having multiple vesicles entrapped within one vesicle. Nanotechnology use multivesicular liposomes (DepoFoam) which permit the delivery of therapeutic agents into the target site. This reduces the frequency of administration, maintains the therapeutic concentration for longer period of time, and minimized the side effects. Recently, several DepoFoam drugs have been approved for the treatment of many diseases including cancer for example DepoCyt® has been approved for the effective treatment of lymphomatous meningitis [70].

On the basis of chemical nature of lipids liposomes are classified into cationic, anionic, and neutral liposomes. These liposomes are widely used in cancer targeting [79]. Cationic liposomes are basically positively charged liposomal formulations that have the capacity to entrap with negatively charged DNA as well as the cell membranes [82]. They are generally composed of cationic lipids such as 1,2-dioleoyl-3-triethylammonium-propane (DOTAP), N-[1-(2,3-Dioleyloxy) Propyl]-N,N,N-trimethylammonium chloride (DOTMA), N,N-dioleyl-N,N-dimethylammonium chloride (DODAC) and dioctadecylamidoglycylspermine (DOGS), etc. It has also been studied that anionic liposomes displayed higher stability in solution due to less aggregation in comparison to the neutrally charged liposomes. These anionic liposomes are fabricated with the help of anionic lipids such as dimyristoyl-phosphatidyl glycerol (DMPG), dipalmitoyl-phosphatidyl glycerol (DPPG), dioleoy-phosphatidyl glycerol (DOPG), 1,2-distearoyl-sn-glycero-3-phosphatidic acid (DSPA), 1,2-distearoyl-sn-glycero-3-phosphatidylglycerol (DSPG), and 1,2-distearoyl-sn-glycero-3-phosphatidylserine (DSPS), etc. Moreover, it is also described that the anionic liposomes improved the process of endocytosis as compared to both cationic liposomes and neutral liposomes, respectively [56]. Neutral phospholipids such as phosphatidylcholine (PC), 1,2-dipalmitoyl-sn-glycero-3-phosphocholine (DPPC), 1,2-distearoyl-sn-glycero-3-phosphoethanolamine (DSPE), 1,2-dioleoyl-sn-glycero-3-phosphoethanolamine (DOPE), 1,2-dioleoyl-sn-glycero-3-phosphatidylcholine (DOPC) are used in the preparation of neutral liposomes [64, 77].

Liposomes are further classified on the basis of advanced modification such as plain, PEGylated, targeted, stimuli-sensitive, immuno, and theranostic liposomes. The long-circulating liposomes (stealth liposomes) can stay in the blood or systemic circulation for a longer duration than the non-modified liposomes. But these liposomes showed non-specific distribution to normal cells which cause toxicity and reduce the therapeutic effect of anticancer drug. Thus to avoid this problem, targeted drug delivery system (DDS) has been designed. Recently, advanced liposomes have been developed such as targeted liposomes (folate, mannose, transferrin, hyaluronic acid, and asialoglycoprotein), stimuli-sensitive liposomes (pH, temperature, magnetic field, hypoxia, and photo), and immuno liposomes. Targeted liposomes are fabricated on the basis of ligand-receptor interactions. Several receptors are overexpressed on the surface of cancer cells which are recognized as the marker to target cancer. Moreover, several characteristics of TME such as acidic pH, elevated temperature, and hypoxia have been employed using SDDS. Immuno liposomes are specialized liposomes which are carrying Ab that attached to their surfaces and able to accumulate in the area within the body where an attached Ab recognizes and binds its antigen [4, 8]. Currently, advanced technology has been used to develop theranostic liposomes which are exploited for the diagnosis and treatment of cancer. Classification of liposomes on the basis of chemical nature of phospholipids, advanced modification, and applications of liposomes are shown in Table 1.

3 Application of Liposomes in Cancer Therapy

Recently, several strategies such as targeted, stimuli-sensitive, theranostic, and immunoliposomes are developed for effective treatment of cancer. Several applications of liposomes in cancer therapy have been discussed below [53].

3.1 Ligand Targeted Liposomes

Generally, liposomes undergo passive and active targeting. The higher accumulation of drug inside the tumor cells (because of EPR effects) occur in passive targeting. Additionally, active targeting approaches (ligand-appended liposomes) have been employed as a third generation of liposomes. Functionalization of liposomes can provide a site-specific delivery like cell organelles (because of high affinity for particular receptors) [30, 72].

3.1.1 Folate-Targeted Liposomes

The folate receptor (FR) is overexpressed on the tumor cells, such as ovary, uterus, testis, colon, kidneys, and lung cancer. Folic acid conjugates are linked via γ-carboxyl group to drug molecules and enter into the folate overexpressing cancer cells. Receptor-mediated endocytosis has been taken place because of high affinity of folate conjugates for cancer cell surfaces. Folic acid functionalization helps the

Table 1 Classification of liposomes on the basis of nature of phospholipids, advanced modification, and applications of liposomes

Type of liposomes	Chemicals and solvents	Method of preparations	Drug/active agents/biomolecules	Uses	References
Cationic	1-Palmitoyl-2-stearoyl-sn-glycero-3-phosphocholine (HSPC), DPPC, unsaturated (POPC and DOPC), DOTAP, dimethyldidodecylammonium bromide (DDAB), Cholesterol (Chol)	Thin-film hydration	Doxorubicin (DOX)	Breast cancer	[11]
	1,2-dioleoyl-3 trimethyl ammonium propane (DOTAP),phosphatidylcholine(PC), 1,2-distearoyl-sn-glycero-3-phosphoethanolamine-N-(aleimide(polyethyleneglycol)-2000) (Mal-PEG) and Cholesterol (Chol)	Thin-film method	Trastuzumab	Prostate cancer	[81]
	PEG, DC-Chol/ dioleoyl phosphatidylethanolamine (DOPE)	Thin-film hydration	KSP siRNA	Ovarian cancer	[40]
	DOTAP/DOPE/Chol	Thin-film hydration method	EGFR siRNA	Glioma	[93]
Anionic	Cholesterol and Distearoyl phosphatidylcholine (DSPC)	Ultrafiltration centrifugation	siRNA	Pulmonary fibrosis	[59]
	Phospholipids and Chol	Thin-film hydration method	Antigen	Autoimmune diseases	[3]
	Dipalmitoyl phosphatidylcholine (DPPC); 1,2-distearoyl-sn-glycero3-phosphoethanolamine-N, Chol, Diethylenetriaminepentaacetic acid anhydride (DTPA)	Film hydration method	Paclitaxel (PTX) and Vinorelbine (VNB)	Lung cancer	[35]
	HSPC, DOTAP, DSPE-PEG2000, Soybean phosphatidylcholine	Film hydration method	Celecoxib (CXB), curcumin (CUR), and DOX	Breast cancer	[47]
Neutral	DSPC, stearyl amine (SA)	Film hydration method	Tamoxifen (TMX)	Breast cancer	[77]
Stealth	DPPC, DSPC, and DSPE-MPEG-2000, Cholesterol	Thin-film hydration method	Sirolimus	Prostate cancer	[55]
	Egg-PC, cholesterol and DSPE-PEG2000	Thin-film hydration method	quercetin	Cervical cancer	[75]
Folate-targeted	Folic acid, 1-Ethyl-(3-dimethylaminopropyl) carbonyldiamine hydrochloride (EDC·HCL), 1-hydroxybenzotriazole (HOBT) and 4-dimethylaminopyridine (DMAP), 1,2-ethanedithiol (EDT)	Ethanol injection-ammonium sulfate gradient method	DOX and Astragaloside IV	Triple negative breast cancer	[101]
Transferrin-targeted	Transferrin, (DPPC), Chol, DSPE-PEG2000, and polyethylene glycol-distearoyl phosphatidylethanolamine-maleimide (DSPE-PEG2000-MAL)	Film dispersion method	Dioscin	Tumor targeting	[92]
Hyaluronic acid- targeted	Hyaluronic acid (HA), DOPE, DOTAP, and DSPE-PEG2000	Film dispersion method	5-fluorouracil (5-FU)	colorectal cancer	[50]

(continued)

Table 1 (continued)

Type of liposomes	Chemicals and solvents	Method of preparations	Drug/active agents/biomolecules	Uses	References
Mannose-targeted	Soya phosphatidyl choline (Pho), Chol	Film dispersion method	PTX	Colon Cancer	[96]
Asialoglycoprotein-targeted	Asialoglycoprotein, (DSPE-PEG2000), egg yolk lecithin; DSPE-PEG2000-GAL, Chol	Lipid film hydration	Norcantharidin	Hepatocellular carcinoma	[46]
pH sensitive	DPPC, DC-Chol, Chol, pH-sensitive octylamine grafted poly aspartic acid (PASP-g-C8)	Reverse-phase evaporation method	Cytarabine	Tumor targeting	[106]
	DOTAP, DOPE, DSPE-PEG2000, heparin	Thin-film hydration method	DOX and Epacadostat	Melanoma	[9]
Immuno liposomes	Lipids soy phosphatidylcholine (SPC), DSPEPEG-2000, CHOL	Thin-film hydration method	Docetaxel (DTX)	Prostate cancer	[15]
	SPC, cholesterol, and DSPE-PEG2000, (2′-Benzothiazolyl)-7-diethylaminocoumarin	Film dispersion/post-insertion method	Calatase and programmed death ligand 1 monoclonal antibodies (aPDL1s)	Melanoma immunotherapy	[23]

targeted delivery of bioactive and diagnostic agents to cancer cells in the presence of normal cells. Folate mediated targeting is better to approach for cancer targeting [48, 67]. Park et al. [61] developed folate-targeted pH-sensitive liposomes loaded with DTX and doxycycline (anti-calpain-2) for non-small cell lungs cancer (NSCLC). The poor diagnosis of NSCLC is due to tumor-associated macrophages (TAMs). Additionally, calpain-2 (CAPN2) is overexpressed in NSCLC and is involved in tumor growth. This study is used to improve antitumor efficacy and reduce side effects associated with the treatment. Cytotoxicity assay demonstrated that the FR-mediated targeting (FRβ) in M2 TAMs and NSCLC cells effectively suppressed tumor growth. This combinational therapy showed synergistic anticancer effects by suppressing CAPN2 expression [61]. Some methods such as non-specific contrast agents, and other treatment approaches are inadequate in the treatment and early prognosis of tumors. Thus, there is necessity to fabricate more specific, targeted approach to overwhelm the limitation of non-targeted DDS. The folate conjugated PTX and VNB encapsulated Tc-99m radiolabeled liposomes were formulated and evaluated. In vivo therapeutic efficacy of liposomes was determined by bio-distribution study. In vitro study was performed in LLC1 cell lines to determine tumor growth inhibition. Active and passive targeted liposomes possesses higher cellular uptake and co-drug loaded liposomes display higher cytotoxicity than free drug combination (in LLC1 cells). Outcomes of bio-distribution studies in tumor-bearing C57BL/6 mice indicated that greater liposomes were observed in cancerous tissue (compared to non-cancerous tissue). Additionally, histopathological assay demonstrated that the FR-targeted liposomes not only suppress tumor growth but also inhibited lungs metastasis. Hence, these liposomes have immense potential in cancer prognosis and therapy [36]. In another study, diacid metabolite of norcantharidin (DM-NCTD)-loaded folic acid-functionalized, polyethylene glycolated (DM-NCTD/FA-PEG) liposomes were prepared as clinically effective approach for hepatocellular carcinoma (HCC). In vitro cytotoxicity study of DM-NCTD/FA-PEG liposomes displayed higher cytotoxicity on H22 (hepatoma cells) compared to non-targeted PEG liposomes. Additionally, bio-distribution study of DM-NCTD liposomes showed improved tumor targeting efficiency. The DM-NCTD/FA-PEG liposomes exhibit higher efficiency to inhibit tumor (in H22 tumor-bearing mice) compared to free DM-NCTD or DM-NCTD/PEG liposomes, in vivo. Thus, FR-targeted liposomes could have immense potential in HCC patients [48].

3.1.2 Transferrin Targeted Liposomes

Transferrin receptors (TfR) are overexpressed on cancer cells which are used for tumor targeting. Several hurdles like blood brain barrier (BBB) complexity and multidrug resistance (MDR) are related to the treatment of gliomas (brain tumor). To overcome above challenges, Tf-functionalized vincristine and tetradrine (TET) entrapped liposomes were prepared [76]. In vitro assay of Tf-functionalized liposomes displayed the improvement in transportability across BBB, enhanced cellular uptake, inhibited the MDR, and blocked the invasion of tumor cell and vasculogenic mimicry (VM) channels. It also induces apoptosis of C6 cells and C6/ADR cells via up-regulating caspases proteins. In vivo studies suggested that Tf-functionalized liposomes could enhance localization of drugs into brain tumor tissues and increase treatment efficiency. Hence, the developed liposomes have higher therapeutic effects in gliomas [85]. D-alpha-tocopheryl PEG 1000 succinate mono-ester (TPGS) liposomes bearing DTX and QDs were formulated and evaluated for diagnosis and treatment of brain tumors (brain theranostic). TPGS-liposomes were characterized for their particle size, polydispersity index, morphology, drug entrapment efficiency, and in vitro release study. The size of liposomes was found to be below 200 nm with higher entrapment efficiency (71%). Outcomes of animal studies demonstrated that the TfR-targeted liposomes are considered an effective brain-targeting strategy. Targeting efficacy of these liposomes is because of its higher permeability and smaller size (nano-size) [83]. In another study, Tf and rabies virus glycoprotein (RVG) peptide conjugated liposomes were developed for brain tumor therapy. These liposomes have homogenous particle size and capability to prevent enzymatic degradation of plasmid DNA. RVG-Tf-conjugated liposomes are exhibiting higher ability to transfect cells than non-functionalized liposomes. In vitro studies (triple co-culture BBB model) of these liposomes revealed the efficient translocation ability and higher permeability through BBB. Thus, the alterations of liposomes with brain-targeting ligands are used as potential strategies for gene delivery [14]. Fernandesa et al. [16] developed Tf-conjugated DTX bearing liposomes for prostate cancer (PC) treatment. The anticancer effects of Tf-conjugated DTX bearing liposomes (LIP-DTX-TF) were evaluated on PC cell lines. MTT assay (in PC-3 and PNT2 cell lines) showed more cytotoxicity of LIP-DTX-TF to PC-3 cells than the conventional formulation. Thus, these liposomal systems have promising potential for PC treatment [16].

3.1.3 Mannose Receptors Targeted Liposomes

Mannose receptor (MR) is highly expressed on tumor cell surface. It plays an important role in the transportation of glucose (energy source for metabolically active cancer cells) into the cancer cells. Mannose-functionalized nanocarriers are used for cancer cell targeting. It may facilitate receptor-mediated intracellular transport of therapeutics. The MR is overexpressed especially in resistant colon cancer cells (HCT8/ADR). Hence, MR-targeted liposomes can enhance the transport of chemotherapeutic drugs intracellularly. Kang et al. [34] prepared mannose-functionalized DOX and dihydroartemisinin (DHA) bearing liposomes for effective targeting of cancer cells. The co-delivery of DOX (anticancer effect) and DHA (anti-MDR effect) in HCT8/ADR cells could overcome the MDR effect and provide promising approach for anti-MDR colon cancer therapy [34]. MR is responsible for tumor growth, invasion, proliferation, metastasis in the TME, and overexpressed in TAMs. Chlorogenic acid (CHA) has been identified as a potent immunomodulators which induced polarization of TAMs from an M2 (pro-tumergenic) to an M1 (anti-tumerogenic) phenotype. MR-targeted CHA-entrapped liposomes have been used for targeting TAMs through MR. Results of in vitro and in vivo studies determined the anticancer effects of CHA-entrapped mannosylated liposomes. They also exhibit a desirable particle size, stability, and targetability. Additionally, CHA-entrapped mannosylated liposomes displayed higher suppression in tumor growth (in G422 gliomas). Hence, CHA-entrapped mannosylated liposomes could provide a promising potential for TAMs-targeted cancer immunotherapy [99]. In another study, targeted mannose-6-phosphate (M6P)-functionalized liposomes bearing anticancer drugs have been formulated for cancer treatment. The behavior of M6P-functionalized and non-functionalized liposomes was compared in MCF-7 cells and HDF normal cells. They also prepared calcein-entrapped liposomes and displayed the higher uptake of M6P-functionalized liposomes in MCF-7 cells as compared to HDF cells. Hence, the developed liposomes have higher strength and selectivity toward tumor cells [51].

3.1.4 Hyaluronic Acid Targeted Liposomes

Hyaluronic acid (HA) is a glycosaminoglycan that is used for cancer cell targeting. HA-functionalized nanocarriers have been used to target CD44 receptors and improve tumor targeting [65, 66]. Combination of phototherapy and chemotherapy was used to enhance the tumor targeting. The HA hexadecylamine (HA-C16) (HA derivative) was functionalized on the surface of liposomes and biocompatible magnetic nanoparticles and anticancer drug (DTX) was entrapped into it. These liposomes show enhanced therapeutic efficacy (IC50 $= 0.69 \pm 0.1036$ μg/mL) compared to radiation and DTX monotherapy. Glioblastoma multiforme (GBM) is a type of brain malignancy. The distinction between tumor and healthy brain cells remains challenging for GBM, thus there is necessity to develop an effective approach to target specific GBM cells. The developed HA-conjugated liposomes (HALMPs) are showing potentiality for targeting GBM cells over other brain cells (because of higher CD44 receptor expression). Thus, HA-conjugated liposomes show targeted delivery of bioactive through CD44 receptors in GBM cells [22]. In another study, HA-functionalized liposomes containing DOX and PTX were developed to improved cancer chemotherapy. This combination displayed synergistic effect and reduced drug resistance. The in vitro study exhibited that the release of both drugs over 72 h were higher in pH 5.5 phosphate buffer than pH 7.4 (PBS). In vitro MTT assay showed highest cytotoxicity and synergistic effects in comparison to free drug or single drug loaded liposomes. Additionally, cell uptake study suggested that higher internalization ability of HA-DOX-Lip compared to HA-unconjugated DOX-Lip and free DOX. Hence, the HA-conjugated dual-drug loaded liposomes exhibit promising potential for targeted tumor therapy [84].

3.1.5 Asialoglycoprotein Receptor (ASGPR) Targeted Liposomes

ASGPR is overexpressed in HCC which is widely used in targeted DDS. The use of natural ligands (asialofetuin), and synthetic ligand with galactosylated or lactosylated residues, such as galactosylated cholesterol and galactosylated polymers have achieved significant targeting efficacy to the liver [10]. Lactoferrin (Lf) is a unique targeting ligand for HCC cells because of its specific binding with ASGPR. In one study, lactoferrin (Lf)-modified DOX-loaded PEGylated liposomes (Lf-PLS) were developed and its targeting efficacy and antitumor effect to HCC were determined. The confocal laser scanning microscopy (CLSM) showed that the Tf-PLS was considerably higher uptaken in ASGPR-positive cells but not in ASGPR-negative cells as compared to non-functionalized PEGylated liposomes (PLS). In vitro cytotoxicity study indicated that the DOX-loaded Lf-PLS displayed stronger anti-proliferative effects on HCC cells. Outcomes of this study suggested that the DOX-loaded Lf-PLS have promising potential in HCC-targeting therapy [94]. Targeted liposomes were designed using polymeric cholesteryl arabinogalactan to target HCC. The in vitro study displayed increased cytotoxicity and specificity towards HepG2 cell lines. These liposomes are specifically localized into liver in comparison to other organs like kidney and heart. These in vitro and in vivo studies showed that the prepared liposomes have high stability and entrapment efficiency due to presence of cholesteryl arabinogalactan. Thus, these targeted liposomes have promising potential in the treatment of HCC [63]. In another study, ASGPR targeted liposomes were developed to determine the anticancer effect and selectivity of co-delivery of DOX and vimentin siRNA to HCC. The aim of this study was to develop galacosylated liposomes bearing drug, known as gal-DOX/vimentin siRNA liposomes (Gal-DOX/siRNA-L). Additionally, animal studies of these liposomes demonstrated that the efficient delivery of drugs to the liver. All findings showed that the Gal-DOX-siRNA-L could be used as effective carrier for the delivery of bioactives and increase transfection efficacy [57].

3.2 Stimuli-Sensitive Liposomes

Liposomes that show trigger responses to various stimuli (pH, temperature, magnetic field, hypoxia, and light) for the delivery of drug to the site of action are known as stimuli-sensitive liposomes. Stimuli-sensitive liposomes are used to attain cancer specific targeting with higher therapeutic effects and lower side effects. The key biomarker for the designing of SDDS is the specific features of TME like lower pH, higher temperature, hypoxia, and overexpression of proteolytic enzymes [86]. The TME responsive strategies can be exploited to target cancer cells. SDDS can elicit the release of bioactive as a result of destabilization caused by several stimuli (such as pH, temperature, and light) [87]. Thus, SDDS has wide application in cancer therapeutics. These DDS have several benefits such as higher accumulation, improved pharmacokinetic parameters enhanced cell uptake, and drug release selectively in cancerous tissues in response to the stimuli [97]. SDDS have been designed to remain the structural integrity of nanocarrier(s) throughout the circulation and triggered the release of bioactive in response to particular stimuli at the target site. These nanocarrier(s) shows faster changes within the TME like disruption and accumulation of system leading to triggered release of bioactive [2, 13]. The fabrication of SDDS and efficacy of these systems were confirmed by preclinical animal experiments [95, 97]. Various stimuli-sensitive DDS have been discussed below.

3.2.1 PH-Sensitive Liposomes

The pH-sensitive liposomes respond to change in the pH of the surrounding disease tissue like cancer cells [19]. The pH of TME is different from the pH of normal cells. Thus, the pH-sensitive liposomes have been employed as favorable strategy for enhancing the localization of drug at tumor site. They show selective release of bioactive in response to pH. They were further modified with ligand to enhance the targetability and enhance the therapeutic efficacy. These liposomes showed pH-dependent disruption of lipid membrane (at acidic pH of endosomes). Therefore, the encapsulated bioactive is internalized into the cell and shows its pharmacological action [58]. Metastasis is the most common hurdle in cancer chemotherapy and leads to failure in melanoma therapy. Indoleamine-2,3 dioxygenase (IDO) overexpressed in the TME lead to the immune escape. The combinational approach (chemotherapeutic agent with IDO inhibitor) was used which was effective for melanoma treatment. Thus, heparin-functionalized pH-sensitive DOX and epacadostat (EPA) bearing liposomes were formulated. Heparin functionalization increases the stability of liposomes, induced the cell uptake and avoids the interaction of B16F10 cells to platelets and extracellular matrix (ECM). In vitro studies of pH-sensitive liposomes showed improved cytotoxicity and apoptosis. Systematic administration of these liposomes exhibited optimum anti-metastasis activity on melanoma. Results of in vitro and in vivo studies suggested that the higher efficiency of heparin-functionalized pH-sensitive liposomes. Reduction in tumor cell invasion, migration, and induction of anti-metastatic immunity showed by DC maturation, CD8 + cytotoxic T lymphocytes (CTLs) activation confirmed the therapeutic efficacy of pH-sensitive liposomes. Thus, heparin-functionalized pH-sensitive DOX and EPA loaded liposomes exhibits higher antitumor immunity and are applicable for the treatment of lungs metastasis in melanoma [9]. Folate-targeted DOX and imatinib (IM) loaded pH-sensitive liposomes (FPL-DOX/IM) were formulated and evaluated for antitumor efficacy. This study triggered the release of bioactive at acidic pH and reversed the drug resistance. Outcomes demonstrated that the FPL-DOX/IM could provide an immense potential against MDR tumors [9]. Han et al. [20] prepared D-α-tocopheryl poly(2-ethyl-2-oxazoline) succinate (TPOS)-modified DTX liposomes. TPOS has ability to prepare pH-responsive nanocarrier(s) and is employed for cancer targeting. Results of in vitro suggested that the TPOS-modified DTX liposomes had shown slower drug release (at pH 7.4), while the higher release ($86.92\% \pm 1.69\%$) at pH 6.4. Cell uptake and flow cytometry study displayed that the TPOS-modified coumarin-6 liposomes

(TPOS-C6-L) had higher cellular uptake (at pH 6.4 than pH 7.4). In vitro cytotoxicity study revealed that cytotoxicity of TPOS-DTX-L (at pH 7.4) was lower than pH 6.8 after 48 h of incubation. The apoptosis assay displayed that the PEG-DTX-L had no remarkable improvement with decreased pH (from 7.4 to 6.8), while TPOS-DTX-L significantly induced apoptosis in HeLa cells with decreased pH (from 7.4 to 6.8). Hence, TPOS-C6-L could be employed as an effective carrier for the designing of pH-sensitive DDS [20].

3.2.2 Temperature-Sensitive Liposomes

Temperature is an external stimuli used to elicit the release of bioactive from temperature-sensitive liposomes (TSLs). TSL is classified into two types: traditional liposomes containing the temperature-sensitive component and liposomes modified with temperature-sensitive polymers. These liposomes show an increase in permeation of membrane at its gel-to-sol crystalline phase TT. These systems show stability at physiological temperature (37 °C) but show a temperature-elicited drug delivery above 37 °C under the influence of stimulus [91]. At hyperthermic temperature (above 40 °C), these liposomes showed rapid release of entrapped drugs. Ruiz et al. [68] developed DOX-loaded TSLs for targeted delivery of drugs to the tumor. These liposomes show a rapid release of DOX in response to mild hyperthermia (HT). The thermosensitivity of TSLs can be affected by DOX crystal loading. The characteristics of DOX crystals were altered through different active loading methods (ammonium sulfate, ammonium EDTA and magnesium sulfate). These studies revealed that the ammonium sulfate or ammonium EDTA loading methods had similar encapsulation efficiency (%EE) into TSLs, while low DOX loading was observed in metal complexation method. Additionally, cryogenic transmission electron microscopy (cryo-TEM) presented a variation in the nature of DOX crystals formed inside TSLs. In vitro DOX uptake through CT26 and PC-3 cells demonstrated that the use of optimum lipid:DOX ratio showed controlled release characteristics in combination with mild HT. Results of animal study indicated that the reduction in tumor growth was found to be highest in TSLs with HT. Hence, developed DOX-loaded TSLs exhibited a pronounced potential in cancer treatment [68]. The TMX and IM loaded TSLs were formulated and evaluated for breast cancer treatment. These liposomes undergo several characterizations like particle size, zeta potential, and drug release below and above the liposomal TT. The % EE of developed liposomes (for TMX and IM) was found to be more than 70% (above TT of lipids, i.e., 39.4 °C). The uptake study demonstrated the faster uptake of TSLs by breast cancer cells (MCF-7 and MDA-MB-231). Hence, the combinational therapy shows synergistic anticancer effect against breast cancer cells (MCF-7, MDA-MB-231). Finding indicated that the TSLs have promising potential for the delivery of TMX and IM for breast cancer treatment [33]. The DOX-entrapped TSLs were prepared for the site-specific delivery of DOX into brain tumors. These liposomes quickly release the encapsulated bioactive inside the tumor in response to HT (>40 °C). HT plays a significant role in the transient opening of the BBB when combined with TSLs. Thus, HT could achieve effective concentration of drugs inside the brain. DOX-entrapped TSLs were intravenously infused over 30 min at a fixed dose (0.94 mg/kg) in anesthetized beagles (age 17 months). Fluorescence microscopy showed an efficient delivery of DOX across the BBB. Additionally, histopathology in haematoxylin and eosin (H and E) stained samples display the localized damage near the probe. Survival studies described that this approach is safe with no normal tissue toxicity [5]. Zhang et al. [102] prepared DTX loaded TSLs for targeted delivery of drug to tumor. In vitro drug release study exhibited higher DTX release at 42 than 37 °C, which could be due to temperature-responsive drug release from the liposomes. Thus, it could be a favorable strategy to deliver drugs at tumor site [102].

3.2.3 Magnetic Field Sensitive Liposomes

Magnetic field (MF) is a significant stimulus that plays a vital role in sustained release of drugs from magnetic field sensitive nanocarriers [32]. The MF is applied locally to increase the accumulation of drug in the tumor, or by varying the MF to induce tumor heating to attain better anticancer efficacy. MF is an external stimulus that facilitates the transfer of anticancer drug to target sites and maintain its concentration in blood up to its complete absorption [2]. Anilkumar et al. (2020) prepared liposomal nanocomposites comprising citric acid-modified iron oxide magnetic nanoparticles (CMNPs) for dual magneto-photothermal cancer therapy. It is prompted by modifying MF (AMF) and near-infrared (NIR) lasers. These CMNPs were entrapped into the cationic liposomes led to form nanometric magnetic liposomes (MLs) for concurrent magnetic hyperthermia (MH) in the presence of AMF and photothermia (PT) induced by NIR-laser exposure. The liposomes with 121 nm size and 87% EE were found to be best for designing of MLs. After the endocytosis of MLs shows inducible thermal effects which led to enhance cytotoxicity of cancer cells in the presence of AMF-NIR lasers. Therefore, developed systems have favorable potential for bimodal MH-PT dual magneto-photothermal cancer therapy [88]. Curcumin loaded PEGylated ML were formulated and evaluated for antitumor efficacy. This study enhances the solubility of lipophilic drugs and provides controlled release of therapeutic agents. In vitro drug release study showed that the carriers displayed no significant release at 37 °C, while it showed rapid release at 45 °C (due to inductive magnetic heating). Moreover, it would show (3 times higher) curcumin release under high frequency MF (HFMF) exposure, than without HFMF exposure at 45 °C. Fluorescence microscopy of PEGylated ML displayed efficient internalization into the MCF-7 cells. Hence, Cur-loaded PEGylated MLs can be used as potential approach for cancer treatment [21].

3.2.4 Hypoxia-Sensitive Liposomes

Tumors are hypoxic in nature which is lack of blood circulations that restrict the drug entry deep inside the tumor cells. This hypoxic condition of tumor causes drug resistance in chemotherapy, radiotherapy, and phototherapy. It is reported in many literature that hypoxia-responsive liposomes may enhance the delivery of bioactive. Hypoxia-responsive liposomes were prepared and evaluated by entrapment of nitroimidazole derivatives into the liposomal membrane. At hypoxia condition of tumor, these derivatives displayed reductive metabolism which facilitates a triggered release of bioactive due liposomal degradation. CLSM and NIR imaging confirmed the oxygen-dependent drug (DOX) release. The in vivo fluorescence microscopy displayed the selective release of drug to hypoxic tumor site. Thus, hypoxia-responsive liposomes revealed selective cytotoxicity in hypoxic cells. Finding suggested that the hypoxia-responsive liposomes provide a better strategy for cancer treatment [43]. Lui et al. (2017) developed hypoxia-responsive ionizable liposomes for the effective transport of siRNA and anticancer drugs into the brain tumor. These liposomes selectively increase cell uptake (siRNA) under hypoxic conditions and low pH to treat gliomas. To achieve the above objectives, malate dehydrogenase lipid molecules were synthesized. The synthesized molecules have nitroimidazole groups which are responsible for the hypoxia mediated specificity and sensitivity. The in vivo studies (histological and microscopic examinations) of brain tissue revealed that the liposomes effectively delivered siRNA into gliomas. Thus, hypoxia-responsive ionizable liposomes provide a wide platform for the delivery of siRNA into the tumor [46].

3.2.5 Light-Sensitive Liposomes

Light is also an extrinsic stimulus that is used for the delivery of bioactives. Light plays a significant role in the activation/inactivation of several biochemical processes which has been known as effective strategy for various biomedical applications. Various factors like wavelength, intensity, pulse duration, and cycle of activated light were employed in biomedical research. Several light radiations such as UV, visible light, and NIR were clinically utilized for the delivery of drugs. The NIR showed deep penetration of bioactive in tumor tissues [107]. Some photosensitizing agents such as chlorins, porphyrin derivatives, phthalocyanines, and porphycenes were employed for prognosis and treatment of tumors. Number of reports displayed that light-triggered drug release is a potential strategy for delivering chemotherapeutic agents to tumor cells. Liposomes loaded with calcein (fluorescence marker), talaporfin sodium (TPS, a water soluble photosensitizer), and gemcitabine (chemotherapeutic drug) were developed and characterized for effective cancer treatment. These liposomes showed site-specific drug release upon irradiation with a NIR laser. Liposomes consist of phospholipids (DSPC/DOPE/cholesterol/PEG2000-DSPE in a molar ratio of 85/10/5/5) and demonstrated higher drug release (70% within 10 min) upon IR irradiation, but no drug release in the absence of NIR-laser irradiation. DOPE is responsible for NIR-laser-triggered drug release, the quantity of DOPE can also affect the leakage of drug in absence of NIR-laser-irradiation at 37 °C (body temperature). Thus, DOPE plays an essential role in the effective delivery of bioactive into the tumor tissue. Gemcitabine is an anticancer drug that is entrapped into the liposomes to establish NIR-triggered drug release. The in vitro cytotoxicity assay showed that the NIR laser with liposomes (entrapped with gemcitabine and TPS) provide highest cytotoxic effects against EMT6/P cells. Outcomes of this study suggested that the light-sensitive liposomes may offer a wide platform for the treatment of cancer [18]. Yang et al. [98] prepared tirapazamine (TPZ) and lipophilic IR780 entrapped liposomes combined with photodynamic therapy (PDT) for synergistic anticancer effect. This PDT employed as non-invasive and effective treatment approach for cancer therapy. The application of PDT is restricted because of its poor selectivity to tumor and hypoxia-induced resistance. Thus, to enhance the effectiveness of PDT, they combine TPZ and lipophilic IR780 entrapped liposomes with PDT. The hydrophilic chemotherapeutic drug (TPZ) was entrapped into the core and lipophilic IR780 was entrapped into the lipid bilayer. IR780 helps to produce ROS and hypoxic microenvironment in local sites. This led to release TPZ radicles, which causes DNA double strands to break. These liposomes cause cell death (by inducing ROS) in response to laser irradiation. In vivo study showed that the TPZ and lipophilic IR780 loaded liposomes have higher anticancer efficacy through combining PDT with chemotherapy. Thus, TPZ and IR780 bearing liposomes have immense potential to combine PDT with hypoxia-activated chemotherapy [98].

3.3 Theranostic Liposomes

Theranostics liposomes have been designed to combine multiple imaging agents with bioactives to improve therapeutic effects in advanced stages of cancer. The liposomes with higher entrapment efficiency (for hydrophilic and lipophilic agents) are formed to co-deliver different diagnostic and therapeutic agents [37]. Thus, this strategy can enhance the imaging capacity along with cancer treatment. Karpuza et al. [35] developed folate-targeted PTX and VNB encapsulated theranostic liposomes for diagnosis and treatment of NSCLC. The particle size, zeta potential, and %EE of these liposomes were found to be around 150–180 nm, −10 mV, and 15% (PTX) and 20% (VNB), respectively. The uptake study suggested that the uptake of drug for targeted liposomes was greater than the untargeted liposomes (in H1299 cells). The targeted liposomes displayed higher cellular internalization rate for H1299 cells than A549 cells. The liposomes showed good biocompatible profile on cancerous cells. Anticancer efficacy of dual-drug loaded liposomes was higher than single drug encapsulated liposomes. Thus, these theranostic nanosized, dual-drug loaded liposomes provide a wide platform for NSCLC treatment [35]. In another study, photosensitizer (verteporfin (VP)) loaded theranostic liposomes have been developed. These liposomes formed by DPPC and the triblock copolymer (F127), and functionalized with 5(6)-carboxyfluorescein (CF) due to its photodynamic potential. Additionally, the cell uptake study showed that VP has 6.0 ns, monoexponential decay by time-resolved fluorescence microscopy. This photoactive approach was used to reduce 98% viability in T98G cells. Thus, it could be concluded that novel theranostic liposomes have immense potential in cancer treatment [12]. Karpuz et al. [36] developed folate-targeted, dual-drug loaded, Tc-99 m radio labeled liposomes for in vivo and ex vivo assessment. The dual-drug loaded liposomes exhibited higher cytotoxicity profiles than blank formulation in LLC1 cells. Bio-distribution study was carried out in NSCLC tumor-bearing C57BL/6 mice. The cell uptake assay demonstrated that the radiolabeled, folate-targeted, dual-drug loaded liposomes (Fol-Lipo/PCX/VNB) showed higher uptake of drug (compared to non-targeted liposomes). Histopathological study displayed that the targeted liposomes could not only suppress the tumor growth but also reduce lung metastasis [36]. In one study, biomimetic system (DOX-QDs-Lip@M,) has been synthesized for the treatment of breast carcinoma. The DOX and quaternary QDs (as imaging agents) were encapsulated into liposomes. These liposomes were prepared by fusing and coating the isolated macrophages on the surface of liposome. This strategy improved blood circulation time of biomimetic system, leading to target tumor cells. Additionally, naturally formed bio-film can stabilize the structure of the liposome, and inhibit drug leakage. Thus, biomimetic liposomes could improve diagnosis and cancer treatment [44]. Zhang et al. [103] developed multifunctional hypoxia-induced theranostic liposomes. They were loaded with photosensitizer (Chlorin e6 (Ce6)), hypoxia-activated prodrug (Tirapazamine (TPZ)), and gene probe for synergistic PDT. The Ce6-mediated PDT upon irradiation with a laser induces hypoxia, leads to disrupt the liposomes, and induced the anti-tumor effects of TPZ for enhanced cancer cell killing. The gene probe helps to detect oncogenic biomarkers which improve diagnosis. The results of in vitro and in vivo assay displayed that the developed system showed enhanced anticancer activity as compared to conventional PDT. This study concluded that the hypoxia-responsive multifunctional liposomal combined PDT can be used as synergetic approach in cancer diagnosis and treatment [103]. Saesoo et al. [69] prepared superparamagnetic iron oxide nanoparticles (SPIONs) conjugated with rituximab (RTX) containing liposomes for brain targeting. The BBB restricts the entry of bioactives into the CNS in the brain. To overcome this problem, targeted nanocarriers are used to transport drugs specifically into cancer cells. The liposomes having theranostic properties enhance the transport of drugs to target site. The results of TEM and VSM (Vibrating Sample Magnetometer) study indicated that the SPIONs loaded liposomes are spherical in shape and have super paramagnetic properties. The BBB model was used to determine targeting and therapeutic efficiency of liposomes. This study concluded that the liposome could use as effective DDS for the delivery of RTX into brain [69]. The chemiluminescent liposomes have also been formulated as a theranostic carrier for the detection of tumor cell under oxidative stress conditions. Hydrogen peroxide (H_2O_2) is one of the fundamental key of oxidative stress and a marker of ROS related diseases (cancer). However, particular exposure and scavenging of overproduced H_2O_2 provide an effective treatment for ROS-associated diseases. This property helps to identify malignant tumor cells utilizing chemiluminescent peroxylate response. Thus, novel contrast agent existing for H_2O_2 (peroxylate) liposomes were developed, which identify H_2O_2 through chemiluminescence response. These liposomes are formed by bis (2, 4, 6-trichlorophenyl) oxalate (TCPO) and curcumin as fluorophore. Here curcumin is used for selective tumor damage and act as photosensitizer which leads to cell death. The cell uptake study suggested that the liposomes improve extravasation into the membrane and enhance the bioavailability of curcumin [52].

3.4 Immunoliposomes

Liposomes are modified with antibody (Ab) where Ab fragments have been widely used as targeting moieties called immunoliposomes. These immunoliposomes have high specificity for their target antigens. Active targeting was achieved by binding liposomes to the tumor cell-specific receptors. Moreover, these Ab conjugated liposomes can enhance the sensitivity of cancer cells to other chemotherapeutics (e.g., through the use of siRNAs). These liposomes are employed for tumor targeting by modifying normal immune system. The system comprises two types of immunity, innate and acquired immunity. Innate immunity includes immune system elements that provide immediate host defense. This immunity includes cells such as neutrophils, monocytes, and macrophages. The acquired immunity involves B and T lymphocytes, and seen in higher animals. Acquired immunity is far slower to grow than innate immunity but retains memory for specific antigens. Both innate and acquired immunity have been examined for immunomodulatory response [26, 27, 30]. Zhang et al. [104] prepared DP7-C-functionalized liposomes to improve immune response and provide anticancer effect of vaccine (a neoantigen-based mRNA). In this study, cholesterol modified cationic peptide DP7 (having transmembrane structure and immune-adjuvant property) has been used. The DOTAP (DOTAP/DP7-C) altered liposomes were utilized for efficient mRNA delivery. The liposomal system was mainly used to enhance the transport of mRNA encoding individualized neoantigens to dendritic cells (DCs) and activate the DCs. This system has dual properties, immunoadjuvant and as a carrier to transport mRNA. In vitro study suggested that the liposomal system could deliver mRNA to different types of DCs. It also promotes the development of antigen-specific lymphocyte reactions that were better than DOTAP/LL2 neoantigen-encoding mRNA complex group. Thus, these liposomes could provide a promising platform for delivery of mRNA [104]. In another study, c(RGDyK)-modified liposomes were developed for targeted spatial delivery into gliomas cells. This study enhanced brain penetration and anti-glioma efficacy. The integrin $\alpha v \beta 3$ receptors are overexpressed on capillary endothelial cells and in glioma cells. This receptor has been employed to target tumor cells via c(RGDyK)-modified liposomes for targeted spatial delivery into glioma cells. The c(RGDyK)-modified liposome encapsulated with AL3810 which brings hypothermia on multiple dosing. This anti-c(RGDyK)-modified liposomes activated IgG and IgM antibody with suitable complements (C3b and C5b-9) cause complement dependent opsonization. This system has some benefits like targetability, higher blood circulation, and anti-glioma efficacy. Thus, the c(RGDyK)-modified liposome has promising potential for cancer therapy [42]. Jacoberger-Foissac et al. [24] reported three vaccine alternatives, with a peptide from HA flu virus protein (as CD4 epitope), HPV16 E7 oncoprotein (as CD8 epitope), and TLR4, TLR2/6 (as adjuvant) in this system. TLR4 liposomes induce a potent Th1-oriented anticancer immunity that leads to suppressing tumor growth and increase the survivability of mice. Thus, TLR4 liposomes were considered as a better treatment against the HPV-transformed orthotopic lung tumor mouse model. Moreover, TLR2/6 liposomes induce weaker Th1-immune response which was not sufficient for the induction of prolonged anticancer activity. Therefore, this study provides a wide platform for strategical development of vaccines [24]. In another study, PEGylated liposomes (PEG-Lip) have been developed to transfer antigens to splenic marginal zone B (MZ-B cells). MZ-B cells are produced from spleen and are responsible for efficient trapping of blood borne pathogens and for induced primary antibody response. The blank PEG-Lip and antigen-loaded-PEG-Lip were introduced intravenously and induced antigen-specific immunity. The liposomes were functionalized with several PEG derivatives having different functional groups via. (methoxy PEG (CH3O-PEG), hydroxy PEG (HO-PEG), or polyglycerol (PG), to activate the complement system and deliver a standard antigen, ovalbumin (OVA), to splenic MZ-B cells by in vitro and in vivo. The in vitro study indicated that the modified HO-PEG-liposomes activated the complement system and assisted the preferential association of HO-PEG-liposomes with MZ–B cells. After a single dose immunization system provides effective antigen delivery to MZ-B cells and improves earlier immunization system. Hence, this approach could deliver antigen to MZ-B cells that induced humoral immune response [80]. Yoshizaki et al. [100] prepared pH-sensitive polymer-modified liposomes (MGlu-HPG-liposomes) for the stimulation of specific immune responses. They used cationic lipids to form MGlu-HPG-liposome which activate DCs and improve anticancer effects. The CpG-DNA (a ligand to toll receptor 9 (TLR9)) expressed in endosomes of DCs) was introduced into polymer-modified liposomes containing cationic lipids. This ligand used to activate antigen-specific immunity by complexation methods. Here TLR9 receptor is expressing in endosomes of DC. The liposomal solution expressed TLR9 receptor more efficiently than free antigen solution. These liposomal formulations induced cytokine production and promoted in vivo antigen-specific immune activities. They also have good anticancer effects than conventional pH-sensitive polymer-modified liposomes. The outcomes suggested that the MGlu-HPG-liposome could be a better approach to design effective vaccine carriers [100]. Liang et al. [45] developed liposomes-coated gold nanocages (Lipo-AuNCs) for targeted delivery of melanoma antigen peptide TRP2 to promote activation of DCs and increase tumor explicit T lymphocytes reactions. The in vivo study established that the targeted antigen-bearing AuNCs could improve anticancer immune response by inhibiting tumor growth and metastasis (in lung metastatic models). Hence, Lipo-AuNCs were a favorable nanoplatform for anticancer immunotherapy and in vivo imaging [45].

Table 2 Marketed liposomal formulation for cancer

Marketed product	Drug used	Application	Manufacturer company	Year
Doxil	Doxorubicin	Ovarian cancer, AIDS-related Kaposi's sarcoma, multiple myeloma	Sequus Pharmaceuticals, USA	1995
Depocyt	Cytarabine	Lymphomatous meningitis	Sky Pharma Inc., La Jolla, CA, USA	1999
Myocet	Doxorubicin	Breast cancer	Elan Pharmaceuticals, Princeton, NJ, USA	2000
Marqibo	Vincristine	Lymphoblastic leukemia	Spectrum Pharmaceuticals, USA	2012
Onivyde	Irinotecan	Pancreatic cancer	Merrimack Pharmaceuticals, USA	2015

4 Marketed Liposomal Products

Liposomes were first discovered by Bangham in 1960s, but at that time it was not marketed for clinical use. Various researches have been done to market the drug loaded liposomes in cancer treatment. This leads to approval of the first DOX-loaded liposomal formulation (Doxil®). Doxil® is a marketed anticancer drug, comprises of DOX which is entrapped in the liposomes. It was approved in 1995 to treat AIDS-related kaposi's sarcoma, ovarian cancer, and breast cancer. Doxil® is consists of high phase-transition-temperature (Tm) phospholipid HSPC, Chol, MPEG-DSPE in a molar ratio of 56:38:5 [6]. Consequently, several products have become available for the treatment of cancer and different disease. Additionally, Marqibo is used to treat lymphoblastic leukemia. It is composed of vincristine sulfate liposomal injection, designed to enhance the pharmacokinetics and pharmacodynamic of vincristine. In 2015, Merrimack Pharmaceuticals marketed a newer anticancer product, i.e., Onivyde. It is composed of liposomal irinotecan injection for intravenous administration. Irinotecan acts by inhibiting topoisomerase-I enzyme (it allow DNA replication inducing single-strand breaks in DNA) and leads to cell death. It is extensively used in the treatment of pancreatic cancer [74]. Some marketed liposomal DDS are described in Table 2.

5 Conclusion and Future Perspectives

Cancer is caused by several genetic alterations and uncontrolled growth of abnormal cells. There are several types of anticancer agents commonly employed in cancer therapy using liposomes, including DOX, PTX, DTX, gemcitabine (GEM), vincristine (VCR), etc. Selective receptors present on cancer cells provide an opportunity for targeted DDS. Targeted liposomes are able to identify specific molecular systems on cell surfaces. This strategy, known as active targeting, is able to enhance the specificity of DDS interaction with target tumors. Liposomes have multifunctional properties based therapeutic strategies for achieving desirable outcomes. The liposomes exhibit potential strategies for the treatment of cancer such as surface modification (via ligands and PEGylation) and stimuli-sensitive characteristics. These advanced liposomes could improve the applicability of liposomes in cancer therapeutics. Additionally, the ligand-modified liposomes selectively target the bioactive to the cancer site and increase the efficiency of bioactives. Moreover, better understanding of molecular biology can help in the selection of existing and novel targeting moiety for cancer treatment. In future, researchers will focus on the fabrication of safe and effective liposomes with improved efficacy for cancer treatment.

Acknowledgements Sarjana Raikwar (SRF), Pritish Kumar Panda (SRF), Pooja Das Bidla (SRF), and Shivani Saraf (SRF) are highly obliged to Indian Council of Medical Research (ICMR, New Delhi) for rendering funding assistance. Ankit Jain gratefully acknowledges financial support from the DBT-RA program (Govt. of India) in Biotechnology and Life Sciences.

Conflict of Interest Authors declare no conflict of interest.

References

1. Akbarzadeh A et al (2013) Liposome: classification, preparation, and applications Nanoscale Res Lett 8:102. https://doi.org/10.1186/1556-276X-8-102
2. Arias JL, Clares B, Morales ME, Gallardo V, Ruiz MA (2011) Lipid-based drug delivery systems for cancer treatment. Curr Drug Targets 12:1151–1165
3. Benne N et al (2020) Atomic force microscopy measurements of anionic liposomes reveal the effect of liposomal rigidity on antigen-specific regulatory T cell responses. J Control Release 318:246–255
4. Bozzuto G, Molinari A (2015) Liposomes as nanomedical devices. Int J Nanomedicine 10:975–999. https://doi.org/10.2147/ijn.s68861
5. Bredlau AL et al (2018) Localized delivery of therapeutic doxorubicin dose across the canine blood–brain barrier with hyperthermia and temperature sensitive liposomes. Drug Deliv 25:973–984
6. Bulbake U, Doppalapudi S, Kommineni N, Khan W (2017) Liposomal formulations in clinical use: an updated review. Pharmaceutics 9:12, UNSP 12. https://doi.org/10.3390/pharmaceutics9020012
7. Çağdaş M, Sezer AD, Bucak S (2014) Liposomes as potential drug carrier systems for drug delivery. In: Application of nanotechnology in drug delivery. InTech
8. Chen Y et al (2018) Co-delivery of doxorubicin and imatinib by pH sensitive cleavable PEGylated nanoliposomes with folate-mediated targeting to overcome multidrug resistance. Int J Pharm 542:266–279
9. Chen Y et al. (2020) Co-delivery of doxorubicin and epacadostat via heparin coated pH-sensitive liposomes to suppress the lung metastasis of melanoma. Int J Pharm 584:119446

10. Cho C, Cho K, Park I, Kim S, Sasagawa T, Uchiyama M, Akaike T (2001) Receptor-mediated delivery of all trans-retinoic acid to hepatocyte using poly (L-lactic acid) nanoparticles coated with galactose-carrying polystyrene. J Control Release 77:7–15
11. Chowdhury N et al (2020) Targeted delivery of doxorubicin liposomes for Her-2+ breast cancer treatment. AAPS PharmSciTech 21:1–12
12. de Oliveira DCS et al (2020) Theranostic verteporfin-loaded lipid-polymer liposome for photodynamic applications. J Photochem Photobiol B Biol 212:112039
13. Deshpande PP, Biswas S, Torchilin VP (2013) Current trends in the use of liposomes for tumor targeting. Nanomedicine 8:1509–1528
14. dos Santos Rodrigues B, Arora S, Kanekiyo T, Singh J (2020) Efficient neuronal targeting and transfection using RVG and transferrin-conjugated liposomes. Brain Res 1734:146738
15. Eloy JO et al (2020) EGFR-targeted immunoliposomes efficiently deliver docetaxel to prostate cancer cells. Colloids Surf B Biointerfaces 194:111185
16. Fernandes MA et al (2021) Transferrin-functionalized liposomes for docetaxel delivery to prostate cancer cells. Colloids Surf Physicochem Eng Aspects 611:125806
17. Fu M et al (2020) Combination of targeted daunorubicin liposomes and targeted emodin liposomes for treatment of invasive breast cancer. J Drug Target 28:245–258
18. Fuse T, Tagami T, Tane M, Ozeki T (2018) Effective light-triggered contents release from helper lipid-incorporated liposomes co-encapsulating gemcitabine and a water-soluble photosensitizer. Int J Pharm 540:50–56
19. Goenka S, Sant V, Sant S (2014) Graphene-based nanomaterials for drug delivery and tissue engineering. J Control Release 173:75–88
20. Han S, Sun R, Su H, Lv J, Xu H, Zhang D, Fu Y (2019) Delivery of docetaxel using pH-sensitive liposomes based on D-α-tocopheryl poly (2-ethyl-2-oxazoline) succinate: comparison with PEGylated liposomes. Asian J Pharm Sci 14:391–404
21. Hardiansyah A, Yang M-C, Liu T-Y, Kuo C-Y, Huang L-Y, Chan T-Y (2017) Hydrophobic drug-loaded PEGylated magnetic liposomes for drug-controlled release. Nanoscale Res Lett 12:355
22. Hayward SL, Wilson CL, Kidambi S (2016) Hyaluronic acid-conjugated liposome nanoparticles for targeted delivery to CD44 overexpressing glioblastoma cells. Oncotarget 7:34158
23. Hei Y et al (2020) Multifunctional immunoliposomes combining catalase and PD-L1 antibodies overcome tumor hypoxia and enhance immunotherapeutic effects against melanoma. Int J Nanomed 15:1677
24. Jacoberger-Foissac C et al (2020) Liposomes as tunable platform to decipher the antitumor immune response triggered by TLR and NLR agonists. Eur J Pharm Biopharm 152:348–357
25. Jain A, Jain S (2016). Ligand-mediated drug-targeted liposomes liposomal delivery systems: advances and challenges. https://doi.org/10.4155/FSEB2013.14.251
26. Jain A, Jain SK (2014) Brain targeting using surface functionalized nanocarriers In: Singh B, Jain NK, Katare OP (eds) Drug nanocarriers, pp 203–255
27. Jain A, Jain SK (2015) Ligand-appended BBB-targeted nanocarriers (LABTNs). Crit Rev Ther Drug Carrier Syst 32:149–180. https://doi.org/10.1615/CritRevTherDrugCarrierSyst.2015010903
28. Jain A, Jain SK (2016) In vitro release kinetics model fitting of liposomes: an insight. Chem Phys Lipids 201:28–40. https://doi.org/10.1016/j.chemphyslip.2016.10.005
29. Jain A, Jain SK (2016c) Liposomes in cancer therapy. In: Nanocarrier systems for drug delivery. Nova Science Publishers, pp 1–42. https://www.novapublishers.com/catalog/product_info.php?products_id=59761&osCsid=e7d370318f328e75748328a1e44e48aa
30. Jain A, Jain SK (2016d) Multipronged, strategic delivery of paclitaxel-topotecan using engineered liposomes to ovarian cancer. Drug Dev Ind Pharm 42:136–149. https://doi.org/10.3109/03639045.2015.1036066
31. Jain A, Tiwari A, Verma A, Saraf S, Jain S (2019) Combination cancer therapy using multifunctional liposomes. Crit Rev Ther Drug Carrier Syst. https://doi.org/10.1615/CritRevTherDrugCarrierSyst.2019026358

32. Jhaveri A (2016) Magnetic field-responsive nanocarriers. In: Smart pharmaceutical nanocarriers. World Scientific, pp 267–308
33. Jose A, Ninave KM, Karnam S, Venuganti VVK (2019) Temperature-sensitive liposomes for co-delivery of tamoxifen and imatinib for synergistic breast cancer treatment. J Liposome Res 29:153–162
34. Kang X-j et al (2017) Codelivery of dihydroartemisinin and doxorubicin in mannosylated liposomes for drug-resistant colon cancer therapy. Acta Pharmacol Sin 38:885
35. Karpuz M, Silindir-Gunay M, Kursunel MA, Esendagli G, Dogan A, Ozer AY (2020) Design and in vitro evaluation of folate-targeted, co-drug encapsulated theranostic liposomes for non-small cell lung cancer. J Drug Deliv Sci Technol 57:101707
36. Karpuz M et al. (2021) Diagnostic and therapeutic evaluation of folate-targeted paclitaxel and vinorelbine encapsulating theranostic liposomes for non-small cell lung cancer. Eur J Pharm Sci 156:105576
37. Kelkar SS, Reineke TM (2011) Theranostics: combining imaging and therapy Bioconj Chem 22:1879–1903
38. Lai F, Schlich M, Sinico C, Fadda AM (2021) Liposomes as brain targeted delivery systems. In: Nanomedicines for brain drug delivery. Springer, pp 29–59
39. Lasic DD (1998) Novel applications of liposomes. Trends Biotechnol 16:307–321
40. Lee J, Ahn HJ (2018) PEGylated DC-Chol/DOPE cationic liposomes containing KSP siRNA as a systemic siRNA delivery carrier for ovarian cancer therapy. Biochem Biophys Res Commun 503:1716–1722
41. Lee W, Im H-J (2019) Theranostics based on liposome: looking back and forward. Nucl Med Mol Imaging (2010) 53:242–246
42. Li J et al (2020) A pentapeptide enabled AL3810 liposome-based glioma-targeted therapy with immune opsonic effect attenuated. Acta Pharmaceutica Sinica B
43. Li Y, Lu A, Long M, Cui L, Chen Z, Zhu L (2019) Nitroimidazole derivative incorporated liposomes for hypoxia-triggered drug delivery and enhanced therapeutic efficacy in patient-derived tumor xenografts. Acta Biomater 83:334–348. https://doi.org/10.1016/j.actbio.2018.10.029
44. Liang B, Deng T, Li J, Ouyang X, Na W, Deng D (2020) Biomimetic theranostic strategy for anti-metastasis therapy of breast cancer via the macrophage membrane camouflaged superparticles. Mater Sci Eng C 115:111097
45. Liang R et al (2017) Liposomes-coated gold nanocages with antigens and adjuvants targeted delivery to dendritic cells for enhancing antitumor immune response. Biomaterials 149:41–50
46. Liu H-M et al (2017) Hypoxia-responsive ionizable liposome delivery siRNA for glioma therapy. Int J Nanomed 12:1065
47. Liu J, Sun Y, Liu X, Yang Y, Widjaya AS, Long Z, Jiang Y (2020) Efficiency of different treatment regimens combining anti-tumor and anti-inflammatory liposomes for metastatic breast cancer. AAPS PharmSciTech 21:1–12
48. Liu M-C, Liu L, Wang X-R, Shuai W-P, Hu Y, Han M, Gao J-Q (2016) Folate receptor-targeted liposomes loaded with a diacid metabolite of norcantharidin enhance antitumor potency for H22 hepatocellular carcinoma both in vitro and in vivo. Int J Nanomed 11:1395
49. Liu X et al (2017) Asialoglycoprotein receptor-targeted liposomes loaded with a norcantharimide derivative for hepatocyte-selective targeting. Int J Pharm 520:98–110
50. Mansoori B et al (2020) Hyaluronic acid-decorated liposomal nanoparticles for targeted delivery of 5-fluorouracil into HT-29 colorectal cancer cells. J Cell Physiol 235:6817–6830
51. Minnelli C et al (2018) Selective induction of apoptosis in MCF7 cancer-cell by targeted liposomes functionalised with mannose-6-phosphate. J Drug Target 26:242–251
52. Mohammadi SS, Vaezi Z, Shojaedin-Givi B, Naderi-Manesh H (2019) Chemiluminescent liposomes as a theranostic carrier for detection of tumor cells under oxidative stress. Anal Chim Acta 1059:113–123
53. Mu L-M et al (2017a) Dual-functional drug liposomes in treatment of resistant cancers. Adv Drug Del Rev 115:46–56

54. Mu LM et al (2017) Dual-functional drug liposomes in treatment of resistant cancers. Adv Drug Deliv Rev 115:46–56. https://doi.org/10.1016/j.addr.2017.04.006
55. Nandi U, Onyesom I, Douroumis D (2020) Anti-cancer activity of sirolimus loaded liposomes in prostate cancer cell lines. J Drug Deliv Sci Technol:102200
56. Neves LF, Duan J, Voelker A, Khanal A, McNally L, Steinbach-Rankins J, Ceresa BP (2016) Preparation and optimisation of anionic liposomes for delivery of small peptides and cDNA to human corneal epithelial cells. J Microencapsul 33:391–399
57. Oh HR, Jo H-Y, Park JS, Kim D-E, Cho J-Y, Kim P-H, Kim K-S (2016) Galactosylated liposomes for targeted co-delivery of doxorubicin/vimentin siRNA to hepatocellular carcinoma. Nanomaterials 6:141
58. Paliwal SR, Paliwal R, Vyas SP (2015) A review of mechanistic insight and application of pH-sensitive liposomes in drug delivery. Drug Deliv 22:231–242. https://doi.org/10.3109/10717544.2014.882469
59. Pan T et al (2021) Suppressing Sart1 to modulate macrophage polarization by siRNA-loaded liposomes: a promising therapeutic strategy for pulmonary fibrosis. Theranostics 11:1192
60. Panda P, Saraf S, Tiwari A, Verma A, Raikwar S, Jain A, SK J (2019) Novel strategies for targeting prostate cancer curr drug del. https://doi.org/10.2174/1567201816666190821143805
61. Park YI et al (2021) pH-sensitive multi-drug liposomes targeting folate receptor β for efficient treatment of non-small cell lung cancer. J Control Release 330:1–14
62. Paszko E, Senge M (2012) Immunoliposomes. Curr Med Chem 19:5239–5277
63. Pathak P et al (2016) Design of cholesterol arabinogalactan anchored liposomes for asialoglycoprotein receptor mediated targeting to hepatocellular carcinoma: in silico modeling, in vitro and in vivo evaluation. Int J Pharm 509:149–158
64. Payne NI, Ambrose CV, Timmins P, Ward MD, Ridgway F (1986) Proliposomes: a novel solution to an old problem J Pharm Sci 75:325–329
65. Prajapati SK, Jain A, Jain A, Jain S (2019a) Biodegradable polymers and constructs: a novel approach in drug delivery. Eur Polym J 120:109191. https://doi.org/10.1016/j.eurpolymj.2019.08.018
66. Prajapati SK, Jain A, Shrivastava C, Jain AK (2019) Hyaluronic acid conjugated multi-walled carbon nanotubes for colon cancer targeting. Int J Biol Macromol 123:691–703. https://doi.org/10.1016/j.ijbiomac.2018.11.116
67. Raikwar S, Vyas S, Sharma R, Mody N, Dubey S, Vyas SP (2018) Nanocarrier-based combination chemotherapy for resistant tumor: development, characterization, and ex vivo cytotoxicity assessment. AAPS PharmSciTech 19:3839–3849
68. Ruiz A, Ma G, Seitsonen J, Pereira SG, Ruokolainen J, Al-Jamal WT (2020) Encapsulated doxorubicin crystals influence lysolipid temperature-sensitive liposomes release and therapeutic efficacy in vitro and in vivo. J Control Release 328:665–678
69. Saesoo S et al. (2018) Characterization of liposome-containing SPIONs conjugated with anti-CD20 developed as a novel theranostic agent for central nervous system lymphoma. Colloids Surf B Biointerfaces 161:497–507
70. Salehi B et al (2020) Multivesicular liposome (DepoFoam) in human diseases. Iran J Pharm Res IJPR 19:9
71. Saraf S, Jain A, Hurkat P, Jain SK (2016) Topotecan liposomes: a visit from a molecular to a therapeutic platform. Crit Rev Ther Drug Carrier Syst 33:401–432. https://doi.org/10.1615/CritRevTherDrugCarrierSyst.2016015926
72. Saraf S, Jain A, Tiwari A, Verma A, Panda PK, Jain SK (2020a) Advances in liposomal drug delivery to cancer: an overview. J Drug Deliv Sci Technol 56:101549. Artn 101549. https://doi.org/10.1016/J.Jddst.2020.101549
73. Saraf S et al (2020) Targeting approaches for the diagnosis and treatment of cancer frontiers in anti-cancer. Drug Discovery 11(11):105
74. Saraswat AL, Maher TJ (2020) Development and optimization of stealth liposomal system for enhanced in vitro cytotoxic effect of quercetin. J Drug Deliv Sci Technol 55:101477

75. Sawant RR, Torchilin VP (2012) Challenges in development of targeted liposomal therapeutics. AAPS J 14:303–315. https://doi.org/10.1208/s12248-012-9330-0
76. Shafaa M (2020) Preparation, characterization and evaluation of cytotoxic activity of Tamoxifen bound liposomes against breast cancer cell line. Egypt J Biomed Eng Biophys 21:19–31
77. Sharma A, Sharma US (1997) Liposomes in drug delivery: progress and limitations. Int J Pharm 154:123–140
78. Shi G, Guo W, Stephenson SM, Lee RJ (2002) Efficient intracellular drug and gene delivery using folate receptor-targeted pH-sensitive liposomes composed of cationic/anionic lipid combinations. J Control Release 80:309–319
79. Shimizu T et al (2020) Complement activation induced by PEG enhances humoral immune responses against antigens encapsulated in PEG-modified liposomes. J Control Release
80. Sicard G, Paris C, Giacometti S, Rodallec A, Ciccolini J, Rocchi P, Fanciullino R (2020) Enhanced antisense oligonucleotide delivery using cationic liposomes grafted with trastuzumab: a proof-of-concept study in prostate. Cancer Pharm 12:1166
81. Simões S, Filipe A, Faneca H, Mano M, Penacho N, Düzgünes N, Pedroso de Lima M (2005) Cationic liposomes for gene delivery. Expert Opinion Drug Deliv 2:237–254
82. Sonali et al (2016) Transferrin liposomes of docetaxel for brain-targeted cancer applications: formulation and brain theranostics. Drug Deliv 23:1261–1271
83. Song M, Liang Y, Li K, Zhang J, Zhang N, Tian B, Han J (2019) Hyaluronic acid modified liposomes for targeted delivery of doxorubicin and paclitaxel to CD44 overexpressing tumor cells with improved dual-drugs synergistic effect. J Drug Deliv Sci Technol 53:101179
84. Song X-l et al (2017) Targeting vincristine plus tetrandrine liposomes modified with DSPE-PEG2000-transferrin in treatment of brain glioma. Eur J Pharm Sci 96:129–140
85. Torchilin VP (2007) Targeted pharmaceutical nanocarriers for cancer therapy and imaging. AAPS J 9:E128–E147
86. Trédan O, Galmarini CM, Patel K, Tannock IF (2007) Drug resistance and the solid tumor microenvironment. J Natl Cancer Inst 99:1441–1454
87. Ts A, Lu Y-J, Chen J-P (2020) Optimization of the preparation of magnetic liposomes for the combined use of magnetic hyperthermia and photothermia in dual magneto-photothermal cancer therapy. Int J Mol Sci 21:5187
88. Verma A et al (2020) Liposomes for advanced drug delivery. In: Advanced biopolymeric systems for drug delivery. Springer, pp 317–338. https://doi.org/10.1007/978-3-030-46923-8_12
89. Vyas A, Jain A, Hurkat P, Jain A, Jain SK (2015) Targeting of AIDS related encephalopathy using phenylalanine anchored lipidic nanocarrier. Colloids Surf B Biointerfaces 131:155–161. https://doi.org/10.1016/j.colsurfb.2015.04.049
90. Wang J, Ayano E, Maitani Y, Kanazawa H (2017) Tunable surface properties of temperature-responsive polymer-modified liposomes induce faster cellular uptake ACS. Omega 2:316–325
91. Wang Y et al (2020) Transferrin modified dioscin loaded PEGylated liposomes: characterization and in vitro antitumor effect. J Nanosci Nanotechnol 20:1321–1331
92. Wei L, Guo X-Y, Yang T, Yu M-Z, Chen D-W, Wang J-C (2016) Brain tumor-targeted therapy by systemic delivery of siRNA with transferrin receptor-mediated core-shell nanoparticles. Int J Pharm 510:394–405
93. Wei M et al (2015) Lactoferrin-modified PEGylated liposomes loaded with doxorubicin for targeting delivery to hepatocellular carcinoma. Int J Nanomed 10:5123
94. Xiao W, Zeng X, Lin H, Han K, Jia H-Z, Zhang X-Z (2015) Dual stimuli-responsive multi-drug delivery system for the individually controlled release of anti-cancer drugs. Chem Commun 51:1475–1478
95. Xiong M et al (2017) Mannosylated liposomes improve therapeutic effects of paclitaxel in colon cancer models. J Microencapsul 34:513–521
96. Yang K, Feng L, Liu Z (2016) Stimuli responsive drug delivery systems based on nano-graphene for cancer therapy. Adv Drug Del Rev 105:228–241

97. Yang Y et al (2019) Near-infrared light triggered liposomes combining photodynamic and chemotherapy for synergistic breast tumor therapy. Colloids Surf B Biointerfaces 173:564–570
98. Ye J et al (2020) Targeted delivery of chlorogenic acid by mannosylated liposomes to effectively promote the polarization of TAMs for the treatment of glioblastoma. Bioactive Mater 5:694–708
99. Yoshizaki Y, Yuba E, Sakaguchi N, Koiwai K, Harada A, Kono K (2017) pH-sensitive polymer-modified liposome-based immunity-inducing system: effects of inclusion of cationic lipid and CpG-DNA. Biomaterials 141:272–283
100. Yue G, Wang C, Liu B, Wu M, Huang Y, Guo Y, Ma Q (2020) Liposomes co-delivery system of doxorubicin and astragaloside IV co-modified by folate ligand and octa-arginine polypeptide for anti-breast cancer RSC. Advances 10:11573–11581
101. Zhang H et al (2014) Preparation, characterization, and pharmacodynamics of thermosensitive liposomes containing docetaxel. J Pharm Sci 103:2177–2183. https://doi.org/10.1002/jps.24019
102. Zhang K, Zhang Y, Meng X, Lu H, Chang H, Dong H, Zhang X (2018) Light-triggered theranostic liposomes for tumor diagnosis and combined photodynamic and hypoxia-activated prodrug therapy. Biomaterials 185:301–309
103. Zhang R et al (2020) DP7-C-modified liposomes enhance immune responses and the antitumor effect of a neoantigen-based mRNA vaccine. J Control Release 328:210–221
104. Zhao Y, Cai F, Shen X, Su H (2020) A high stable pH-temperature dual-sensitive liposome for tuning anticancer drug release. Synthetic Syst Biotechnol 5:103–110
105. Zhu L, Torchilin VP (2013) Stimulus-responsive nanopreparations for tumor targeting. Integr Biol (Camb) 5:96–107. https://doi.org/10.1039/c2ib20135f

Chapter 7
Chitosan Based Nanocomposites for Drug Delivery Application

Malihe Pooresmaeil and Hassan Namazi

Abbreviations

5-ASA	5-Aminosalicylic acid
5-FU	5-Fluorouracil
ADM	Adriamycin
AFM	Atomic force microscopy
AgNPs	Silver NPs
AuNPs	Gold NPs
BET	Brunauer-Emmett-Teller
BSA	Bovine serum albumin
CAM	Chorioallantoic membrane
CFX	Ciprofloxacin hydrochloride
CFNP	Calcium ferrite NPs
CLSM	Confocal laser scanning microscopy
CS	Chitosan
cisP	Cisplatin
CNT	Carbon nanotube
COS CS	Oligosaccharide
CLX	Chlorhexidine
CNC	Nanocrystalline cellulose
CUR	Curcumin

M. Pooresmaeil · H. Namazi (✉)
Research Laboratory of Dendrimers and Nano-Biopolymers, Faculty of Chemistry, University of Tabriz, P.O. Box 51666, Tabriz, Iran
e-mail: namazi@tabrizu.ac.ir

H. Namazi
Research Center for Pharmaceutical Nanotechnology (RCPN), Tabriz University of Medical Science, Tabriz, Iran

S. Gopi et al. (eds.), *Nanotechnology for Biomedical Applications*,
Materials Horizons: From Nature to Nanomaterials,
https://doi.org/10.1007/978-981-16-7483-9_7

DEX	Dexamethasone
DFT	Density Functional Theory
DLS	Dynamic light scattering
DOP	Dopamine
DOX	Doxorubicin
DS	Diclofenac sodium
DSC	Differential scanning calorimetry
CPT	Camptothecin
FA	Folic acid
FE-SEM	Field emission scanning electron microscopy
FeO	Iron oxide
FT-IR	Fourier transform infrared
GC	Gentamicin
GO	Graphene oxide
GQDs	Graphene quantum dot
Hap	Hydroxyapatite
HA	Hyaluronic acid
HRTEM	High resolution transmission electron microscopy
HNT	Halloysite nanotube
HPLC	High-performance liquid chromatography
IBU	Ibuprofen
IC50	Half-maximal inhibitory concentration
IDM	Indomethacin
INH	Isoniazid
IONPs	Iron oxide NPs
LBL	Layer-by-layer
LDH	Layered double hydroxide
LH	Lidocaine hydrochloride
MMT	Montmorillonite
MC	Minocycline
MOF	Metal-organic framework
Mox	Moxifloxacin
Ms	Magnetization saturation
MTT	(3-(4,5-Dimethylthiazol-2-yl)-2,5-diphenyltetrazolium bromide
MWCNTs	Multi-walled carbon nanotubes
MWt	Molecular weights
MZ	Metronidazole
NFZ	Nitrofurazone
NIR	Near-infrared
NPs	Nanoparticles
OCMC	Ocarboxymethyl chitosan
PAA	Polyacrylic acid
PAI	Photoacoustic imaging
PCL	Polycaperlactone
PEG	Poly (ethylene glycol)

PHT	Phenytoin sodium
PRD	Prednisolone
PTX	Paclitaxel
PVA	Polyvinyl alcohol
PX	Piroxicam
RAFT	Reversible addition-fragmentation chain transfer
RBCs	Red blood cells
rGO	Reduced graphene oxide
RFP	Rifampicin
ROS	Reactive oxygen species
SA	Sodium alginate
SAED	Selected area electron diffraction
SEM	Scanning electron microscope
SIF	Simulated intestinal fluid
SPR	Surface plasmon resonance
SS	Sumatriptan Succinate
ss	Sodium salicylate
SWCNTs	Single-walled carbon nanotubes
TC	Tetracycline
TR	Triclosan
TGA	Thermal gravimetric analysis
TPP	Tripolyphosphate
TEM	Transmission electron microscopy
TiO_2	Titanium dioxide
UV-Vis	Ultraviolet-visible
VC	Vancomycin
VH	Vancomycin hydrochloride
VSM	Vibrating sample magnetometer
WVTR	Water vapor transmission rate
XRD, MTX	Methotrexate
ZnO	Zinc oxide

1 Introduction

Drug releases in an explosive manner are one of the main issues in traditional drug delivery systems (DDSs) that leads to several concerns. Additionally, the traditional DDSs have several other disadvantages that the wasting a large portion of the drug, poor oral bioavailability, and harmful effects for the healthy cells because of the passive targeting of the drug are some of them [1]. Therefore the main challenge in the drug delivery area is the engineering of a smart nanocarrier that could competently encapsulate the drug molecules at high loading amount, meanwhile controlling their release. By considering these, in recent years the interests are focused on designing the new DDSs with more efficiency that can control the release behavior [2]. On the other hand, one of the other main aims of drug delivery technologies is the optimization of the drug delivery to the body with safe and non-toxic engineered materials [3]. Up to now various systems like dendrimers, hyperbranched polymers, nanocomposites, organic or inorganic materials, porous materials, and, etc. have been introduced as DDSs [4–10]. Among them, the polymeric nanocomposites comprised nanoparticles (NPs) and polymer recently gained remarkable interest owing to having the feature of both NPs and polymer.

The small size of the NPs provides their easily traveling all over the body via the bloodstream without recognizing by the immune system, permeating through the blood vessels, penetrating tissues, and delivering their cargo to the targeted cells, hence, these materials with various beneficial physicochemical properties were received a lot of attention in the biomedical area. Moreover, these materials can play a main role in drug delivery due to their discrimination potentiality toward unhealthy and healthy cells. However, due to some limitations, their surface is commonly engineered with various polymers like naturally occurring polymers for tuning these properties. Modification of the NPs maintains the previous prerequisites, besides, it induces the biodegradable and biocompatibility properties to them. In fact, the engineering of NPs with polymers not only improves their blood circulation time and biocompatibility but also improves their drug encapsulation efficiency and provides binding sites for attachment of different agents like targeting or imaging agents [1, 11]. Hence modified NPs with biopolymers have proved to be promising carriers and have attracted immense interest for drug delivery and developing more effective and safer formulations. Eco-friendly, facile availability, reactivity at a moderate level, biocompatibility, mild gelling ability, biodegradability, low cost, and safety are some of the outstanding benefits of the biopolymer-based DDSs in the pharmaceutical industry [2]. Several types of biopolymers have been used for nanocomposites fabrication, which among them the chitosan (CS) biopolymer has received extensive attention because of its interesting features and caused more development due to the sustained drug delivery release pattern in the system based on it.

CS is a linear polycationic polymer with natural origins. This polymer is the second most abundant biopolymer after cellulose in nature [12]. In the structure of CS the β-(1 → 4)-linked D-glucosamine (de-acetylated unit) and N-acetyl-D-glucosamine (acetylated unit) are randomly distributed [13]. The structure of CS is provided in Fig. 1a. In an aqueous medium, the CS has the pKa value of about 6.0–6.5, thus, its charged state and physiochemical features are considerably affected by the ambient medium pH [14]. CS is prepared via deacetylation of the present chitin in crustaceans, marine, and mollusks shells. This polymer has many beneficial medicinal properties like biodegradability, non-toxicity, antioxidant, antitumor, antifungal, analgesic, mucoadhesive, and hemostatic features, which increase its importance in biomedical applications. The biodegradability of CS arose from the formation of the hydrogen bonds and then metabolizing by human keratin enzymes like isoenzyme [15, 16]. Also, the antimicrobial character of CS leads to its use as a CS-based drug carrier without any concern about bacterial infection [17]. The crystallinity degree of CS is a function of the deacetylation degree. CS has poor solubility in water,

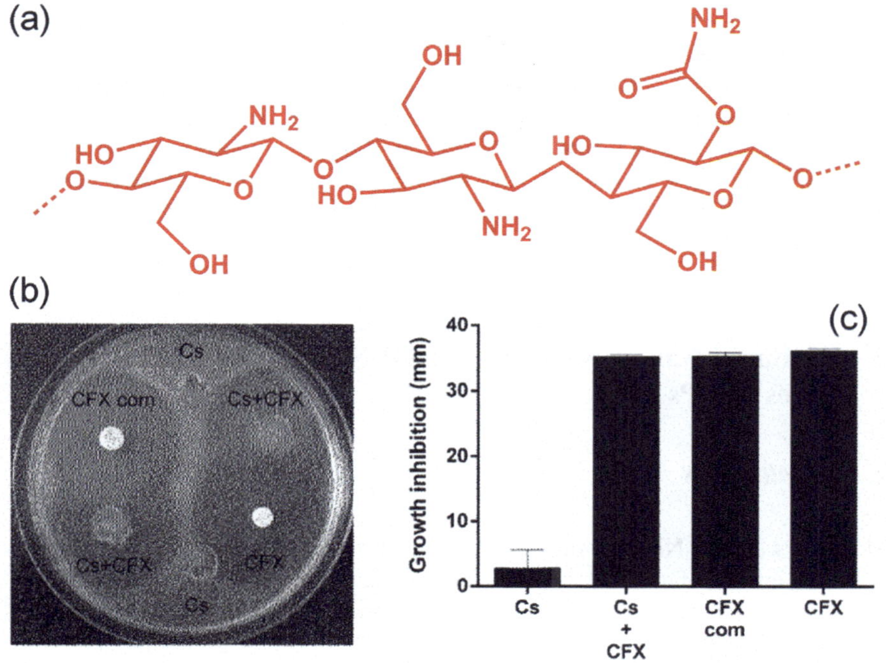

Fig. 1 The chemical structure of CS (**a**), a digital image of the CFX loaded CS films antibacterial activity (**b**), measured growth inhibition diameters (**c**). Reproduced with permission from [24]

however, in an acidic environment, protonation of the amino-functional groups of the residues D-glucosamine leads to its dissolving [18]. The composites based on CS also can be used for wastewaters treatment because of their possibility for the formation of complexes between the amino groups of CS with numerous species. The ability for environmental stimuli answering, owing to the reversible solubility varies by changing the pH value converted the CS-based systems to a smart polymer. In fact in the CS-based system, the ionization degree of amine groups is changeable in response to the acidic of the media. On the other hand, CS with an abundant availability, having a high density of hydroxyl, amine, and ether functional groups in the structure, and non-toxicity introduced as an ideal candidate for modification of the different NPs via both of the covalent or non-covalent methods [11]. Besides, the present amino groups in the structure of CS offer a great density of the positive charge in acidic mediums, which contribute to the cellular internalization and endosomal escape, and also, it provides the CS dissolving possibility in acidic aqueous solutions [19]. Topical ocular drug delivery, wound healing, bioadhesives implantation, injection, tissue engineering, blood anticoagulants, wound dressings, nonviral vectors to genes delivery, and stem cell technology are some of the biomedicinal areas that this biopolymer attracted considerable attention because of its positive characters [12]. Additionally, the CS is more efficient for the avoidance and cure of dental plaques due to its inherent antibacterial features [20]. By considering the above-mentioned facts this chapter reports the significant and basic principles of the CS-based nanocomposites which were prepared via each of the different types of NPs; metal NPs, metal oxide NPs, silica NPs, hydroxyapatite (HAp) NPs, layered double hydroxide (LDH) NPs, graphene oxide (GO) and graphene quantum dots (GQDs) NPs, carbon nanotube (CNT) NPs, CS NPs, hybrid NPs as a nano-theranostic system for drug delivery applications. Hence, we expect that due to the ongoing potential of CS in drug delivery area, this chapter is helpful for a researcher that works in this field to develop new DDSs with improved features.

2 CS-Based Nanocomposites as a DDSs with Different Types of NPs

2.1 Metal NPs

Metal NPs are the NPs that only have metals in their structure. Over the past few years, the interest of material scientists for research in the metal NPs area has dramatically been widespread owing to their unique physicochemical characteristics for instance antibacterial activity, catalytic activity, and magnetic properties. There are numerous types of metal NPs which among them the gold NPs (AuNPs) and silver NPs (AgNPs) are the commonly studied metal NPs in designing of DDSs. Silver ions have both bactericidal and bacteriostatic performance, and therefore can be utilized as an antimicrobial agent. Hence, recently, the use of AgNPs is growing to eliminate the provided concerns that arose from the excessive use of antibiotics that produce resistance to the microorganismal [21, 22]. By considering this, in 2019 Pereira et al. in situ synthesized the AgNPs on the CS microspheres and then evaluated their potential for the antibacterial performance and controlled delivery of ibuprofen (IBU). Because both CS and AgNPs have antimicrobial activity, it is expected that the formation of the nanocomposite from two of these components (CS-AgNP) can significantly develop anti-bactericidal properties. The antibacterial test exhibited that the microspheres possess antibacterial activity against both *S. aureus* and *E. coli* bacteria. Respectively about 10.9% and 24.8% of IBU as a non-steroidal medicine incorporated in the neat and AgNPs containing CS microspheres. Furthermore study of the IBU releases behavior for the CS microspheres with and without AgNPs displayed their pH-dependent drug release profile. The observed release pattern; higher IBU release rate at pH 1.2 than pH 7.4 could be attributed to the special nature of IBU; so, IBU is a weak acid (pKa = 5.2) with improved solubility in an alkaline medium. Therefore, a more amount of IBU releases occurred in simulated intestinal fluid (SIF). Therefore release results exhibited that the AgNPs can promote an extension of the polymer matrix and facilitate the absorption of a higher amount of water molecules, hence, support and releasing a greater drug quantity [21]. In another research, the use of gamma irradiation as a safe crosslinking agent obtained the crosslinked hydrogel composite membranes based on

CS-loaded $AgNO_3$, polyvinyl alcohol (PVA), and vitamin E. Thereafter its biomedical applications was investigated. The influence of several affecting parameters on hydrogel membranes preparation, for instance, plasticizer, copolymer concentration, $AgNO_3$ concentration, irradiation dose, and vitamin E were investigated in detail. The Fourier transform infrared (FT-IR) approved occurring the crosslinking reaction between the PVA and CS. A homogenous one-phase interpenetrating morphology was observed in the canning electron microscope (SEM) analysis of the nanocomposite. In vitro antimicrobial test exhibited a significant antimicrobial performance for PVA-Cs-Ag composed hydrogel membranes in particular toward *Streptococcus mutans* owing to the presence of AgNP in membranes. Consequently, the current research displayed that the PVA/CS/$AgNO_3$-Vit.E hydrogel composite membranes have acceptable features for use as wound dressing materials [23]. Similarly, the antibacterial activity and the ability of Ag/CS film for controlled delivery of the ciprofloxacin hydrochloride (CFX) were studied by *I. Aranaz* and coworkers. In this way, the antimicrobial-loaded films were simply created via the casting-solvent evaporation method by using tripolyphosphate (TPP) as a crosslinker. Then, the CS films were evaluated as a carrier of CFX to the skin at pH 7.4, either in combination with AgNPs or alone. In all of the synthesized samples; the CFX was released at a two-step procedure, so, the initial burst release was followed by a more controlled release stage. Antimicrobial activity of CFX, CS films, CFX based CS films, nanocomposite films, and CFX nanocomposite films were tested against *P. aeruginosa* via the disk diffusion method. As it is reported in Fig. 1b, c, the inhibition zone diameter of the CFX loaded CS films was analogous to that of the CFX which indicates that the CFX diffused and stayed active whereas the CS film alone doesn't show antibacterial performance. Additionally, the nanocomposite films exhibited no microbial inhibition. Therefore the antimicrobial performance of CFX nanocomposite films may be related to the CFX activity because no differences with CFX CS films were found. A probable reason could be the unlike ionic properties of Mueller Hinton agar media compared to the phosphate-buffered saline (PBS) media. Additionally, the wide variation in silver susceptibility of *P. aeruginosa* ranging from 8 to 70 mg mL^{-1}, which hamper the comparison of results among experiments pointing to the need for standards in this field could be another possible explanation [24].

Namazi and coworkers also in situ synthesized the AgNPs (in different percentages) inside the physically crosslinked CS hydrogel beads via TPP as the crosslinker in the facile and economic procedure. It is reported that the appearance of the peaks at 2θ values of about 38°, 44°, and 64° of the nanocomposites XRD spectrum approved the formation of AgNPs in the hydrogels. The swelling test results displayed that the AgNPs enhanced the swelling capacity of the hydrogel bead nanocomposite. Furthermore, the observations approved that the concentration of AgNPs influenced the antibacterial activity of the prepared systems against *Escherichia coli* and *Staphylococcus aureus* bacteria which was approved via observed inhibition zone diameter. IBU was loaded as a model drug in the prepared nanocomposite hydrogel beads. More evaluation of the release pattern exhibited that in comparison with neat CS hydrogel beads the time of IBU release from CH/Ag beads is extended owing to needing a longer path for IBU for migration from AgNPs having hydrogels to the buffer medium. Based on the findings, the fabricated CH/Ag nanocomposite hydrogels can be promising candidates for controlled drug delivery without any concerning bacterial infections [25]. DDSs with imaging capabilities are a good option in cancer treatment. In this regard bioimaging via CS-based nanocomposites having AgNPs was also studied by using the CS conjugated with near-infrared (NIR) photoluminescent silver sulfide quantum dots (Ag_2S QDs) capped by long-chain carboxylic acid. In this way, at first, the oleic acid-capping Ag_2S QDs were prepared, reacted with N-hydroxysuccinimide, and eventually conjugated with CS at the amino-functional groups. NIR photoluminescent Ag_2S QDs exhibited good biocompatibility and tunable optical properties. Doxorubicin (DOX) as an anticancer drug was entrapped in core–shell nanostructured and Ag_2S (DOX)@CS nanospheres was obtained. In vitro and in vivo tests exhibited that the nanospheres released the DOX at the acidic pH of the tumor cells and have a great antitumor efficiency. In the case of DOX load and release should be mentioned that the strong hydrophobic interaction between the oleoyl groups at pH $\geq$ 7.0, the DOX is entrapped in the nanospheres, while, the chains expand due to the repulsion between protonated amine groups increased the DOX in an acidic medium. HeLa cells viability after treatment with different concentrations of free and DOX loaded Ag_2S@CS nanospheres confirmed the biocompatibility of the carrier and its ability for cancer cells killing after drug loading on it. By doing several further investigations the authors mentioned that the NIR photoluminescent Ag_2S QDs serve as fluorescent tags and for in vitro and in vivo tracking of the nanosphere spreading in a body via the strong NIR signal [14].

AuNPs are other NPs with biocompatible, bio-inert, superior optic, and electronic properties, thus their insertion into the CS structure could be induced new beneficial features to each newly prepared system. Therefore Chen et al. designed and fabricated a new type of hydrogels based on CS via a physical crosslinking method regarding in situ reductions of $AuCl_4^-$ ions by CS. Figure 2a shows the mechanism of CS-Au hydrogel formation. Transmission electron microscopy (TEM) of freeze-dried CS-Au hydrogel approved the formation of Au NPs via the appearance of the small dark spots which are related to the Au NPs. In the swelling test, the CS-Au hybrid hydrogel displayed exceptional water-absorbing. Around 9.77% of doxorubicin hydrochloride (DOX·HCl) was loaded in it. The DOX release behavior was investigated in different conditions and the achieved results (Fig. 2b) indicated that the CS-Au hydrogel is a promising sustained drug release system in which its performance is influenced by external stimuli, like pH or temperature. As it is reported in Fig. 2c the over 100% of C6 cells viability after treatment with CS-Au hybrid hydrogel indicated its biocompatibility. But the free DOX and DOX loaded CS-Au hydrogel both showed cytotoxicity to the C6 cells. All of the obtained results approved that the physically crosslinked CS-Au hybrid hydrogel combines the merits of both CS and AuNPs, and confirmed that the CS-Au hybrid hydrogel is an ideal smart local drug delivery platform [26].

Additionally, because of the inherent features of Au NPs including optical property and surface plasmon resonance (SPR), they can be used as a contrasting agent. By considering this special advantage, the AuNPs were prepared in the green procedure

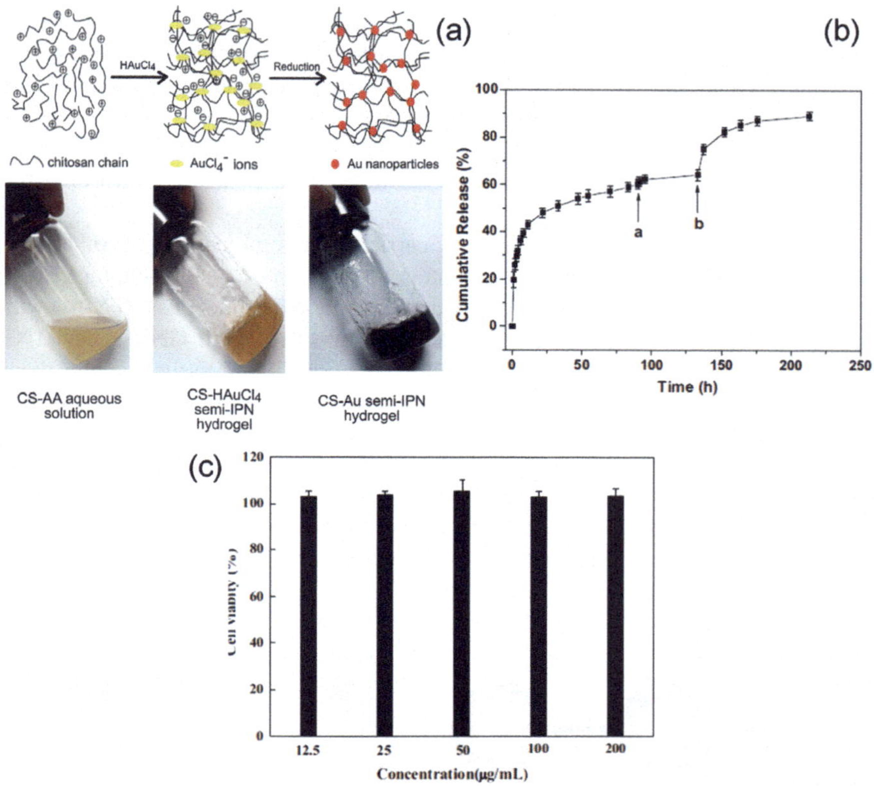

Fig. 2 Mechanism of the CS-Au hydrogel formation (**a**), cumulative DOX·HCl release behavior from drug-loaded CS-Au hydrogel at 37 °C and pH 7.4 (Point a: increase temperature to 70 °C for 1 h, then re-incubate the hydrogel in PBS at 37 °C; Point b: the pH value of PBS was changed to 3.0) (**b**), and C6 cell viability for the treated cells with hydrogel (**c**). Reproduced with permission from [26]

by using the CS oligosaccharide (COS) as a stabilizing and reducing agent and then loaded with paclitaxel (PTX). Subsequently, the efficiency of the prepared system was studied to reveal its capability in drug delivery and photoacoustic imaging (PAI) of MDA-MB-231 cells. Field emission scanning electron microscopy (FE-SEM) analysis exhibited spherical in shape for PTX-loaded COS-stabilized gold nanoparticles (PTX-COS AuNPs), moreover, the dynamic light scattering (DLS) determined its average particle size around 61.86 nm. The pH-sensitive and sustained PTX release profiles, as well as, strong cytotoxic effect toward MDA-MB-231 cells in the apoptosis induction with enhanced reactive oxygen species (ROS) formation, and altered mitochondrial membrane potential level approved the efficiency of the prepared system for use in biomedicine. The cellular uptake of PTX-COS AuNPs was confirmed via flow cytometry, as well as, via fluorescence microscopy. Also, PTX-COS AuNPs were investigated as an optical contrast agent for PAI. The results of this work displayed the promising potential of PTX-COS AuNPs in molecular imaging and drug delivery. The overall pattern for the AuNPs biosynthesis using the COS, PTX loading on COS AuNPs, and the probable mechanism for cellular internalization of PTX-COS AuNPs in MDA-MB-231 cancer cells are provided in Fig. 3a. This study shows that the PTX-COS AuNPs can use as a contrast agent for PAI and is a good candidate for future applications in the pharmaceutical industry [27].

In the line of investigations about nanocarrier with combination photothermal-chemotherapy ability, one injectable nanocomposite hydrogel incorporating PEGylated gold nanorods (GNRs) and PTX-loaded CS polymeric micelles (PTX-M) was prepared (PTX-M/GNR/gel). After intratumoral injection, both GNRs and PTX-M were simultaneously carried and deposited in the tumor tissue via the thermosensitive hydrogel matrix. Exposure to the laser irradiation caused the GNR-mediated photothermal damage confined to the tumor with sparing the adjacent normal tissue. Synergistically, the co-delivered PTX-M exhibited prolonged tumor retention with the sustained anticancer drug release to efficiently kill the residual tumor cells that evade the photothermal ablation because of the heterogeneous heating in the tumor site. Summation of the outcomes showed that the combination

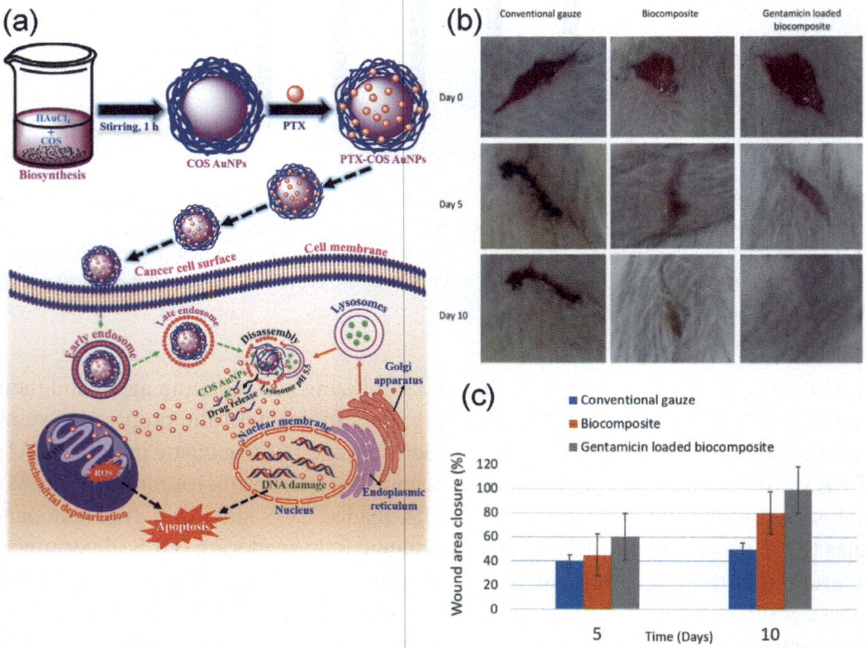

Fig. 3 The schematic procedure for the gold NPs biosynthesis by using COS, PTX loading on COS AuNPs, and the probable mechanism for the MDA-MB-231 cancer cells internalization of PTX-COS AuNPs (**a**), the obtained image of the treated wounds with commercial gauze, CS/PEG/ZnO, and gentamicin-loaded CS/PEG/ZnO (**b**), and the determined wound area closure (**c**). Reproduced with permission from [27, 37]

of photothermal-chemotherapy presents more influences on suppressing the tumor recurrence and prolonging the survival in the Heps-bearing mice, in comparison with the photothermal therapy alone [28].

Noble metal NPs are another metal NPS which received more attention due to their reduced toxicity, special chemical, optical, and electronic properties when compared to other materials. Palladium is one example of noble metal NPs with the ability to use in various areas like an antimicrobial agent, catalyst, biosensor, and dental appliances. For example, the CS/Pd nanocomposite was designed and fabricated via an economic chemical reduction way and then was evaluated for the delivery of 5-Fluorouracil (5-FU) and curcumin (CUR) in a conjugated form or separately. At first, the CS containing palladium solution was crosslinked via TPP. In the following, the palladium NPs deposited CS was achieved via in situ reductions of the palladium acetate as a result of the sodium borohydride dropwise addition to the above solution. The loading amount of the CUR and 5-FU was calculated at about 96.33% and 95.93% respectively. The study of the drug releases kinetic exhibited that the release profile is fitted with zero-order kinetics and denotes the constant release of drugs from drug carriers with prolonged drug release which leads to the reduction in the amount of the administered doses. Eventually, the drug-loaded nanocomposites exhibited more cytotoxicity on colon cancer cells; half-maximal inhibitory concentration (IC50) respectively was found to be 21.5 μg/mL, 18.3 μg/mL, and 14.6 μg/mL for CUR, 5-FU, and 5-FU + CUR encapsulated nanoformulations. which demonstrates the efficiency of the composite system for inhibiting cancer cells growth. In conclusion, dual CUR and 5-FU loaded CS/Pd are promising vehicles for multiple therapeutic agent delivery and achieving efficient therapy [29]. A review of the published works in the metal NPs/CS DDSs area shows that a wide range of drugs could be delivered with metal NPs/CS DDSs, besides, the biocompatibility of the CS is preserved even after combination with metal NPs. Because of the high volume of published research on the study of the CS-based nanocomposites having metal NPs for avoiding prolongation, the results of other published cases are summarized in Table 1.

2.2 *Metal Oxide NPs*

It is reported that the metal oxide NPs like titanium dioxide (TiO_2), iron oxide (FeO), and zinc oxide (ZnO) NPa have the exceptional capability for use in a different areas of medicine for example in biosensing, photothermal therapy, antibacterial, and anticancerous drug delivery, as well as in cell imaging because of their minimum toxicity, biocompatibility, and low price [20, 30–32]. However, because of their inherent tendency for agglomeration, these materials need to be modified via appropriate surface stabilizing agents to solve these drawbacks [33]. Additionally, the functionalization of the metal oxide NPs with modifier agents can reduce the probably some toxic effects which arose from their direct application [34]. Among the various used agents for the metal oxide NPs modification, the CS is a commonly used safe

Table 1 Summary of a selection of the studies CS-based nanocomposites for drug delivery application

Short description of DDS	Type of NPs	Carried drug	Release test pHs	The mentioned direction of applicability	Studied cell in biological assays	References
CSN films	Metal NPs	Moxifloxacin (Mox)	pH 6.8	Antibacterial	–	[109]
CSSN films	Metal NPs	Mox	pH 6.8	Antibacterial	–	[110]
CS-pullulan-silver-nanocomposite (CSPN)	Metal NPs	Mox	pH 6.8	Antibacterial	–	[111]
CS-gold nanocomposite (CGNC)	Metal NPs	5-FU	pH 7.4, and 5.3	Tumor therapy	SiHa	[112]
CS-g-glycolic acid and gold nanoflower nanohybrid	Metal NPs	Cyclophosphamide	pH 7.4	Tumor therapy	SP2/0 mouse myeloma	[113]
CS/gold (CS-Au)	Metal NPs	5-FU	pH ~5	Tumor therapy	MCF-7, and VERO	[114]
CS-g-poly(acrylamide)/Zn (CPA-Zn)	Metal NPs	Ofloxacin	pH 2.2, 7.4, and 9.4	Chlamydial infections, treating mycobacterial infections, antimicrobial	–	[115]
CS/Au based micelles	Metal NPs	Paclitaxel	pH 7.4	Tumor therapy	MCF7	[116]
Fe_3O_4@APTES/CS/TG	Metal oxide NPs	CUR	pH 3.4, and 7.4	Tumor therapy	–	[117]
Magnetic CMC/β-CD/CS hydrogel	Metal oxide NPs	MTX	pH 1.2, 5.3, and 7.4	Tumor therapy	–	[118]
Fe_3O_4@C/CMC/CS	Metal oxide NPs	DS	SGF, SIF, and SCF	Anti-inflammatory drug	–	[119]
CS-PEO-hyd-ZnO(Im) mats	Metal oxide NPs	Imipenem/cilastatin sodium drug and hydrocortizon	pH 7.2–7.4	–	–	[120]

(continued)

Table 1 (continued)

Short description of DDS	Type of NPs	Carried drug	Release test pHs	The mentioned direction of applicability	Studied cell in biological assays	References
PTX-NFS-BSA-CS-FA	Metal oxide NPs	PTX	pH 5.4, and 7.4	Tumor therapy	MCF 7	[121]
OA-Fe_3O_4@CS-PEG	Metal oxide NPs	DOX	pH 5.0, and 7.4	Tumor therapy	HepG2	[122]
CFNP-CS	Metal oxide NPs	Ampicillin	pH 4, 7.4, and 9	Antibiotic	–	[123]
Fe_3O_4/CS nanocomposites	Metal oxide NPs	Gemcitabine	pH 7.4	Tumor therapy	Leukemia	[124]
CS-CDpoly-MNPs	Metal oxide NPs	5-FU	pH 1.2, 6.8, and 7.4	Tumor therapy	–	[125]
CS/PAA/Fe_3O_4	Metal oxide NPs	5-FU	pH 7.4	Tumor therapy	–	[126]
PAA-g-CS/ MIONs	Metal oxide NPs	DOX	pH 5.4, and 7.4	Tumor therapy	–	[127]
CipChCuM	Metal oxide NPs	Ciprofloxacin	pH 7.4	Antibacterial	–	[128]
Fe_3O_4/CS/INH	Metal oxide NPs	Isoniazid	pH 7.4	Antibiotic	–	[129]
ZnO/CMC/CS	Metal oxide NPs	5-FU	pH 1.2, 6.8, and 7.4	Tumor therapy	–	[130]
SPION coated palmitoyl chitosan	Metal oxide NPs	PTX	pH 7.4	Tumor therapy	MCF-7	[131]
CS-dextran sulfate (DS) nanoparticles coated iron oxide	Metal oxide NPs	5-ASA	pH 7.0	Inflammatory treatment	–	[132]

(continued)

Table 1 (continued)

Short description of DDS	Type of NPs	Carried drug	Release test pHs	The mentioned direction of applicability	Studied cell in biological assays	References
Chitosan/ZnO	Metal oxide NPs	IBU	pH 7.4		–	[133]
Magnetic CS	Metal oxide NPs	5-hydroxytryptophan (5-HTP)	pH 3.5, and 8.5	Chemical precursor in the biosynthesis of important neurotransmitters serotonin	–	[134]
CS/$ZnFe_2O_4$	Metal oxide NPs	Lidocaine	pH 4.8, 5.5 and 7.4	Anesthetic drug	Fibroblast	[135]
Magnetic κ-carrageenan/CS	Metal oxide NPs	MTX	pH 5.3, and 7.4	Tumor therapy	–	[136]
HNT/CTS/PCN	Silica NPs	Phenytoin sodium (PHT)	SGF, and SIF	Antiepileptic	–	[137]
Hal-CTS	Silica NPs	Aspirin (Asp)	pH 1.2, and 7.4	Pain relief	–	[138]
HNT@CUR-Au/CS	Silica NPs	CUR	pH 5.5, and 7.4	Antitumor	MCF-7	[139]
PCB and PCBSF	Silica NPs	cisP	pH 4.0 and 7.4	Antitumor	HEK-293 and HCT116	[140]
PVA/Cs-MMT	Silica NPs	Nitrofurazone (NFZ)	pH 7.4	Antibiotic	–	[141]
CS/Pectin/clay	Silica NPs	Diclofenac	pH 6.8	Non-steroidal anti-inflammatory	–	[142]
Palygorskite/CS (Pal/CS)	Silica NPs	Diclofenac	pH 6.8	Non-steroidal anti-inflammatory	–	[143]

(continued)

Table 1 (continued)

Short description of DDS	Type of NPs	Carried drug	Release test pHs	The mentioned direction of applicability	Studied cell in biological assays	References
CS/MMT	Silica NPs	Vancomycin (VC) and Gentamicin (GC)	pH 7.4	Antibiotics	Fibroblast (NIH/3T3), and osteoblast (SaOS-2)	[144]
CS-MMT	Silica NPs	IBU	pH 5.4, and 7.8	–	–	[145]
MCM-48/chitosan nanocomposite (MCM-48/CH)	Silica NPs	IBU	pH 1.2, and 7.4	–	–	[146]
PTX-MMT-CS	Silica NPs	PTX	pH 1.2 and 7.4	Antitumor	COLO-205 cells	[147]
CS-silica/CpG oligodeoxynucleotide (ODN)	Silica NPs	CpG ODN	pH 5.5	Cancer, allergies, and infectious diseases treatment	293XL-hTLR9, and PBMCs	[148]
MSN-BCS-FA	Silica NPs	DOX	pH 5.0, 6.5, and 7.4	Antitumor	HepG-2 cells	[149]
CS-Cloisite	Silica NPs	Paracetamol and theophylline	pH 2.2	–	–	[150]
CS-g-LA/MMT	Silica NPs	IBU	pH 7.4	–	HL-60 Leukemia	[151]
Alg-CS/5-FU/Mt	Silica NPs	5-FU	pH 1.2, 7.4, and 10	–	–	[152]
Cs-CMS-MMT	Silica NPs	CUR	pH 4.5, and 7.4	Anti-biofilm activity	–	[153]
Chitosan-clay	Silica NPs	Propranolol	pH 0.1 M HCl and 6.8	–	–	[154]
CS-MMT	Silica NPs	Vitamin B12	pH 7.4	–	–	[155]

(continued)

Table 1 (continued)

Short description of DDS	Type of NPs	Carried drug	Release test pHs	The mentioned direction of applicability	Studied cell in biological assays	References
CS-PLA/MMT	Silica NPs	PTX	pH 1.2, and 7.4	–	–	[156]
PLGA-loaded CH-BGscaffolds	Silica NPs	Dexamethasone (DEX)	PBS	–		[157]
MMT:CS	Silica NPs	CLX	pH 1.2, 4.2, and 6.8	Bisguanide antiseptic	–	[158]
(CS-MAS)	Silica NPs	Nicotine	pH 6.0	For smoking cessation	–	[159]
Chitosan-PVA/Na^{+}MMT	Silica NPs	5-FU	7.4 pH	Antitumor	–	[160]
CS/CPL cryogels	Silica NPs	Indomethacin (IDM)	pH 1.2, and 7.4	Non-steroidal anti-inflammatory	–	[161]
CS-Si-CUR	Silica NPs	CUR	pH 4, 7.4, and 9.2	Antimicrobial	–	[162]
Cur-Ga NF	GO NPs	CUR and gallic acid codelivery	pH 5.4	Anticancer, antimicrobial, antioxidant, and anti-inflammatory	A549 and KB	[163]
CS-g-PMAA/GO	GO NPs	DOX	pH 4.0, 5.4, and 7.4	Tumor therapy	MCF7	[164]
PEO/CS/GO	GO NPs	DOX	pH 5.4, and 7.3	Tumor therapy	A549	[165]

(continued)

Table 1 (continued)

Short description of DDS	Type of NPs	Carried drug	Release test pHs	The mentioned direction of applicability	Studied cell in biological assays	References
GO-CS	GO NPs	Proanthocyanidins	–	Antiviral, anti-inflammatory, cardioprotective, antidiabetic, anticancer, and anti-aging	HEK 293	[166]
CS-GN, CS-GP, and CS-G-IF	GO NPs	Ifosfamide	–	Tumor therapy	–	[167]
CNP/CS films	GQDs NPs	Bovine serum albumin (BSA)	–	Wound healing	–	[168]
FA-COS/MHNTs	Hybrid NPs	Camptothecin (CPT)	pH 5, 6.8, and 7.4	Tumor therapy	Caco-2	[169]
30%MFe_2O_4/Silica/Chitosan (M = Ni, Cu, Co and Mn)	Hybrid NPs	cisP	pH 5.6	Tumor therapy	MCF-7	[170]
Chitosan–MGQD nanocomposite	Hybrid NPs	Lidocaine hydrochloride (LH) and BSA	pH 7.4	–	–	[171]
MagSi@Chi-g-NIPAAm	Hybrid NPs	DOX	pH 4.0, and 7.4	Tumor therapy	HeLa	[172]
MNS-SO_3H-MTX-CH	Hybrid NPs	MTX and prednisolone (PRD)	pH 7.4	Anticancer and anti-inflammatory	HD-MY-Z, EJ and K-562 the	[173]
CS/QD/MNP	Hybrid NPs	Insulin	pH 5.3, and 7.4	Treating Type II diabetes	L02	[174]
CS/Fe_3O4@$mSiO_2$	Hybrid NPs	DOX	pH 4.0, 5.8, and 7.5	Tumor therapy	HepG2	[175]

(continued)

Table 1 (continued)

Short description of DDS	Type of NPs	Carried drug	Release test pHs	The mentioned direction of applicability	Studied cell in biological assays	References
MQ-MSN/PNIPAM-g-CS	Hybrid NPs	Adriamycin (ADM)	pH 5.0, 6.5, and 7.4	Tumor therapy	HepG2	[176]
CS/cobalt ferrite/titanium oxide nanofibrous scaffolds	Hybrid NPs	DOX	pH 5.3, and 7.4	Tumor therapy	B16F10 melanoma	[177]
Glycolic acid-functionalized CS-Au-Fe_3O_4	Hybrid NPs	Cyclophosphamide	pH 7.4	Tumor therapy	L929	[178]
mGO-CS/SA	Hybrid NPs	DOX	pH 5.0, and 7.4	Tumor therapy	A549	[179]
ZnO-QD-CS-folate	Metal GQD NPs	DOX	pH 5.3, and 7.4	Tumor therapy	Fibroblasts	[180]
CS-OM-HA	Hybrid NPs	BSA	–	Bone tissue engineering	MG 63 osteoblast	[181]
CS/GN/WN nanocomposite (CGW)	Hybrid NPs	CUR and DOX codelivery	pH 5.4, and 7.4	Tumor therapy	–	[182]
CS/CUR@GQDs@MIL-88(Fe)	Hybrid NPs	CUR	pH 5.0, and 7.4	Tumor therapy	MDA-MB 231	[183]
f-GE-g-(CS/MMT-PANI)	Hybrid NPs	DOX	SIF and SGF	Tumor therapy	–	[184]

biopolymer which modification introduces the additional properties that are desired for biofunctionalization.

ZnO NPs is one example of the investigated metal oxide NPs in the structure of the drug carriers owing to its specific features like antibacterial performance toward pathogenic bacterial species, as well as, displaying excellent pharmacognostic performance toward deadly diseases like diabetes and cancer, inflammation, and itching relief [20, 35, 36]. Therefore the contribution of nanotechnology in the disease treatment field like in the area of bio-dental application is significant. In 2020 Javed et al. evaluated the performance of the CS capped metallic oxide NPs in restorative dentistry from the ZnO NPs release viewpoint. So, this research group fabricated the uncapped and CS-coated ZnO NPs via an easy co-precipitation method. Analysis obtained the size of the uncapped and CS-coated ZnO NPs respectively about 20–25 nm and 25–30 nm. Besides the zeta potential value of ZnO NPs was increased from 12.4 ± 0.5 mV to $+ 21.3 \pm 1.2$ mV after coating with CS. A series of in vitro biological assays displayed that the antidiabetic, antibacterial, and antioxidant activity of ZnO NPs improved as a result of capping with CS biopolymer. About 90% cytotoxicity was detected for ZnO-CS NPs toward brine shrimps. Also, efficient secondary caries remediation was observed via the incorporation of ZnO NPs and ZnO-CS NPs within the dentine bonding systems. Moreover, a notable decrease in *Streptococcus mutans* and *Lactobacillus acidophillus* strains was obtained, when boosted via CS capped ZnO NPs reinforced dental adhesive discs. To conclude the mentioned positive features with detected improved properties, especially high release pattern, more resistance to solubility and water sorption, and minor shear bond strength values variation exhibited that the CS-coated NPs can be an auspicious theranostic means for combating a wide range of human pathogens [20].

In the following researches for the preparation of the CS-based nanocarrier having ZnO, in 2020 Masud et al. designed and fabricated the CS-ZnO NPs, sodium tri polyphosphate (STPP) crosslinked poly (ethylene glycol) (PEG) to evaluate its wound healing efficiency. In this way, initially, the ZnO NPs were synthesized via the co-precipitation method. The performed analysis determined the average size of the synthesized ZnO NPs about 50 nm. In the next step, the synthesized ZnO NPs were embedded within the CS/PEG bionanocomposite to improve the antimicrobial activity of nanocomposite, and semi-porous was obtained. Eventually, the capability of the system for gentamicin delivery and its effect on antibacterial activity were tested. An optimum 76% of gentamicin was loaded in the prepared nanocomposite. The biocompatibility of the prepared bionanocomposite was confirmed after doing the cytotoxicity analysis against Vero cells and BHK 21 cells. In vitro antibacterial activity test indicated the improved antibacterial performance for gentamicin-loaded bionanocomposites toward *E. coli* and *S. enterica* that could be related to the combined influence of the ZnO NPs and gentamicin. For further confirmation of the obtained results from the in vitro analysis, the in vivo test was performed. As it is reported in Fig. 3b, c the results of the in vivo test displayed that the gentamicin-loaded bionanocomposites have better healing ability in comparison with traditional hydrogel wound dressing without any scar creation. The performed studies confirmed

CS/PEG/ZnO based release system can be a potential candidate for wound dressing applications with sustained drug release [37].

Currently, the utilization of the targeted carriers is one of the other popular used ways for improving the bioavailability of drugs. Fe_3O_4 NPs with high magnetic properties and good biocompatibility indicated its exceptional properties for use in this area of medicine [33]. Therefore, in 2018, Wang et al. used an easy in situ hybridization technique for fabrication of the magnetic CS hydrogel (MCH) that remotely changed drug release from passive release to pulsatile release in the presence of a low frequency alternating magnetic field (LAMF). Avoiding the MNPs aggregation is the important advantage of the MNPs synthesis within the hydrogel matrix. Chelation of the iron ions by amino groups of CS is a reason for avoiding the MNPs aggregation. In the presence of LAMF with 15 min ON–15 min OFF cycles, the fraction release presented a zigzag shape and pulsatile release pattern with a fast answer. Elastic modulus and strength of MCH were enhanced respectively as high as 265% and 416%. Both hydrophobic rifampicin (RFP) drug and hydrophilic adriamycin (ADM) drug releases from the hydrogel were investigated in vitro. The cumulative release of the drug from MCH was enhanced by 67.2%. The results of (3-(4,5-dimethylthiazol-2-yl)-2,5-diphenyltetrazolium bromide (MTT) assay and cell morphology analysis toward MG-63 cells exhibited that the MCH has exceptional biocompatibility; cell viability was 87% even at treatment dosage up to 200 mg/mL and no acute adverse influence on them. In particular, based on the results the MCH has the potential applications as an implantable platform for localized drug delivery stimulated by a non-invasive low frequency alternating magnetic field [38]. In the other similar research work consistent with the targeted drug delivery capability of the magnetic NPs, the pH and temperature-responsive hydrogel (Fig. 4a) was prepared through the surface reversible addition-fragmentation chain transfer (RAFT) copolymerization. In this regard magnetic nanocarrier was fabricated through simultaneous RAFT polymerization of two stimuli-responsive acrylic monomers; N-isopropyl acrylamide and acrylic acid onto the functionalized CS in the existence of Fe_3O_4 NPs. In the following, the fabricated CS hydrogel was used as a platform for in situ fabrication of magnetic Fe_3O_4 nanoparticles on its. The vibrating sample magnetometer (VSM) analysis approves the in situ growth of the magnetic NPs in the hydrogel network (Fig. 4b). In vitro drug delivery studies indicated that about 89% of DOX was loaded in the nanocomposite, also, as is reported in Fig. 4c it observed that about 82% of loaded DOX was released in a sustained-release way from the nanocomposites. The achieved results signify that the CS/Fe_3O_4 nanocomposite could be a hopeful drug carrier for sustained and controlled release of anticancer drugs and addressing the associated problems with traditional drug administration ways [33].

A further study for introducing the potential of the CS-based nanocomposites having Fe_3O_4 NPs for targeted delivery, the Fe_3O_4/CS/isoniazid magnetic NPs (Fe_3O_4/CS/INH-MNPs) were fabricated as a stimuli-responsive platform for delivery of isoniazid (INH) as a main antituberculosis drug. In the structure of the prepared system, the TPP and Fe_3O_4 magnetic NPs acted as crosslinkers to create intermolecular cross-linkages for the CS chains. Thermal gravimetric analysis (TGA) analysis determined the average mass content of Fe_3O_4 in NPs by about 53.68%. The in vitro

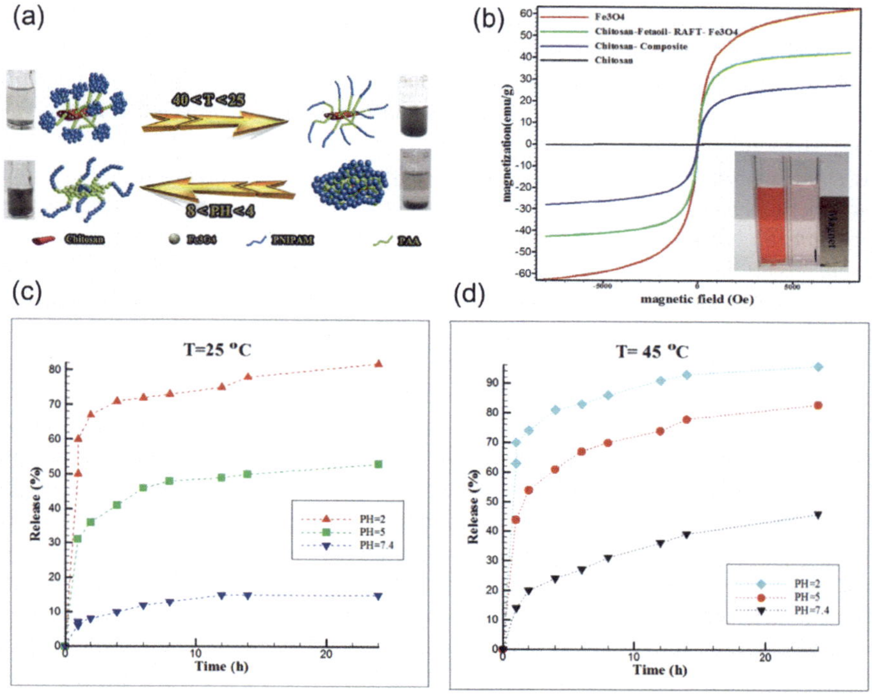

Fig. 4 Representation schematic for the CS/Fe_3O_4 nanocomposite response to temperature and pH variation in aqueous solution (**a**), measured magnetization curves (**b**), and DOX release behavior from the drug-loaded CS/Fe_3O_4 at different conditions (**c, d**). Reproduced with permission [33]

INH release evaluation in two different pH values; pH 7.4 and pH 5.7 that mimic endosomal and lysosomal conditions showed that the release value increased as the pH value of PBS medium reduced, which was consistent with the pH sensitivity of the drug carrier structure. Cytotoxic test of Fe_3O_4/CS/INH-MNPs was performed on OCI-LY3 cells and 90% of OCI-LY3 cells viability was determined even at a concentration of 100 μg/mL which confirmed the biocompatibility of the fabricated system. The summation of the obtained results exhibited that the prepared system is expected to be an appropriate candidate in targeted, pH-dependent, and favorable tuberculosis therapy with outstanding sustained release and magnetic sensitivity [39]. In another interesting research regarding the targeted anticancer drug delivery behavior of magnetic NPs, the pH-responsive drug delivery platforms were prepared via the κ-carrageenan-crosslinking of the magnetic CS with different molecular weights (MWt) for controlled release of sunitinib as a model of anticancer drug. So, firstly Fe_3O_4 NPs were synthesized in the existence of CS with various molecular weights via in situ method. Subsequently, as-synthesized magnetic CS NPs were crosslinked with κ-carrageenan. The XRD analysis indicated that the size of magnetic carriers is influenced by changing the molecular weight of used CS. As well as it was detected that the sunitinib encapsulation efficiency and its release pattern were affected via the size of the magnetic NPs. So, respectively 78.42%, 69.57%, and 62.38% of sunitinib loading amount were calculated by high, medium, and low molecular weights of magnetic CS platforms. pH-responsive releases of sunitinib from the fabricated magnetic CS/κ-carrageenan agents were confirmed via observation of the higher amount of drug release in pH 4.5 compared with pH 7.4. As well as the sunitinib release was molecular weight-dependent, so it is mentioned above, the carriers prepared in the presence of CS with higher MWt results in a fast release rate for sunitinib because of the enhanced surface area of magnetic NPs. The achieved results could be evidence that the prepared magnetic carriers have the potential for use in tumor therapy due to the pH-sensitive targeting of tumor cells without affecting the healthy adjacent tissues and open up several exciting opportunities [40].

The capability of the magnetic CS-based systems was also investigated for the targeted delivery of methotrexate (MTX). In this regard, the MTX-PEG-CS-IONPs-Cy5.5 nanocomposites were prepared via the functionalization of the CS decorated iron oxide NPs (CS-IONPs) surface with polyethylene glycolated methotrexate (MTX-PEG) and then cyanin (Cy5.5) as a near-infrared fluorescent dye. The prepared system displayed the self-targeting/kill capability, superparamagnetic character, and fluorescence property for simultaneous cancer therapy and diagnosis. Briefly, at the first step, the IONPs were prepared via the chemical co-precipitation technique. In the next step, the fabricated CS-IONPs were covered via a non-solvent-aided coacervation process followed by a crosslinking. In the end, the pre-prepared MTX-PEG and Cy5.5 dye were grafted on the CS-IONPs surface through carbodiimide chemistry. MTX is an anticancer drug that could be a tumor-specific targeting ligand for targeted directing of these NPs to tumor cells overexpressing folate (FA) receptors due to the high similarity between the structure of FA and MTX. The stability of the synthesized system was indicated owing to the observation of no obvious particle size change in dynamic light scattering (DLS) after 24 h. From the in vitro and in vivo output, i.e., the achieved higher tumor accumulation, controlled drug MTX, more cell uptake, reduced side effects, and enhanced therapeutic efficacy the MTX-PEG-CS-IONPs-Cy5.5 could be proposed as a valuable platform for targeted and controlled drug injection [41].

In continuous research about the development of nanocarriers functionalized with FA, a desired folic acid (FA)-modified reduction-responsive magnetic CS nanocapsules (FA-RMCNCs) successfully were prepared from the FA modified thiolated CS with a facile sonochemical technique. Coumarin 6 (C6) was loaded into the FA-RMCNCs via dissolution in the hydroxyl silicone oil before sonicating. SEM analysis exhibited the spherical morphology for the as-synthesized FA-RMCNCs with the size of 200–350 nm. The obtained 27.25 emu/g value for Ms of the FA-RMCNCs approved the magnetic responsive property of the system. The selective internalization by the HeLa cells via folate-mediated endocytosis which was observed via confocal laser scanning microscopy (CLSM) technique discovered that the fabricated FA-RMCNCs with core–shell architecture could be an efficient system for drug delivery. Moreover, the reductant-triggered coumarin 6 release proposed that the FA-RMCNCs have greater drug release reduction-responsivity. The results of this research disclosed the capability of the FA-RMCNCs as magnetic/reduction dual-responsive, folate-receptor-mediated targeting nanoplatforms in the controlled delivery of hydrophobic drugs and their targeted transferring. Accordingly, the prepared FA-RMCNCs with reduction/magnetic dual-responsive and folate-receptor-mediated targeting capability can be a very hopeful candidate for targeted transferring and triggered release of hydrophobic drugs [42]. Additionally, the survey of the published research works showed that drug delivery by using the magnetic biopolymeric systems could be an effective way and address the low cargo capacity of hydrophobic drugs and improve their delivery. Keeping these in mind, in one similar research work, Jardim designed and investigated the magneto-responsive nanovehicles based on $MnFe_2O_4$ NPs layer-by-layer (LBL) modified with CS and sodium alginate (SA) for targeted delivery of CUR. Briefly, $MnFe_2O_4$ magnetic NPs (MNPs) were prepared via thermal decomposition, coated with citrate, and subsequently modified with the layer-by-layer (LbL) assembly of CS as polycation and SA as polyanion in the polyelectrolyte multilayers shape. The step-by-step change in the magnetization saturation (Ms) from 41.1 to 15.3 emu/g verified the efficacy of the LbL method to excellently alteration of the polymeric shell thickness on the NPs. After each deposition, the recorded charge reversals (± 30 mV) established the formation of polyelectrolyte adsorption and a stable LbL assembly. On the other hand, it is expected that the combination of cationic and anionic characteristics can be helpful and improve the carrier distribution and internalization by tumor cells. ~100 μg/mg of CUR was payloaded in the prepared drug delivery agent. MTT assays after 72 h treatment of the MCF-7 cells with the fabricated system exhibited enhancing of the system performance as a result of capping via biopolymers. Achieved results developed the applicability of the prepared nanovehicles as an appropriate approach for remotely delivering drugs for medical purposes, upon the usage of an external alternating magnetic field in order the enhance the carrier efficiency biocompatibility [34].

Similarly, the pH-sensitive release behavior of the CS-based nanocomposite containing $Mn_{1-x}Zn_xFe_2O_4$ ((Mn, Zn) ferrite) MNPs was evaluated for DOX delivery as a commonly used anticancer drug for tumor therapy. In the first step, the $Mn_{0.9}Zn_{0.1}Fe_2O_4$ MNPs with superparamagnetic behavior; MS value of about 56.1 emu/g was fabricated via chemical co-precipitation technique. In the following the synthesized $Mn_{0.9}Zn_{0.1}Fe_2O_4$ MNPs were encapsulated in PLGA-coated CS polymer. A decrease in the saturation magnetization to about 13.2 emu/g confirmed the encapsulation process. The analysis obtained the thickness of the formed polymeric shell layer about 50 nm. A pH-sensitive release pattern was observed for the fabricated system, so, around 57% of DOX released within the first 5 h, however, the amount of DOX release was around 46% after 25 h. In the MTT test, the DOXPLGA@CS@$Mn_{0.9}Zn_{0.1}Fe_2O_4$ exhibited enhanced anticancer performance toward HeLa cells and acceptable therapeutic effect. The summation of the obtained results is good proven for showing the suitability of the designed system [43]. In order to expand research in the targeted anticancer drug delivery field and with the aim of introducing a new CUR delivery system, in 2020, a CUR-loaded hybrid NPs of vanillin-CS coated with superparamagnetic calcium ferrite was fabricated and evaluated by K. Srirama et al. The biocompatible and superparamagnetic nature of calcium ferrite NPs (CFNP) convert it's to an effective candidate in the magnetic targeting drug delivery. In this interesting research at first, the CS modified vanillin was synthesized via the Schiff-base reaction to improve the hydrophobic drug loading capacity. Afterward, the CFNP NPs were added to the modified CS. In the end, the vanillin-CS-CFNP was prepared via ionic gelation. The optimum loading of the CUR as a plant-based anticancer drug was calculated at about 98.3% which was determined after optimization of the system. The CUR release study was done at different pH, magnetic fields, and initial CUR loading concentrations. Biological tests toward L929 fibroblast cells approved the biocompatibility of the hybrid material. Moreover, the performed cytotoxicity test toward MCF-7 breast cancer cell line exhibited the antitumor property of the CUR-loaded hybrid nanocarrier. Eventually, the inverted microscopy images of the treated MCF-7 cells with two different concentrations of CFNP, vanillin ligand, and CUR-loaded CS-vanillin with CFNP are shown in Fig. 5a. From Fig. 5a, it is detected that the nontreated MCF-7 cells are highly viable. But the cell viability is reduced after treatment with the samples in a concentration-dependent manner. The output of this research indicated the anticancer property of CFNP, CUR, and CUR-loaded CS-vanillin with CFNP toward the MCF-7 cancer cell lines [44].

TiO_2 NPs are one of the other metal oxide biocompatible NPs which have attracted more attention owing to their special chemical and physical properties. In this regard, in one interesting research, a series of nanohybrid composites of dopamine (DOP)/CS (CS)@TiO_2 were fabricated via sol–gel technique and their performance was studied for delivery of DOP as a model drug. In the following the ultraviolet–visible (UV–Vis) absorbance and electrochemical determination, were used to determine the drug release pattern from the composites. The overall process of the synthesis consisted of two steps; (1) fabrication of the DOP/CS (10–40 wt %) composites via solubilization of the DOP in the CS aqueous solution. (2) coating of the DOP/CS composites via different amounts of TiO_2 (10–50 wt.%) which titanium tetra n-butoxide was a precursor. Several characterization analyses like FT-IR, FE-SEM, and XRD approved the successful formation of DOP/CS@TiO_2 nanohybrid composites in detail. The Brunauer–Emmett–Teller (BET) analysis determined the specific surface area of the nanohybrid composite, i.e., DOP/CIS/TiO_2(20, 30) (B2) about 68 m^2/g which is anticipated to be indicative of adsorption in macropores. In vitro DOP delivery studies exhibited that the incorporation of the TiO_2 within DOP/CIS composites significantly improved the drug loading amount and leads to sustained drug releases. So, the whole DOP drug was released after 10 min in pH 7.4 for the DOP/CS composites without TiO_2, but, after coating the TiO_2 the entrapped drug was released after 16 h. The achieved results are very pleasing and opened a good and promising way for the development of the oral administration of DOP [3]. Because of the high volume of published research on the study of the CS-based nanocomposites having metal oxide NPs for avoiding prolongation the results of other published cases are summarized in Table 1.

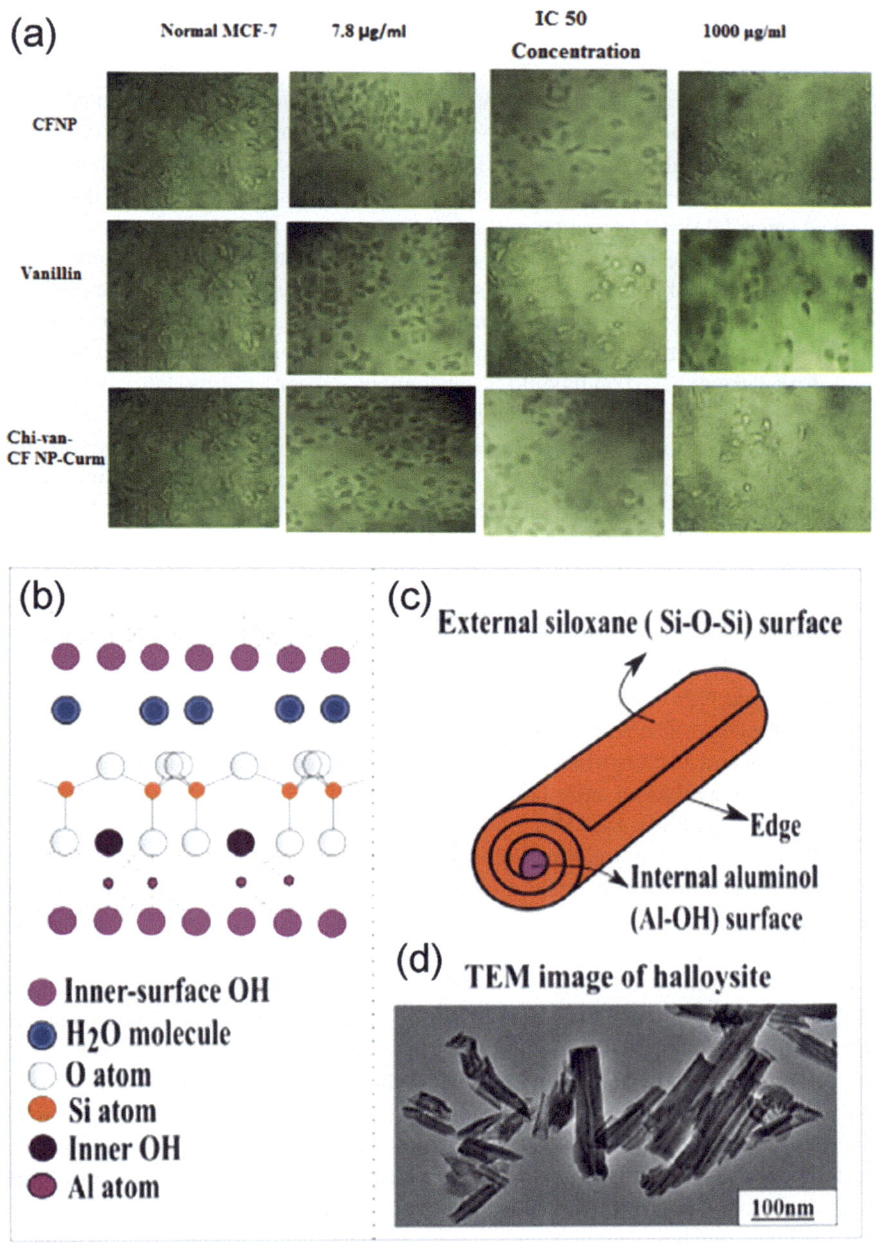

Fig. 5 Phase inverted microscopy images of control and treated MCF-7 cells at two different concentrations (**a**), the structure of HNT (**b**, **c**), and the TEM image of HNT (**d**). Reproduced with permission from [44, 48]

2.3 Silica NPs

Silica-based NPs offer improved hydrophilicity, high surface area, good permeability, and toughness to the nanocomposites [45, 46]. There are several types of silica-based NPs which halloysite, montmorillonite (MMT), mesoporous silica nanoparticles (SiO_2), clay are some of the CS-based nanocomposite having silica that their application as DDSs is presented in the following.

Halloysite is one member of the clay minerals family with exceptional efficiency like natural abundance, hollow spiral nanostructure, tuning drug release, high mechanical feature, high surface area, more availability, good thermal properties, the potential of functionalization, chemical stability, and biocompatibility, hence, the nano-drug carriers based on it could be a beneficial carrier for drug delivery applications. Additionally, the dual electric charge nature of halloysite nanotube (HNT) is one exceptional feature of it. So, in a wide pH range of 3–8, the inner surface (lumen) has a positive charge but its outer surface has a negative charge which provides the possibility of the opposite charge molecules adsorbing on its outer or inner surfaces [47]. As it is shown in Fig. 5b–d this material could be observed in nature as a hydrated mineral formed from rolled aluminosilicate sheets with the $Al_2Si_2O_5$ $(OH)_4.nH_2O$ formula. Depending on the geological deposits, the size of the halloysite nanotubes is varied. As the halloysite mineral was heated it converted to the molecular formula of $Al_2Si_2O_5(OH)_4$ [48].

Silica-based NPs could be modified via organic materials to intercalate the drugs into their structure. The modification with polymers results in enhanced drug entrapping capacity and also controls the drug release behavior [49]. By establishing these, in 2020 norfloxacin as an antibiotic drug was loaded in HNT via vacuum operation and sonication. Initially, the HNT powder was vacuum dried for obtaining the dried HNT. Dring leads to the water molecules removing from the lumen of the halloysites and enhancing the lumen space for drug loading. In the next step, the norfloxacin was loaded in the dried HNT. Some used characterization analyses like FT-IR and XRD confirmed the loading of norfloxacin into nanotubes via showing the characteristic peaks of HNT and loaded norfloxacin. In the TEM analysis, the norfloxacin-loaded HNT exhibited the tubular structure of HNT and also the existence of globular structures in its inner lumen that could be a good sign for successfully loading norfloxacin into the lumen of the HNT. In the following the solvent casting and freeze-drying were used for the fabrication of the halloysite/CS nanocomposites which drug-loaded halloysites were embedded in there. The authors reported that they were used from pre-loading step for loading of norfloxacin into the halloysite lumen owing to the hollow tubular structure of the halloysite. The in vitro norfloxacin release evaluation displayed that halloysite/CS nanocomposites can sustain the norfloxacin release. The sustained release during the 6 h could be attributed to the presence of HNT in the nanocomposite structure. An enhanced antimicrobial effect was observed for the nanocomposites toward both gram-positive and gram-negative bacteria by detecting the inhibition zones around the discs. The value of the inhibition zones diameters was dependent on the amount of the loaded antimicrobial agent. As well as, the prepared nanocomposite certified the biocompatibility against 3 T3 cells in the MTT assay [48]. Similarly in the field of the wound dressing, HNT was functionalized with different amount of poly (lactic-co-glycolic acid) (PLGA) with a negative charge under the acidic condition and CS as a polycationic polysaccharide by using the LbL method to develop a new system with advantages of both inorganic and organic materials; CS@PLGA@ APTES/Intercalated Etched HNT (CPMNT). The procedure of the nanocomposite preparation and minocycline (MC) loading on it is shown in Fig. 6a. After characterization of the prepared nanocomposites, some of the needing tests for proposing each of the prepared systems as a wound healing agent like water uptake, biodegradability, water vapor transmission, and photostability were

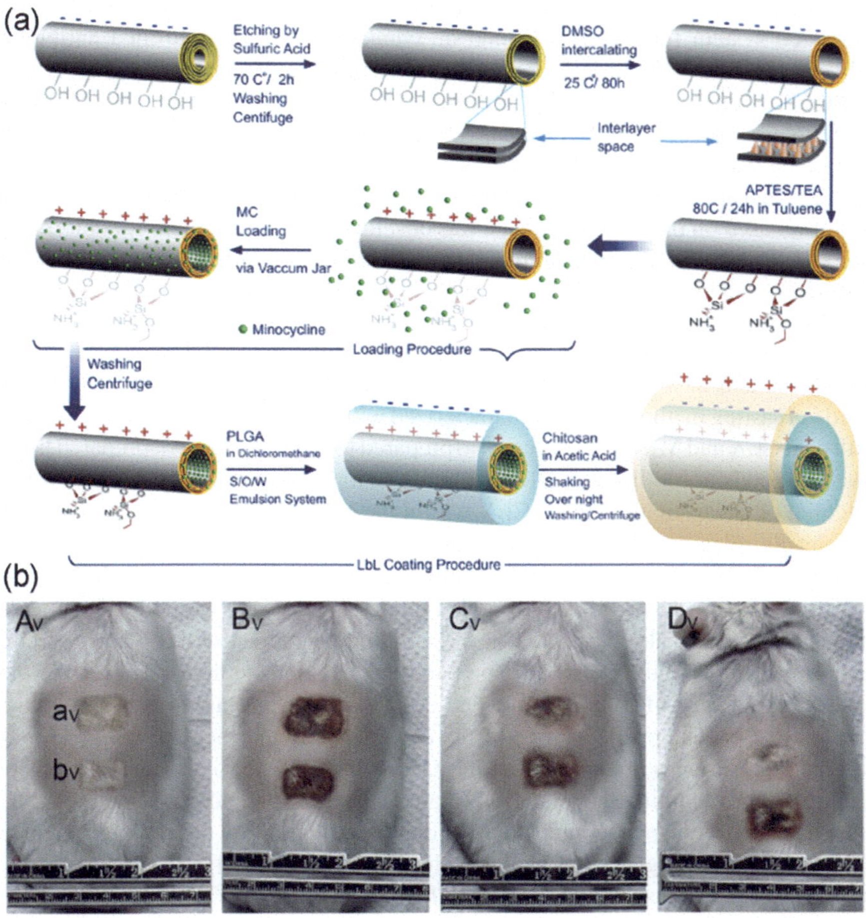

Fig. 6 Representation of CPMNT fabrication and MC loading within the HNT lumen (**a**), the digital image of treated wounds with F/C40P60MNT-MC (av) and PVA/HNT (bv) at (Av) 0 day, (Bv) 3 days, (Cv) 6 days and, (Dv) 12 days of post-treatment (**b**). Reproduced with permission from [47]

performed and the output was confirmed that the composite film can absorb wound secretions, penetrated properly to wound site, and did not degrade. Additionally, the protein adsorption assay exhibited blood compatibility for the optimized composite (C40P60MNT). In the following the in vitro pH-controlled and targeted delivery of MC as a broad-spectrum antibiotic from the tetracycline family was detected which was well fitted with Korsmeyer-Peppas kinetic model. The in vitro antibacterial assay exhibited a suitable antibacterial activity for the C40P60MNT-MC nanocomposites against both the gram-negative and gram-positive bacteria; *Staphylococcus aureus* and *pseudomonas aeruginosa*. The in vivo wound healing test was performed in rats to complete the in vitro studies, it was detected that in comparison with the control sample, the MC loaded C40P60MNT composite film has faster healing of the burn wound (Fig. 6b). According to the achieved results the suitability of the fabricated system for use as a wound dressing system with controlled antibiotic delivery capability for chronic wounds like the burn wounds suggested [47].

MMT is another natural clay mineral which due to its suitable features like swelling and water uptake, mechanical and thermal behavior, and bioadhesion commonly was used for modification of some features of polysaccharides to use in biomedicinal applications [50, 51]. In one remarkable research, a new wound healing platform based on the modified MMT was developed as a new system for chlorhexidine (CLX) delivery. The main aim of this research was to limit the CLX cytotoxicity towards human fibroblasts. Therefore, in this way, the CLX was intercalated between the MMT layers, and the intercalated product (MMT-CLX) was obtained. TGA and UV–Vis spectrophotometry obtained the amount of the loaded CLX in MMT-CLX about 32% and 27% respectively. Afterward, the CS/MMT-CLX films were fabricated via casting method and then characterized with several techniques. Films loaded with neat CLX and MMT/CLX were also prepared for comparison. All prepared films revealed good antibiofilm and antimicrobial activities. Results demonstrated that the potential of encapsulation of the CLX within the MMT and CS nanocomposite can prolong its releases while preserving the antimicrobial and antibiofilm activities which confirmed the suitability of the fabricated system for localized CLX delivery in wound dressing systems [52]. In another interesting research work, the CS-silicate biocomposites were also fabricated via the solid–liquid interaction of CS and a certain type of montmorillonite (VHS), the prepared nanocomposite then was used for the controlled delivery of 5-aminosalicylic acid (5-ASA). Interlayer exchange of the CS with the exchangeable cations of the clay mineral, and then the CS precipitation on VHS surfaces proposed as an interaction mechanism for the nanocomposite synthesis. For obtaining the best performance, the 5-ASA loading capacity and its release pattern of the nanocomposite were compared with those of the 5-ASA loaded VHS or CS alone. Higher drug loading value for 5-ASA/CS/VHS (23% mass/mass) and slower drug release in comparison with both of the 5-ASA/VHS and 5-ASA/CS displayed a better performance for the nanocomposite than individual components. The obtained result could be related to a synergic effect between the polysaccharide and clay mineral. Additionally, the presence of CS in the nanocomposite structure prevented matrix disintegration (Fig. 7a). Results conclude that the prepared biocomposites are promising supports for modified formulations of 5-ASA [53].

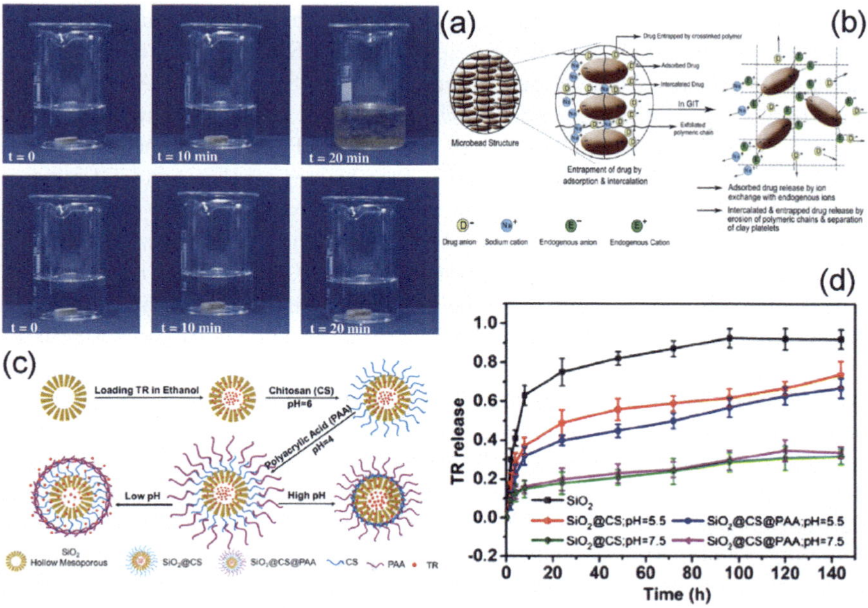

Fig. 7 Digital photo of 5-ASA/VHS (top) and 5-ASA/CS/VHS (bottom) matrices in the acidic environment at three different hydration time intervals (**a**), microbead structure and mechanism of drug release (**b**), schematic of the SiO_2@CS@PAA preparation, and the TR release pattern at two different pH values (**c**), and curves of TR release from TR-loaded SiO_2, SiO_2@CS, and SiO_2@CS@PAA NPs (**d**). Reproduced with permission from [49, 53, 54]

In the contribution of clay-polymer based systems as a drug carrier, in 2019 the new nanocomposites were designed and prepared by using the (Optigel (OPT) and Laponite (LAP)) as members of the clay family and SA and CS as biopolymers via ionotropic gelation method. The performance of the prepared systems was evaluated for oral controlled delivery of diclofenac sodium (DS). DS-LAP-SA, DS-OPT-SA, and DS-LAP-CS microbeads were prepared in three different combinations by varying proportions of clays and polymers in this research. FT-IR and differential scanning calorimetry (DSC) analysis confirmed the presence of DS in the nanocomposite structure via the appearance of its corresponding peaks. Exceptional structural and mechanical properties, in vitro drug loading efficiency, release behavior, and also swelling behavior was performed to obtain the optimum formulation. High DS loading amount and a controlled DS release pattern were detected for DS-OPT-SA clay and polymer combination. A schematic of the microbead structure and drug release mechanism is provided in Fig. 7b. The prolonged DS release was observed for the optimized formulation in rats. In vivo ulcer induction activity and anti-inflammatory potential were studied for optimized formulation and respectively the longer anti-inflammatory and biocompatibility effect was observed. These results are in accordance with the fact that the prepared microbeads effectively could be used for oral drug administration to improve the therapeutic efficacy [49]. In another study for further improving the CS-based systems as a drug carrier and obtaining the system with more pH-controlled drug delivery capability, the CS and polyacrylic acid (PAA) multilayer encapsulated hollow mesoporous silica NPs (SiO_2@CS@PAA) was fabricated. In this regard, the prepared hollow mesoporous SiO_2 NPs were loaded with triclosan (TR). Then the SiO_2@CS and SiO_2@CS@PAA NPs respectively were prepared via the coating with CS and PAA. Step by step synthesis and TR releases are presented in Fig. 7c. The change in the obtained values for the zeta potential showed success in the synthesis; about −10.4, +13.5, and −13.1 mV were measured for the zeta potential of the SiO_2, SiO_2@CS, and SiO_2@CS@PAA nanoparticles. As it is shown in Fig. 7d comparing the in vitro TR release results in pH 5.5 and pH 7.4 displayed that the PAA decoration can avoid the burst release of TR with pH sensitivity [54].

CS-based nanocomposites having Si NPs were also studied in the case of antibiotic controlled release. In this way, in the one interesting research the antimicrobial electrospun CS-polyethylene oxide (CS-PEO) nanofibrous mats having cefazolin, fumed silica (F. silica), and cefazolin-loaded fumed silica NPs were designed, prepared via electrospinning, and then their biomedical applications were studied. A diameter of about 70 ± 15 nm, 90 ± 20, and 160 ± 30 was determined respectively for the CS-PEO-1% F. silica-0.5% cefazolin, CS-PEO-1% F. silica, and CS-PEO nanofibrous mats. Both of the CS-PEO mats containing 2.5% cefazolin and 1% F. and silica-0.50% cefazolin displayed 100% anti-bactericidal activities against both *S. aureus* and *E. coli* bacteria in the antibacterial assay. Comparing the hydrophilicity, tensile strength, and cefazolin release pattern during 15 days proposed that the CS-PEO-1% F. silica-0.50% cefazolin with high tensile strength and the more sustained release is the best nanocomposite for tissue/device biomedical applications. Also, the wound healing capability of the CS-PEO-F. silica-cefazolin mat was assessed as a wound dressing scaffold toward the wounded skins of the female Wistar rats and it was detected that after ten days the wounded skins of the rats were completely healed using this mat [45].

The nanocomposites based on CS and Si-based NPs were also evaluated as a drug carrier for avoiding vaginal candidiasis. Candida albicans is the main important causative agent of leukorrhea in reproductive stage women. In this regard, the ionic gelation approach via TPP was used for the fabrication of the Cal A loaded CS NPs. Cal A is a drug-like molecule. On the other hand, marine sponge derived spicules (Spi) were successfully isolated, characterized using several techniques, and Cal A was loaded in it. About 99.9% of silica dioxide was in the composition of sponge spicules. Respectively around 65% and 38% of Cal A loading amount was determined for Cal@CS and Cal@Spi composites. In the following, both Cal A load CS and Spi were evaluated using in vitro methods. Additionally, the vaginal candidiasis induced animal model was used to study the efficiency of the drug composites. The achieved results presented a promising anticandidal performance for Cal@CS than Cal@Spi. Therefore, these types of formulations with improved antifungal activity and lesser toxic influences in comparison with oral or systemic administration can be a good way for human treatment [55].

Omega-3 fatty acids have valuable health outcomes like improvement of brain development and reduced cardiovascular diseases, as well as, have an anti-inflammatory, anti-thrombotic, antimicrobial, and antioxidant activity. Therefore, obtaining the maximum health benefit and decreasing the oxidation of omega-3 rich oils (flaxseed or fish oils) is valuable in biomedicine. In this object, a new series of SA/CS-based nanocomposite microspheres was designed and prepared. Moreover, the nanocomposite microsphere containing the CUR as a natural antioxidant with an average particle size of 153 nm was synthesized through a three-step route (oil-in-water (o/w) emulsification, gelation, and then microencapsulation. Several microscopy analyses illustrated a spherical shape for the fabricated microspheres. FT-IR and XRD analysis exhibited the existence of omega-3 rich oils in the core of the fabricated microspheres. Microencapsulation efficiency, the release profile of oils, oxidative stability, antibacterial activities, and antioxidant performance were evaluated for the fabricated microspheres. Microspheres having CUR scavenged the free radicals at a higher speed than without CUR. A more sustained release at a neutral environment in comparison with the acidic condition for SA and SA-CS-based microspheres recommend that a combination between CS and SA may be better for omega-3 rich oils encapsulation. The results of this work are in accordance with the fact that the prepared microspheres in this work could be used as safe and efficient edible platforms for hydrophobic nutraceuticals like omega-3 rich oils with a wide spectrum of antibacterial performance [56].

In the area of biomedicine, especially in drug delivery, the multifunctional nanocomposites with both diagnosis and therapy ability are valuable. Taking into account a dual-stimuli-responsive and luminescent nanocomposite based on mesoporous silica, decatungstoeuropate, and poly (N-isopropylacrylamide)-CS was prepared. Briefly, the prepared silica NPs were coated with pH/Thermo dual-responsive poly (N-isopropylacrylamide)-CS, and then the copolymer was functionalized with luminescent decatungstoeuropate. The prepared nanocarrier displayed a good red luminescence at different pHs and temperatures. About 36.0% of DOX was loaded in it. In vitro drug release assay specified that the DOX releases was pH and thermo dependent, so, acidic condition and high temperatures conditions were satisfactory for the fast release of DOX. Finally, the HeLa cell viability of over 90% and 33.6% respectively after h treatment with neat and DOX loaded composites confirmed the applicability of the prepared system for anticancer drug delivery [57].

In continuing efforts to design and preparation of the CS-based nanocomposites having Si NPs, in 2020 the Oliveira et al. fabricated the hydrogel nanocomposite of poly(N-2-vinil-pirrolidone)(PVP) having clay and CS NPs for delivery of glucantime drug. The gamma irradiation in ^{60}Co source with doses of 25 kGy was used for the crosslinking. SEM images revealed a super porous structure for the prepared hydrogels. A higher gel fraction was observed for the system with 1.0% of clay. Comparing the release profile of the glucantime from the PVP/clay and PVP/CS/clay hydrogels validated a more sustained release profile for PVP/CS/clay which is a beneficial feature for each DDS [58]. Because of the high volume of published research on the study of the CS-based nanocomposites having silica NPs for avoiding the prolongation the results of other published cases are summarized in Table 1.

2.4 *Hydroxyapatite*

HAp with a molecular stoichiometric formula of $Ca_{10}(PO_4)_6(OH)_2$, is inorganic material commonly used in tissue engineering because of its special properties, like osteoconductivity, bioactivity, noninflammatory, nonimmunogenicity, and biocompatibility. The structure of the commercially accessible HAp is alike to the mineral phase of human bone. The addition of the HAp into the polymeric system increases its mechanical strength. Nanosized HAp has low crystallinity, compatibility with soft tissues, highly active surface, and can uniformly disperse within the polymer matrices. Owing to these advantages, nano-HAp has received more attention for use as a drug carrier in a controlled manner [59, 60]. Moreover, the smaller size of nano-HAp in comparison with red blood cells leads to its capability for easy transfuse in the blood cycle [61]. However, unfortunately, the weak interactions between HAp NPs and drugs lead to its burst release. Therefore, it is hypothesized that the combination of HAp with polysaccharides be an operative route for dissolving this drawback [62].

With the aim of obtaining a new oral drug delivery vehicle based on HAp and CS, a new ternary system comprised of CS/hyaluronic acid (HA)/HAp hydrogels were designed and fabricated via gamma rays irradiation as a crosslinking agent. After the characterization of the prepared system by using the FT-IR and SEM analysis, its capability for 5-FU oral delivery was studied. Additionally, the prepared hydrogels were investigated from the swelling viewpoint. The results of in vitro drug delivery studies exhibited that the addition of HAp and HA in the structure of hydrogel improved the drug loading amount (from 3.2 to 12.74 mg/g) and release capability. The results exhibited that the HA/CS/HAp hydrogel could be proposed as an appropriate candidate for drug transporting in the human body [59]. Similarly, irradiation was also used for the preparation of the SA/CS/HAp nanocomposite because of its safe nature. In this regard, the porous SA/CS/HAp nanocomposite hydrogel having different values of HAp; 0.6, 2.0, 3.5, and 5.0% wt/v were synthesized by using gamma radiation as a crosslinking agent and a free radical initiator. Gel fraction analysis exhibited that an increase in the HAp content increased the crosslinking yield. This is because of the high tendency of HAp for the formation of H-bonding with -OH groups and the creation of the crosslinked structure. After evaluation of the success in the synthesis, the efficiency of the prepared samples as a DOX delivery vehicle was investigated in the oral delivery route. Additionally, the influence of initial DOX concentration, as well as, the HAp % on drug release behavior was studied. A decrease in the drug release percentage was observed as a result of an increase in drug concentration which may be owing to the irregular drug spreading within the particles. Also, a higher HAp percentage resulted in a higher DOX release that may be related to the dissolving of the HAP at lower pH that lead to the releasing of the loaded biomolecules or drugs. Overall, this investigation gives insight into the fact that the drug releases occur as a result of water penetration in the polymeric network of hydrogels, hydrogel swelling, drug dissolving, and finally drug diffusion through the hydrated polymer matrix into the release medium [61].

In recent years, development of the biodegradable composites with the capability to suppress or remove the pathogenic micro-biota or modulate the inflammatory answer has received great deals of attention due to its importance in limiting/repairing periodontal tissue destruction. In view of the above background and for periodontal treatment, a non-steroidal anti-inflammatory drug-loaded biodegradable CS/PVA/HAp electrospun (e-spun) composite nanofibrous films and mats were fabricated. The effect of heat treatment on the morphology of fibers and films was also studied. The smooth surface was detected for the control e-spun fibers, however, after heat treatment, the smoothness and uniformity of fibers were missed. Because of heat treatment, the thermal stability of both films and fibers was enhanced, but the pore size and fiber diameter was decreased. Also, the differences in the in vitro drug release patterns for the heat treated and non-heat treated (control) nanofibrous films and mats having different concentrations of piroxicam (PX) were studied. In vitro PX release pattern evaluation at physiological pH 7.4 for not heat treated and heat treated e-spun fibers, films and mats were analyzed. The burst PX release profile was detected for both films and fibers which was followed by later sustained-release patterns. Also, the release rate of control was slightly higher than heat treated e-spun mats. In vitro cytocompatibility assay toward VERO cell line of epithelial cells indicated the non-cytotoxic nature for all of the synthesized materials. The current observations suggested the applicability of the CS and HAp based systems as a potential candidate for periodontal regeneration [63]. In similar work for study the potential of HAp as a local antibiotic delivery agent, in 2021 He et al. designed and fabricated the CS-coated polytrimethylene carbonate (PTMC)/polylactic acid (PLLA)/oleic acid-modified HAp (OA-HAp)/vancomycin hydrochloride (VH) microsphere scaffold via a double emulsion procedure (W/O). A schematic of CS-coated core/shell structure microsphere synthesis is provided in Fig. 8a. Surface modification of HAp can enhance the interaction between the NPs and the polymer, as well as, it can overcome its poor dispersibility in the polymer matrix. More studies approved that the insertion of PLLA, VH, and OA-HAp into PTMC microspheres improved the surface properties scaffold (Fig. 8c). Evaluation of the composite microsphere scaffold surface showed its suitability for osteoblast adhesion. For evaluation of the biocompatibility of the scaffold, the osteoblast cells were co-cultured on the scaffolds. As it is presented in Fig. 8b the study of the osteoblast fluorescence images depicted that all of the scaffolds have good biocompatibility and could be utilized as supporting materials for osteoblast. Based on the findings of this work the authors proposed that the CS-PTMC/PLLA/OA-HA/VH microsphere scaffold could be a talented candidate for bone tissue engineering usages [64].

Ciprofloxacin is another antibiotic that the effect of the HAp NPs insertion in the structure of the carrier was investigated on its release behavior. In this way, the HAp NPs were synthesized with a size of about 20–30 nm; TEM and XRD results. The energy dispersive spectroscopy (EDS) approved the existence of HAp NPs in nanocomposites structure. Comparing the ciprofloxacin release behavior for the neat and HAp containing CS/κ-carrageenan complex exhibited that the introduction of the HAp leads to obtaining a drug transporting system with a more sustained release manner. So, for the pristine CS/κ-carrageenan complex about 98% of ciprofloxacin

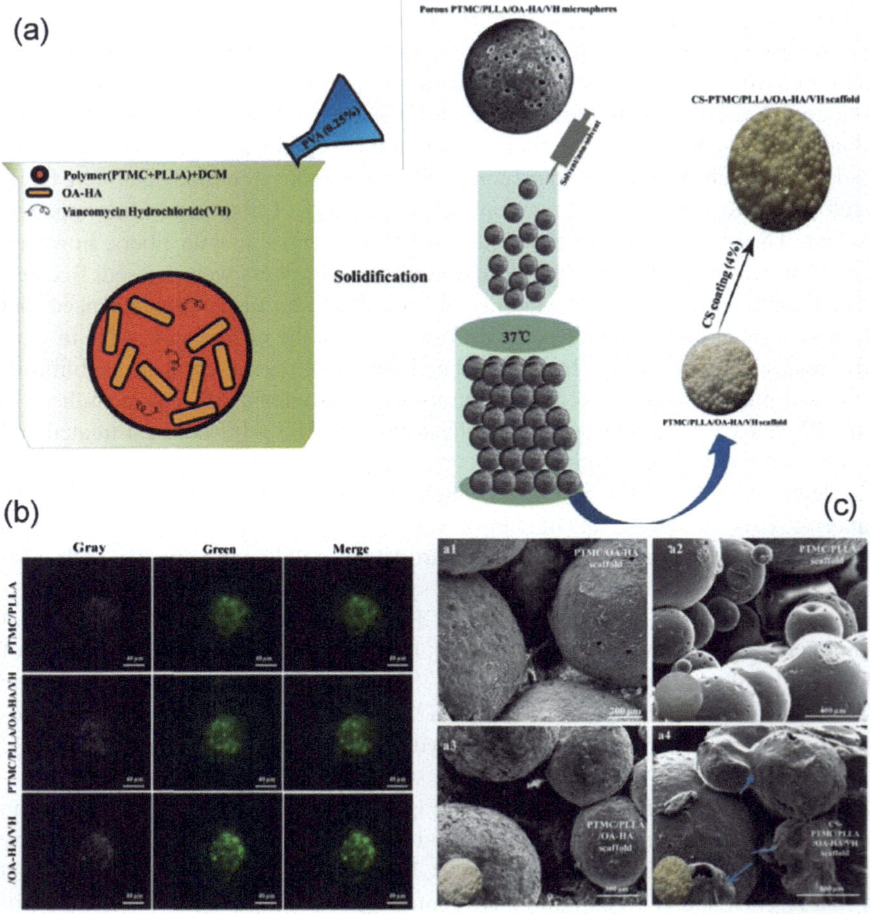

Fig. 8 Schematic of CS-coated core/shell structure microsphere synthesis (**a**), Osteoblast fluorescence detection (**b**), and SEM images of scaffolds (**c**). Reproduced with permission from [64]

was released during 120 h, while, the release amount was about 66% and 52% respectively for the hydrogel nanocomposites containing a low and high amount of HAp. It is hypothesized that the obtained sustained ciprofloxacin release profile for the CS/κ-carrageenan complex hydrogel as a resulting of the HAp embedding in the hydrogel network is a confirmation for prolonged-release behavior of the nanocomposites [62].

HAp based nanocomposite was also investigated as an orthopedic implant material after combination with CS and then their Vitamin B12 delivery ability was investigated. A further study was performed to evaluate the effect of the Ca-deficient hydroxyapatite (CDHA)-polymer interaction on the Vitamin B12 release profile of the monolithic membrane fabricated via CDHA/CS nanocomposite. In this consideration, the kinetics of release was evaluated to explore the effect of the different parameters for example ex-situ and in situ ways, and various CDHA values. The higher diffusion exponent (n) value was measured for the in situ synthesized membranes. Moreover, the n value of the in situ synthesized membranes increased with increasing the amount of CDHA, but it stayed in the range of below 10 wt.%. The obtained results specified that the mechanism of drug diffusion is changed via the CDHA-CS interaction and is affected via both the amount of the CDHA in the membrane and synthesis route. Conversely, the prepared membranes through the in situ process exhibited a lower permeability (P) value, and the value of P decreased and then increased as a result of CDHA amount increasing respectively in the lower range and higher 10 wt.%. The summation of the observed results displayed that based on the CDHA nanofiller concentration and the synthesis process, the CDHA as nanofillers could act as either diffusion barrier or diffusion improver for the CDHA/CS membranes [65].

In one interesting research effort in the bone tissue engineering area, a biocompatible scaffold based on HAp nanocomposite was designed and prepared through a multi-step method for bone tissue engineering. In this regard, a CS-graft-poly(acrylic acid-coacrylamide)-n-HAp was prepared via freeze-drying method and used as a matrix of the scaffold. The output of the crystallinity, elemental analysis, mechanical properties, and pores size displayed that the prepared system has the capability for using as a scaffold. A maximum of 97.6% of celecoxib loading was obtained for the sample with the highest amount of HAp. Celecoxib as a model of poorly water-soluble drug with anti-inflammatory, antipyretic, analgesic, and activities was capsulated via the hydrogen and ionic bonding within the fabricated scaffolds. The schematic of the CS-graft-poly(AA-co-AAm)/n-HAp interaction with celecoxib, and drug release profile is presented in Fig. 9a. Further study on the celecoxib delivery displayed two regions for drug release; a low initial burst release which followed up to 14 days with a sustained release (Fig. 9b). Performing the MTT assays toward HUGU fibroblastic cells exhibited an improvement in the biocompatibility of scaffolds as a result of the HAp incorporation in their structure (Fig. 9c). The outcome concludes the efficiency of the fabricated nanocomposite scaffolds by proposing implants and drug delivery platforms in bone tissue engineering encouraging clinical applications [66].

In similar research from the drug type viewpoint, the CS modified HAp nanocarriers were fabricated and its potential for celecoxib delivery was studied. After the characterization of the prepared system, the in vitro biological analyses were performed to investigate the influence of celecoxib-loaded NPs on colon cancer cell morphology, proliferation, cellular uptake, apoptosis, and the cytoskeleton. The experimental test results displayed narrow and small hydrodynamic size distributions, high entrapment drug efficiencies (approximately 59% of entrapment efficiency obtained via high-performance liquid chromatography (HPLC) analysis), hemocompatibility, and sustained celecoxib release pattern. Notable apoptosis, antiproliferation, and time-dependent cytoplasmic internalization of celecoxib-loaded HAp-CS NPs were detected after treating the HCT 15 and HT 29 colon tumor cells with

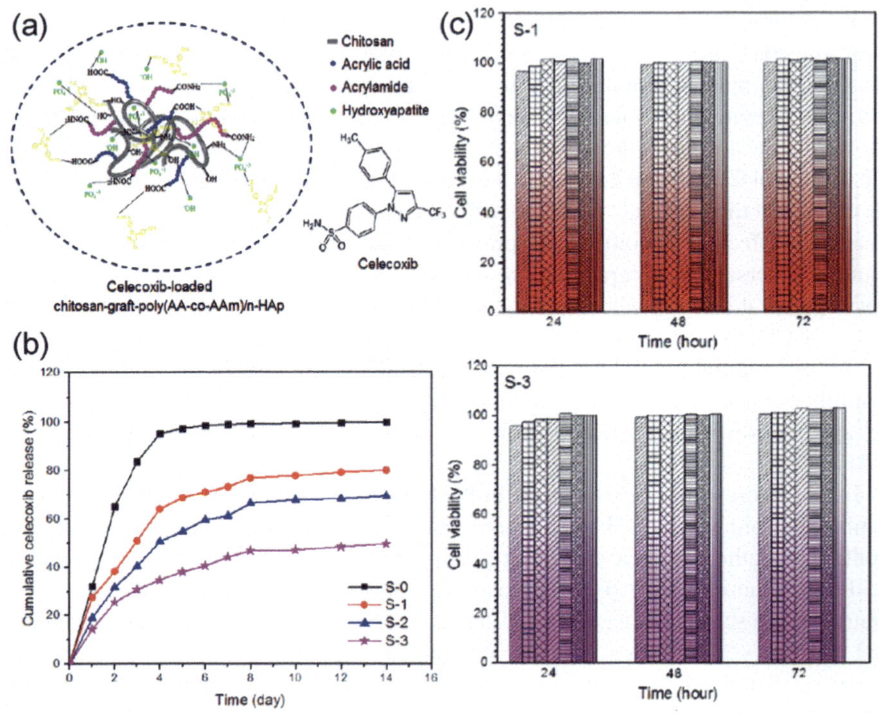

Fig. 9 The schematic of celecoxib loading on CS-graft-poly(AA-co-AAm)/n-HAp (**a**), drug release pattern for the nanocomposite scaffolds (**b**), and obtained cell viability in MTT assay (**c**). Reproduced with permission from [66]

it. Eventually, the in vivo investigations revealed enhanced tumor growth inhibition after treatment with celecoxib-loaded HAp-CS nanocomposite compared to the cells treated with free celecoxib. The results of this study showed that HAp-CS nanocomposite is a safe and efficient tool for the delivery of therapeutic agents to colon cancer [60].

2.5 *LDH NPs*

LDH is one member of the inorganic nanomaterials that belong to the group of hydrotalcite-like compounds. In the lamellar structure of LDH, there are exchangeable anions with positive layers having Mg^{2+} and Al^{3+} ions which are surrounded by hydroxyl groups [67–69]. The two-dimensional LDHs are usually represented with the common formula of $[M^{2+}{}_{1-x}M^{3+}{}_{x}(OH)_2](A^{n-})^{x/n}\cdot mH_2O$, which respectively M^{3+}, M^{2+}, and A^{n-} are trivalent cations, divalent cations, and interlayer anionic molecule [70]. Additionally, the present H_2O molecules between the LDH layers lead to coupling each layer with another layer through the H-bonding. Exceptional biocompatibility, low price, large surface area, and great surface positive charge are some of the positive features of these anionic clay for biomedicinal applications. An easy degradation of LDHs in the acidic medium is a reason for excellent biodegradability and low toxicity of their even at higher dosage [67].

Today because of the increasing trend of cancer development, obtaining the new targeted drug carriers is valuable. Given these premises, in 2020, Anirudhan and coworkers designed and fabricated the new dual-responsive system comprising of isocyanato modified layered double hydroxide (LDH-NCO), folic acid conjugated thiolated chitosan (TCS), and AuNPs for targeted and efficient chemophotothermal therapy of breast tumors. In fact, the used FA is a targeting ligand with the enhanced binding capability to the folate receptors that are overexpressed on some of the cancer cells. For this aim, briefly, at first, the FA was conjugated with TCS. In the second step, the surface of the previously prepared LDH (Mg/Al) was functionalized via 3-isocyanatopropyl triethoxysilane and dimethylformamide that lead to introducing the NCO groups on the LDH surface and enhancing the reaction sites. In the continue, the LDH-NCO was functionalized with modified CS, and LDH-NCO-TCS nanocomposite was prepared. The formation of SH bond on the surface of LDH-NCO-TCS by using the FA-Cys was an aiding tool for its facile interaction with AuNPs. Eventually, the LDH-NCO-TCS/AuNp nanocomposite was achieved. After the chemical characterization of the synthesized drug carrier, its capability was evaluated for DOX delivery. In vitro drug delivery test exhibited that the maximum of 94.6% DOX was loaded via static and hydrogen bonding interactions. Evaluation of the obtained DOX release pattern showed the stimuli-responsive DOX release pattern with more releases of it in pH 5.5 compared to 7.4. In the following, the in vitro analysis illustrated that the presence of AuNPs tolerates high temperature via the NIR light irradiation and improves the photothermal treatment. MCF7 cells were selected for the evaluation of the DOX loaded LDH-NCO-TCS/AuNPs capability in cancer cell killing. The negligible hemolytic activity was observed in the blood compatibility studies. Additionally, the notable cell killing was detected in G0/G1 phase for the treated MCF7 cells with DOX loaded LDH-NCO-TCS/AuNp via flow cytometry analysis (Fig. 10a, b). These results open an insight that the combination of cancer therapy is a promising way for maximizing the therapeutic efficacy and extended this approach [67]. Similarly, in one interesting research work, the pH-sensitive bionanocomposite beads containing the FA intercalated LDH (Mg/Al) and CS were prepared and its FA release mechanism was described in detail. In this regard, in one co-precipitation procedure, the FA was intercalated in the LDH structure via anion exchange and co-precipitation approaches. FT-IR and XRD analyses confirmed the presence of FA between the LDH layers respectively through the presence of functional group peaks belonging to the FA and enhancing the LDH layer space. In the following the FA intercalated LDH (LDH-FA) was combined with different amounts of CS via the ultrasonication in a greenway and the nanocomposite beads were obtained. Study of the FA releases in the simulated condition to the gastrointestinal tract; pH 1.2, 6.8, and 7.4 displayed a good shield effect of polymer in the acidic pH, as well as, the rate of FA releases was decreased with the addition of CS concentration. So, the FA

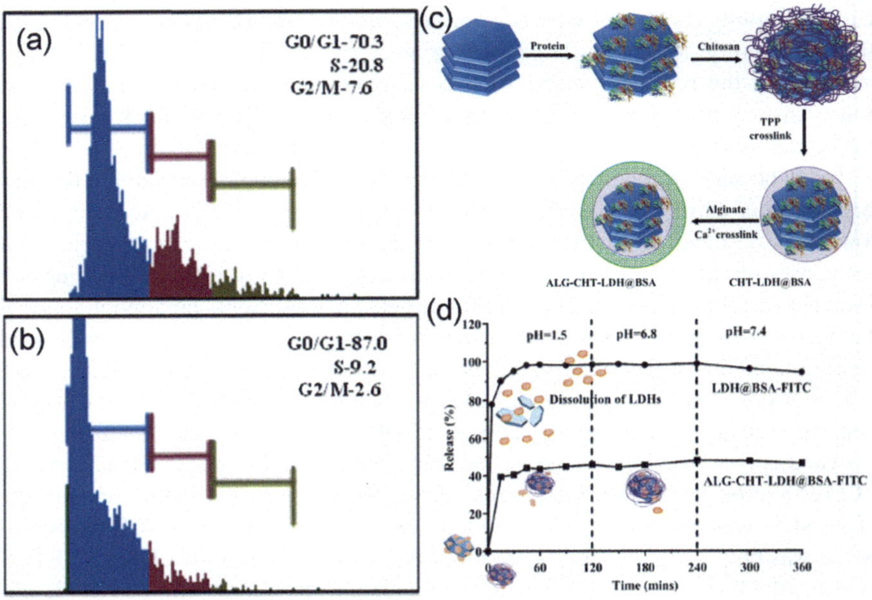

Fig. 10 The quantified status of the MCF7 cells for nontreated (**a**) and LDH-NCO-TCS/AuNp treated (**b**) cells after 24 h incubation at 37 °C, a schematic illustration of the SA-CS-LDH NPs synthesis (**c**), and profiles of BSA-FITC releases from LDH@BSA-FITC and ALG-CHT-LDH@BSA-FITC in simulated oral delivery condition (**d**). Reproduced with permission from [67, 72]

releases assay exhibited 81% of FA releases in the stomach condition for LDH-FA hybrid within 2 h. But the present CS in the structure of nanocomposite prolonged the release rate, so, 68.69% and 57.09% of FA were released after 8 h respectively for LDH-FA/Cs1 and LDH-FA/Cs2. The achieved results from the kinetic study exhibited that the FA releases could be controlled via LDH surface diffusion or intraparticle diffusion. Fitting the obtained experimental results with linear form of the several kinetic models and the calculated R^2 value for them exhibited that the several kinetic models and their related mechanisms like ion exchange, diffusion-controlled of a flat surface by heterogeneous sites, concurrent leaching of the FA anions, and LDH intraparticle diffusion may occur phenomena in the FA release from LDH structure. Findings demonstrated that the release of the intercalated drugs in the structure of the LDH hybrid was controlled via its modification [70].

In the continued efforts for obtaining a novel bio-hybrid DDS based on LDH and CS merit, Mg/Al-NO_3 LDH intercalated either with a drug-loaded phospholipid bilayer (BL) or IBU was fabricated and afterward embedded in CS beads via dropwise addition of the acidic mixture in alkaline solution. That several characterization methods were used for clarifying the presence of BL and IBU in the interlayer space of LDH, as well as, the encapsulation of LDH in the beads. So, the presence of LDH on the surface of the bead was clarified via SEM. The evaluation of the release pattern was performed for sodium ibuprofen, as well as, for the association of 17 β -estradiol inside the negatively charged BL which was encapsulated in the LDH/CS hybrid structures. The performance of the fabricated new biohybrids demonstrated the ability of these types of systems as sustained-release platforms for a wide range of neutral and anionic drugs [71].

Besides the drug delivery, the LDH NPs with low cytotoxicity and good biocompatibility have been shown their exceptional capability to enhance the immune response via protein antigen delivery. However, the enzyme/acidic degradation in the stomach and low bioavailability in the small intestine are two main limitations of LDHs for use as the oral delivery of vaccine platforms. By considering this, in one interesting research work, and with the aim of achieving the improved performance for LDHs, Yu et al. fabricated the LDH encapsulated with SA-CS (SA-CS-LDH) to improve oral protein delivery of vaccine (BSA). Summary at first the antigen was loaded in LDH NPs and the LDH@antigen was obtained. In the next step, the fabricated LDH@antigen was coated by using the CS through crosslinking with TPP. The formation of the CS layer was done due to its helpful character for enhancing the adsorption and uptake of LDH@antigen in the small intestine. In the following alginate coated on the previously prepared hybrid via the crosslinking with $CaCl_2$. The SA coat was used to inhibit the acidic degradation in the stomach. The schematic of the SA-CS-LDH NPs synthetic process is shown in Fig. 10c. DLS analysis obtained the size of the SA-CS-LDH@BSA about 327 nm. The coating with polymers was confirmed via TEM analysis. In the following, the protein release pattern was investigated for the neat LDHs and SA-CS-LDH at various pHs. Comparing the obtained release profiles in the simulated oral delivery way (Fig. 10d) exhibited that the SA-CS-LDH NPs can partially protect from the protein release at the acidic condition (pH 1.2). Flow cytometry was performed against macrophage and Caco2 cells to evaluate the cellular uptake efficiency respectively for the macrophages and intestine cells. From this analysis, it was detected that the CS-coated LDHs considerably improve the internalization of the protein into the Caco2 and macrophage cells [72].

2.6 GO and GQDs

Recently GO as an oxidized carbon layer with advanced properties, like a large two-dimensional plan surface area, abundant oxygen-containing functional groups, hydrogen bonding interactions, more biocompatibility, water solubility, and low price is converted to one of the ideal materials for delivering a variety of drugs and received tremendous attention in the biomedical applications. High drug loading capacity owing to the hydrophobic interaction, π-π stacking, hydrogen bond formation, and the capability for the plasma membrane traversing are other positive features of GO as a drug carrier [19, 73–76]. Although the GO sheets provide a stable suspension in water, the poor solubility of the GO sheets in a biological medium, particularly when drugs were loaded in it is one main driving force for GO surface modification. Besides the surface modification, GO can be used as a filler in the structure of the nanocomposites. Moreover, the GO has exceptional photothermal properties at NIR radiation and it can absorb the light radiation and change its energy into heat that is used for the bacterial pathogens killing [15, 77–80].

In recent years, chemotherapy looks to introduce and develop new methods of cancer-treating via manipulating the nanocarriers with pharmacokinetic features. Among the recently introduced methods, codelivery is one of the popular methods for cancer therapy with high efficiency. The synergistic influence of improved cancer cell inhibition and the low side effects are two main advantages of combinatorial therapy. Therefore today several research groups were tried to display the codelivery benefit for cancer therapy via designing and preparation of the new systems for the codelivery of anticancer drugs. Of these, graphene with its special structure, having a π-conjugated system, and large surface area could encapsulate more than one therapeutic agent and therefore is a hopeful candidate in the formulation of co DDSs. For example, 5-FU and CUR-loaded CS/reduced graphene oxide (CS/rGO) nanocomposite was fabricated through the facile chemical method; crosslinking with TPP. Drug delivery tests exhibited a higher entrapment efficiency; about >90% for CS/rGO nanocarrier. Both CUR and 5-FU peaks were observed in the XRD spectrum of the composite at 8.8° and 28.6° respectively that approve the preparation of dual-drug-loaded composite. It is demonstrated that the release mechanism follows from the three suitable regions for each drug; in the case of 5-FU the first region fitted well to the Higuchi kinetics, the second region follows the first order, whereas the third region of the release profile obeys the Korsmeyer-Peppas model. However, in the case of CUR; the first region obeys from zeroth-order kinetics, the second region supported the Higuchi model, and the third region follows first-order kinetics. The observed release patterns depended on the interaction between CS/rGO and each drug molecule. About 80.3% of NIH 3T3 mouse embryonic fibroblast cells viability after treating with 40 μg/mL of drug-free CS/rGO clearly showed the safety of CS/rGO against normal cells, moreover, an IC50 of 23.8 μg/mL approved the efficient cytotoxicity for dual-drug-loaded nanocomposite against HT-29 colon tumor cells. Summation of the obtained results exhibited that the CS/rGO is a hopeful carrier for multiple therapeutic deliveries in clinical applications [81]. In the area of cancer therapy, a major challenge is the ability for producing the drug carrier with high biocompatibility, and controlled drug delivery features which are often undertaken by biocompatible stimuli-responsive polymers. In one interesting research work, the nanocarrier of functionalized GO with CS and SA was fabricated via an electrostatic layer-by-layer self-assemble procedure. The functionalization of GO was confirmed via zeta potential and atomic force microscopy (AFM) techniques. The AFM analysis exhibited that the thickness of pure GO increased from 2 nm to about 60 nm after functionalization with CS and SA. Moreover, the zeta potential of −33.25 mV, 24.98 mV, and −26.52 mV was calculated respectively for GO, GO-CS, and GO-CS/SA which proposes that the SA was crosslinked successfully with GO-CS. Additionally, notable stability of GO-CS/SA in PBS buffer, distilled water, and cell culture medium having 10% serum (Fig. 11a) verified the successful modification of GO, as well as, its potential for biomedicinal applications. 70.16% of DOX was loaded in GO-CS/SA. As it is reported in Fig. 11b a notable pH-sensitive drug release

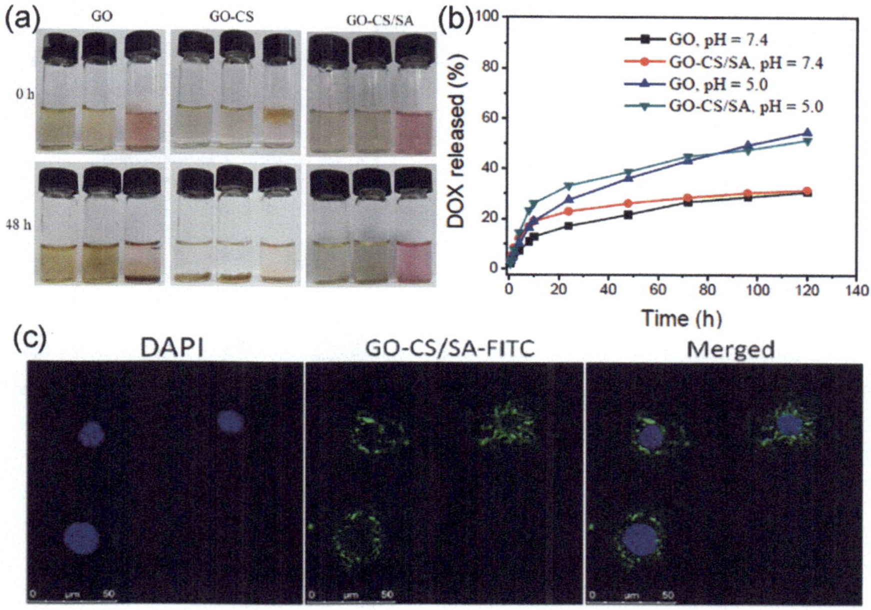

Fig. 11 The obtained digital images of GO, GO-CS, and GO-CS/SA in purified water (left), phosphate buffer saline (PBS, middle), and PRMI-1640 cell culture with 10% serum (right) (**a**), obtained DOX release profiles from drug-loaded GO and GO-CS/SA in PBS buffer (pH 7.4, 37 °C) and (pH 5.0, 37 °C) (**b**), fluorescence image of GO-CS/SA intracellular MCF-7 cells uptaking after 2 h treatment with GO-CS/SA (**c**). Reproduced with permission from [77]

pattern was observed for DOX loaded system, so, the percentage of DOX release in pH 7.4 was higher than pH 5.0. Additionally, the in vitro biological studies approved that the DOX loaded GO-CS/SA nanocomposites can internalize in MCF-7 cancer cells, localized in the cytoplasm, and released the DOX within theirs. Therefore, the above-mentioned results and the observed cytotoxicity output approved the ability of GO-CS/SA nanoparticles for antitumor drug delivery to the cancerous cells [77].

Similarly, in investigated drug type, another research work highlighted the potential of GO-based functionalized CS polyelectrolyte for targeted and pH-dependent delivery of DOX. In this regard at first, the GO was changed to amine-functionalized GO (AGO) via reaction with triethyl tetramine (TETA). AGO acted as a cationic polyelectrolyte. Afterward in the presence of AGO, the AA and itaconic acid monomers were grafted to the hydroxyl group of CS in the CS conjugated with FA (FA-CS) form by using the ethyleneglycol dimethacrylate as crosslinker, and chemically modified CS (CMCS) was formed. A schematic of the FA-CMCS/AGO synthesis is provided in Fig. 12a. The 95.0% of DOX was loaded in FA-CMCS/AGO and released at a higher rate at pH 5.3 compared with physiological conditions; pH 7.4. L929 cell viability after treatment with FA-CMCS/AGO approved its biocompatibility. Additionally, the viability of HeLa and MCF7 cells was studied via MTT assay. The obtained results displayed that the prepared drug carrier has the potential for clinical applications for

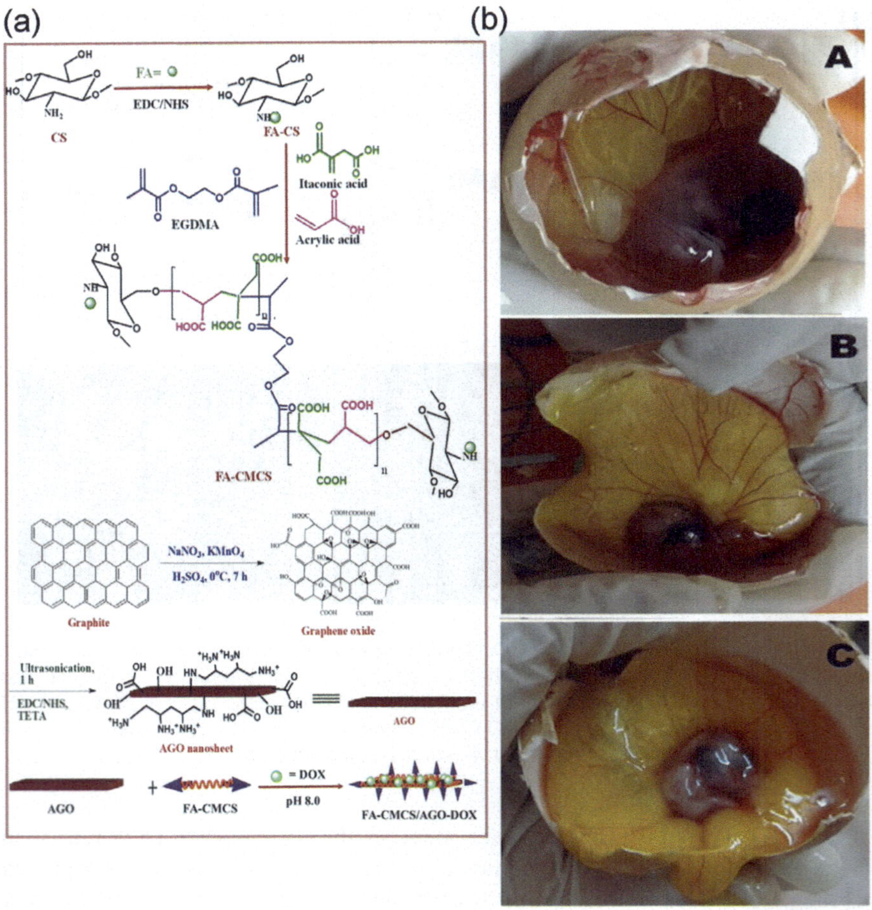

Fig. 12 Graphic illustration of the FA-CMCS/AGO synthesis and DOX loading on it (**a**) and CAM assay image for (A) control, (B) FA-CMCS/AGO, and (C) FA-CMCS/AGO-DOX (**b**). Reproduced with permission from [82]

targeted drug delivery. Study the cellular uptake via CLSM exhibited the passage of the FA-CMCS/AGO-DOX within the cancerous cell's cytoplasm under the effect of the FA targeting ligand. As a supplementary test the in vivo anti-angiogenesis activities of the FA-CMCS/AGO and FA-CMCS/AGO-DOX were studied via chorioallantoic membrane (CAM) assay. As it is shown in Fig. 12b, respectively 5.5, 5.2, and 2.2 g percentages were found for the hemoglobin percentage of control, FA-CMCS/AGO, and FA-CMCS/AGO-DOX. The results prove the non-toxicity of the synthesized nanocomposite and releases of the DOX from the drug-loaded carrier without the premature drug release. Therefore the outstanding work showed that the FA-CMCS/AGO is a talented carrier for controlled and targeted anticancer drug delivery [82].

HA is another targeted ligand for CD44 which is used in the structure of the DDSs to recognize the particular recognition of cancer cells and enhance the performance of anticancer drug delivery. Low cost with better stability and also the water solubility are some advantages of the HA compared to the other targeting counterparts, like antibodies or peptides. By considering this, initially, the GO was modified with CS to increase its biocompatibility and then was conjugated by HA to be a targeted drug carrier. Targeted drug delivery in the systems having HA is due to the special recognition of HA to the overexpressed transmembrane glycoprotein CD44 on the different cancer cell surfaces. The SNX-2112 as an anticancer drug; the Hsp90 inhibitor was loaded into GO-CS-HA about exceeded 110%. Evaluation of the drug release showed considerably higher SNX-2112 release in an acidic condition in comparison with the physiological conditions. In the lower concentration, the GO-CS-HA had negligible influence on the red blood cells (RBCs) lysis and blood coagulation and exhibited low cytotoxicity toward NHBE and A549 cells. The observation of the biocompatibility for the GO-CS-HA/SNX-2112 against normal human bronchial epithelial cells (NHBE cells), as well as, inhibiting and killing the A549 cells confirmed the efficiency of GO-CS-HA/SNX-2112 for cancer therapy. In vivo histological examinations and blood property analyses were also performed to investigate the cytotoxicity of the fabricated materials against the main organs in SD rats. The in vivo test results displayed that although an inflammatory answer was established in the short-term, however, any severe long-term injury was not observed after treatment with GO-CS-HA/SNX-2112. Consequently, GO-CS-HA presented excessive potential as an efficient and non-toxic drug carrier with low side effects for tumor therapy [19].

The nanocomposites based on CS are the value systems for wound healing owing to the CS distinctive features like the increased speed of the tissue regeneration, stimulating blood coagulation, enhancing O_2 transmission rate, avoiding microorganisms, and increasing the speed of epithelialization. In view of the importance of CS-based nanocomposites having the GO NPs as a wound healing system, in 2020 Nowroozi and coworkers fabricated the porous CS/GO/CUR nanocomposite via the freeze-drying method and proposed it as 3D scaffolds for wound dressing. It is expected that the hydrophilic carboxyl, epoxide, and hydroxyl functional groups on the GO surface improve their biological interactions and improve the scaffold hydrophilicity that is approved via the swelling ratio and water vapor transmission rate (WVTR) results. In this regard, 0–1.5%wt. of GO/CUR nanocomposite incorporated into CS matrix to obtain the appropriate nanocomposite. The FE-SEM image before and after cell seeding exhibited that the cells are well attached and proliferated on the surface of the scaffolds. A study of the CUR releases in PBS solution (pH 7.4) at 37 °C by UV–Vis spectra exhibited that the CUR release from all of the scaffolds in the initial 12 h has more rate and then decreased slowly. The morphology and viability of NIH/3T3 cells are studied via FE-SEM and WST assay kit analysis. The achieved FE-SEM images exhibited the good spreading of NIH/3T3 fibroblasts on the nanocomposite scaffolds. The measured 95% of NIH/3T3 cells viability confirmed the biocompatibility of the prepared nanocomposite. The outcome evidenced that the fabricated CS/GO/Cur scaffolds are excellent candidates for wound dressing applications due to the mentioned features, as well as, cell growth and proliferation extensively. Therefore the outcomes direct that the CS/GO/CUR nanocomposites have the potential to be applied as a new wound dressing platform [15].

Migraine is an extreme headache that associates with nausea, vomiting, and sensitivity to light. In 2020 Jafari et al. established that CS-GO nanocomposite could be used for controlled release of the Sumatriptan Succinate (SS) drug that is a commonly used drug for migraine attacks. treatment. In this way, the nanocomposite hydrogel beads of CS/TPP/GO were prepared with different amounts of GO (0–20%wt.) via physical interactions between GO nanosheets and CS hydrogel. After characterization of the prepared system via some of the common analytical methods, the amount of drug loading was optimized by considering the influence of the effective factors including pH, temperature, adsorbent amount, and contact time. After doing the in vitro tests the Langmuir isotherm ($R^2 = 0.9976$) was determined as an adsorption mechanism with the highest adsorption capacity of 45.4 mg/g. A study of the antibacterial activity on standard strains of *E. coli* and *Staphylococcus* displayed that the modification of the GO with CS increases the antimicrobial activity and the prepared beads have antibacterial activity. Eventually, the obtained HEK 293 cells viability after 48 h treatment with CS/TPP/GO/SS beads displayed that the CS/TPP/GO hydrogel beads containing SS had no important cytotoxicity to HEK 293 cells. Findings support that the CS-based nanocomposites could be considered as a promising vehicle for improving drug efficacy and migraine therapy [83].

CS-based nanocomposites having GO NPs were also investigated for local biomimetic delivery owing to their keeping local higher concentration and avoiding side effects. With the aim of evaluating the biomimetic local pulsatile delivery of Teriparatide, in one interesting example in 2020 Wang et al. at first, designed and suggested the concept of Teriparatide dose and time manner on bone marrow mesenchymal stem cells which resulted from ovariectomized rats (OVX-BMSCs). Afterward, the NIR light-sensitive, CS/rGO hydrogel films were prepared via electrodeposition. The photothermal conversion was utilized for achieving the local delivery of Teriparatide to develop regeneration of osteoporotic bone defects via Teriparatide biomimetic local pulsatile delivery. As it is presented in Fig. 13a, the critical-sized calvarial bone defect models in osteoporotic rats were used for the evaluation of pulsatile delivery of Teriparatide potential to more approve the influences for the intermittent administration of Teriparatide on osteogenesis and angiogenesis in vivo. In fact, CS/rGO with the stimuli-responsive feature was applied for drug delivery, which can speedily obtain light-to-heat conversion under the low-power NIR irradiation to trigger delivery systems (Fig. 13b, c). Neat CS/rGO hydrogel film was utilized as a control vehicle. Each day for 10 min NIR light was irradiated locally to the calvarial defects centers. As well, a temperature detector was administered avoiding the superheat damaging and the inactivation of Teriparatide via keeping the local temperature around 45 °C (Fig. 13b). 1 h after NIR irradiation, the serum Teriparatide level was determined (Fig. 13d). The serum Teriparatide level displayed a pulsed fluctuation within the 1-week detecting cycle. In comparison with the control group, the blood concentration of Teriparatide was kept at a relatively more level in the constant release group. In the continue, the histological assay and micro-computed

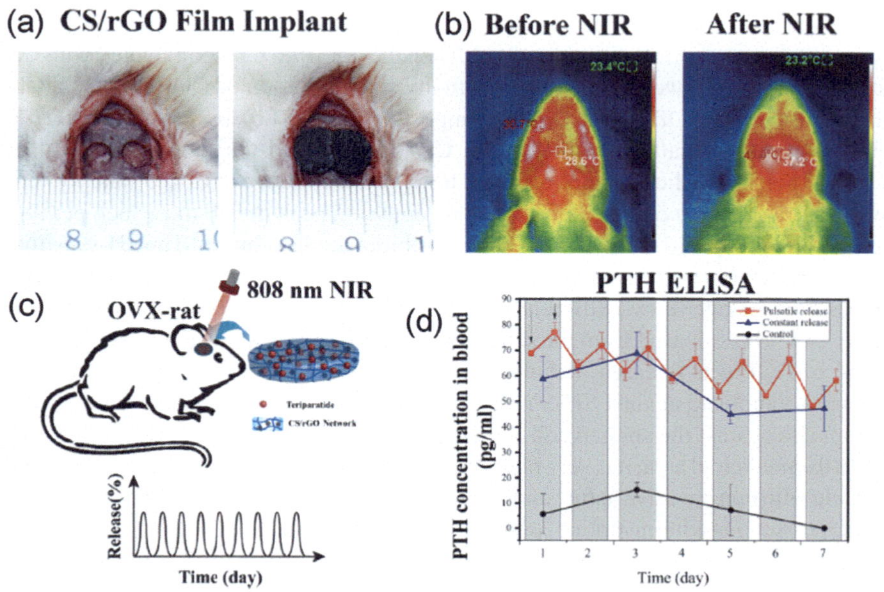

Fig. 13 Implanted CS/rGO hydrogel films into calvarial defect models of osteoporotic rats (**a**), achieved photothermal conversion via low-power NIR irradiation (**b**), representation graphs of the in vivo experimental design (**c**), and the measured serum levels of Terissparatide by using the PTH ELISA kit (**d**). Reproduced with permission from [84]

tomography (Micro-CT) clarify that the NIR light-responsive CS/rGO films obtain a satisfactory influence for bone as a complex tissue generation via biomimetic Teriparatide delivery. Furthermore, a greater density of blood vessels is detected between the center of the defect region and freshly formed bone. Combining the results of this research display the development of the physiological pulsatile secretion of Teriparatide through the stimuli-responsive influence of the biomaterial and provides a novel approach for improving osteoporotic bone regeneration [84].

Up to now numerous materials proposed as drug carriers but easily functionalization with other materials, water solubility, and good cell permeability makes carbon dots (CDs) an excellent candidate for drug delivery. Bottom-up and top-down methods are two common ways for GQDs synthesis. Another advantage of the CDs in drug delivery is their capability for using as makers or tracers for monitoring the drug release metabolism. In fact, CDs containing systems help us to find the followed path via drug encapsulated nanocarriers to obtain information about how these drug molecules are delivered to the target site [85–87]. Because the structure of GQDs is completely composed of "carbon" atoms they are safer than inorganic quantum dots that have toxic metal ions in their structure [88, 89].

Specific drug delivery has a significant value in medicine because of its negligible side effects on normal cells and also needing a low dosage of the drug. Hence in 2020 Mathew et al. designed and developed the CS/CDs-based nanocomposite to evaluate its efficacy for sustained delivery of dopamine. In this way, firstly CDs were fabricated via carbonization of CS. Then the prepared CDs conjugated with CS and CS/CD matrix were achieved. Eventually, encapsulation of DOP in the CS/CD matrix formed a DOP@CS/CD nanocomposite. The 97% of IC-21 and SH-SY5Y cell lines viability after treatment with CS/CD approved the biocompatibility of CSLCD for biomedicinal application. The photoluminescence of the final system was detected via the observation of the one emission peak at 550 nm which permits to use of the prepared CDs having systems as bioimaging vehicles. The pH-sensitive release behavior of the CS/CD was clarified via a comparison of the obtained drug release percentage in two different pHs. So, at pH 4.4 around 60% of the encapsulated DOP was released from the drug-loaded nanocomposite, while, at pH 7.0 only 4.5% of DOP was released. Results of the MTT assay to find the neurotoxicity of the nanocomposite against SH-SY5Y differentiated cells revealed the safety of the composite against the studied cell line [85].

In the medicinal area, it is very pleasing and important to fabricate the drug delivery vehicle with non-invasive nature and controlled drug release behavior. In compression with intravenous administration, oral delivery is most usually used owing to the non-invasive nature and the fact that avoids patient pain and discomfort. By considering this goal, in 2018 GQDs were prepared from the pyrolysis of citric acid and then used as a crosslinker for CS. Afterward, the sodium salicylate (ss) was loaded in the crosslinked CS and the CS-GQD/ss was fabricated. Finally, the ss-loaded CS-GQD was coated by using the carboxymethylcellulose (CMC) as a pH-sensitive biopolymer and CS-GQD/ss@CMC hydrogel beads were achieved. Due to the presence of the GQDs in the structure of the final carrier, the nanocomposite showed an emission peak at its PL spectrum. The observed fluorescent properties could be a reason that the synthesized bionanocomposite beads potentially could be utilized as a fluorescent bioimaging agent. The ss release profile of CS-GQD/ss@CMC bionanocomposite was studied in SGI, the obtained in vitro drug release profile in SGT conditions confirmed the efficiency of the CS-GQD/ss@CMC hydrogel beads as a controlled oral drug carrier that related to the protection capability of CMC due to its certain structure. Cytotoxicity assay against HT29 cells exhibited low toxicity against human colon cancer cells, therefore the prepared CS-GQD/ss@CMC potentially can be applied as a safe carrier for oral SS delivery [88]. The efficiency of the CDs/CS-based system was also investigated for transdermal drug delivery. In this regard, Justin et al. fabricated the biodegradable microneedles for electrically stimulated and tracked transdermal delivery of BSA. In the following the structure, electrical, mechanical, biodegradation, and crystallinity, of CS-GQD nanocomposites were evaluated. It was found that the composites had 0.25–2 wt.% GQDs considerably improved the electrical conductivity, although, their biodegradation rate and mechanical properties are similar to the composites having 1 wt.% of GQDs. The authors mentioned that the prepared microneedle arrays have 1 wt.% GQDs are strong and can withstand inserted force into the body. Additionally, the drug release tests approved that the prepared nanocomposite is able to deliver large and small molecular weight drugs. Finding displayed that the prepared multifunctional nanocomposites offer a universal platform for tracked and iontophoretic delivery of therapeutics with a different range of molecular weight in an enhanced and controlled drug delivery [90]. Because of the high volume of published research on the study of the CS-based nanocomposites having GO NPs for avoiding the prolongation the results of other published cases are summarized in Table 1.

2.7 CNT NPs

CNTs with interesting mechanical and structural properties could be utilized in the fabrication of the composites that have the potential of using in tissue engineering and gene or drug delivery vehicles [91]. Nanotubes have the ability to enter the living cell without leading any damage or without leading their death. Although the mechanism of CNTs penetration over the cell membrane is unclear, however, the survey of the literature displayed that the functionalized-CNTs are initially absorbed and admitted onto the surface of the membrane with an axis parallel to the membrane plane. Therefore it has been seen that they inhibit the growth of numerous cells; for example the Chinese Hamster Ovary cells, human promyelocytic leukemia cells (HL-60), and endothelial and human T cells [17]. More research on the published works shows that the incorporation of the CNTs in the polymer composites structure improves their structural and mechanical feature. Moreover, it was observed that the CNTs have positive influences on cell proliferation and differentiation as well. Because of the interesting properties like excellent mechanical, electronic, and thermal properties recently the multi-walled carbon nanotubes (MWCNTs) have been extensively used as potential nanofillers for improvement of properties of biopolymers in comparison with single-walled carbon nanotubes (SWCNTs) [12].

Taking these into account Azeem et al. used from MWCNTs nanofillers to design and fabricate the nanocomposite of a polyelectrolytic complex SA-CS with acceptable stability without the conventional concerns in the use of a variety of crosslinkers. In fact, in the structure of the fabricated system, the COOH-MWCNTs acted as a reinforcing material to enhance the crosslinking density and prolong the drug release. The synthesis of nanocomposite was performed under the ultrasonic waves via a one-step solution casting procedure. The authors aimed to ultrasonic waves simplify the interactions between counterions. The formation of H-bonding between the hydroxyl groups of SA and reactive amino and primary and secondary hydroxyl groups at C-2, C-3, and C-6 of CS is a reason for polyelectrolyte complex formation. After characterization with several common techniques, the drug delivery tests displayed that about 88% of IBU was successfully incorporated in the formulation. As well, it was observed that the drug release kinetic followed by diffusion of IBU from the synthesized system in a first-order, non-Fickian, and intraparticle diffusion manner [2]. In another research work in order to exhibition of the MWCNTs positive role in drug delivery, pH and thermo-responsive nanocomposites based on CS were prepared via the creation of an imine bond as a result of the reaction between the aldehyde groups of the 1,3,5-Triazine-2,4,6-tribenzaldehyde derivative (TRIPOD) as crosslinker with amino groups of CS in the presence of MWCNTs (MWCNTs 0.1 and 0.3% (w/w)).

The authors of this research work mentioned that up to 2017 there is no report for the fabrication of CS-based hydrogels via the Schiff-base linking via the tris-aldehyde derivative of 1,3,5-triazine as crosslinker. The schematic of the hydrogel nanocomposite fabrication is presented in Fig. 14. The use of MWCNTs improved the stability and mechanical properties of the polymeric hydrogel. The outcome of the swelling assay displayed a temperature and pH-responsive swelling pattern and was strongly influenced by the amounts of MWCNTs and crosslinker. Similarly, the metronidazole (MZ) release pattern exhibited that the temperature, pH, and the wt.% of MWCNTs were strongly affected the observed drug release behavior. For all of the samples, it was detected that the MZ release in SGF was much higher than SIF. Additionally, in comparison to the pure hydrogel, the presence of MWCNT-COOH in the nanocomposites structure obtained the better drug-releasing, so, with increasing the MWCNT-COOH content the drug release was more controllable and has a slower release rate [12].

In the continuous effort for the preparation of the CS-based nanocomposites having MWCNTs in 2018 the new MWCNTs/gelatin-CS nanocomposite films were fabricated via the solution casting method and then the effect of MWCNTs on water uptake, physicomechanical, and thermal features of the nanocomposites were evaluated. Ciprofloxacin as a usual antibiotic drug was loaded into the nanocomposite matrix. Both antimicrobial susceptibility and dissolution tests were used to evaluate the effect of MWCNTs on ciprofloxacin release pattern. The results showed a higher antibacterial activity for ciprofloxacin loaded MWCNTs/gelatin-CS nanocomposite in comparison with neat gelatin-CS composite films against all microorganisms [91].

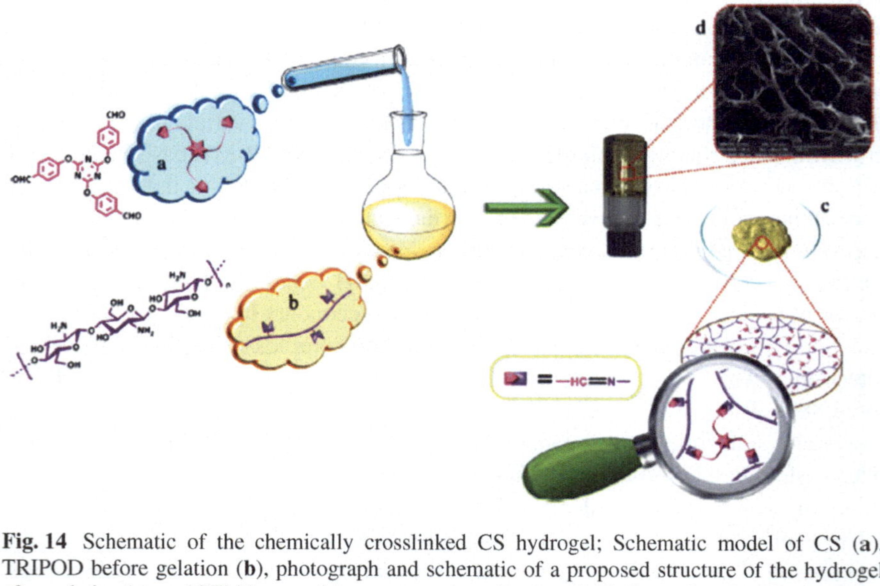

Fig. 14 Schematic of the chemically crosslinked CS hydrogel; Schematic model of CS (**a**), TRIPOD before gelation (**b**), photograph and schematic of a proposed structure of the hydrogel after gelation (**c**), and SEM image of the cross-section of hydrogel (**d**). Reproduced with permission from [12]

2.8 CS NPs

CS NPs is one member of the organic nanomaterials family with a large surface area and a certain amount of zeta potential which provides a superior activity. These materials have potential applications in biomedical for loading vaccines, drugs, and genes. Owing to the special character of the NPs, the CS NPs displayed a more antibacterial performance in comparison with traditional CS. Emulsion droplet coalescence, emulsion crosslinking, precipitation, reverse micelle, ionotropic gelation, molecular self-assembly, and template polymerization are some of the commonly investigated methods for the CS NPs preparation [92].

In recent years, hydrogels emerged as a promising candidate for drug delivery applications. Owing to this, recently the application of these materials in medicine is being developed. In this way, nanocomposite hydrogels based on PVA, 5% CS NPs, and different amounts of tetracycline (TC) as antibiotics were fabricated via freezing–thawing cycles. FT-IR analysis approved the formation of the nanocomposites via the appearance of the related peak of each constituent and exhibited the interaction between PVA, CS NPs, and tetracycline. Antibacterial tests displayed a great antibacterial activity toward *S. aureus* and *E. coli* respectively as gram-positive and gram-negative bacteria. TC release was evaluated at 37 °C and pH 7.4. It was found that the release mechanism fitted in Peppas-Korsmeyer and Higuchi models. The viability of the Caucasian fetal foreskin fibroblasts (HFFF2) cells was obtained higher than 60% after incubation for 72 h with prepared nanocomposite. The observed results clarified that the nanocomposite hydrogel having 5% of TC could be proposed as an efficient candidate for wound dressing and drug transporting for colon-specific drugs [18].

In another study and with the aim of preparing the new antibacterial DDS based on nanotechnology, the CS NPs were prepared through the ionic gelation method; the interaction between amino groups in CS and TPP. The synthesized CS NPs were selected as a nanocarrier for the TC, ciprofloxacin, and gentamycin. A schematic of the CS crosslinking with TPP is shown in Fig. 15a. The selected drugs were loaded in the prepared CS NPs. In the following, the samples of cotton (100%) and (50:50) cotton/polyester blended fabrics were treated with various amounts of drug/CSNPS as superior antibacterial compounds with minimum toxicity to induce antibacterial activity. The SEM displayed the insertion of free and antibiotics loaded CS NPs in the structure of the prepared fabrics. Finally, the evaluation of the antibacterial activity via an inhibition zone determination confirmed the antibacterial performance of these fabrics [92].

One of the more usual forms of leishmaniasis which arose from various types of Leishmania parasites is cutaneous leishmaniasis. So today this disease is considered as a second and fourth reason for mortality and morbidity. By considering the therapeutic influence of paromomycin (PM) known as aminosidine, on leishmaniasis Leishmania parasite Esfandiari et al. designed and fabricated the PM-loaded into mannosylated CS (MCS) NPs via dextran (PM-MCS-dex-NPs) and ionic gelation. The Zeta potential and particle size of PM-MCS-dex-NPs were determined at about

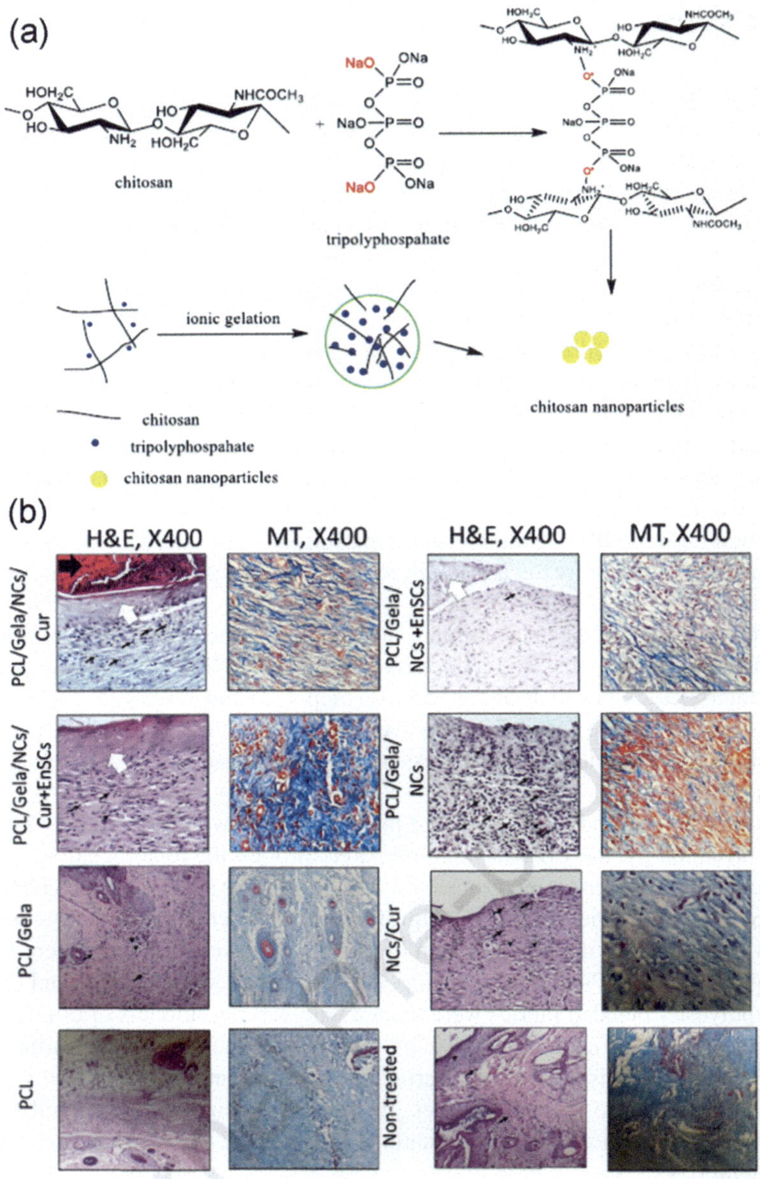

Fig. 15 Schematic of the CS NPs formation (**a**) and H&E and MT stained microscopic sections of the healed incisions in the various treatment groups at 14th days post-wounding (**b**) (Thick black arrows, thick white arrows, thin arrows, and arrowhead respectively are related to the crusty scab, epidermal layer, mononuclear inflammatory cells (MNICs), and polymorphonuclear inflammatory cells (PMNICs)). Reproduced with permission from [92, 94]

+31 mV and 246 nm respectively. Mannosylation percentage of CS quantitatively was calculated about 17% via CHNO elemental analysis. 83.5% of PM encapsulated in the prepared nanocomposite. The more release rate in the acidic environment and the higher PM-MCS-dex-NPs uptake inside THP-1 cell confirmed that the proposed new targeted construct denoted a strong anti-leishmanial performance [93].

CS NPs based systems were also used as a wound dressing agent. Additionally, the hybrid electrospun fiber with bioactive molecules in their structure which can deliver the cells within the wound bed leads to obtaining an enhanced therapeutic effect. In this contribution, an electrospun gelatin (Gela) and polycaperlactone (PCL) scaffold having CUR-loaded CS NPs (CS NPs/CUR) were fabricated to accelerate the repair process, and its in vivo wound healing performance was investigated. The biological test outputs showed that the electrospun hybrid scaffold seeded with human endometrial stem cells (EnSCs) has acceptable biocompatibility with the host immune system and wound healing in a full-thickness excisional animal model. All of the typical absorption characteristics of Gela, PCL, CS NPs, and CUR were detected in FT-IR analysis of the prepared structure. More study on the prepared system displayed the good structural and mechanical characteristics, perfect contact angle, degradability, and wettability of hybrid fiber scaffolds including the shape uniformity, pore size, and porosity. Additionally, higher proliferation and cell attachment were observed for PCL/Gela/CS NPs/CUR in comparison with each of PCL/Gela and PCL scaffolds. The observed higher cellular growth for PCL/Gela/CS NPs/CUR fiber mat in comparison with other fiber mats might be related to the bioactive interaction of cells with CUR as well as CS NPs as substrates. Additionally, as it is reported that the wound closure evaluation results approved the high efficiency of PCL/Gela/CS NPs/CUR in wound healing. Reducing the ROS-induced damages owing to the presence of CUR and CS NPs can be the reason for wound healing via promotion of the healing progression. Moreover, the PCL/Gela/NCs/Cur fiber mat exhibited a positive effect on the skin regeneration in the histological analysis for the H&E and MT stained skin wounds (Fig. 15b) [94].

In another similar interesting work, by considering the importance of growth factors to wound healing the HA-based wound dressing platforms were prepared by using the nanocrystalline cellulose (CNC) as reinforcing agents. In this research the CS NPs loaded with GM-CSF (GM-CSF-CS NPs) were developed and the optimized form was used as a filler and CNC-HA/GM-CSF-CS NPs composite was obtained. The observed positive features like appropriate mechanical properties, controlled release of GM-CSF up to 48 h, and high swelling capacity make CNC-HA/GM-CSF-CS NPs composite a brilliant candidate for wound healing. Nearly full wound closure and complete re-epithelialization were observed after 13 days of the wound covering with CNC-HA/GM-CSF-CS NPs composite. Moreover, the positive influences of GM-CSF on wound healing were approved via a lower inflammatory reaction. Improved granulation tissue formation and enhanced re-epithelialization were observed for CNC-HA/GM-CSF-CS NPs composite treated wounds in comparison with CNC-HA/CS NPs. The output of this research work displayed that the CNC-HA/GM-CSF-CS NPs composite can be utilized potentially in clinical for wound therapy [95].

Developing the controlled release of anti-hyperglycemic DDS is another challenge for developing an efficient drug carrier. In this contribution, CNC with a dimension of about 86–237 nm and 5–7 nm respectively length and width were extracted from palm fruit stalks via sulfuric acid hydrolysis. Then a controlled DDS with an average size of 215–310 nm was prepared from the CS NPs loaded with repaglinide (RPG) drug and CNC through ionic gelation with TPP as a crosslinking agent and spacer. RPG drug encapsulation efficiency of the prepared systems was calculated at ~98%. In vitro drug release study showed that the free RPG completely was released after 4 h, however, 64–77% of RPG was released from the RPG-loaded systems at the end of the release time. Additionally, evaluation of the effect of CNC surface oxidation on the drug releases exhibited the more sustained release for the composite with oxidized CNC (OXCNC) which might be related to the introduction of carboxylic groups at the surface of CNC. The formation of hydrogen bonding between OXCNC or CNC and RPG could be the reason for the observed controlled and sustained-release pattern. Therefore the fabricated CS NPs/OXCNC/RPG or CS NPs/CNC/RPG systems could be proposed as a controlled antidiabetic DDS [96]. Besides the antidiabetic drug delivery, the ability of the CS NPs was also investigated for the delivery of rosuvastatin as a drug. In this object, first, the rosuvastatin-loaded CS NPs were fabricated via the ionic gelation method in mean size of 100–150 nm. In the following, different hydrogel films with various ratios of SA:PVA were prepared. Finally, different amounts of rosuvastatin-loaded CS NPs were added into a hydrogel with an optimum ratio of SA:PVA. The tensile test output exhibited that hydrogel film with optimal mechanical properties obtained in the SA:PVA ratio of 7:3 and 3 wt.% of drug-loaded CS NPs. Prepared DDS exhibited that 100% of the loaded rosuvastatin was released within 24 h. The mentioned outputs and higher cell viability after 72 h of treatment of the human fibroblast cells with the prepared system illustrated a great potential of the prepared system for rosuvastatin drug release [97].

2.9 *Hybrid NPs*

Due to the before-mentioned favorable properties of each NPs, using them together in pharmaceutical formulations can be a promising way for obtaining the systems with enhanced performance. Therefore, there are numerous reports about the nanocomposite of CS having two or higher numbers of NPs. For example, theranostic is one type of system which could be prepared via the NPs combination in one system [98–102]. The combination of therapeutic and imaging agents within each system presents a new kind of nano-system recognized as 'theranostic' which for the first time in 2002 was devised by John Funkhouser. In this regard, M. Saviz Baktash et al. designed and reported the fabrication of the new system having combined the imaging ability of MNPs with the potential of CS-grafted GO as a pH-responsive nanocarrier. The grafting procedure was performed in different concentrations and MWt of the CS. In vitro test results displayed that an increase in the CS concentration and a decrease in CS MWt, reduce the DOX loading by ~24%, while, a decrease in MWt enhanced the release percentage is higher than 200%. It is reported that the more contribution of π-π stacking in DOC-CS interactions and fewer hydrogen binding could be reasons for observed phenomena. The MTT assay against normal L929 cell lines discovered the high biocompatibility of MNP/GO/CS nano-system, however, this assays for the cancer cells MCF7 with MNP/GO/CS loaded with DOX presented an upgraded performance for MNP/GO grafted with a low MWt of CS fulfill the requirements of a good drug carrier with desirable characteristics [11]. In another similar research, the fluorescent CS/Ag-Au nanocomposites with a different weight percentage of Ag and Au were prepared via the chemical reduction technique. The appearance of the CS, silver, and gold peaks in the XRD spectrum of nanocomposite displayed success in the nanocomposite fabrication. Additionally, the silver-gold as a filler phase and the CS as the matrix were clear in the obtained microscopic image through the high resolution transmission electron microscopy (HRTEM). The efficiency of the prepared nanocomposite for cellular imaging was confirmed via the presence of one emission peak at 719 nm in the PL spectra. About 97% of 5-FU loaded nanocomposite was also prepared. The authors reported that an increase in particle size from ~5 to ~12 nm could be a sign for 5-FU encapsulation within the nanocomposite structure, this also was confirmed via elemental mapping. A study of the 5-FU release profile in the simulated cancer cells condition exhibited a 67.6% of 5-FU release in 72 h. Eventually, the cytotoxicity analysis of the 5-FU loaded nanocomposite against MCF-7 cells displayed an IC50 value of about 55 μg/mL, showing its appropriateness for the cancer treatment [103]. The 5-FU delivery via the CS-based nanocomposites having the nanohybrid was also studied via CS/Ag/MWCNT nanocomposites to compare the release profile and cytotoxicity of the CS/Ag/MWCNT nanocomposites system with CS/Ag towards MCF-7 cell line. Experimental works displayed that the presence of the Ag NPs induced antibacterial activity to the system, as well as, Ag NPs are toxic to cancerous cells. Using the XRD pattern the triclinic structure of 5-FU, the semi-crystalline nature of CS, the face-centered cubic structure of Ag, and graphitic reflection of carbon from carbon nanotube were detected for the nanocomposites containing them. 5-FU loaded respectively about 95% and 96% in the prepared CS/Ag/MWCNT nanocomposites and CS/Ag nanocomposite. The existence of 5-FU reflections in the corresponding selected area electron diffraction (SAED) pattern approved the successful loading of 5-FU in the nanocomposites. The comparative study showed a prolonged-release profile for 5-FU loaded C/Ag/MWCNT which is because of the relatively strong binding of 5-FU to MWCNT. Cell viability assay against MCF-7 cell lines exhibited better cytotoxicity for CS/Ag/MWCNT when compared to the CS/Ag system. Summation of the obtained results displayed that the addition of MWCNT to the nanocomposite structure is productive and improves the release behavior of 5-FU, as well as, the cytotoxicity of the system [17].

In the following research on the preparation of a new DDS for 5-FU delivery and by considering the advantage of co-drug delivery, Nejadshafiee et al. successfully fabricated a multifunctional magnetic bio-metal–organic framework (Fe_3O_4@Bio-MOF) covered with FA-CS conjugate (FC) for cancer-targeted delivery of 5-FU and CUR. In this regard the zinc was used as a metal ion and CUR was used as an organic linker and the Bio-MOF nanocomposite was prepared in the presence of amine-functionalized Fe_3O_4 nanoparticles (Fe_3O_4@NH_2 MNPs). MNPs). Afterward, about 60% of the 5-FU was capsulated in the magnetic Bio-MOF and the obtained system was then coated with FC network (Fe_3O_4@Bio-MOF-FC). The step-by-step schematic of the Fe_3O_4@Bio-MOF-FC synthesis is reported in Fig. 16a. The obtained in vitro drug release profile displayed the controlled release of 5-FU

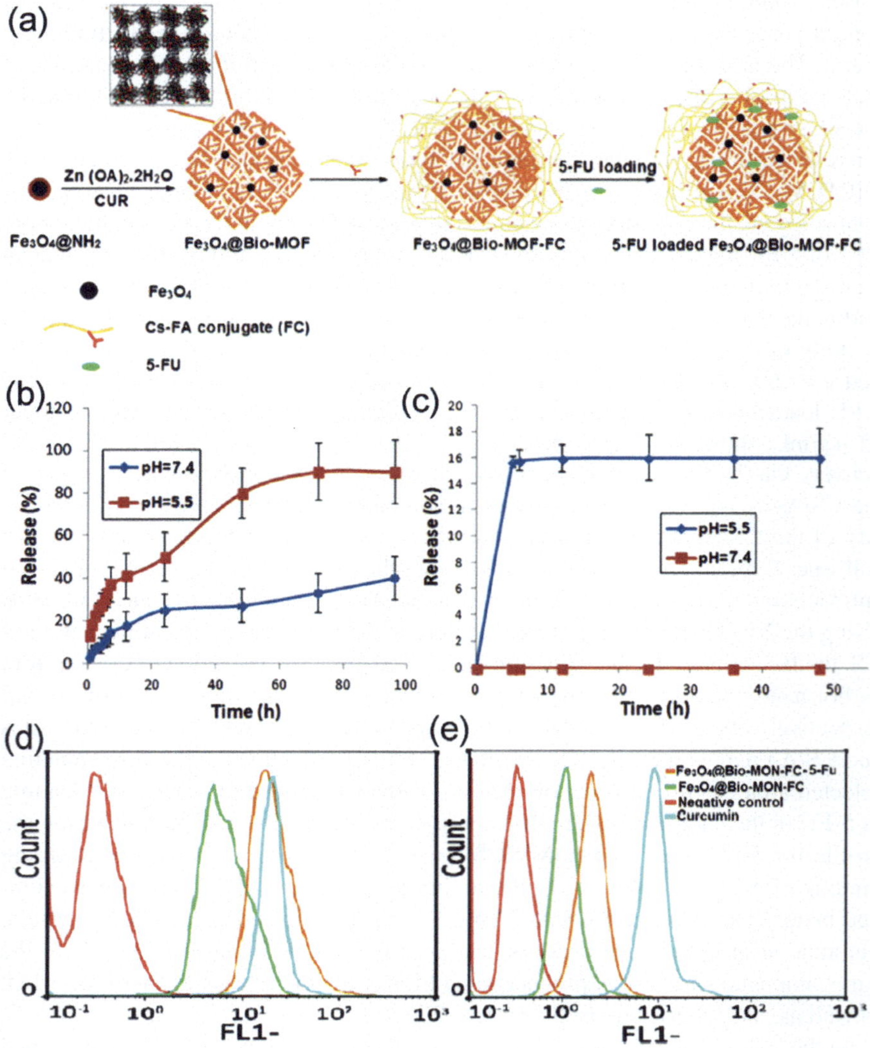

Fig. 16 Step by step procedure for fabrication of 5-FU loaded Fe_3O_4@Bio-MOF-FC (**a**), 5-FU (**b**), and CUR (**c**) release behavior from 5-FU-loaded Fe_3O_4@Bio-MOF-FC, flow cytometric analysis of the samples for cellular internalization in MDA-MB-231 (**d**) and NIH-3T3 (**e**). Reproduced with permission from [104]

and CUR in acidic pH that reveal the selectivity and high efficiency of the carrier in tumor microenvironments (Fig. 16b, c). As it is reported in Fig. 16d, e the cell uptake studies exhibited the selective uptaking of the 5-FU-loaded Fe_3O_4@Bio-MOF-FC via folate receptor-positive MDA-MB-231 cells. Additionally, Fe_3O_4@Bio-MOF-FC exhibited biocompatibility, while the drug-loaded nanocarrier exhibited selective and more toxicity toward the tumor cells in comparison with normal cells. In vitro and in vivo MRI investigations presented negative signal enhancement in the tumor that confirm the capability of the fabricated nanocarrier for use as a diagnostic agent. Due to the mentioned positive features, the prepared nanocarrier could be supposed as a promising vehicle in breast cancer treatment [104].

Taking the antibacterial property of AgNPs into account and the specific structure of GO, Namazi, and coworkers easily fabricated the pH-responsive bionanocomposite hydrogel beads based on GO-Ag nanohybrid and CS for controlled release of DOX. The in situ formation of Ag NPs on the GO sheets was confirmed through the FT-IR, TEM, SEM, and XRD, methods. Afterward, the capsulation of GO-Ag nanohybrid with CS in different percentages of GO-Ag was occurred via its crosslinking with TTP. The swelling test displayed that the incorporation of CS/GO-Ag in the structure of hydrogel beads considerably increased the swelling value. As well as the in vitro DOX load and release study exhibited that the increase in the GO-Ag nanohybrid content leads to a more controlled and sustained DOX release pattern. Additionally, the antimicrobial activity against both gram-positive and gram-negative bacteria was enhanced by increasing the percentage of GO-Ag in the nanocomposite structure. Based on the output of this work, the fabricated CS/GO-Ag nanocomposite hydrogel beads could be presented as a new biocompatible drug delivery vehicle for the controlled release of anticancer drugs [13]. In the field of using the hybrid NPs in drug delivery, in 2019, Lin et al. open up a novel core–shell MOFs/CDs@OCMC NPs for cancer diagnosis and treatment. In fact, the present MIL100 (Fe-BTC) in the structure of the nanocomposite could be utilized for magnetic resonance imaging owing to having a large amount of Fe^{+3}, as well as, it has a large specific surface area that is an advantage for each drug carrier. The fluorescence feature of the system was provided via the insertion of CDs into the synthesized MOFs structure. Ocarboxymethyl chitosan (OCMC) was covered on MOFs/CDs to increase safety and introduce pH sensitivity to the system. The observed magnetic resonance imaging (MRI) and fluorescence imaging (FOI) for the prepared composite confirmed that the MOFs/CDs@OCMC can be utilized for FOI/MRI dual-mode imaging. The output of the MTT test indicated the biocompatibility of MOFs/CDs@OCMC particles, hence, it is suitable as a drug carrier. DOX was loaded about 50 mg/g in the prepared carrier and released in a pH-triggered manner; the rate of DOX release at pH 3.8 was faster than pH 7.4. The findings proposed that the fabricated system is an ideal material for the association of treatment and diagnosis [105]. As previously mentioned the design and fabrication of the new system with combinatorial therapeutic efficacy are useful for future clinical applications. Therefore in one interesting research for obtaining the system with combinatorial therapeutic efficacy, the Au NPs were synthesized in the presence of porous silicon nanoparticles (PSiNPs) as host matrix and PSiNPs@Au nanocomposites were obtained. In the following the incorporation of PSiNPs@Au into thermosensitive CS hydrogels produced the CS/PSiNPs@Au nanocomposite hydrogels with in situ gelation property at the physiological condition. CS/PSiNPs@Au nanocomposite exhibited a long-term stable photothermal effect with more local temperature under NIR laser irradiation. Moreover, CS/PSiNPs@Au nanocomposite hydrogels showed a long-term sustained release of DOX in acidic tumor medium under the NIR laser irradiation. A single intratumoral injection of DOX/CS/PSiNPs@Au precursors into tumor-bearing mice offered an important synergistic chemo-photothermal therapeutic efficacy under repeated NIR laser radiation. The summation of the results showed that the developed multifunctional CS/PSiNPs@Au nanocomposite hydrogels with potential application in localized cancer therapy for use in future clinics [106]. In the case of targeted cancer fluorescence imaging and drug delivery, a CS-coated with rGO and implanted with Fe_3O_4 NPs with enhanced antibiofilm and antioxidant properties were prepared. After the characterization of the prepared system, the in vivo toxicity analysis against A549 and MCF7 cells approved the biocompatibility of the system. 0.448 mg/mL^{-1} of DOX was loaded in the rGO/Fe_3O_4/CS nanocomposite. Multimodal imaging and cellular uptake discovered the advantage of the FA-conjugated nanocomposite as a drug carrier due to remarkably improving the DOX co-localization to the cytoplasmic region of cancer cells over-express folate receptors. The summation of the obtained results supports the potential of nanocomposite in targeted chemotherapy [107]. Similarly, a magnetic nanocomposite having GO, superparamagnetic iron oxide nanoparticles (SPIONs) as NPs, as well as, PVA, and CS as biocompatible polymers were fabricated to evaluate its potential as imaging and drug delivery agent. In this regard, initially, the synthesized SPIONs were covered via CS and the stable SPIONs were obtained. Afterward, these NPs were deposited on GO. In the end, PVA is used as a coat on the nanocomposite surface to lead to the stability of nanocomposites for further in vivo applications. After the characterization of the fabricated system, its capability for 5-FU load and release was investigated. The superparamagnetic nature of nanocomposite was beneficial for its application in drug delivery to the human body. The outline of in vitro drug delivery tests showed a high 5-FU entrapment efficiency and faster and more 5-FU release in acidic pH, so, up to about 91.9% of 5-FU was released in pH 5.8, while, the 5-FU release amount was about 80.6%in pH 7.4 [108]. Similarly in order to take the advantage of the magnetic hybrid systems for delivery of other anticancer drugs, in 2020 the Abdel-Bary and coworkers evaluated the potential of a series of CS functionalized with magnetite (M), silicon dioxide (S), and GO NPs for cisplatin (cisP) load and release. The fabricated nanocomposites were evaluated by XRD, SEM, FT-IR, and EDS analysis. A near-spherical morphology was detected for the studied nanocomposites in TEM analysis. The in vitro drug release results exhibited the highest loading percentage for CS/M and CS/M/S/GO nanocomposites (87% and 84% respectively), while the potential of its release was the highest for CS/M composite in comparison with the other ones (91%) at pH 6.5. The obtained experimental data were further evaluated and approved via Density Functional Theory (DFT) calculations [1]. A comparison on the obtained systems indicated that the prepared nanocomposites with hybrid NPs can be considered as hopeful candidates for more medical application to investigate the in vivo and clinical application of theirs. Because of the high volume of published research on the study of the CS-based nanocomposites having hybrid NPs for avoiding the prolongation, the results of other published cases are summarized in Table 1.

3 Conclusion

Over the past several decades because of the inherent advantages of NPs like non-toxicity, high surface area, the possibility for surface modification, low cost, etc. the NPs exhibited special merits for use as a drug carrier. However, they have some limitations which lead to the combination of them with other materials like polymers and the formation of nanocomposites. CS is one member of the biopolymer family with biodegradability, antibacterial activity, low cost, availability, and biocompatibility advantages that are commonly used for nanocomposites fabrication which is investigated as a drug carrier. One of the main properties of nanocomposites is the fact that these materials show some new positive features besides the merits of both consisting portions. Consequently, it is concluded that CS-based nanocomposites unquestionably are versatile materials for use in biomedical applications. Taken together, this chapter summarizes and highlights the new advances in DDSs based on CS nanocomposites; especially the CS-based nanocomposites having each of the metal NPs, metal oxide NPs, silica NPs, HAp NPs, LDH NPs, GO and GQDs NPs, CNT NPs, CS NPs, hybrid NPs under nine categories. This chapter also focused on the delivery of various drugs such as anti-inflammatory, antibiotics, anticancer, proteins, and antidiabetic for therapeutic applications. Therefore we hope and expect that this chapter will be able to provide useful insights for future researchers using CS-based nanocomposites in the drug delivery area.

Acknowledgements This research is supported by a research grant from the University of Tabriz (number SAD/2761-991003).

Conflicts of Interest The authors declare that there are no conflicts of interest.

References

1. Abdel-Bary AS, Tolan DA, Nassar MY, Taketsugu T, El-Nahas AM (2020) Chitosan, magnetite, silicon dioxide, and graphene oxide nanocomposites: synthesis, characterization, efficiency as cisplatin drug delivery, and DFT calculations. Int J Biol Macromol
2. Bibi A, Akhtar T, Akhtar K, Farooq M, Shahzad MI (2020) Alginate-chitosan/MWCNTs nanocomposite: a novel approach for sustained release of Ibuprofen. J Polym Res 27(12):1–16
3. Safari M, Ghiaci M, Jafari-Asl M, Ensafi AA (2015) Nanohybrid organic–inorganic chitosan/dopamine/TiO2 composites with controlled drug-delivery properties. Appl Surf Sci 342:26–33

4. Namazi H, Jafarirad S (2011) Application of hybrid organic/inorganic dendritic ABA type triblock copolymers as new nanocarriers in drug delivery systems. Int J Polym Mater 60(9):603–619
5. NAMAZI H, BAHRAMI SA, Entezami AA (2005) Synthesis and controlled release of biocompatible prodrugs of beta-cyclodextrin linked with PEG containing ibuprofen or indomethacin
6. Namazi H, Babazadeh M, Sarabi A, Entezami A (2001) Synthesis and hydrolysis of acrylic type polymers containing nonsteroidal antiinflammatory drugs. J Polym Mater 18(3):301–311
7. Adeli M, Namazi H, Du F, Hönzke S, Hedtrich S, Keilitz J, Haag R (2015) Synthesis of multiarm star copolymers based on polyglycerol cores with polylactide arms and their application as nanocarriers. RSC Adv 5(20):14958–14966
8. Namazi H, Motamedi S, Namvari M (2011) Synthesis of new functionalized citric acid-based dendrimers as nanocarrier agents for drug delivery. BioImpacts BI 1(1):63
9. Namazi H, Toomari Y, Abbaspour H (2014) Fabrication of triblock ABA type peptide dendrimer based on glutamic acid dimethyl ester and PEG as a potential nano drug delivery agent. BioImpacts BI 4(4):175
10. Toomari Y, Namazi H, Entezami AA (2015) Fabrication of biodendrimeric β-cyclodextrin via click reaction with potency of anticancer drug delivery agent. Int J Biol Macromol 79:883–893
11. Baktash MS, Zarrabi A, Avazverdi E, Reis NM (2020) Development and optimization of a new hybrid chitosan-grafted graphene oxide/magnetic nanoparticle system for theranostic applications. J Mol Liquids 114515
12. Karimi AR, Tarighatjoo M, Nikravesh G (2017) 1, 3, 5-Triazine-2, 4, 6-tribenzaldehyde derivative as a new crosslinking agent for synthesis of pH-thermo dual responsive chitosan hydrogels and their nanocomposites: swelling properties and drug release behavior. Int J Biol Macromol 105:1088–1095
13. Rasoulzadehzali M, Namazi H (2018) Facile preparation of antibacterial chitosan/graphene oxide-Ag bio-nanocomposite hydrogel beads for controlled release of doxorubicin. Int J Biol Macromol 116:54–63
14. Tan L, Huang R, Li X, Liu S, Shen Y-M, Shao Z (2017) Chitosan-based core-shell nanomaterials for pH-triggered release of anticancer drug and near-infrared bioimaging. Carbohyd Polym 157:325–334
15. Nowroozi N, Faraji S, Nouralishahi A, Shahrousvand M (2020) Biological and structural properties of graphene oxide/curcumin nanocomposite incorporated chitosan as a scaffold for wound healing application. Life Sci 264:118640
16. Namazi H, Mosadegh M (2011) Bio-nanocomposites based on naturally occurring common polysaccharides chitosan, cellulose and starch with their biomedical applications. Recent Devel Bio-nanocomposites Biomed Appl 379–397
17. Nivethaa E, Dhanavel S, Rebekah A, Narayanan V, Stephen A (2016) A comparative study of 5-Fluorouracil release from chitosan/silver and chitosan/silver/MWCNT nanocomposites and their cytotoxicity towards MCF-7. Mater Sci Eng, C 66:244–250
18. Parsa P, Paydayesh A, Davachi SM (2019) Investigating the effect of tetracycline addition on nanocomposite hydrogels based on polyvinyl alcohol and chitosan nanoparticles for specific medical applications. Int J Biol Macromol 121:1061–1069
19. Liu X, Cheng X, Wang F, Feng L, Wang Y, Zheng Y, Guo R (2018) Targeted delivery of SNX-2112 by polysaccharide-modified graphene oxide nanocomposites for treatment of lung cancer. Carbohyd Polym 185:85–95
20. Javed R, Rais F, Fatima H, ul Haq I, Kaleem M, Naz SS, Ao Q (2020) Chitosan encapsulated ZnO nanocomposites: fabrication, characterization, and functionalization of bio-dental approaches. Mater Sci Eng C:111184
21. Pereira AKdS, Reis DT, Barbosa KM, Scheidt GN, da Costa LS, Santos LSS (2020) Antibacterial effects and ibuprofen release potential using chitosan microspheres loaded with silver nanoparticles. Carbohydrate Res 488:107891
22. Namazi H, Fard AMP (2011) Preparation of gold nanoparticles in the presence of citric acid-based dendrimers containing periphery hydroxyl groups. Mater Chem Phys 129(1–2):189–194

23. Nasef SM, Khozemy EE, Kamoun EA, El-Gendi H (2019) Gamma radiation-induced crosslinked composite membranes based on polyvinyl alcohol/chitosan/AgNO3/vitamin E for biomedical applications. Int J Biol Macromol 137:878–885
24. Aranaz I, Harris R, Navarro-García F, Heras A, Acosta N (2016) Chitosan based films as supports for dual antimicrobial release. Carbohyd Polym 146:402
25. Yadollahi M, Farhoudian S, Namazi H (2015) One-pot synthesis of antibacterial chitosan/silver bio-nanocomposite hydrogel beads as drug delivery systems. Int J Biol Macromol 79:37–43
26. Chen R, Chen Q, Huo D, Ding Y, Hu Y, Jiang X (2012) In situ formation of chitosan–gold hybrid hydrogel and its application for drug delivery. Colloids Surf B 97:132–137
27. Manivasagan P, Bharathiraja S, Bui NQ, Lim IG, Oh J (2016) Paclitaxel-loaded chitosan oligosaccharide-stabilized gold nanoparticles as novel agents for drug delivery and photoacoustic imaging of cancer cells. Int J Pharm 511(1):367–379
28. Zhang N, Xu X, Zhang X, Qu D, Xue L, Mo R, Zhang C (2016) Nanocomposite hydrogel incorporating gold nanorods and paclitaxel-loaded chitosan micelles for combination photothermal–chemotherapy. Int J Pharm 497(1–2):210–221
29. Dhanavel S, Nivethaa E, Narayanan V, Stephen A (2017) In vitro cytotoxicity study of dual drug loaded chitosan/palladium nanocomposite towards HT-29 cancer cells. Mater Sci Eng, C 75:1399–1410
30. Ahmadian Y, Bakravi A, Hashemi H, Namazi H (2019) Synthesis of polyvinyl alcohol/CuO nanocomposite hydrogel and its application as drug delivery agent. Polym Bull 76(4):1967–1983
31. Rakhshaei R, Namazi H, Hamishehkar H, Kafil HS, Salehi R (2019) In situ synthesized chitosan–gelatin/ZnO nanocomposite scaffold with drug delivery properties: higher antibacterial and lower cytotoxicity effects. J Appl Polym Sci 136(22):47590
32. Namazi H, Pooresmaeil M, Hasani M (2020) Oxidized starch/CuO bio-nanocomposite hydrogels as an antibacterial and stimuli-responsive agent with potential colon-specific naproxen delivery. Int J Polym Mater Polym Biomater 1–10
33. Hosseinzadeh S, Hosseinzadeh H, Pashaei S, Khodaparast Z (2019) Synthesis of stimuli-responsive chitosan nanocomposites via RAFT copolymerization for doxorubicin delivery. Int J Biol Macromol 121:677–685
34. Jardim KV, Palomec-Garfias AF, Andrade BYG, Chaker JA, Báo SN, Márquez-Beltrán C, Moya SE, Parize AL, Sousa MH (2018) Novel magneto-responsive nanoplatforms based on MnFe2O4 nanoparticles layer-by-layer functionalized with chitosan and sodium alginate for magnetic controlled release of curcumin. Mater Sci Eng C 92:184–195
35. Farhoudian S, Yadollahi M, Namazi H (2016) Facile synthesis of antibacterial chitosan/CuO bio-nanocomposite hydrogel beads. Int J Biol Macromol 82:837–843
36. Namazi H, Hasani M, Yadollahi M (2019) Antibacterial oxidized starch/ZnO nanocomposite hydrogel: synthesis and evaluation of its swelling behaviours in various pHs and salt solutions. Int J Biol Macromol 126:578–584
37. Masud RA, Islam MS, Haque P, Khan MNI, Shahruzzaman M, Khan M, Takafuji M, Rahman MM (2020) Preparation of novel chitosan/poly (ethylene glycol)/ZnO bionanocomposite for wound healing application: effect of gentamicin loading. Materialia 12:100785
38. Wang Y, Li B, Xu F, Han Z, Wei D, Jia D, Zhou Y (2018) Tough magnetic chitosan hydrogel nanocomposites for remotely stimulated drug release. Biomacromol 19(8):3351–3360
39. Zhao C, Liu X, Zhang X, Yan H, Qian Z, Li X, Ma Z, Han Q, Pei C (2017) A facile one-step method for preparation of Fe3O4/CS/INH nanoparticles as a targeted drug delivery for tuberculosis. Mater Sci Eng C 77:1182–1188
40. Karimi MH, Mahdavinia GR, Massoumi B (2018) pH-controlled sunitinib anticancer release from magnetic chitosan nanoparticles crosslinked with κ-carrageenan. Mater Sci Eng C 91:705–714
41. Lin J, Li Y, Li Y, Wu H, Yu F, Zhou S, Xie L, Luo F, Lin C, Hou Z (2015) Drug/dye-loaded, multifunctional PEG–chitosan–iron oxide nanocomposites for methotraxate synergistically self-targeted cancer therapy and dual model imaging. ACS Appl Mater Interfaces 7(22):11908–11920

42. Zhong S, Zhang H, Liu Y, Wang G, Shi C, Li Z, Feng Y, Cui X (2017) Folic acid functionalized reduction-responsive magnetic chitosan nanocapsules for targeted delivery and triggered release of drugs. Carbohyd Polym 168:282–289
43. Montha W, Maneeprakorn W, Buatong N, Tang I-M, Pon-On W (2016) Synthesis of doxorubicin-PLGA loaded chitosan stabilized (Mn, Zn) Fe2O4 nanoparticles: biological activity and pH-responsive drug release. Mater Sci Eng C 59:235–240
44. Kamaraj S, Palanisamy UM, Mohamed MSBK, Gangasalam A, Maria GA, Kandasamy R (2018) Curcumin drug delivery by vanillin-chitosan coated with calcium ferrite hybrid nanoparticles as carrier. Eur J Pharm Sci 116:48–60
45. Fazli Y, Shariatinia Z (2017) Controlled release of cefazolin sodium antibiotic drug from electrospun chitosan-polyethylene oxide nanofibrous Mats. Mater Sci Eng, C 71:641–652
46. Namazi H, Jafarirad S (2011) Controlled release of linear-dendritic hybrids of carbosiloxane dendrimer: the effect of hybrid's amphiphilicity on drug-incorporation; hybrid–drug interactions and hydrolytic behavior of nanocarriers. Int J Pharm 407(1–2):167–173
47. Mohebali A, Abdouss M (2020) Layered biocompatible pH-responsive antibacterial composite film based on HNT/PLGA/chitosan for controlled release of minocycline as burn wound dressing. Int J Biol Macromol 164:4193–4204
48. Barman M, Mahmood S, Augustine R, Hasan A, Thomas S, Ghosal K (2020) Natural halloysite nanotubes/chitosan based bio-nanocomposite for delivering norfloxacin, an anti-microbial agent in sustained release manner. Int J Biol Macromol 162:1849–1861
49. Raut SY, Gahane A, Joshi MB, Kalthur G, Mutalik S (2019) Nanocomposite clay-polymer microbeads for oral controlled drug delivery: development and in vitro and in vivo evaluations. J Drug Delivery Sci Technol 51:234–243
50. Namazi H, Belali S (2016) Starch-g-lactic acid/montmorillonite nanocomposite: synthesis, characterization and controlled drug release study. Starch-Stärke 68(3–4):177–187
51. Namazi H, Dadkhah A, Mosadegh M (2012) New biopolymer nanocomposite of starch-graft polystyrene/montmorillonite clay prepared through emulsion polymerization method. J Polym Environ 20(3):794–800
52. Ambrogi V, Pietrella D, Nocchetti M, Casagrande S, Moretti V, De Marco S, Ricci M (2017) Montmorillonite–chitosan–chlorhexidine composite films with antibiofilm activity and improved cytotoxicity for wound dressing. J Colloid Interface Sci 491:265–272
53. Aguzzi C, Capra P, Bonferoni C, Cerezo P, Salcedo I, Sánchez R, Caramella C, Viseras C (2010) Chitosan–silicate biocomposites to be used in modified drug release of 5-aminosalicylic acid (5-ASA). Appl Clay Sci 50(1):106–111
54. Qiu Q, Quan Z, Zhang H, Qin X, Wang R, Yu J (2020) pH-triggered sustained drug release of multilayer encapsulation system with hollow mesoporous silica nanoparticles/chitosan/polyacrylic acid. Mater Lett 260:126907
55. Arumugam G, Rajendran R (2020) Callophycin A loaded chitosan and spicules based nanocomposites as an alternative strategy to overcome vaginal candidiasis. Int J Biol Macromol
56. Hamed SF, Hashim AF, Hamid HAA, Abd-Elsalam KA, Golonka I, Musiał W, El-Sherbiny IM (2020) Edible alginate/chitosan-based nanocomposite microspheres as delivery vehicles of omega-3 rich oils. Carbohydrate Polym 116201
57. Huang N, Wang J, Cheng X, Xu Y, Li W (2020) Fabrication of PNIPAM-chitosan/decatungstoeuropate/silica nanocomposite for thermo/pH dual-stimuli-responsive and luminescent drug delivery system. J Inorg Biochem 211:111216
58. Oliveira MJ, Estefânia OS, Lúcia MB, Regina M, Amato VS, Lugão AB, Parra DF (2014) Influence of chitosan/clay in drug delivery of glucantime from PVP membranes. Radiat Phys Chem 94:194–198
59. Taşdelen B, Erdoğan S, Bekar B (2018) Radiation synthesis and characterization of chitosan/hyraluronic acid/hydroxyapatite hydrogels: drug uptake and drug delivery systems. Materials Today: Proceedings 5(8):15990–15997
60. Venkatesan P, Puvvada N, Dash R, Kumar BP, Sarkar D, Azab B, Pathak A, Kundu SC, Fisher PB, Mandal M (2011) The potential of celecoxib-loaded hydroxyapatite-chitosan nanocomposite for the treatment of colon cancer. Biomaterials 32(15):3794–3806

61. Abou Taleb MF, Alkahtani A, Mohamed SK (2015) Radiation synthesis and characterization of sodium alginate/chitosan/hydroxyapatite nanocomposite hydrogels: a drug delivery system for liver cancer. Polym Bull 72(4):725–742
62. Mahdavinia GR, Karimi MH, Soltaniniya M, Massoumi B (2019) In vitro evaluation of sustained ciprofloxacin release from κ-carrageenan-crosslinked chitosan/hydroxyapatite hydrogel nanocomposites. Int J Biol Macromol 126:443–453
63. Farooq A, Yar M, Khan AS, Shahzadi L, Siddiqi SA, Mahmood N, Rauf A, Manzoor F, Chaudhry AA, Ur Rehman I (2015) Synthesis of piroxicam loaded novel electrospun biodegradable nanocomposite scaffolds for periodontal regeneration. Mater Sci Eng C 56:104–113
64. He J, Hu X, Cao J, Zhang Y, Xiao J, Chen D, Xiong C, Zhang L (2020) Chitosan-coated hydroxyapatite and drug-loaded polytrimethylene carbonate/polylactic acid scaffold for enhancing bone regeneration. Carbohydrate Polym 253:117198
65. Liu T-Y, Chen S-Y, Li J-H, Liu D-M (2006) Study on drug release behaviour of CDHA/chitosan nanocomposites—effect of CDHA nanoparticles. J Control Release 112(1):88–95
66. Saber-Samandari S, Saber-Samandari S (2017) Biocompatible nanocomposite scaffolds based on copolymer-grafted chitosan for bone tissue engineering with drug delivery capability. Mater Sci Eng C 75:721–732
67. Anirudhan T, Sekhar VC (2020) Fabrication of functionalized layered double hydroxide/chitosan nanocomposite with dual responsive drug release for the targeted therapy of breast cancer. Eur Polymer J 139:109993
68. Pooresmaeil M, Nia SB, Namazi H (2019) Green encapsulation of LDH (Zn/Al)-5-Fu with carboxymethyl cellulose biopolymer; new nanovehicle for oral colorectal cancer treatment. Int J Biol Macromol 139:994–1001
69. Nia SB, Pooresmaeil M, Namazi H (2020) Carboxymethylcellulose/layered double hydroxides bio-nanocomposite hydrogel: a controlled amoxicillin nanocarrier for colonic bacterial infections treatment. Int J Biol Macromol 155:1401–1409
70. Mallakpour S, Hatami M (2019) Fabrication and characterization of pH-sensitive bionanocomposite beads havening folic acid intercalated LDH and chitosan: drug release and mechanism evaluation. Int J Biol Macromol 122:157–167
71. Lerner DA, Bégu S, Aubert-Pouëssel A, Polexe R, Devoisselle J-M, Azaïs T, Tichit D (2020) Synthesis and properties of new multilayer chitosan@ layered double hydroxide/drug loaded phospholipid bilayer nanocomposite bio-hybrids. Materials 13(16):3565
72. Yu X, Wen T, Cao P, Shan L, Li L (2019) Alginate-chitosan coated layered double hydroxide nanocomposites for enhanced oral vaccine delivery. J Colloid Interface Sci 556:258–265
73. Pooresmaeil M, Namazi H (2018) β-Cyclodextrin grafted magnetic graphene oxide applicable as cancer drug delivery agent: synthesis and characterization. Mater Chem Phys 218:62–69
74. Pooresmaeil M, Namazi H (2018) Surface modification of graphene oxide with stimuli-responsive polymer brush containing β-cyclodextrin as a pendant group: preparation, characterization, and evaluation as controlled drug delivery agent. Colloids Surf B 172:17–25
75. Namvari M, Namazi H (2015) Preparation of efficient magnetic biosorbents by clicking carbohydrates onto graphene oxide. J Mater Sci 50(15):5348–5361
76. Kabiri R, Namazi H (2016) Synthesis of cellulose/reduced graphene oxide/polyaniline nanocomposite and its properties. Int J Polym Mater Polym Biomater 65(13):675–682
77. Lei H, Xie M, Zhao Y, Zhang F, Xu Y, Xie J (2016) Chitosan/sodium alginate modificated graphene oxide-based nanocomposite as a carrier for drug delivery. Ceram Int 42(15):17798–17805
78. Kazempour M, Namazi H, Akbarzadeh A, Kabiri R (2019) Synthesis and characterization of PEG-functionalized graphene oxide as an effective pH-sensitive drug carrier. Artif Cells Nanomed Biotechnol 47(1):90–94
79. Hashemi H, Namazi H (2018) Sonochemically synthesized blue fluorescent functionalized graphene oxide as a drug delivery system. Ultrason Sonochem 42:124–133
80. Hashemi H, Namazi H (2018) Blue fluorescent graphene oxide hybrid: synthesis, characterization, and application as a drug delivery system. J Drug Deliv Sci Technol 48:355–362

81. Dhanavel S, Revathy T, Sivaranjani T, Sivakumar K, Palani P, Narayanan V, Stephen A (2020) 5-Fluorouracil and curcumin co-encapsulated chitosan/reduced graphene oxide nanocomposites against human colon cancer cell lines. Polym Bull 77(1):213–233
82. Anirudhan T, Sekhar VC, Athira V (2020) Graphene oxide based functionalized chitosan polyelectrolyte nanocomposite for targeted and pH responsive drug delivery. Int J Biol Macromol 150:468–479
83. Jafari Z, Rad AS, Baharfar R, Asghari S, Esfahani MR (2020) Synthesis and application of chitosan/tripolyphosphate/graphene oxide hydrogel as a new drug delivery system for Sumatriptan Succinate. J Mol Liquids 315:113835
84. Wang X, Guo W, Li L, Yu F, Li J, Liu L, Fang B, Xia L (2020) Photothermally triggered biomimetic drug delivery of Teriparatide via reduced graphene oxide loaded chitosan hydrogel for osteoporotic bone regeneration. Chem Eng J 127413
85. Mathew SA, Praveena P, Dhanavel S, Manikandan R, Senthilkumar S, Stephen A (2020) Luminescent chitosan/carbon dots as an effective nano-drug carrier for neurodegenerative diseases. RSC Adv 10(41):24386–24396
86. Rakhshaei R, Namazi H, Hamishehkar H, Rahimi M (2020) Graphene quantum dot cross-linked carboxymethyl cellulose nanocomposite hydrogel for pH-sensitive oral anticancer drug delivery with potential bioimaging properties. Int J Biol Macromol 150:1121–1129
87. Pooresmaeil M, Namazi H, Salehi R (2020) Synthesis of photoluminescent glycodendrimer with terminal β-cyclodextrin molecules as a biocompatible pH-sensitive carrier for doxorubicin delivery. Carbohydrate Polym 246:116658
88. Javanbakht S, Shaabani A (2019) Encapsulation of graphene quantum dot-crosslinked chitosan by carboxymethylcellulose hydrogel beads as a pH-responsive bio-nanocomposite for the oral delivery agent. Int J Biol Macromol 123:389–397
89. Javanbakht S, Nazari N, Rakhshaei R, Namazi H (2018) Cu-crosslinked carboxymethylcellulose/naproxen/graphene quantum dot nanocomposite hydrogel beads for naproxen oral delivery. Carbohyd Polym 195:453–459
90. Justin R, Román S, Chen D, Tao K, Geng X, Grant RT, MacNeil S, Sun K, Chen B (2015) Biodegradable and conductive chitosan–graphene quantum dot nanocomposite microneedles for delivery of both small and large molecular weight therapeutics. RSC Adv 5(64):51934–51946
91. Sharmeen S, Rahman AM, Lubna MM, Salem KS, Islam R, Khan MA (2018) Polyethylene glycol functionalized carbon nanotubes/gelatin-chitosan nanocomposite: an approach for significant drug release. Bioactive Mater 3(3):236–244
92. El-Alfy EA, El-Bisi MK, Taha GM, Ibrahim HM (2020) Preparation of biocompatible chitosan nanoparticles loaded by tetracycline, gentamycin and ciprofloxacin as novel drug delivery system for improvement the antibacterial properties of cellulose based fabrics. Int J Biol Macromol 161:1247–1260
93. Esfandiari F, Motazedian M, Asgari Q, Morowvat M, Molaei M, Heli H (2019) Paromomycin-loaded mannosylated chitosan nanoparticles: synthesis, characterization and targeted drug delivery against leishmaniasis. Acta Tropica 197:105072
94. Zahiri M, Khanmohammadi M, Goodarzi A, Ababzadeh S, Farahani MS, Mohandesnezhad S, Bahrami N, Nabipour I, Ai J (2020) Encapsulation of curcumin loaded chitosan nanoparticle within poly (ε-caprolactone) and gelatin fiber mat for wound healing and layered dermal reconstitution. Int J Biol Macromol 153:1241–1250
95. Dehkordi NK, Minaiyan M, Talebi A, Akbari V, Taheri A (2019) Nanocrystalline cellulose–hyaluronic acid composite enriched with GM-CSF loaded chitosan nanoparticles for enhanced wound healing. Biomed Mater 14(3):035003
96. Abo-Elseoud WS, Hassan ML, Sabaa MW, Basha M, Hassan EA, Fadel SM (2018) Chitosan nanoparticles/cellulose nanocrystals nanocomposites as a carrier system for the controlled release of repaglinide. Int J Biol Macromol 111:604–613
97. Afshar M, Dini G, Vaezifar S, Mehdikhani M, Movahedi B (2020) Preparation and characterization of sodium alginate/polyvinyl alcohol hydrogel containing drug-loaded chitosan nanoparticles as a drug delivery system. J Drug Deliv Sci Technol 56:101530

98. Pooresmaeil M, Namazi H (2019) Preparation and characterization of polyvinyl alcohol/β-cyclodextrin/GO-Ag nanocomposite with improved antibacterial and strength properties. Polym Adv Technol 30(2):447–456
99. Karimzadeh Z, Javanbakht S, Namazi H (2019) Carboxymethylcellulose/MOF-5/Graphene oxide bio-nanocomposite as antibacterial drug nanocarrier agent. BioImpacts BI 9(1):5
100. Pooresmaeil M, Javanbakht S, Nia SB, Namazi H (2020) Carboxymethyl cellulose/mesoporous magnetic graphene oxide as a safe and sustained ibuprofen delivery bio-system: synthesis, characterization, and study of drug release kinetic. Colloids Surf A Physicochem Eng Aspects 124662
101. Pooresmaeil M, Namazi H (2020) pH-sensitive ternary Fe3O4/GQDs@ G hybrid microspheres; synthesis, characterization and drug delivery application. J Alloys Compounds 846:156419
102. Pooresmaeil M, Namazi H, Salehi R (2020) Simple method for fabrication of metal-organic framework within a carboxymethylcellulose/graphene quantum dots matrix as a carrier for anticancer drug. Int J Biol Macromol 164:2301–2311
103. Nivethaa E, Dhanavel S, Narayanan V, Stephen A (2017) Chitosan stabilized Ag-Au nanoalloy for colorimetric sensing and 5-Fluorouracil delivery. Int J Biol Macromol 95:862–872
104. Nejadshafiee V, Naeimi H, Goliaei B, Bigdeli B, Sadighi A, Dehghani S, Lotfabadi A, Hosseini M, Nezamtaheri MS, Amanlou M (2019) Magnetic bio-metal–organic framework nanocomposites decorated with folic acid conjugated chitosan as a promising biocompatible targeted theranostic system for cancer treatment. Mater Sci Eng C 99:805–815
105. Lin C, Sun K, Zhang C, Tan T, Xu M, Liu Y, Xu C, Wang Y, Li L, Whittaker A (2020) Carbon dots embedded metal organic framework@ chitosan core-shell nanoparticles for vitro dual mode imaging and pH-responsive drug delivery. Microporous Mesoporous Mater 293:109775
106. Xia B, Zhang W, Tong H, Li J, Chen Z, Shi J (2019) Multifunctional chitosan/porous silicon@ Au nanocomposite hydrogels for long-term and repeatedly localized combinatorial therapy of cancer via a single injection. ACS Biomater Sci Eng 5(4):1857–1867
107. Viswanathan K, AlSalhi MS, Sandhanasamy D, Kasi G, Ayyakannu A, Marimuthu G (2020) Chitosan overlaid Fe3O4/rGO nanocomposite for targeted drug delivery, imaging, and biomedical applications. Sci Rep (Nature Publisher Group) 10(1)
108. Aliabadi M, Shagholani H (2017) Synthesis of a novel biocompatible nanocomposite of graphene oxide and magnetic nanoparticles for drug delivery. Int J Biol Macromol 98:287–291
109. Shah A, Yameen MA, Fatima N, Murtaza G (2019) Chemical synthesis of chitosan/silver nanocomposites films loaded with moxifloxacin: their characterization and potential antibacterial activity. Int J Pharm 561:19–34
110. Shah A, Buabeid MA, Arafa E-SA, Hussain I, Li L, Murtaza G (2019) The wound healing and antibacterial potential of triple-component nanocomposite (chitosan-silver-sericin) films loaded with moxifloxacin. Int J Pharm 564:22–38
111. Shah A, Ashames AA, Buabeid MA, Murtaza G (2020) Synthesis, in vitro characterization and antibacterial efficacy of moxifloxacin-loaded chitosan-pullulan-silver-nanocomposite films. J Drug Deliv Sci Technol 55:101366
112. Chandran PR, Sandhyarani N (2014) An electric field responsive drug delivery system based on chitosan–gold nanocomposites for site specific and controlled delivery of 5-fluorouracil. RSC Adv 4(85):44922–44929
113. Kumari S, Singh RP (2012) Glycolic acid-g-chitosan-gold nanoflower nanocomposite scaffolds for drug delivery and tissue engineering. Int J Biol Macromol 50(3):878–883
114. Nivethaa E, Dhanavel S, Narayanan V, Vasu CA, Stephen A (2015) An in vitro cytotoxicity study of 5-fluorouracil encapsulated chitosan/gold nanocomposites towards MCF-7 cells. RSC Adv 5(2):1024–1032
115. Pathania D, Gupta D, Kothiyal N, Eldesoky G, Naushad M (2016) Preparation of a novel chitosan-g-poly (acrylamide)/Zn nanocomposite hydrogel and its applications for controlled drug delivery of ofloxacin. Int J Biol Macromol 84:340–348
116. Pourjavadi A, Bagherifard M, Doroudian M (2020) Synthesis of micelles based on chitosan functionalized with gold nanorods as a light sensitive drug delivery vehicle. Int J Biol Macromol 149:809–818

117. Shafiee S, Ahangar HA, Saffar A (2019) Taguchi method optimization for synthesis of Fe3O4@ chitosan/tragacanth gum nanocomposite as a drug delivery system. Carbohydrate Polym 222:114982
118. Naderi Z, Azizian J, Moniri E, Farhadyar N (2020) Synthesis and Characterization of carboxymethyl cellulose/β-cyclodextrin/chitosan hydrogels and investigating the effect of magnetic nanoparticles (Fe3O4) on a novel carrier for a controlled release of methotrexate as drug delivery. J Inorg Organomet Polym Mater 30(4):1339–1351
119. Sun X, Shen J, Yu D, Ouyang X-k (2019) Preparation of pH-sensitive Fe3O4@ C/carboxymethyl cellulose/chitosan composite beads for diclofenac sodium delivery. Int J Biol Macromol 127:594–605
120. Fazli Y, Shariatinia Z, Kohsari I, Azadmehr A, Pourmortazavi SM (2016) A novel chitosan-polyethylene oxide nanofibrous mat designed for controlled co-release of hydrocortisone and imipenem/cilastatin drugs. Int J Pharm 513(1–2):636–647
121. Bano S, Afzal M, Waraich MM, Alamgir K, Nazir S (2016) Paclitaxel loaded magnetic nanocomposites with folate modified chitosan/carboxymethyl surface; a vehicle for imaging and targeted drug delivery. Int J Pharm 513(1–2):554–563
122. Xie P, Du P, Li J, Liu P (2019) Stimuli-responsive hybrid cluster bombs of PEGylated chitosan encapsulated DOX-loaded superparamagnetic nanoparticles enabling tumor-specific disassembly for on-demand drug delivery and enhanced MR imaging. Carbohyd Polym 205:377–384
123. Bilas R, Sriram K, Maheswari PU, Begum KMS (2017) Highly biocompatible chitosan with super paramagnetic calcium ferrite (CaFe2O4) nanoparticle for the release of ampicillin. Int J Biol Macromol 97:513–525
124. Arias JL, Reddy LH, Couvreur P (2012) Fe 3 O 4/chitosan nanocomposite for magnetic drug targeting to cancer. J Mater Chem 22(15):7622–7632
125. Ding Y, Shen SZ, Sun H, Sun K, Liu F, Qi Y, Yan J (2015) Design and construction of polymerized-chitosan coated Fe3O4 magnetic nanoparticles and its application for hydrophobic drug delivery. Mater Sci Eng, C 48:487–498
126. Amini-Fazl MS, Mohammadi R, Kheiri K (2019) 5-Fluorouracil loaded chitosan/polyacrylic acid/Fe3O4 magnetic nanocomposite hydrogel as a potential anticancer drug delivery system. Int J Biol Macromol 132:506–513
127. Kurdtabar M, Koutenaee RN, Bardajee GR (2018) Synthesis and characterization of a novel pH-responsive nanocomposite hydrogel based on chitosan for targeted drug release. J Polym Res 25(5):119
128. Mabrouk M, Abd El-Wahab RM, Beherei HH, Selim MM, Das DB (2020) Multifunctional magnetite nanoparticles for drug delivery: preparation, characterisation, antibacterial properties and drug release kinetics. Int J Pharmaceutics 587:119658
129. Qin H, Wang C, Dong Q, Zhang L, Zhang X, Ma Z, Han Q (2015) Preparation and characterization of magnetic Fe3O4–chitosan nanoparticles loaded with isoniazid. J Magn Magn Mater 381:120–126
130. Sun X, Liu C, Omer A, Lu W, Zhang S, Jiang X, Wu H, Yu D, Ouyang X-k (2019) pH-sensitive ZnO/carboxymethyl cellulose/chitosan bio-nanocomposite beads for colon-specific release of 5-fluorouracil. Int J Biol Macromol 128:468–479
131. Mansouri M, Nazarpak MH, Solouk A, Akbari S, Hasani-Sadrabadi MM (2017) Magnetic responsive of paclitaxel delivery system based on SPION and palmitoyl chitosan. J Magn Magn Mater 421:316–325
132. Saboktakin MR, Tabatabaie R, Maharramov A, Ramazanov MA (2010) Synthesis and characterization of superparamagnetic chitosan–dextran sulfate hydrogels as nano carriers for colon-specific drug delivery. Carbohyd Polym 81(2):372–376
133. Yadollahi M, Farhoudian S, Barkhordari S, Gholamali I, Farhadnejad H, Motasadizadeh H (2016) Facile synthesis of chitosan/ZnO bio-nanocomposite hydrogel beads as drug delivery systems. Int J Biol Macromol 82:273–278
134. dos Santos MJ, Santos M-KMS, Dias DJS, Chaker JA, Sousa MH (2015) One-step synthesis of magnetic chitosan for controlled release of 5-hydroxytryptophan. J Magn Magn Mater 380:117–124

135. Liu H, Liu J, Xie X, Li X (2020) Development of photo-magnetic drug delivery system by facile-designed dual stimuli-responsive modified biopolymeric chitosan capped nano-vesicle to improve efficiency in the anesthetic effect and its biological investigations. J Photochem Photobiol B Biol 202:111716
136. Mahdavinia GR, Mosallanezhad A, Soleymani M, Sabzi M (2017) Magnetic-and pH-responsive κ-carrageenan/chitosan complexes for controlled release of methotrexate anti-cancer drug. Int J Biol Macromol 97:209–217
137. Jamshidzadeh F, Mohebali A, Abdouss M (2020) Three-ply biocompatible pH-responsive nanocarriers based on HNT sandwiched by chitosan/pectin layers for controlled release of phenytoin sodium. Int J Biol Macromol 150:336–343
138. Li X, Ouyang J, Yang H, Chang S (2016) Chitosan modified halloysite nanotubes as emerging porous microspheres for drug carrier. Appl Clay Sci 126:306–312
139. Rao KM, Kumar A, Suneetha M, Han SS (2018) pH and near-infrared active; chitosan-coated halloysite nanotubes loaded with curcumin-Au hybrid nanoparticles for cancer drug delivery. Int J Biol Macromol 112:119–125
140. Kouser R, Vashist A, Zafaryab M, Rizvi MA, Ahmad S (2018) pH-responsive biocompatible nanocomposite hydrogels for therapeutic drug delivery. ACS Appl Bio Mater 1(6):1810–1822
141. Feiz S, Navarchian AH (2019) Poly (vinyl alcohol) hydrogel/chitosan-modified clay nanocomposites for wound dressing application and controlled drug release. Macromol Res 27(3):290–300
142. Cheikh D, García-Villén F, Majdoub H, Zayani MB, Viseras C (2019) Complex of chitosan pectin and clay as diclofenac carrier. Appl Clay Sci 172:155–164
143. Yahia Y, García-Villén F, Djelad A, Belaroui LS, Sanchez-Espejo R, Sassi M, López-Galindo A, Viseras C (2019) Crosslinked palygorskite-chitosan beads as diclofenac carriers. Appl Clay Sci 180:105169
144. Kimna C, Deger S, Tamburaci S, Tihminlioglu F (2019) Chitosan/montmorillonite composite nanospheres for sustained antibiotic delivery at post-implantation bone infection treatment. Biomed Mater 14(4):044101
145. Abdeen R, Salahuddin N (2013) Modified chitosan-clay nanocomposite as a drug delivery system intercalation and in vitro release of ibuprofen. J Chem
146. Abukhadra MR, Refay NM, Nadeem A, El-Sherbeeny AM, Ibrahim KE (2020) Insight into the role of integrated carbohydrate polymers (starch, chitosan, and β-cyclodextrin) with mesoporous silica as carriers for ibuprofen drug; equilibrium and pharmacokinetic properties. Int J Biol Macromol
147. Bothiraja C, Thorat U, Pawar A, Shaikh K (2014) Chitosan coated layered clay montmorillonite nanocomposites modulate oral delivery of paclitaxel in colonic cancer. Mater Technol 29(sup3):B120–B126
148. Chen S, Zhang H, Chinnathambi S, Hanagata N (2013) Synthesis of novel chitosan–silica/CpG oligodeoxynucleotide nanohybrids with enhanced delivery efficiency. Mater Sci Eng C 33(6):3382–3388
149. Chen C, Yao W, Sun W, Guo T, Lv H, Wang X, Ying H, Wang Y, Wang P (2018) A self-targeting and controllable drug delivery system by fabricating with multi-stimuli responsive chitosan-based thin film layer on mesoporous silica nanoparticles. Int J Biol Macromol
150. Cojocariu A, Profire L, Aflori M, Vasile C (2012) In vitro drug release from chitosan/cloisite 15A hydrogels. Appl Clay Sci 57:1–9
151. Depan D, Kumar AP, Singh RP (2009) Cell proliferation and controlled drug release studies of nanohybrids based on chitosan-g-lactic acid and montmorillonite. Acta Biomater 5(1):93–100
152. Azhar FF, Olad A (2014) A study on sustained release formulations for oral delivery of 5-fluorouracil based on alginate–chitosan/montmorillonite nanocomposite systems. Appl Clay Sci 101:288–296
153. Jahanizadeh S, Yazdian F, Marjani A, Omidi M, Rashedi H (2017) Curcumin-loaded chitosan/carboxymethyl starch/montmorillonite bio-nanocomposite for reduction of dental bacterial biofilm formation. Int J Biol Macromol 105:757–763

154. Khlibsuwan R, Pongjanyakul T (2016) Chitosan-clay matrix tablets for sustained-release drug delivery: effect of chitosan molecular weight and lubricant. J Drug Deliv Sci Technol 35:303–313
155. Liu K-H, Liu T-Y, Chen S-Y, Liu D-M (2008) Drug release behavior of chitosan–montmorillonite nanocomposite hydrogels following electrostimulation. Acta Biomater 4(4):1038–1045
156. Nanda R, Sasmal A, Nayak P (2011) Preparation and characterization of chitosan–polylactide composites blended with Cloisite 30B for control release of the anticancer drug paclitaxel. Carbohyd Polym 83(2):988–994
157. Nazemi K, Azadpour P, Moztarzadeh F, Urbanska A, Mozafari M (2015) Tissue-engineered chitosan/bioactive glass bone scaffolds integrated with PLGA nanoparticles: a therapeutic design for on-demand drug delivery. Mater Lett 138:16–20
158. Onnainty R, Onida B, Páez P, Longhi M, Barresi A, Granero G (2016) Targeted chitosan-based bionanocomposites for controlled oral mucosal delivery of chlorhexidine. Int J Pharm 509(1–2):408–418
159. Pongjanyakul T, Khunawattanakul W, Strachan CJ, Gordon KC, Puttipipatkhachorn S, Rades T (2013) Characterization of chitosan–magnesium aluminum silicate nanocomposite films for buccal delivery of nicotine. Int J Biol Macromol 55:24–31
160. Reddy AB, Manjula B, Jayaramudu T, Sadiku E, Babu PA, Selvam SP (2016) 5-Fluorouracil loaded chitosan–PVA/Na+ MMT nanocomposite films for drug release and antimicrobial activity. Nano-micro Lett 8(3):260–269
161. Dinu MV, Cocarta AI, Dragan ES (2016) Synthesis, characterization and drug release properties of 3D chitosan/clinoptilolite biocomposite cryogels. Carbohyd Polym 153:203–211
162. Gaware SA, Rokade KA, Kale S (2019) Silica-chitosan nanocomposite mediated pH-sensitive drug delivery. J Drug Deliv Sci Technol 49:345–351
163. Sattari S, Tehrani AD, Adeli M, Soleimani K, Rashidipour M (2020) Fabrication of new generation of co-delivery systems based on graphene-g-cyclodextrin/chitosan nanofiber. Int J Biol Macromol 156:1126–1134
164. Abbasian M, Roudi M-M, Mahmoodzadeh F, Eskandani M, Jaymand M (2018) Chitosan-grafted-poly (methacrylic acid)/graphene oxide nanocomposite as a pH-responsive de novo cancer chemotherapy nanosystem. Int J Biol Macromol 118:1871–1879
165. Ardeshirzadeh B, Anaraki NA, Irani M, Rad LR, Shamshiri S (2015) Controlled release of doxorubicin from electrospun PEO/chitosan/graphene oxide nanocomposite nanofibrous scaffolds. Mater Sci Eng C 48:384–390
166. Figueroa T, Aguayo C, Fernández K (2020) Design and characterization of chitosan-graphene oxide nanocomposites for the delivery of proanthocyanidins. Int J Nanomed 15:1229
167. Shariatinia Z, Mazloom-Jalali A (2019) Chitosan nanocomposite drug delivery systems designed for the ifosfamide anticancer drug using molecular dynamics simulations. J Mol Liq 273:346–367
168. Kandra R, Bajpai S (2020) Synthesis, mechanical properties of fluorescent carbon dots loaded nanocomposites chitosan film for wound healing and drug delivery. Arab J Chem 13(4):4882–4894
169. Dramou P, Fizir M, Taleb A, Itatahine A, Dahiru NS, Mehdi YA, Wei L, Zhang J, He H (2018) Folic acid-conjugated chitosan oligosaccharide-magnetic halloysite nanotubes as a delivery system for camptothecin. Carbohyd Polym 197:117–127
170. Jermy R, Ravinayagam V, Alamoudi W, Almohazey D, Elanthikkal S, Dafalla H, Rehman S, Chandrasekar G, Baykal A (2020) Tuning pH sensitive chitosan and cisplatin over spinel ferrite/silica nanocomposite for anticancer activity in MCF-7 cell line. J Drug Deliv Sci Technol 101711
171. Justin R, Chen B (2018) Multifunctional chitosan–magnetic graphene quantum dot nanocomposites for the release of therapeutics from detachable and non-detachable biodegradable microneedle arrays. Interface Focus 8(3):20170055
172. Pon-On W, Tithito T, Maneeprakorn W, Phenrat T, Tang I-M (2019) Investigation of magnetic silica with thermoresponsive chitosan coating for drug controlled release and magnetic hyperthermia application. Mater Sci Eng C 97:23–30

173. Popova M, Trendafilova I, Szegedi Á, Momekova D, Mihály J, Momekov G, Kiss LF, Lázár K, Koseva N (2018) Novel SO3H functionalized magnetic nanoporous silica/polymer nanocomposite as a carrier in a dual-drug delivery system for anticancer therapy. Microporous Mesoporous Mater 263:96–105
174. Shen J-M, Xu L, Lu Y, Cao H-M, Xu Z-G, Chen T, Zhang H-X (2012) Chitosan-based luminescent/magnetic hybrid nanogels for insulin delivery, cell imaging, and antidiabetic research of dietary supplements. Int J Pharm 427(2):400–409
175. Wu J, Jiang W, Shen Y, Tian R (2017) Synthesis and characterization of mesoporous magnetic nanocomposites wrapped with chitosan gatekeepers for pH-sensitive controlled release of doxorubicin. Mater Sci Eng C 70:132–140
176. Gui R, Wang Y, Sun J (2014) Encapsulating magnetic and fluorescent mesoporous silica into thermosensitive chitosan microspheres for cell imaging and controlled drug release in vitro. Colloids Surf B 113:1–9
177. Radmansouri M, Bahmani E, Sarikhani E, Rahmani K, Sharifianjazi F, Irani M (2018) Doxorubicin hydrochloride-Loaded electrospun chitosan/cobalt ferrite/titanium oxide nanofibers for hyperthermic tumor cell treatment and controlled drug release. Int J Biol Macromol 116:378–384
178. Kumari S, Singh RP (2013) Glycolic acid functionalized chitosan–Au–Fe3O4 hybrid nanoparticle based nanohybrid scaffold for drug delivery. Int J Biol Macromol 54:244–249
179. Xie M, Zhang F, Peng H, Zhang Y, Li Y, Xu Y, Xie J (2019) Layer-by-layer modification of magnetic graphene oxide by chitosan and sodium alginate with enhanced dispersibility for targeted drug delivery and photothermal therapy. Colloids Surf B 176:462–470
180. Yuan Q, Hein S, Misra R (2010) New generation of chitosan-encapsulated ZnO quantum dots loaded with drug: synthesis, characterization and in vitro drug delivery response. Acta Biomater 6(7):2732–2739
181. Kar S, Kaur T, Thirugnanam A (2016) Microwave-assisted synthesis of porous chitosan–modified montmorillonite–hydroxyapatite composite scaffolds. Int J Biol Macromol 82:628–636
182. Omidi S, Pirhayati M, Kakanejadifard A (2020) Co-delivery of doxorubicin and curcumin by a pH-sensitive, injectable, and in situ hydrogel composed of chitosan, graphene, and cellulose nanowhisker. Carbohydrate Polym 231:115745
183. Pooresmaeil M, Namazi H (2020) Facile preparation of pH-sensitive chitosan microspheres for delivery of curcumin; characterization, drug release kinetics and evaluation of anticancer activity. Int J Biol Macromol 162:501–511
184. Mohanty S, Sarangi S, Roy GS (2019) Formulation and evaluation of graphene grafted chitosan/polyaniline nanocomposites for controlled release of anticancer drug doxorubicin. Int J App Pharm 11(3):138–143

Chapter 8
Targeted Drug Delivery of Nanoparticles

Hayretin Tonbul and Yılmaz Capan

Nanoparticle drug delivery is a very attractive field that can provide to develop more potent and specific drugs. Packaging therapeutic cargos into nanoparticles improve their biocompatibility, bioavailability, and safety profiles. Moreover, somehow targeting these particles to a disease site can detect the presence of the disease site, block a function there, or specifically accumulate a drug to it. In last decade, one of the main research areas for nanoparticle drug delivery system is investigating targeting strategies to improve benefits of them. In this chapter, we review the targetable nano drug delivery system and possible targeting strategies including passive targeting and cellular-subcellular active targeting.

1 Overview of Targetable Nanoparticle Drug Delivery Systems

Nanoparticle drug delivery systems or nanomedicines could be defined as clinical and biomedical applications of engineered nanometer-sized materials. These nanosized materials, e.g., nanoparticles show unique properties as biologically, magnetically, and optically that are generally not found in the bulk sample of the same material [1] and could be inorganic (quantum dots, iron oxide, silica, etc.) or organic (polymeric nanoparticles, dendrimers, liposomes, etc.) based [2].

H. Tonbul (✉)
Faculty of Pharmacy, Department of Pharmaceutical Technology, Inonu University, Malatya, Turkey
e-mail: hayrettin.tonbul@inonu.edu.tr

Y. Capan
Faculty of Pharmacy, Department of Pharmaceutical Technology, Lokman Hekim University, Ankara, Turkey

S. Gopi et al. (eds.), *Nanotechnology for Biomedical Applications*,
Materials Horizons: From Nature to Nanomaterials,
https://doi.org/10.1007/978-981-16-7483-9_8

1.1 Polymeric Nanoparticles

Depend on the methods, polymeric nanoparticles can be prepared as nanocapsules or nanospheres and they are one of the most widely researched drug carriers. Synthetic polymers such as polycaprolactone (PCL), polyacrylate, polylactic acid (PLA), polyglycolic acid (PGA), and their copolymer poly lactic-co-glycolic acid (PLGA) [3], and natural polymers such as chitosan, gelatin, and sodium alginate could be used in drug delivery application of polymeric nanoparticles [4]. Polymeric nanoparticles are able to carry a wide range of therapeutics and have good surface modification potential. Their tunable size and easily functionalized and modified surface ensure to use them in targeted drug delivery systems.

1.2 Liposomes

Liposomes are bilayered vesicular systems formed from phospholipids which can be unilamellar and multilamellar. One of the key features of liposomes is the ability to carry both water-soluble and lipid-soluble drugs at the same time [5]. They are also promising biocompatible and biodegradable drug carriers having potential flexibility for internal and surface modifications [6]. Liposomes have a relatively long history of investigation and there are certain number of liposome formulations already using in the clinic that are generally passively targeted [7].

1.3 Polymeric Micelles

Polymeric micelles are spherical amphiphilic copolymer formed self-assembly in an aqueous environment. These supramolecular structures have gathered considerable attention in drug delivery due to their low toxicity, good biocompatibility, prolonged blood circulation time, and ability to enrapt high amount of drug in their micellar core [8]. Also, their stealth properties induced by hydrophilic polymeric brush on the micellar surface and opportunity for the chemical modification of the core–shell structure make them a promising candidate for drug targeting [9].

1.4 Dendrimers

Dendrimers are novel synthetic tree-shaped drug delivery systems having improved chemical and physical features. They have also predictable size, molecular weight, and shape according to generation number [10]. Although cationic dendrimers have toxicity problems, PEGylation of the surface neutralizes the charge and make them biocompatible [11]. Moreover, surface modification with ligands may also reduce their toxicity and make them suitable for targeted drug delivery with controlled release [12].

1.5 Solid Lipid Nanoparticles

Solid lipid nanoparticles (SLNs) were developed as an alternative to emulsions, liposomes, and polymeric nanoparticles as a colloidal system for controlled drug delivery [13]. The potential advantages of SLNs over other nanoparticle drug delivery systems are having higher drug entrapment capacity, high biocompatibility, and being easy to produce formulation [14]. Also, the surface of SLNs can be modified by conjugating different ligands to improve their targeting ability [15].

1.6 Ceramic Nanoparticles

Ceramic nanoparticles are primarily made from phosphates, carbides, oxides, metals, and metalloids such as silicon, alumina, and titanium. Their chemical inertness, high heat, and pH resistance make them useful in a wide range of drug delivery applications [16]. Their only drawback is the possibility of safety problems due to lack of biodegradation and slow dissolution [17]. However, their robustness and opportunity to synthesize them with very small size may make them a perfect candidate for targeted drug delivery.

2 Targeting Strategies in Nanoparticle Drug Delivery Systems

Targeted drug delivery systems divide two major categories in literature which are passive and active targeting [18]. The main goal of both strategies is to deliver the nanocarrier and their cargo preferentially to the diseased site while minimizing the accumulation of cargo in healthy cells and organs.

2.1 Passive Targeting

The term "passive targeting" in literature has been widely used in nanoparticle drug delivery systems to define findings related to the accumulation of nanomedicines in tumor tissue. The first observation related to passive targeting was reported by

Matsumura and Maeda in 1986 that show spontaneous accumulation of administered nanocarrier in the solid tumor due to leaky vasculature and decreased lymphatic recovery [19]. This study forms the basis of the term "enhanced permeability and retention" (EPR) effect that is closely related to passive targeting.

Pathophysiology of tumor tissue should be discussed to better understand this relationship. Sustained angiogenesis, neovasculature, and formation of new blood vessels, in tumor tissue to provide oxygen and nutrient supply to tumor cells and remove the metabolic waste product is one of the hallmarks of metabolically active cancer [20]. Such blood vessels highly abnormal and often exhibit a disorganized and chaotic course that structurally different from healthy vessels [21]. Moreover, these vessels may exhibit interendothelial gaps that size vary from a few nanometers to several hundred nanometers that increase the leakiness of tumor vasculature [22, 23]. Another pathophysiologic condition of tumor tissue is poor lymphatic drainage of solid tumors due to compression of lymphatic vessels of solid tumors by the high density of rapidly proliferating cancer cells that cause the collapse of the vessels leading to poor infiltration of lymphocytes [24]. Different from healthy tissues collapsed lymphatic vessels not able to drain fluid from tumor tissue efficiently and cause elevated interstitial fluid pressures inside solid tumor [25].

The EPR effect suggests that administered drug-loaded nanoparticles do not penetrate into normal healthy vascular walls while penetrating incomplete tumor vascular walls through interendothelial gaps. Meanwhile, retention of nanoparticles longer periods there due to impaired lymphatic drainage. Thereby, two nanoparticle criteria critical for passive targeting: (i) size of nanoparticle must be smaller than the cutoff size of tumor interendothelial gaps, and (ii) blood circulation time of nanoparticles should be extended as high as possible [26].

2.2 Factors Affecting Passive Targeting

Size, shape, elasticity, surface charge, and chemistry of nanoparticles play a vital role in the success of passive targeting. Among these, particle size of nanoparticles is the main factor due to passive targeting is mostly achievable during diffusion mediated transport [27]. The size of nanomaterial should be in the range of 40–200 nm to be most favorable for diffuse through tumor mass, extend circulation time and decrease renal clearance [28]. The shape of nanoparticles also plays a vital role in determining their in vivo characteristics for passive targeting. In a study, sphere-shaped nanoparticles were taken up 5 times greater rate than rod-shaped particles [29]. In another study, the impact of nanoparticles' elasticity was investigated for their biological effects, tumor uptake, and blood circulation. In vivo results show that softer nanoparticles (10 kPa) have higher blood circulation time than harder nanoparticles (3000 kPa) [30].

Surface charge and chemistry is another main factor for passive targeting that play a vital role in blood circulation time. To better understanding for impact of surface properties of nanoparticles, the physiology of the mononuclear phagocyte system

(MPS) should be discussed. In blood circulation, nanoparticles, just like other large particles such as bacteria, interact with plasma proteins and adsorb many of them that lead to a process called opsonization. The specificity and amount of these plasma proteins may vary and depends on the surface properties of nanoparticles [31]. One or more of these plasma proteins may interact with specific receptors on surface hepatocytes in the liver and macrophages and play a vital role in the clearance of nanoparticles [32]. Consequently, blood circulation time of nanoparticles, as well as the success of passive targeting, decreases dramatically with the increase in the opsonization process. Modifying the surface of nanoparticles and make the surface hydrophilic, "look like water", and slightly anionic or neutral is preferred to reduce opsonization [33]. To achieve this, the surface of the nanocarriers is modified with water-soluble polymers and gain the nanoparticle "stealth" behavior. Polyethylene glycol (PEG) is generally used for this purpose and this process is called PEGylation in literature [34].

2.3 Limitations of Passive Targeting

Passive targeting mainly depends on the EPR effect that relies on pathophysiologic features of solid tumors, including impaired lymphatic drainage and leaky vasculature. Although these features may theoretically result in the accumulation of nanoparticles in tumor tissue, it is important to mention that physiologic conditions substantially vary not only between tumors but also within tumors [35]. The extent of the vascular permeability, lymphatic drainage, blood perfusion rates, extracellular matrix density, and composition may vary between tumors and even different parts of the same tumor. Hence, this heterogenicity may lead to different outcomes in different tumors and irregular drug distribution in the same tumor and cause developing resistance to drugs and lessen therapeutic effect [26]. Moreover, the EPR effect is mainly derived based on mouse tumor model studies, and the extent of EPR in humans is controversial and subject to debate [36–38].

Current data show that only a small percentage of the total administered nanoparticles reaches the targeting site [39]. Thereby nanoparticles need better tools for targeting the nanoparticles to their needed site of action.

2.4 Active Targeting

Besides the passive targeting, nanoparticles must be efficiently targeted to desired tissue and subsequent cellular/subcellular locations in order to use all advantages of the unique properties of nanoparticle drug delivery systems [41, 42]. To achieve this, active targeting, also called ligand-mediated targeting, of nanoparticles is crucial (Fig. 1). Generally, active targeting involves the use of one or more targeting moieties such as antibodies, peptides, and some small molecules used to modify the surface

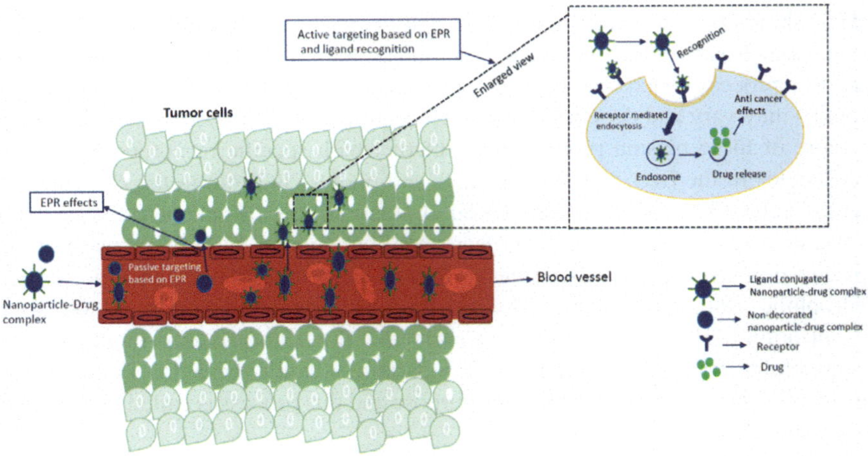

Fig. 1 Schematic representation of passive and active targeting. After entering the blood circulation, nanoparticles passively targeted and accumulated in tumor tissue by the EPR effect. Active targeting is achieved by modifying the surface of nanoparticles with a targeting agent and internalize the cell via overexpressed receptors on the surface of the tumor cells and EPR effects. Reprinted from [40]

of nanocarrier that interact specifically with receptor or antigens that are either overexpress or uniquely expressed on diseased tissue [43]. Active targeting strategies can be divided into two groups: (i) cellular targeting and (ii) subcellular targeting.

2.4.1 Cellular Targeting

The field of cancer research is one of the most researched areas for the implantation of active targeting of nanoparticles and identification of cell surface markers for nanoparticle targeting. Mostly investigated markers include programmed death ligand-1 (CD274), cell adhesion glycoprotein (CD44), myeloid antigen (CD13), vascular endothelial growth factor receptor (VEGFR), integrins, folate receptor protein, epidermal growth factor receptor (HER2), and somatostatin receptors (SSTRs) that overexpressed on the surface of the cancer cells [44]. Generally, specific ligands are used to targets these receptors and mostly investigated are listed below.

Polysaccharides and Proteins

Antibodies are Y-shaped proteins and they are tens of kilodaltons in size and have high specificity [45]. Antibodies are one of the most investigated and have the longest history in receptor targeting among all ligands [46]. Actually, apart from targeting agent propose of antibodies, many of them have also therapeutic effect [47]. The development of different antibody-associated targeted delivery systems for the different disease has become possible with recent advances in antibody engineering strategies. In cancer research, antibody-associated targeted nanoparticle delivery systems hold big potential for chemotherapy, chemoprevention, and early diagnosis of cancer [48]. However, it should be noted that the large size of antibodies may limit their density on the surface of nanoparticles during surface modification [49].

Transferrin is an iron-binding glycoprotein that is vital for iron transportation [50]. Transferrin receptors are overexpressed in specific tissues and cells that can be targeted by transferrin modified nanoparticles. Moreover, active targeting is also achieved with either conjugation of transferrin with drugs or genetically by infusion of therapeutic proteins or peptides into the structure of transferrin [51].

Hyaluronic acid is a polysaccharide that also capable of binding cell surfaces for cellular targeting. Hyaluronic acid is a natural linear polymer that is one of the main components of the extracellular matrix along with collagen [52]. Hyaluronic acid can specifically bind CD44 receptor that is overexpressed on various type of cancer cell including breast, colon, lung, prostate, head, and neck [53]. Not only nanoparticle surface modification agent but also hyaluronic acid drug conjugates have been widely researched and applied for tumor targeting [54].

Peptides

Using peptides as an active targeting agent has several advantages such as good stability, ease of conjugation on the surface of nanocarriers due to their small size, showing no or limited recognition by the mononuclear phagocyte system, and low cost [55, 56]. Moreover, it is possible that chemically modifying the amino acid sequences of peptides to prevent enzymatic degradation, enhance the peptide stability and improve the effectiveness of binding to several receptors. Selected peptides can be efficiently conjugated to nanoparticles and drugs for active targeting [56].

Aptamers

Aptamers are short single-stranded nucleic acid (RNA or DNA) sequences comprising several nucleotides. They can fold to three-dimensional structures and bind to targeted cell surface that overexpresses specific receptors for different aptamers [49]. Aptamers have several advantages including easy modification, small size, high stability in rough chemical and physical environment, economic and rapid production, lack of batch to batch variation, high flexibility, and low immunogenicity that make them good candidates for active targeting agent [57]. However, they are also some limitations such as exposure to serum degradation, relatively low specificity, and short half-life [58].

Small Molecules

One of the most investigated small molecules for the active targeting agent is folic acid that specifically binds the folate receptor. Folic acid or vitamin B9 is a poorly immunogenic, stable, and water-soluble vitamin that is vital for several cell functions including methylation, DNA synthesis, and replication, cell division, and growth [59]. Folate or folic acid receptors is overexpressed on several solid tumor cells such as brain, breast, ovarian, cervical, lung, colorectal, and kidney and macrophages that make them attractive targets for many nanoparticle drug delivery systems via receptor-mediated endocytosis [60–62].

Another small molecule using as an active targeting agent is anisamide that is a low molecular benzamide derivative known to bind sigma receptors overexpressed in cancer cells [63]. Although using anisamide as targeting ligand relatively new, there are some studies show that anisamide decorated nanocarriers increase tumor cell internalization. However, current studies suggest that further studies are needed to identify anisamide's exact mechanism of action [64, 65].

There are also some other small molecules that are investigated use them as a targeting ligand by research groups. For example, Kataoka group investigated that phenylboronic acid strongly attached to N-acetylneuraminic acid that are the main components of sialic acid. It is also known that sialic acid epitopes are overexpressed in many tumor cells. Their results show that phenyboronic acid-mediated targeting of nanoparticles offers a highly translational approach for cancer diagnosis and therapy [66].

2.4.2 Subcellular Targeting

Although several studies prove that cell surface markers effective for active targeting and improve the drug efficacy tremendously in a wide range of cell types, this may not sufficient to deliver the drugs to specific subcellular locations, i.e., organelles, and achieve the ultimate target [67–70]. In order to target the nanoparticles to organelles, arrange the nanoparticle size to match the biologic properties of subcellular location, biocompatibility, appropriate particle surface modifications, and charge are important aspects to consider [71]. Targeting the organelles which are responsible for vital mechanisms to cell survival including cellular energy level, protein synthesis, replication, and digestion are the most important four aspects in subcellular-based therapeutic strategies. Therefore, mitochondria, nucleus, endoplasmic reticulum, and golgi apparatus are primary targets in subcellular targeting (Fig. 2).

Targeting to Mitochondria

Mitochondria is one of the most vital organelles that responsible for several critical roles in cellular homeostasis and physiology including the regulation of reactive

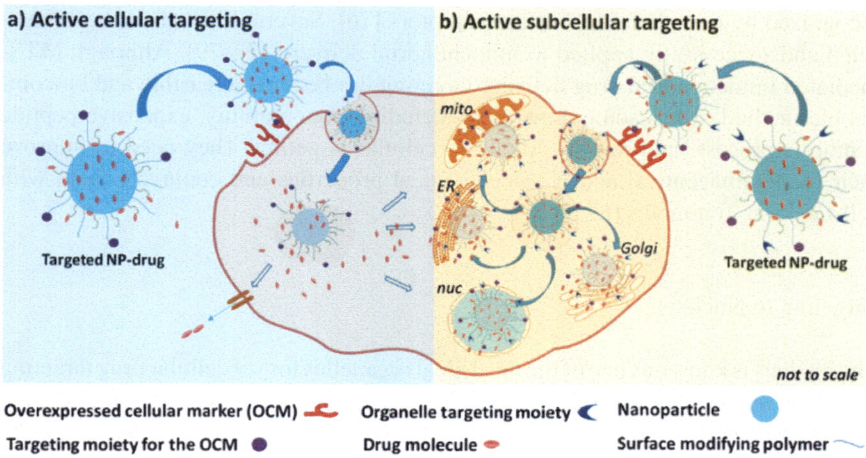

Fig. 2 Schematic representation of active cellular (**a**) and subcellular (**b**) targeting of nanoparticle formulations. In active cellular targeting, nanoparticles bind the overexpressed cell surface marker such as folate and epidermal growth factor receptors via affinity of a targeting moiety such as antibodies, peptides, and specific small molecules. The nanoparticles are internalized via endocytosis and the carried drugs are released in the cytosol and diffuse to subcellular sites of action. In active subcellular targeting, similar to cellular targeting, nanoparticles bind the cell surface, internalize the cells. However, in this approach, the nanoparticles themselves localize to specific subcellular compartments such as mitochondria (mito), nucleus (nuc), endoplasmic reticulum (ER), and golgi apparatus (Golgi) and release the drug in it. Subcellular targeting minimizes adverse off-target drug effects. Reprinted from [67]

oxygen species, provide energy to the cell in the form of ATP via oxidative phosphorylation, lipid synthesis, cell proliferation, control the intracellular calcium ion concentration, and apoptosis [72, 73]. Therefore, it enormously contributes the cellular mortality management and many clinically approved drugs directly affect mitochondria by triggering apoptosis. As a consequence, mitochondria would be privileged pharmacological target, and targeting mitochondria with nanoparticle drug delivery systems hold big potential to control mitochondrial dysfunction-related diseases [68]. However, the highly complex nature of mitochondria makes it difficult to develop such nanoparticle drug delivery systems. Targeting mitochondria and deliver the drugs on it depends on both monitoring of mitochondrial entry and intracellular trafficking routes.

One of the most investigated methods to overcome the difficulties is utilizing delocalized lipophilic cations as ligand including triphenyphophonium, nitrogen containing heterocycles, quaternary ammonium salts, rodamine, and dequalinum [68, 71]. They have a unique ability to both traverse phospholipid membrane and accumulate mitochondrial matrix due to their high lipophilicity and stable cationic charge [74, 75].

Another method is developing specific mitochondrial targeting sequences or mitochondrial targeting ligands (MTSs). MTSs consist of 20–40 amino acids that are recognized by mitochondrial surface receptors [76]. Several MTSs have been identified and successfully applied to mitochondrial delivery [77–79]. Although MTS-mediated mitochondrial drug delivery targeting has become a feasible and biocompatible method, it has some drawbacks including poor stability, expensive peptide synthesis process, and low specificity to cellular targeting. They need to improve their biopharmaceutical and physicochemical properties and conjugate them with cell-targeting fragments [67, 76].

Targeting to Nucleus

The nucleus is known as one of the most ideal organelles for subcellular drug targeting due to it has central position for metabolism, cell cycle, and cell reproduction. However, there are major obstacles that need to consider for nucleus targeting. The nucleus is separated from the cytosol by the double-layered lipid membrane that is the biggest barrier need to nanoparticles overcome [74]. This membrane is perforated by nuclear pore complex (NPC) with transverse diameter of ~70 nm that acts as a selective portal between cytoplasm and nucleus. Moreover, the inner walls of NPCs are tethered with phenylalanine glycine nucleoporins that reduce the inner diameter to even ~40 nm [80]. Although, ions and small molecules access to the nucleus via passive diffusions, macromolecules larger than 10 nm requires active transport.

One strategy that may be used to target the nucleus is regulate the nanoparticle size and make them smaller than 10 nm and passively diffuse the nucleus [81]. However, this approach obviously cannot apply all drug carriers and the charge and shape of the nanoparticle may also hinder passive diffusion even if a sub 10 nm sized particle is obtained [76].

In the nucleus, there is also an active transport system mediated by nuclear transport receptors that recognize and bind to specific basic peptide sequences exposed on cargo known as nuclear localization signal (NLS). NLS binds to protein cargo and translocates cargos, even those larger than 10 nm, into the nucleus. Recent studies show that specific peptides such as transactivator of transcription (TAT), nuclear factor kappa B (NF-κB), and KRRRR peptide can be used to modify nanoparticles to target them nucleus via NLS [76, 82, 83].

In addition to improving the physicochemical properties of nanoparticles to pass through NPCs, using cell membrane penetrating peptides to open the nuclear membrane may be another effective method for nuclear targeting. Some research groups have successfully synthesized a series of cell membrane peptides to modify the nanoparticles that could effectively penetrate the cell and nucleus membrane and directly damage to DNA [84].

In conclusion, nuclear delivery of nanoparticles is one of the most promising approaches for subcellular targeting but a strict physicochemical arrangement of nanoparticles including size, shape, surface charge, and surface modification with specific ligands is needed.

Targeting to Endoplasmic Reticulum (ER) and Golgi Apparatus (GA)

The ER and GA have attracted attention due to their important roles in endocytosis and having a large intracellular surface. ER is the largest organelle that controls the biosynthesis, folding, and assembly of proteins and other biological macromolecules and playing a vital role in the homeostasis of cells [85]. Golgi apparatus is closely linked to the endoplasmic reticulum that plays important roles in cell secretory pathways [86]. Both ER and GA valuable targets for the delivery of chemotherapeutics and induce cell death [76].

One of the mechanisms for targeting these subcellular units is regulating the endocytosis pathway of nanoparticles. Caveola mediated endocytosis pathway can directly transport nanoparticles into the ER and GA [76]. There are several factors that play role in the endocytosis pathway selection of molecules and nanoparticles can internalize the cell via caveolae-mediated endocytosis by making the necessary physicochemical adjustment on nanoparticles [87, 88].

Another approach to designing ER and GA targeted nanoparticle system is to avoid discharging of nanoparticles by exocytosis via enhancing their retention time in target substructures. For example, Xeu and coworkers reported a pH-responsive photothermal ablation agent that modified with bovine serum albumin to obtain nanoparticles. Results show that these nanoparticles accumulated in GA of cancer cells during endocytosis due to their hypertrophic morphology. Moreover, nanoparticles can be activated for effective photothermal therapy in response to the acidic media of the GA [89]. Also, there are some studies that show successful ER delivery with nanoparticle drug delivery systems [90–92]. However, it should be noted that subcellular targeting of the ER and GA is still in infancy when compared with the mitochondria and nucleus targeting.

3 Conclusion

Nanoparticle drug delivery systems are the modern form of drug delivery that could minimize side-effects, improve bioavailability and reduce both dosage and dosage frequency. Moreover, targeting them to a specific area can tremendously improve the overall therapeutic outcomes of drugs. When targeting strategies are considered, it is seen that using only passive targeting, which is the oldest method that first showed up more than 30 years ago, is generally insufficient. There are several passively targeted nanomedicine products on market and their clinical results show that although they improve the overall therapeutic outcomes, the expected effect is not at the desired level. It is clear that cellular and subcellular targeted drug delivery of nanoparticles is an ever-expanding area of drug development and holds big potential to solving problems where passive targeting is not enough. However, many obstacles remain to be overcome before both cellular and subcellular active targeting nanoparticles reach their full clinical potential.

References

1. Auffan M et al (2009) Towards a definition of inorganic nanoparticles from an environmental, health and safety perspective. Nat Nanotechnol 4(10):634–641
2. Khalid K et al (2020) Advanced in developmental organic and inorganic nanomaterial: a review. Bioengineered 11(1):328–355
3. Banik BL, Fattahi P, Brown JL (2016) Polymeric nanoparticles: the future of nanomedicine. Wiley Interdiscipl Rev Nanomed Nanobiotechnol 8(2):271–299
4. Zhang Z et al (2013) Polymeric nanoparticles-based topical delivery systems for the treatment of dermatological diseases. 5(3):205–218
5. Akbarzadeh A et al (2013) Liposome: classification, preparation, and applications. Nanoscale Res Lett 8(1):102
6. Jain A, Jain (2018) Advances in tumor targeted liposomes. Curr Mol Med 18(1):44–57
7. Bulbake U et al (2017) Liposomal formulations in clinical use: an updated review. Pharmaceutics 9(2):12
8. Zhang Y, Huang Y, Li S (2014) Polymeric micelles: nanocarriers for cancer-targeted drug delivery. AAPS PharmSciTech 15(4):862–871
9. Mahmud A et al (2007) Polymeric micelles for drug targeting. J Drug Target 15(9):553–584
10. Scott RWJ, Wilson OM, Crooks RM (2005) Synthesis, characterization, and applications of dendrimer-encapsulated nanoparticles. J Phys Chem B 109(2):692–704
11. Kesharwani P et al (2017) Chapter 11—dendrimers in targeting and delivery of drugs, in nanotechnology-based approaches for targeting and delivery of drugs and genes. In: Mishra V et al (eds). Academic Press, pp 363–388
12. Mandal AK (2021) Dendrimers in targeted drug delivery applications: a review of diseases and cancer. Int J Polym Mater Polym Biomater 70(4):287–297
13. Bhatia S (2016) Natural polymer drug delivery systems : nanoparticles, plants, and algae. Springer International Publishing: Imprint: Springer, Cham, p. 1 online resource (X, 225 pages 31 illustrations, 3 illustrations in color)
14. Tekade RK et al (2017) Solid Lipid nanoparticles for targeting and delivery of drugs and genes. In: Nanotechnology-based approaches for targeting and delivery of drugs and genes, pp 256–286
15. Rostami E et al (2014) Drug targeting using solid lipid nanoparticles. Chem Phys Lipid 181:56–61
16. Thomas SC et al (2015) Ceramic nanoparticles: fabrication methods and applications in drug delivery. Curr Pharm Des 21(42):6165–6188
17. Emerich DF, Thanos CG (2007) Targeted nanoparticle-based drug delivery and diagnosis. J Drug Target 15(3):163–183
18. Rosenblum D et al (2018) Progress and challenges towards targeted delivery of cancer therapeutics. Nat Commun 9(1):1410
19. Matsumura Y, Maeda H (1986) A new concept for macromolecular therapeutics in cancer chemotherapy: mechanism of tumoritropic accumulation of proteins and the antitumor agent smancs. Cancer Res 46(12 Pt 1):6387–6392
20. Nagy JA et al (2010) Heterogeneity of the tumor vasculature. Semin Thromb Hemost 36(3):321–331
21. Nagy JA et al (2009) Why are tumour blood vessels abnormal and why is it important to know? Br J Cancer 100(6):865–869
22. Hobbs SK et al (1998) Regulation of transport pathways in tumor vessels: role of tumor type and microenvironment. Proc Natl Acad Sci USA 95(8):4607–4612
23. Nagy JA, Dvorak AM, Dvorak HF (2012) Vascular hyperpermeability, angiogenesis, and stroma generation. Cold Spring Harbor Persp Med 2(2)
24. Lobatto ME et al (2015) Atherosclerotic plaque targeting mechanism of long-circulating nanoparticles established by multimodal imaging. ACS Nano 9(2):1837–1847
25. Fredman G et al (2015) Targeted nanoparticles containing the proresolving peptide Ac2–26 protect against advanced atherosclerosis in hypercholesterolemic mice. Sci Transl Med 7(275)

26. Narum SM et al (2020) Chapter 4—passive targeting in nanomedicine: fundamental concepts, body interactions, and clinical potential. In: Chung EJ, Leon L, Rinaldi C (eds) Nanoparticles for biomedical applications. Elsevier, pp 37–53
27. Patel JK, Patel AP (2019) Passive targeting of nanoparticles to cancer. In: Pathak YV (ed) Surface modification of nanoparticles for targeted drug delivery. Springer International Publishing, Cham, pp 125–143
28. Liechty WB, Peppas NA (2012) Expert opinion: responsive polymer nanoparticles in cancer therapy. Eur J Pharm Biopharm 80(2):241–246
29. Chithrani BD, Chan WCW (2007) Elucidating the mechanism of cellular uptake and removal of protein-coated gold nanoparticles of different sizes and shapes. Nano Lett 7(6):1542–1550
30. Anselmo AC et al (2015) Elasticity of nanoparticles influences their blood circulation, phagocytosis, endocytosis, and targeting. ACS Nano 9(3):3169–3177
31. Verhoef JJF, Anchordoquy TJ (2013) Questioning the use of PEGylation for drug delivery. Drug Deliv Transl Res 3(6):499–503
32. Yan X, Scherphof GL, Kamps JA (2005) Liposome opsonization. J Liposome Res 15(1–2):109–139
33. Attia MF et al (2019) An overview of active and passive targeting strategies to improve the nanocarriers efficiency to tumour sites 71(8):1185–1198
34. Suk JS et al (2016) PEGylation as a strategy for improving nanoparticle-based drug and gene delivery. Adv Drug Deliv Rev 99(Pt A):28–51
35. Prabhakar U et al (2013) Challenges and key considerations of the enhanced permeability and retention effect for nanomedicine drug delivery in oncology. Cancer Res 73(8):2412–2417
36. van der Meel R, Lammers T, Hennink WE (2017) Cancer nanomedicines: oversold or underappreciated? Expert Opin Drug Deliv 14(1):1–5
37. Björnmalm M et al (2017) Bridging bio-nano science and cancer nanomedicine. ACS Nano 11(10):9594–9613
38. Crockford A et al (2014) Implications of intratumour heterogeneity for treatment stratification. J Pathol 232(2):264–273
39. Wilhelm S et al (2016) Analysis of nanoparticle delivery to tumours. Nat Rev Mater 1(5)
40. Jahan S, Karim ME, Chowdhury EH (2021) Nanoparticles targeting receptors on breast cancer for efficient delivery of chemotherapeutics. Biomedicines 9(2)
41. Singh R, Lillard JW Jr (2009) Nanoparticle-based targeted drug delivery. Exp Mol Pathol 86(3):215–223
42. Bazak R et al (2015) Cancer active targeting by nanoparticles: a comprehensive review of literature. J Cancer Res Clin Oncol 141(5):769–784
43. Pearce AK, O'Reilly RK (2019) Insights into active targeting of nanoparticles in drug delivery: advances in clinical studies and design considerations for cancer nanomedicine. Bioconjug Chem 30(9):2300–2311
44. Jahan ST et al (2017) Targeted therapeutic nanoparticles: an immense promise to fight against cancer. J Drug Deliv 2017:9090325
45. Kim KS et al (2018) Multifunctional trastuzumab-chlorin e6 conjugate for the treatment of HER2-positive human breast cancer. Biomater Sci 6(5):1217–1226
46. Alibakhshi A et al (2017) Targeted cancer therapy through antibody fragments-decorated nanomedicines. J Control Release 268:323–334
47. Phillips GDL et al (2008) Targeting HER2-positive breast cancer with trastuzumab-DM1, an antibody-cytotoxic drug conjugate. Can Res 68(22):9280–9290
48. Attarwala H (2010) Role of antibodies in cancer targeting. J Nat Sci Biol Med 1(1):53–56
49. Yoo J et al (2019) Active targeting strategies using biological ligands for nanoparticle drug delivery systems. Cancers 11(5)
50. Tonbul H et al (2019) Combination drug delivery with actively-targeted PLGA nanoparticles to overcome multidrug resistance in breast cancer. J Drug Deliv Sci Technol 54:101380
51. Qian ZM et al (2002) Targeted drug delivery via the transferrin receptor-mediated endocytosis pathway 54(4):561–587

52. Larrañeta E et al (2018) Synthesis and characterization of hyaluronic acid hydrogels crosslinked using a solvent-free process for potential biomedical applications. Carbohyd Polym 181:1194–1205
53. Kapoor A, Kumar S (2014) Cancer stem cell: A rogue responsible for tumor development and metastasis. Indian J Cancer 51(3):282–289
54. Fang Z et al (2019) Hyaluronic acid-modified mesoporous silica-coated superparamagnetic Fe(3)O(4) nanoparticles for targeted drug delivery. Int J Nanomed 14:5785–5797
55. Chen K, Conti PS (2010) Target-specific delivery of peptide-based probes for PET imaging. Adv Drug Deliv Rev 62(11):1005–1022
56. Roveri M et al (2017) Peptides for tumor-specific drug targeting: state of the art and beyond. J Mater Chem B 5(23):4348–4364
57. Proske D et al (2005) Aptamers–basic research, drug development, and clinical applications. Appl Microbiol Biotechnol 69(4):367–374
58. Jo H, Ban C (2016) Aptamer-nanoparticle complexes as powerful diagnostic and therapeutic tools. Exp Mol Med 48(5):e230–e230
59. Zwicke GL, Mansoori GA, Jeffery CJ (2012) Utilizing the folate receptor for active targeting of cancer nanotherapeutics. Nano Rev 3
60. van Dam GM et al (2011) Intraoperative tumor-specific fluorescence imaging in ovarian cancer by folate receptor-alpha targeting: first in-human results. Nat Med 17(10):1315–1319
61. Nogueira E et al (2016) Folate-targeted nanoparticles for rheumatoid arthritis therapy. Nanomedicine 12(4):1113–1126
62. Son J et al (2018) Folate-modified PLGA nanoparticles for tumor-targeted delivery of pheophorbide a in vivo. Biochem Biophys Res Commun 498(3):523–528
63. Fitzgerald KA et al (2016) A novel, anisamide-targeted cyclodextrin nanoformulation for siRNA delivery to prostate cancer cells expressing the sigma-1 receptor. Int J Pharm 499(1–2):131–145
64. Dasargyri A et al (2016) Findings questioning the involvement of Sigma-1 receptor in the uptake of anisamide-decorated particles. J Control Release 224:229–238
65. Dasargyri A, Kümin CD, Leroux J-C (2017) Targeting nanocarriers with anisamide: fact or artifact? 29(7):1603451
66. Deshayes S et al (2013) Phenylboronic acid-installed polymeric micelles for targeting sialylated epitopes in solid tumors. J Am Chem Soc 135(41):15501–15507
67. Nag OK, Delehanty JB (2019) Active cellular and subcellular targeting of nanoparticles for drug delivery. Pharmaceutics 11(10)
68. Ma X et al (2016) Future of nanotherapeutics: targeting the cellular sub-organelles. Biomaterials 97:10–21
69. Sakhrani NM, Padh H (2013) Organelle targeting: third level of drug targeting. Drug Des Devel Ther 7:585–599
70. Rajendran L, Knolker HJ, Simons K (2010) Subcellular targeting strategies for drug design and delivery. Nat Rev Drug Discov 9(1):29–42
71. Huang JG, Leshuk T, Gu FX (2011) Emerging nanomaterials for targeting subcellular organelles. Nano Today 6(5):478–492
72. Kausar S, Wang F, Cui H (2018) The role of mitochondria in reactive oxygen species generation and its implications for neurodegenerative diseases. Cells 7(12):274
73. Chen Z et al (2015) Hierarchical targeted hepatocyte mitochondrial multifunctional chitosan nanoparticles for anticancer drug delivery. Biomaterials 52:240–250
74. Torchilin VP (2006) Recent approaches to intracellular delivery of drugs and DNA and organelle targeting 8(1):343–375
75. Breunig M, Bauer S, Goepferich A (2008) Polymers and nanoparticles: intelligent tools for intracellular targeting? Eur J Pharm Biopharm 68(1):112–128
76. Fu XL et al (2020) Precise design strategies of nanomedicine for improving cancer therapeutic efficacy using subcellular targeting. Signal Transduction Targeted Therapy 5(1)
77. Jiang L et al (2015) Overcoming drug-resistant lung cancer by paclitaxel loaded dual-functional liposomes with mitochondria targeting and pH-response. Biomaterials 52:126–139

78. Kawamura E et al (2013) Intracellular observation of nanocarriers modified with a mitochondrial targeting signal peptide. J Biosci Bioeng 116(5):634–637
79. Mukhopadhyay A et al (2005) Bacterial signal peptide recognizes HeLa cell mitochondrial import receptors and functions as a mitochondrial leader sequence. Cell Mol Life Sci 62(16):1890–1899
80. Alber F et al (2007) The molecular architecture of the nuclear pore complex. Nature 450(7170):695–701
81. Huo SD et al (2014) Ultrasmall gold nanoparticles as carriers for nucleus-based gene therapy due to size-dependent nuclear entry. ACS Nano 8(6):5852–5862
82. Sun YL et al (2020) Novel polymeric micelles as enzyme-sensitive nuclear-targeted dual-functional drug delivery vehicles for enhanced 9-nitro-20(S)-camptothecin delivery and antitumor efficacy. Nanoscale 12(9):5380–5396
83. Cheng Y et al (2019) A multifunctional peptide-conjugated AIegen for efficient and sequential targeted gene delivery into the nucleus 58(15):5049–5053
84. Zhang L et al (2016) Functionalized cell nucleus-penetrating peptide combined with doxorubicin for synergistic treatment of glioma. Acta Biomater 42:90–101
85. Xiao H et al (2017) A new endoplasmic reticulum-targeted two-photon fluorescent probe for imaging of superoxide anion in diabetic mice. Biosens Bioelectron 91:449–455
86. Zhang H et al (2013) An Off-On COX-2-specific fluorescent probe: targeting the golgi apparatus of cancer cells. J Am Chem Soc 135(31):11663–11669
87. Manzanares D, Cena V (2020) Endocytosis: the nanoparticle and submicron nanocompounds gateway into the cell. Pharmaceutics 12(4)
88. Foroozandeh P, Aziz AA (2018) Insight into cellular uptake and intracellular trafficking of nanoparticles. Nanoscale Res Lett 13
89. Xue FF et al (2017) A smart drug: a pH-responsive photothermal ablation agent for Golgi apparatus activated cancer therapy. Chem Commun 53(48):6424–6427
90. Nam JS et al (2016) Endoplasmic reticulum-localized iridium(III) complexes as efficient photodynamic therapy agents via protein modifications. J Am Chem Soc 138(34):10968–10977
91. Sher YP et al (2019) Endoplasmic reticulum-targeting sequence enhanced the cellular immunity of a tumor-associated antigen L6-based DNA vaccine. Am J Cancer Res 9(9):2028–2036
92. Feng Z et al (2018) Enzymatic assemblies disrupt the membrane and target endoplasmic reticulum for selective cancer cell death. J Am Chem Soc 140(30):9566–9573

Chapter 9
Biomedical Applications of Nano-Biosensor

Mamta Bishnoi, Deepika, Nishi Mody, and Ankit Jain

1 Introduction

Biosensors are individualized to identify a particular chemical species or a biochemical molecule (DNA sequence, protein biomarkers, genes, cells, or pathogens) concerning proportional concentration, bio-composition, structure, and functionalities in analytical samples. Nano-biosensors work selectively, reversibly, and repetitively, and are embodied with a biologically active element like isolated enzymes, tissues, whole cells, and immune systems with the physicochemical transducer to generate electrical, thermal, or optical indications [35, 149]. Depending on their intended use and nature of the biochemical entities, biosensors are also called glucometers, immunosensors, resonant mirrors, optrodes, chemical canaries, biochips, and biocomputers. The invention of glucose biosensor in 1962, was the first milestone on the pathway of biosensors development led by Leland Clark and Lyons, then in 1969 Guilbault and Montalvo Jr. came out with urea biosensor based on potentiometric applications elaborated in Table 1 [30]. Nowadays, biosensor applications

M. Bishnoi
School of Pharmaceutical Sciences, Delhi Institute of Pharmaceutical Sciences and Research, New Delhi 110017, India

Deepika
School of Pharmaceutical Sciences, Om Sterling Global University, Hisar, Haryana 125001, India

N. Mody
Department of Pharmaceutical Sciences, Dr. Harisingh Gour Central University, Sagar, MP 470003, India

A. Jain (✉)
Department of Materials Engineering, Indian Institute of Science, Bangalore, Karnataka 560012, India
e-mail: ankitjain@iisc.ac.in; ankitjainsagar@gmail.com

S. Gopi et al. (eds.), *Nanotechnology for Biomedical Applications*,
Materials Horizons: From Nature to Nanomaterials,
https://doi.org/10.1007/978-981-16-7483-9_10

Table 1 Historical overview of biosensor development

Year	Event	References
1916	First reported immobilized invertase enzyme adsorbed on activated charcoal and $Al(OH)_3$	[44]
1956	The invention of the first oxygen electrode by clark	[110]
1962	First glucose detection biosensor with an amperometric electrode	[110]
1969	First urea biosensor with urease enzyme and potentiometric biosensor	[50]
1970	The invention of ion-sensitive field-effect transistor (ISFET)	[43]
1975	First marketed glucose biosensor	[151]
1975	First cell/microbe-based immunosensor	[41]
1983	Evolution of immunosensor with surface plasmon resonance (SPR)	[83]
1984	First glucose biosensor with glucose oxidase and ferrocene having amperometric transducer	[24]
1987	Launching of "The Medi Sense Exac Tech™"	[161]
1990	SPR-based biosensor for quick analysis of interaction kinetics and affinities by Pharmacia Biacore (Uppsala, Sweden)	[118]
1992	Portable clinical analyzer (PCA) by i-STAT for glucose, sodium, chloride, potassium, urea, and nitrogen	[45]
1996	Launching of Glucocard	[108]
1998	Launch of Life Scan Fast Take biosensor	[2]
2001	Market launching of The QuickVue H. pylori Test (detects IgG antibodies specific to *Helicobacter pylori* in human serum, plasma, or whole blood)	[14]
2003	The development of commercial biosensors to detect *Escherichia coli* O157 in stool specimens	[14]
2004	The first digital ovulation test for human chorionic gonadotrophin in urine launched by Clearblue®	[127]
2004	Enzyme-linked immunospot test "T-SPOT®.TB Test" for Mycobacterium tuberculosis	[94]
2006	Dexcom (a US-based company), released an "Artificial pancreas" with a 7 days life	
2008	ANP technologies launched NIDS®, handheld biothreat assay, and reader for biothreat detection	[120]
2009	Launched rapid diagnosis of Influenza A + B point-of-care (POC) with brand name QuickVue®	[14]
2009	FDA approved xTAG® respiratory virus panel (developed by Luminex Corporation, USA) for respiratory viral infections	[120]
2010	GeneXpert designed by Cepheid, the USA for the identification of various pathogens	[120]
2012	FDA approved the first over-the-counter OraQuick, in-home self-testing HIV kit	[40]
2013	The launching of the first continuous *glucose* monitoring system "Guardian™"	[66]

(continued)

Table 1 (continued)

Year	Event	References
2013	For online respiration rate measurement "The Ra-TOX®" Online Toxicity Analyzer launched	[120]
2016	The first G5 Mobile-based Glucose Monitoring System (CGM) developed by Dexcom	[22]
2020	'ID NOW Rapid Isothermal System' was launched by Abbott for the identification of COVID-19	[156]
2020	IRIDICA BAC BSI assay for bloodstream infections	[120]

are found in various fields, earlier their uses were limited to research and laboratories. The term biosensor was coined by Karl Camman in 1977 and defined by the IUPAC (International Union of Pure and Applied Chemistry) standard in 1997 as "*A biosensor is a self-contained integrated device, which is capable of providing specific quantitative or semi-quantitative analytical information using a biological recognition element (biochemical receptor), which is retained in direct spatial contact with a transduction element*" [20]. We can classify biosensors in three generations; in the first generation, the resultant product of the reaction is converted into signal and amplified electrical response analyzed by the transducer, in the second generation-specific 'mediators' are plotted to link the resultant product with the transducer to induce a response, while in third generation, analyte's presence indicated by the direct electrone transfer to electrode without the dispersion of product or mediator.

2 Components of Biosensor

Finally, biosensors comprise (i) biomolecule (ii) transducer (iii) electronic components, and (iv) readout/display unit. Practically in a biosensor, a biomolecule/bioreceptor unit senses the biochemical/chemical entity then the transducer transforms it into a quantifiable signal (Fig. 1) [20, 124].

(i) A biomolecule is selected considering its specificity to the analyte and has a direct link with the selective analyte during operation. In biosensors, bioreceptors (e.g., enzymes, aptamers, antibodies, whole cells, and DNA) are either adsorbed or immobilized as a layer on the electrodes against a specific analyte [20].

(ii) A transducer analyzes the specific biochemical reaction due to electrochemical signal generated. It may be calorimetric (exothermic and endothermic chemical reaction), potentiometric (shifting in electrical potential), amperometric (electronic movement in a redox reaction), optical (the light emission and absorbance difference during the reaction), piezo-electric (reactant and product mass consideration), pyroelectric (analyte's current variations),

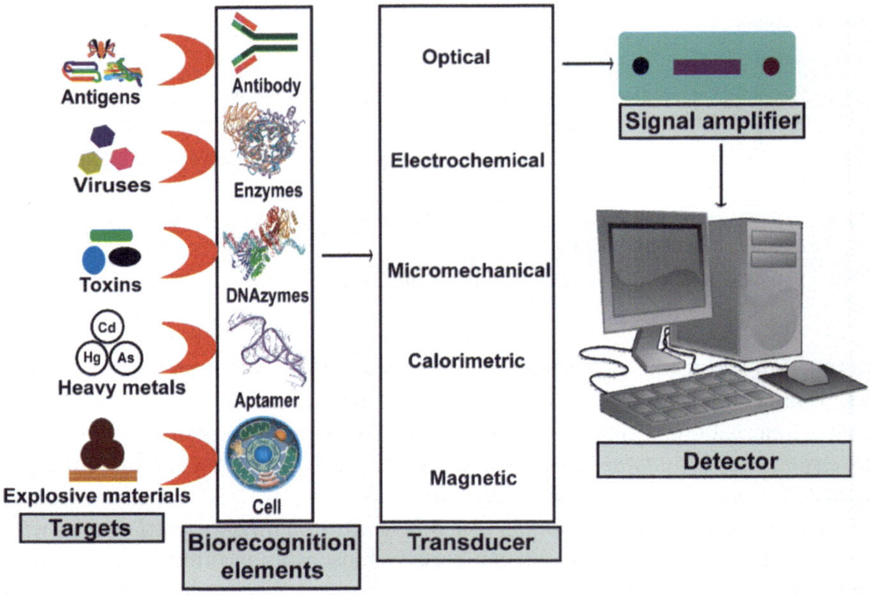

Fig. 1 Schematic showing a biosensor and its fundamental components [124]. Adapted under CC BY 4.0 license

impedance (reactants and product's variations in the conductivity or resistivity), acoustic (changes in mass density, viscoelastic, elastic, electric, or dielectric characteristics of chemical layer) and barometric (pressure measurement of bacterial cell's respiration) [137], (iii) Electronic components consists of complex electronic circuitry which converts the electronic indication into a physical parameter after amplification and converts the signals analog form into the digital form so-called as signal processor and (iv) Finally, a display unit to show the results to the operator [131].

3 Nano-Biosensor

As the name itself depicts nano-biosensor means incorporation of nano-sized materials (like nanoparticles, magnetic nanomaterials, metal nanoparticles, oxide nanoparticles, nanotubes, nanorods, nanomembrane, nanowires, carbon materials, quantum dots, and metallo phthalocyanine) (<100 nm) into the formulation of these [29]. Nano-biosensors are classified on various grounds as demonstrated in Table 2, while biosensors are classified according to the analyte to be determined and the kind of transducer involved for signal detection. Nano-biosensors can be further classified as per carbon-based nanomaterial used in their fabrication, explored in Table 3 [5, 37, 112].

Table 2 Classification of nano-biosensors

Type of biological element used	Type of nanomaterial used	Type of transducer used	According to the performance	According to implantable
1. Enzyme based 2. Antibody-based 3. Biological tissue-based 4. Micro-organism based 5. Cell-based 6. Nucleotide based 7. Sensory receptor-based 8. Amino acid-based	1. Carbon-based nano-biosensor a. Carbon Nanotube (CNT) b. Graphene c. Nanowire- silicone Nanowire d. Nanomembrane 2. Polymer-based A. Natural a. Dextran b. Chitosan c. Gelatin B. Artificial a. Polyaniline (PANI) b. Poly (3,4 ethylene dioxythiophene) (PEDOT) c. Polypyrrole d. Poly(L-ornithine) e. Poly (amido-amine) f. Poly (L-lysine) 3. Hybrid nanomaterial Conducting polymer nanotube 4. Organic and inorganic nanomaterial a. Ion-channel based sensing	1. Electrochemical biosensors a. Potentiometric b. Amperometric c. Voltammetric d. Conductometric and impedimetric 2. Optical biosensor a. Fluorescence-based b. SPR based c. Optical fiber-based d. Cantilever biosensor e. Colorimetric f. Raman and FTIR 3. Piezoelectric Biosensor a. Surface Acoustic waves b. Quartz crystal Microbalance (QCM) 4. Magnetic Biosensor 5. Thermal detection biosensor 6. Ion- sensitive Biosensor	1. High Performance biosensor (FET) 2. Giant Magnetic Resistance sensor (GMR)	1. Miniaturized device 2. Implantable biosensor

Table 3 Differentiation among carbon-based nano-biosensors

Types properties	Carbon nanotubes	Graphene	Nanowire	Nanomembrane
History	Launched in 1960 by Roger Bacon, actually discovered by Sumio Lijima in 1991	Discovered in 2004 as monolayer graphene	First reported in 1964 with silicon and liquid Au droplet later in 2001 by the Lieber group	Invented by Decher in the 1990s later in the 2000s, Geim and Novoselov discovered graphene, a type of nanomembrane
Structure	1-Dimensional (D) seamless rolled cylindrical tube with fullerene hemispheres as a cap	2-D single-layer materials of graphite-like honeycomb lattice having atomically sp2-hybridized carbon	1-D field-effect transistor (FET)-based devices	2-D layers or sheets with artificial freestanding/free-floating structure
Types	Single-walled (SW-armchair, zigzag, chiral), Double-walled (DW), and Multiwalled (MW)	Single-layer 2D graphene nanosheets (GNS), graphene oxide (GO), reduced graphene oxide (rGO), graphene nanoribbons (GNRs), and graphene quantum dots (GQD)	Top-down SiNW and Bottom-up SiNW	Inorganic, organic, hybrid and synthetic biological nanomembrane
Size	Dia.: SWCNT (0.7 nm), MWCNT (2 to 25 nm) with 0.34 nm b/w tubule sheets with several μm length	Very thin (atomic thickness 0.345 nm with 1-foot-long length	Round, rectangular, and triangular shape with approx. 40 nm broad and 20–200 nm long	Thickness ranging from ~10 to below 100 nm

(continued)

Table 3 (continued)

Types properties	Carbon nanotubes	Graphene	Nanowire	Nanomembrane
Characteristics	High thermal conductivity, chemical stability, aspect ratio (surface-to-volume), and low overvoltage	High mechanical strength, elasticity, purity, response time, surface area, extraordinary conductivity (1500–2500 $Wm^{-1} K^{-1}$), quantum Hall effect, and inexpensive than CNT	High catalytic activity, sensitivity, excitation intensity (SiNWs at 750 nm), and maximum interband photoluminescence in the range of 500–1100 nm	Easily fabricated, withstand to heat, electron irradiation, and easy self-supported in air/vacuum or b/w two different fluids (including gas/liquid, exhibit Extremely large surface-to-volume ratios
Analysis techniques	By TEM, SEM, XRD	XRD, X-Ray Fluorescence	By TEM, AFM	By STM, AFM, X-ray photoelectron, Raman spectroscopy, and UV–vis optical reflection spectrometry
Biomedical applications	Detection of sugars, neurotransmitters, amino acids, immunoglobulin, albumin, insulin, cancer biomarkers, cells, and microorganisms	Detection of glucose, NADH, dopamine, H_2O_2, hemoglobin, cholesterol, DNA, ascorbic and uric acid, horseradish peroxidase, gases, catechol, heavy metal ions, and proteins	Detection of types of prostate cancer, cardiac biomarkers, MicroRNA, protein–DNA interactions, Dengue, Influenza A, H_3N_2, H1N1, and HIV detection	Detection of proteins, DNA and RNA, biomarker, gene sequencing, pathogenic bacteria, virus, and prions particles, scaffolds for tissue engineering, wearable and implantable artificial kidney and lung, drug-loading and delivery systems

3.1 Supermacy of Nano-Biosensor Over Conventional Biosensors

Nano-biosensor generally has a small compact size, improved sensitivity, linearity, shock bearing ability, high electric conductivity, versatility, improved precision, accelerated detection time, and reproducibility [20, 23]. So, nano-biosensors can detect the very minute or nano-sized amount of a substance otherwise that would be harmful. These sensors work at the atomic scale in an energy-efficient way. They have shape-dependent properties also like metallic nanoparticles with an increased surface area that helps in enhanced adsorption as shown in Fig. 2 [10, 20]. The activity of nano-biosensors makes them liable to error and thus limitations are also associated with them.

4 Biomedical Applications of Nano-Biosensors

The role of nano-biosensors emerged in the medical field as a boom owing to their biocompatibility, flexibility, and sensitivity. Nano-biosensors are used in the medical field in clinical applications for diagnosis of diabetes, detection of sugars (glucose), cholesterol, urea, steroids, blood, hormones, biotoxins, carcinogens, pathogens (Avian influenza virus, human papillomavirus), various heart diseases/disease biomarkers (cancers, HIV, Hepatitis, dengue fever), and in genetic monitoring and drug discovery as shown in Fig. 3.

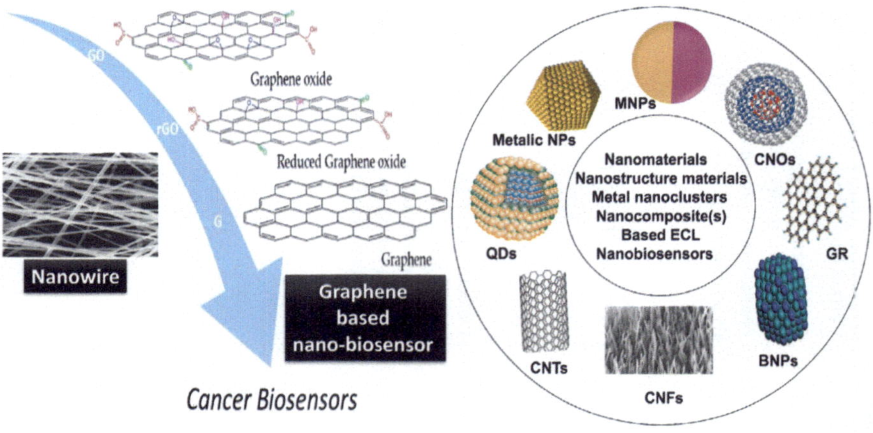

Fig. 2 Different nanomaterials used in nano-biosensors: Graphene-based nano-biosensor [36], Nanowire [3], Magnetic nanoparticles (MNPs), Carbon nano-onions (CNOs), Graphene (GR), Bimetallic nanoparticles (BNPs), Carbon nanofiber (CNFs), Carbon nanotubes (CNTs), Quantum dots (QDs), and Nanoparticles (NPs) [124]. Adapted under CC BY 4.0 license

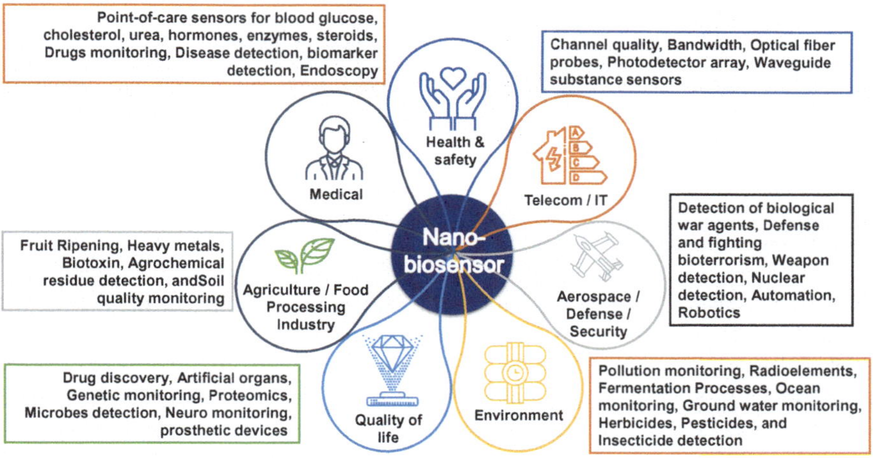

Fig. 3 Biomedical applications of nano-biosensor

4.1 Nano-Biosensors in the Exploration of Various Compounds in the Health Monitoring System

Detection of drugs (e. g. 17β-estradiol, acetylcholine, histamine, catechol), ethanol, phenol and phenolic compounds (e.g., chlorophenol, 3,4-dimethylphenol (DMP), 4-chloro-3-methylphenol, 4-Cl-2-MP), caffeic acid, artesunate, H_2S, BPA, 3-Microperoxidase (MP), 4-MP, captopril, clenbuterol, H_2O_2, concanavalin A (Con A), OPs pesticides, paraoxon-ethyl and chlorpyrifos, herbicides, gases and electrolytes are carried out with the help of nano-biosensors.

4.2 Nano-Biosensor's Applications in Agriculture, Industry, Environment, Research and Development, Military, and Defense Systems

These are broadly used in the detection of enzymes (e.g., acetylcholinesterase, horseradish peroxidase, tyrosinase), microbes (like E. coli, Salmonella, S. aureus), and pesticides. Recent techniques make them feasible in biotechnology, tissue engineering [139], drug discovery, delivery systems [100], and 3-D Printing also [76]. In the defense services nano-biosensors do amazing discoveries like detection of noxious chemicals and biological artilleries [117, 157], and in rockets and vehicles [73]. They are also employed in food industry (detection of microbes, etc.), textile industry [21], and civil structures [57].

4.2.1 In Health Monitoring Systems

Different nanoforms are used in nano-biosensors like nanomembrane, quantum dots, CNTs, gold nanoparticles (GNPs), metallic and magnetic nanoparticles in the examination of clinical samples.

Glucose Detection

Glucose biosensors are used not only in the examination of clinical samples but in the food processing sector as a quality control parameter also. In a diabetic situation, quick and accurate glucose detection is very important. Generally, glucose biosensors are made up of glucose oxidase enzymes. These glucose biosensors can be ranked into three generations, the first generation detects the glucose by reduction with consumption of hydrogen peroxide as a natural oxygen substrate, second generation biosensors detect glucose with redox mediators having specificity and selectivity to oxygen present in the biological fluids while the third generation based on direct interchange of electrons between the electrode and enzyme to speed up the analysis without the help of mediator or reagents [98]. In 1962 Clark and Lyons developed the first glucose biosensor with glucose oxidase (GOx) enzyme [151]. The GOx reduces the glucose into gluconic acid with the release of electrons and protons which further react with oxygen and result in hydrogen peroxide formation. This process converts the reduced enzyme back into its original oxidized form. This GOx is reused further in the same sequence like first act upon glucose and then on oxygen simultaneously. Hence, the metal electrode detects exhaustion of oxygen (with glucose) to assess the glucose concentration.

$$C_6H_{12}O_6 + \text{Glucose oxidase (GOx)} - \text{FAD} \rightarrow \text{Gluconolactone} + \text{GOx} - \text{FADH}_2 \quad (1)$$

$$\text{GOx} - \text{FADH}_2 + O_2 \rightarrow \text{GOx} - \text{FAD} + H_2O_2 \quad (2)$$

These biosensors' sensitivity was enhanced by incorporating gold nanoparticles with chitosan so that surface area increased that further improved by Luo et al. [88]. In the next development, small-sized silica sol–gel technology was induced to preserve the catalytic effect of the immobilized enzymes. Various silica compounds called oxysilanes, e.g., tetramethoxysilane (TMOS), tetraethoxysilane (TEOS), and methyltrimethoxy-silane (MTMOS) [113] are utilized for the high sensitivity, linearity, accuracy, and immobilization of enzymes. Later graphene was discovered as a promising material in glucose biosensors for its large surface area to volume ratio, good functionalization with other chemical groups, smart fluorescence quenching, good electron transferability, and economical cost [6]. As immobilized enzymes in graphene-based glucose biosensors could not perform direct electron conductivity and enzymatic catalytic effect, simultaneously so mediator fewer glucose biosensors were developed. In glucose biosensors, various electrochemical mediators like lutetium phthalocyanines, copper phthalocyanines were used, which act synergistically with graphene to strengthen the enzyme and help in reduction of the voltage requirement. Apart from these, various structural configurations of graphene-like GO, rGO, GNRs, nanosheets, multilayers, and nanostructures [90, 91], nanorod, nanotube [92], nanoscrolls [95], and 3D thin film [91] were also utilized to strengthen the specific characteristics of the glucose biosensors. Glucose biosensor's analytical capability is also enhanced by using smart nanocomposite with metals and different biopolymers, e.g., TiO_2 NP [51, 62], micro-sized nickel oxide with hollow and porous spheres [87], MnO_2 nanorods [85], chitosan with GO and nitrogen-doped graphene [18], SWNT with chitosan, GNPs and GOx (for continuous salivary glucose monitoring) [164] and polyaniline, titanium dioxide combination with rGO [154].

GOx is expensive and unstable, so simple, efficient, selective, and economical non-enzymatic sensors were developed. These glucose biosensors had different nanocomposites like MWCNT suspension deposited on glassy carbon electrode (GCE) [145], Pt nanoflowers with graphene [153], Pt nanoflower with MWNT [12], Pt nanoflowers with rGo and nickel oxide [79], Co_3O_4 NFs in the GO hydrogel [34], Cu_2O nanocubes within graphene nanosheets [84], flower-like CuO and flower-like CuO/few-layer graphene nanosheet [165], 2D bismuth oxychloride (BiOCL)-graphene nanohybrid sheets (BiOCl-G NHS) [48], bimetallic Pd and Cu NP embellished with 3-D graphene hydrogel [162], MnO_2/gold NPs with RGo [163], silver NPs with CNT [13], Pt NPs and a 3D graphene hydrogel [141], Ni NPs with polyvinylpyrollidone-graphene nanosheet [86], Cu and CuO NP with graphene [4] and polyaniline nanosheets (PANINS) with nickel oxide NPs [63]. Oxidized Gr and GQD were used in optical glucose biosensors according to their fluorescence quenching property. They were designed as non-enzymatic colorimetric [135], non-enzymatic fluorescence [19] besides enzymatic optical glucose biosensors [75]. Later paper-based optical glucose biosensor developed like microfluidic paper-based analytical devices (μPADs) [11] and with improved performance electrochemical paper-based analytical devices (ePADs) with colorimetric indicators [8]. These are simple, cost-effective, eco-friendly and easily used with and without sophisticated technologies. Optical transducer-based biosensors were used with different colorimetric indicators viz. 3,3'-diaminobenzidine, 4-amino antipyrine, and Methyl-N (3-sulfopropyl)-3-methyl-aniline sodium salt, 4-amino antipyrine, sodium 3,5-dichloro-2-hydroxy-benzenesulfonate, and potassium iodide 3-D printed electrodes. Wearable devices like a contact lens, tattoo-based epidermal device, subcutaneous microneedle based sensors, e.g., the FreeStyle Libre (Abbott), Enlite (Medtronic), and G5 (Dexcom) are the new advancement in the glucose biosensor field [128].

4.2.2 Cholesterol Detection

Cholesterol not only works as a prototype for vitamin D, steroid hormones, and bile acid but as a biomarker in heart diseases, hypertension, strokes, peripheral arterial disease, type 2 diabetes, hypolipoproteinaemia, anemia, septicemia, malnutrition, brain thrombosis, and heart attack also. In a normal person three different levels found in the blood (a) ideal level, when total cholesterol concentration vary between 125 and 200 mg/dL (<5.17 mM) (b) borderline level when b/w 200 and 239 mg/dL (5.17–6.18 mM), and (c) high concentration level 240 mg/dL and above [136]. Generally, more prevalent cholesterol biosensors are fabricated with cholesterol oxidase (ChOx) enzymes by virtue of their stability, sensitivity, and wide linearity. The enzyme ChOx oxidizes the cholesterol into H_2O_2 and cholest-4-en-3-one in the company of oxygen.

$$\text{Cholesterol} + O_2 \rightarrow \text{Cholest} - 4 - \text{en} - 3 - \text{one} + H_2O_2$$

Since then numerous novel cholesterol biosensors were designed and developed, e.g., in the electrochemical category, microfluidic [65], screen-printed electrode with gold NP, Prussian blue NP [55], pencil graphite electrode with boronic acid [39], GCE with magnetic NP, MWCNTs with sol–gel silica-chitosan and poly(thionine) [102], the platinum electrode with polypyrrole film and epoxy resin membrane [150], the paper supported cholesterol biosensor [102], and carbon nanofiber/polymer nanocomposite (BMCP)-based electrode with Cu and Ni [17] were fabricated. In electro-chemiluminescent cholesterol biosensor category, AuNPs/ion liquid/hollowed TiO_2 nano-shell [144], and hemin-functionalized graphene (hemin-GR) [102] were reported. For cholesterol determination in clinical samples, Kaur and Tomar in 2017, fabricated a reagent-less cholesterol biosensor with ZnO–CuO compound matrix on corning glass electrode as an external mediator [64] while in optic cholesterol biosensors, temperature-triggered, hydrogel supported chemically treated electrodes were designed with magnetically immobilized enzymes for cholesterol and glucose [54]. Therefore, in enzyme-based cholesterol biosensors modified electrodes (like carbon and metal nanomaterials, metal complexes, self-assembled monolayers, polymers, porous materials, hydrogels, and sol–gel with different biomaterials) were used. Generally, enzymatic biosensors lack efficiency as their immobilization process and stability are affected by temperature and pH. By considering all these limitations, Pt, Ag, Au/Pt, and Cu_2S electrodes were designed with electrochemical detection by amperometry, voltammetric methods for non-enzymatic cholesterol biosensors. Indicator displacement assay based biosensors in which calixarenes, crown ethers, and cyclodextrins are employed as supra-molecular complexes while methylene blue (MB) and ferrocene as indicators, e.g., GC electrode with Gr, β-cyclodextrin (β-CD) and MB; Nitrogen-doped Gr quantum dots modified with β-CD and MB; GC electrode with Gr, calix [6] arene and MB. Redox agents are also used like mediator in cholesterol biosensors, e.g., copper (I) sulfide nanoplates deposition on copper rode, polymethyl orange (PMO), and Cu/Ni bimetal on carbon nanofibers, Ru, Pi Nanoclusters coated on Polypyrrole and TiO_2 nanotubes on Cu_2O nanoparticles. Cholesterol's interactions with matrix also explored for its detection, e.g., cholesterol and laponite-montmorillonite with modified ITO electrode; altered GC electrode with AuNPs; MWCNTs with p-amino thiophenol and cholesterol [7]. Future research required on new improved electrochemical electrodes or sensors as non-enzymatic biosensors have selectivity issues.

4.2.3 Urea Detection

Renal diseases are also correlated with urea concentrations. The normal urea level ranges from 7 to 20 mg/dL (i.e., 2.5–7.5 mM) in the blood. A rise in urea concentration is correlated with renal disorder whereas a decline in urea concentration is correlated with hepatic failure, nephritic syndrome, and cachexia. Now urea biosensors are fabricated with the nanomaterials such as CNTs, enzymes, nanoparticles, metal nanoparticles, semiconducting, conducting and non-conducting nanoparticles, nanocomposites, redox dyes, and polymers, oxides, zeolites, sol-gels, clays, and nanoclusters to combine a cooperative improved performance. Nanomaterials have properties of high chemical stability, surface-to-volume ratio, and unique structural make-up. The earliest urea biosensor was formulated by Guilbault and Montalvo in 1969 with urease and potentiometric electrodes. Urease (Urs) enzyme hydrolyzes urea into NH^{4+} and HCO^{3-}ions and these ions are measured by ammonium ion-specific electrode (AISE) transducer. In amperometric type biosensors, various polymer-based urea biosensors fabricated like gold nanoparticles with polyester-Boltorn R H40 (H40–Au) [147], poly-*o*-phenylenediamine (PPD) film modified with lanthanide oxide, Yb_2O_3 nanodisks with nafion membrane [32], polypyrrole film sheathing at Pt electrode [58], polyaniline-MWCNT with disk-shaped graphite electrode and Pt wire [93], cystamine/PAMAM dendrimer-grafted MWCNT/Urease on Au electrode [39], porous polytetrafluoroethylene with disuccinimidyl cross-linkers, e.g., disuccinimidyl glutarate (DSG) [68], and PET/ITO coated with vanadium pentoxide (V_2O_5) by a sol–gel process [38] used in urine, blood, and spent dialysate samples in clinical studies. All these biosensors provide results with elevated sensitivity and reproducibility with extended functionality and stabilities up to a half year [121]. While in the potentiometric biosensor, Saeedfar et al. designed nano fullerene with immobilized Urs (C_{60}-urease) involving N, N′-dicyclohexylcarbodiimide (DCC) or N-(3-dimethyl aminopropyl)-N′-ethyl carbodiimide hydrochloride (EDC) [126] and fluorescence-based biosensor with jack bean meal and urease with CdS quantum dots using mercaptoacetic acid which catalyzed by gold clusters. These emitting gold clusters by fluorescence quenching quantify urea in the blood [105]. Further detection of creatinine and urea was analyzed by the gold laminated nano-sized film with Kretschmann through SPR response in varying concentrations of non-enzymatic solution [96]. A non-enzymatic electrochemical biosensor of kappa-carrageenan/PVA/nano-eggshell biocomposite was prepared to detect low-level urea (limit 60 μM) in phosphate-buffered solution (PBS) [146].

4.2.4 Detection of Drugs and Their Derivatives

Metal nanocomposites containing biosensors have been fabricated for the investigation of drugs along with metabolites. Drug identifying developed biosensors are, detection of S-nitrosothiol derivatives using Au nanoparticles [16], and Au coating on an ultra-microelectrode [143], adenosine triphosphate by gold NP/GRO nanocomposites [42], anticholinesterase drugs (dementia drugs, e.g., M donepezil) with silver nanoparticles [132], levodopa and uric acid determination by Au–Pd nanoparticles adsorbed on porous SS steel electrode [122], nifedipine, hydroquinone, dopamine, NO, and DNA by metal nanocomposite with Gr and IL [89], progesterone by nanocomposite of GQDs and Fe_3O_4 nanoparticles [9], acetylcholine with Fe_2O_3 nanoparticles and poly(3,4- ethylene dioxythiophene) (PEDOT)-rGO nanocomposite [26], diethylstilbestrol by Au NPs/MWCNTCS [15]. New generation biosensors/wearable biosensors are advanced to identify caffeine from sweat [142] whereas β-lactam antibiotics by detecting changes in local pH by invasive β-lactamase containing microneedle-based biosensor [49].

4.2.5 Detection of Electrolytes, Metabolites, Metals, Amino Acids, Cytokines, and Hormone by Wearable Biosensors

These days, wearable biosensors proved a game-changer in drug detection, pharmacokinetic, pharmacodynamic studies for personalizing drug dosage, doping control, closed-loop therapeutics, drug monitoring for antibiotics, and tracking patients' compliance with prescriptions. For a complete body checkup, BioSenHealth 1.0 device is launched having smart biosensors with an Arduino Mega 2560. It has a body temperature sensor, oximetry sensor, pulse sensor for measuring heartbeat, pulse rate, body temperature, and oxygen level of patient [104]. Biosensor with complementary metal oxide and semiconductor integrated circuits on a silicon chip was designed to analyze DNA testing and drug screening [78]. In the same series, Chauhan and coworkers have fabricated a glycated hemoglobin biosensor with gold NPs embedded on N-doped GNS for the examination of diabetes mellitus [61]. Detection of electrolytes level and other complement factors are necessary for sports analytics (like dehydration/rehydration, electrolyte balance, muscle cramps), early disease diagnosis (to judge autonomic nervous disorders, metabolic alkalosis, immunoglobulin detection), alcohol consumption, and investigation of vaccine effectiveness [159]. Each electrolyte has its significance like Ca^{2+} for kidney stones and hyperparathyroidism, and Na^+ and K^+ for electrolyte balance and muscle cramps [155]. Sweat and urine would be a simple sample for a health checkup as it contains electrolytes, such as Na^+, K^+, NH_4^+, etc., and heavy metal species like zinc, iron, copper, and magnesium [67]. Microfluidic/electrochemical-based sweat biosensors are employed to analyze sodium ions (Na^+) [129], potassium ions (K^+) [82], calcium ions (Ca^{2+}), and ammonium (NH_4^+) [70]. Metabolite (glucose, lactate, uric acid, and urea) monitoring through saliva and sweat analysis is proved very beneficial not only in disease detection (like glucose for diabetes; chloride for cystic fibrosis; urea, uric acid, and creatinine for kidney diseases, e.g., uremia; cholesterol for atherosclerosis) but in sports analytics also (like lactate for muscle fatigue, tissue hypoxia, and exercise intensity; glucose for energy availability and consumption) [159]. The noninvasive biosensors were developed for the salivary analysis of fluoride, Na^+, pH, and uric acid [69] while wearable mouthguard electrochemical biosensors for creatinine, chloride, uric acid, urea, and cholesterol in saliva. For sweat analysis, Zhang et al. established a microfluidic biosensor for creatinine and urea by colorimetric analysis for disease diagnostics and health monitoring in kidney function tests [32]. Further, Choi et al. also developed a skin-interfaced multipurpose microfluidic method for monitoring pH, chloride, glucose, and lactate by colorimetric analysis in newborns for the recognition of cystic fibrosis [33]. Heavy metals detection through an excreted fluid (urine and sweat) is beneficial to monitor overall health status, metal detoxification level (excess or deficiency), and emergence of diseases by heavy metals. Non-invasive biosensors were also used for calcium ion [47, 109], Cu, Cd, Pb, Hg [159], and chloride, sodium, and zinc detection by fluorometric sensing [129]. The role of biosensors in cytokine detection is useful in inflammation monitoring like IL-1β, CRP, TNF-α and in precise treatment like IL-1β detection for IBD treatment [60, 159].

Nano-biosensors also found applications in the identification of hormones like cortisol level in various diseases (e.g., Cushing's syndrome, Addison's disease), in mental health management, or stress levels [148, 159]. Biosensors help in the determination of many diseases like tyrosine, glutamate in liver diseases, traumatic brain injury, and Alzheimer's disease by amino acid level in the biological fluid. It also helps in nutritional management or eating disorders [158, 159].

4.2.6 In Drug Delivery

Microneedle-based patches are not only used in topical drug delivery but also helps in controlled long-term delivery, interstitial fluid biosensing, sampling, disease diagnosis by using thermoresponsive, multi-layered, stretch-triggered, effervescent microneedles [71]. Transdermic polymeric microneedle patches for insulin delivery were designed by stereolithography [115]. An effervescent microneedle-based biodegradable biosensor patch was reported with polymer poly(lactic-co-glycolic) acid that gradually releases the contraception hormone levonorgestrel approximately 1 month [81].

4.2.7 Detection of Diseases

Nano-biosensors help in early detection, treatment, mitigation, and prevention of various diseases like Brucellosis (An infectious disease results from gram-negative bacteria) diagnosed by nanosensor of Pd NPs with an immobilized brucella-specific probe [119], and in many more diseases as follows.

Cancer Detection

Cancer biomarkers are used not only in diagnosis, risk assessment, treatment prediction, pharmacokinetics and pharmacodynamic, recurrence but in tracking of the treatment also. The invention of electrochemical nano-biosensors in the identification of biomarkers gives a sensitivity (improving the detection limit, signal to noise ratio), velocity (examination time), selectivity (greater accuracy and precision), and cost-effectiveness (portability) but still, they need optimized electrode and transducer with special recognition and improved sensitivity of the signals. There is an extensive range of cancer biomarkers including conjugation of DNA or aptamers (circulating tumor DNA/ctDNA); antibodies (Merkel cell polyomavirus); Glycoproteins [mucin 1 (MUC1), and 27–29, mammaglobin (MAM)]; DNA (BRCA1, BRCA2); Proteins (Ki-67, OPN) and proteins (alpha-fetoprotein/AFP, BCR-ABL), etc. are reported. Over the years, nanotechnology-based biosensors are designed with metals, e.g., Au (widely used), Ag (toxic for internal use), and Cu NPs, and with carbon electrodes and biomaterials as per their specific physicochemical features [134]. Li et al. have designed prostate-specific antigen (PSA) sensing biosensors with $Cu_2O@CeO_2$-gold nanocomposites with a minimum analytical limit of 0.03 pg mL^{-1} [114]. In breast cancer, mucin biomarkers-based biosensors like the MUC1 aptamer with SWCNTs [103], DNA aptamer with MUC1, ferrocene, and bovine serum albumin were developed. Biosensors for the exploration of CA15-3 biomarkers include silver nanoparticles, GQD with anti-CA125 on modified GCE [59]. Recently, for circulating miRNA-21, a non-enzymatic sandwich assay was adopted and developed [166]. In the diagnosis of early-stage cancer biosensors, Wang and Dai [152] and Farzin et al. reported CNTs supported immunosensors with specificity and accuracy [46]. However, biosensors with nanocomposite structure were developed like CNT, metal, magnetic, and silica-based nanoparticles, conducting polymers and QDs but still few challenges like automation of the device, miniaturization for point-of-care testing, and availability at affordable prices with high resolution and precision must be bridged for future growth [133].

Cardiovascular Diseases

Cardiac biomarker troponin biosensor composed of polymer film assembled on the gold surface widely applied to detect myocardial infraction by analyzing the refractive index by SPR technique [111]. To identify early cardiac damage, a myoglobin biosensor was designed with 2-aminoethyl methacrylate hydrochloride and 4-styrene sulfonic acid to judge the interchange between myoglobin and its biosensor [101]. Later Chien et al. developed an injectable biosensor for analyzing oxygen amount in ischemic patients that consist of poly (2-hydroxyethyl methacrylate) and Pd-porphyrin, a probe for near-infrared (NIR) spectroscopy [31].

4.3 In the Exploration of Various Compounds

4.3.1 Detection of Enzymes

Various biosensors with carbon nanocomposites were fabricated for enzymes detection, like MCNT for glucose oxidase and horseradish peroxidase detection [77], CNT paste electrodes for lactate oxide [125], and CNTs-based electrochemical biosensors for malate dehydrogenase [138].

4.3.2 DNA Detection

Recent developments in imaging technology opened the door in the exploration of nanoparticles, viruses, and DNA by smartphones with nanoscale resolution. Miodek et al. have detected the DNA within 0.3 fM detection limit with nanocomposite built from MWCNTs covered with polypyrrole [99]. Smartphones associated fluorescence-based nano-biosensor with the QDs applied for the barcoding of DNA and notating various particles. Under fluorescent light, smartphones help to analyze the color change in DNA, and microbes (bacteria, viruses, etc.) [130].

4.3.3 Detection of Microorganisms

Improved immunological or t-based analytical methods have been generated for the detection of microorganisms. Generally, low level of oxygen helps in the growth and multiplication of dangerous pathogens (e.g., Clostridium botulinum) [97]. Park et al. used a smartphone-associated optical biosensor, a composite of nanoflower with magnetic antibody to detect Salmonella. Park et al. also developed a sensitive E. Coli detection piezoelectric biosensor with BSA and E. coli conjugated beads in QCM system. In a different project, a QCM based system with gold nanoparticles is utilized for Salmonella typhimurium detection within the range of 10–20 CFU/mL [130].

4.3.4 The Sensing of Ethanol

Biosensor's sensitivity for low level oxygen detection utilized in the development of alcohol nano-biosensor. Commercial ethanol biosensors contain a chromagen with two immobilized enzymes alcohol oxidase and peroxidase with a concentration of 10 μL/L ethanol in the gas phase at 50 °C with the exposure of 15 s. Alcohol oxidase oxidizes ethanol into acetaldehyde and H_2O_2 catalyzed by oxygen while peroxidase oxidizes the chromagen for a color change. The microbial biosensor was also adjusted to analyze ethanol and methanol concentration in alcohol-containing beverages during fermentation and yeast cultivation. This biosensor is composed of immobilized yeast or bacteria with an oxygen electrode which gives results in 10 min at 30 °C [97].

$$\text{Methanol} + O_2 \rightarrow H_2O_2 + \text{Formaldehyde}$$

$$\text{Ethanol} + O_2 \rightarrow H_2O_2 + \text{Acetaldehyde}$$

4.3.5 Biosensor for Organic Acid

Fermented food generally has a variable concentration of organic acids, i.e., ascorbic acid, lactic acid, or glutamic acid. The acidity of the curd governs the quality of mozzarella cheese. Now, biosensors are utilized to assess the acidity of lactic acid by platinum electrode coated with the immobilized lactate oxidase in which pyruvate formation is signaled by an amperometric transducer.

$$\text{Lactate} + O_2 \rightarrow \text{Pyruvate} + H_2O_2$$

The H_2O_2 probe detects the concentration of lactate formation in the curd. This method has more sensitivity than the pH probe [72, 97].

4.3.6 Detection of Hydrogen Peroxide (H_2O_2)

Nano-biosensors are used to quantify H_2O_2 that is a by-product of oxidation of alcohol, lysine, glucose, glutamate, cholesterol, amino acid, etc. in a cell or a particular disease. Now, H_2O_2 biosensors are made of metals (Au, Ag, Pd, Ni, Pt, and Cu), bimetal nanoparticles (Pt/Pd, Pt/Cu, Pd/Cu, Pt/Ir Rh/Pd, and Ru/Rh) metal oxides (ZnO and TiO_2) and carbon-based nanocomposites [107]. The development of H_2O_2 sensor with graphene and ZnO nanorods structure was proved useful in self-cleaning, anti-biofouling properties, liquid repellence, and anti-platelet adhesion properties [52]. A biosensor was developed with non-precious metal cobalt and carbon nanocomposite GCEs, polyacrylonitrile nanofibers, and zeolitic imidazole framework for H_2O_2 detection in various biological and chemical reactions [123].

4.3.7 Detection of NO

Nitric oxide (NO) production is associated with biological fluids, physiological systems and it is also involved in the outbreak of various diseases. Commercially various biosensors are developed like electrochemical biosensor systems NO_{550}, amiNO (by Innovative Instruments, Inc. USA), and fluorescence-based biosensor NCNO. rGO is used with metal or metal oxides to detect NO. Later, it was detected with silver nanoparticles on rGO by Ikhsan et al. [56]. For continuous sensing of NO levels in cultured cells, organs, and living mammals for long period, a wireless, biodegradable, flexible sensor was generated with a wide sensing range [80].

4.4 In Agriculture

4.4.1 Acetylcholinesterase Biosensors

In agriculture, it is extensively used for pesticide detection (e.g., carbamate and organophosphate). Acetylcholinesterase (AChE) hydrolyzes acetylthiocholine chloride into thiocholine (TCh), which further oxidized electrochemically. This hydrolysis reaction blocked by carbamate and organophosphorus pesticides as AChE irreversibly complexed with these and results in a low concentration of TCh with oxidation peak. Therefore, the peak of TCh was measured as the concentration of acetylcholine inhibitors. A sensitive, stabilized biosensor with AChE, Au NPs, and Pt electrode was fabricated for the identification of methyl paraoxon, carbofuran, and phoxim [160]. Following the same principle, Nesakuma and coworkers used AChE biosensor for carbosulfan pesticide detection in rice samples, having a detection limit equal to 0.24 nmol/l by cyclic voltammetry [106]. AChE was again used by Hatefi-Mehrjardi with a gold electrode and a monolayer of mercapto propionic acid and choline oxidase for the quantification of carbaryl with improved detection limit detection 5.96 nmol/l [53]. Another AChE biosensor was also developed for carbaryl detection was composed of poly(amidoamine) with cysteamine core having immobilized AChE, Nafion, carbon nanotubes, tin oxide nanoparticles, and chitosan [27]. Another pH based immobilized AChE containing biosensor was used for the analysis of aflatoxin B1 in samples like sesame, pea, and walnuts [140]. For pesticide methamidophos, poly(allylamine) hydrochloride, SiO_2 NPs with immobilized AChE biosensor was developed. The nano-biosensor was also used for assay of pesticides phorate, parathion, chlorpyrifos, methamidophos, and dimethoate with a corresponding limit of detection equal for all chosen pesticides 10^{-7} mol/l [116]. Other AChE based biosensors are used for chlorpyrifos pesticide detection with the MWCNTs/dicyclohexyl phthalate coated on screen-printed electrodes [28], while pesticide methyl parathion detection in fruits is done by AChE crosslinked glutaraldehyde immobilized on SWCNTs wrapped with BSA [74].

4.5 In Environment Monitoring

4.5.1 Tyrosinase Biosensors

Tyrosinase enzyme-based biosensor was reported by various scientists for phenol assay because of their high sensitivity, efficiency, and compatible design [1]. Tyrosinase based nano-biosensors have two general limitations based on stability and sensitivity issues. The immobilization of tyrosinase was reported by Carralero-Sanz et al., on a GCE electrode with electrodeposited gold nanoparticles. Another tyrosinase based electrode biosensor with colloidal gold nanoparticles, incorporated in graphite-Teflon composite exhibited suitable responses for the several phenolic compounds (catechol, phenol, 4-chlorophenol; 4-chloro-2-methyl phenol, 3-methyl phenol, and 4-methyl phenol).

4.5.2 Detection of Pollutants and Toxicants

Bioluminescent (BL) sensing technique is extensively used for the identification of organic chemicals and heavy metals (e.g., lead, zinc, cadmium, copper, and mercury) in water, food, or environmental toxicology and bioterrorism controls. BL bacteria emit light in a hospitable environment which reacts inversely as the amount of the toxicants increased. The weaker the light emitted by BL bacteria, the higher the degree of toxicity was recorded there. To facilitate on-field detection of toxicants, Cevenini et al. designed a real-time, cost-effective, quantitative smartphone-based BL whole-cell toxicity biosensor with the human embryonic kidney cells (Hek293T, a green-emitting luciferase) and dimethyl sulfoxide as a reference toxic compound employed without trained personnel [25].

5 Conclusion

Extensive use of nano-biosensors has revolutionized various sectors viz. medicine, environmental science, food safety, information technology, and defense also. Biomedical benefits due to nano-biosensors applications include disease prevention, diagnosis, and treatment. The role of biosensors expanding with technological advancement to understand, diagnose, analyze complex biological systems and tissue engineering. Still, more research and commercialization are required to tackle various challenges in the understanding of various pathophysiological mechanisms at the molecular level, metabolic pathways, in vitro and in vivo correlations in clinical studies to ensure their smooth transition from research to medical practice.

References

1. Achi F, Bensana A, Bouguettoucha A, Chebli D (2020) Nanobiosensors for detection of phenolic compounds. In: Nanosensor technologies for environmental monitoring. Springer, pp 275–307
2. Aggidis AG, Newman JD, Aggidis GA (2015) Investigating pipeline and state of the art blood glucose biosensors to formulate next steps. Biosens Bioelectron 74:243–262
3. Ali AS (2020) Application of nanomaterials in environmental improvement. In: Nanotechnology and the environment. IntechOpen
4. Alizadeh T, Mirzagholipur S (2014) A Nafion-free non-enzymatic amperometric glucose sensor based on copper oxide nanoparticles–graphene nanocomposite. Sens Actuators B Chem 198:438–447
5. Ambhorkar P, Wang Z, Ko H, Lee S, Koo K-i, Kim K, Cho D-iD (2018) Nanowire-based biosensors: from growth to applications. Micromachines 9:679

6. Ambrosi A et al (2016) Graphene and its electrochemistry–an update. Chem Soc Rev 45:2458–2493
7. Amiri M, Arshi S (2020) An overview on electrochemical determination of cholesterol electroanalysis
8. Arduini F, Cinti S, Caratelli V, Amendola L, Palleschi G, Moscone D (2019) Origami multiple paper-based electrochemical biosensors for pesticide detection. Biosens Bioelectron 126:346–354
9. Arvand M, Hemmati S (2017) Magnetic nanoparticles embedded with graphene quantum dots and multiwalled carbon nanotubes as a sensing platform for electrochemical detection of progesterone. Sen Actuators B Chem 238:346–356
10. Atta NF, Galal A, Ali S (2011) Nanobiosensors for health care. In: Biosensors for health, environment and biosecurity. IntechOpen
11. Avoundjian A, Jalali-Heravi M, Gomez FA (2017) Use of chemometrics to optimize a glucose assay on a paper microfluidic platform. Anal Bioanal Chem 409:2697–2703
12. Badhulika S, Paul RK, Terse T, Mulchandani A (2014) Nonenzymatic glucose sensor based on platinum nanoflowers decorated multiwalled carbon nanotubes-graphene hybrid electrode. Electroanalysis 26:103–108
13. Baghayeri M, Amiri A, Farhadi S (2016) Development of non-enzymatic glucose sensor based on efficient loading Ag nanoparticles on functionalized carbon nanotubes. Sens Actuators B Chem 225:354–362
14. Bahadır EB, Sezgintürk MK (2015) Applications of commercial biosensors in clinical, food, environmental, and biothreat/biowarfare analyses. Anal Biochem 478:107–120
15. Bai J et al (2017) Ultrasensitive sensing of diethylstilbestrol based on AuNPs/MWCNTs-CS composites coupling with sol-gel molecularly imprinted polymer as a recognition element of an electrochemical sensor. Sens Actuators B Chem 238:420–426
16. Baldim V, Ismail A, Taladriz-Blanco P, Griveau S, de Oliveira MG, Bedioui F (2016) Amperometric quantification of S-nitrosoglutathione using gold nanoparticles: a step toward determination of S-nitrosothiols in plasma. Anal Chem 88:3115–3120
17. Bairagi PK, Verma N (2018) Electrochemically deposited dendritic poly (methyl orange) nanofilm on metal-carbon-polymer nanocomposite: a novel non-enzymatic electrochemical biosensor for cholesterol. J Electroanal Chem 814:134–143
18. Barsan MM, David M, Florescu M, Ţugulea L, Brett CM (2014) A new self-assembled layer-by-layer glucose biosensor based on chitosan biopolymer entrapped enzyme with nitrogen doped graphene. Bioelectrochemistry 99:46–52
19. Basiruddin S, Swain SK (2016) Phenylboronic acid functionalized reduced graphene oxide based fluorescence nano sensor for glucose sensing. Mater Sci Eng C 58:103–109
20. Bhattarai P, Hameed S (2020) Basics of biosensors and nanobiosensors. Nanobiosens From Des Appl 1–22
21. Bishnoi A, Rajaraman Ts, Dube CL, Ambegaonkar NJ (2021) Smart nanosensors for textiles: an introduction. In: Nanosensors and nanodevices for smart multifunctional textiles. Elsevier, pp 7–25
22. Boscari F, Galasso S, Acciaroli G, Facchinetti A, Marescotti M, Avogaro A, Bruttomesso D (2018) Head-to-head comparison of the accuracy of Abbott FreeStyle Libre and Dexcom G5 mobile Nutrition. Metab Cardiovasc Dis 28:425–427
23. Bousiakou LG (2018) Current trends in nanobiosensing towards lab-on-a-chip devices for ultrasensitive analyte detection
24. Cass AE et al (1984) Ferrocene-mediated enzyme electrode for amperometric determination of glucose. Anal Chem 56:667–671
25. Cevenini L, Calabretta MM, Tarantino G, Michelini E, Roda A (2016) Smartphone-interfaced 3D printed toxicity biosensor integrating bioluminescent "sentinel cells". Sens Actuators B Chem 225:249–257
26. Chauhan N, Chawla S, Pundir C, Jain U (2017) An electrochemical sensor for detection of neurotransmitter-acetylcholine using metal nanoparticles, 2D material and conducting polymer modified electrode. Biosens Bioelectron 89:377–383

27. Chen D, Jiao Y, Jia H, Guo Y, Sun X, Wang X, Xu J (2015) Acetylcholinesterase biosensor for chlorpyrifos detection based on multi-walled carbon nanotubes-SnO_2-chitosan nanocomposite modified screen-printed electrode. Int J Electrochem Sci 10:10491–10501
28. Chen D, Liu Z, Fu J, Guo Y, Sun X, Yang Q, Wang X (2017) Electrochemical acetylcholinesterase biosensor based on multi-walled carbon nanotubes/dicyclohexyl phthalate modified screen-printed electrode for detection of chlorpyrifos. J Electroanal Chem 801:185–191
29. Chen L, Hwang E, Zhang J (2018) Fluorescent nanobiosensors for sensing glucose. Sensors 18:1440
30. Cherian E, Baskar G (2020) Enzyme Biosensors Based on Enzyme Inhibition for Pesticide and Heavy Metal Detection. Enzym Inhibit Environ Biomed Appl 1:59–72
31. Chien JS et al (2017) Injectable phosphorescence-based oxygen biosensors identify post ischemic reactive hyperoxia. Sci Rep 7:1–10
32. Chirizzi D, Malitesta C (2011) Potentiometric urea biosensor based on urease immobilized by an electrosynthesized poly (o-phenylenediamine) film with buffering capability. Sens Actuators B Chem 157:211–215
33. Choi J et al. (2019) Soft, skin-integrated multifunctional microfluidic systems for accurate colorimetric analysis of sweat biomarkers and temperature. ACS Sens 4:379–388
34. Chung JS, Hur SH (2016) A highly sensitive enzyme-free glucose sensor based on Co_3O_4 nanoflowers and 3D graphene oxide hydrogel fabricated via hydrothermal synthesis. Sens Actuators B Chem 223:76–82
35. Coulet PR, Blum LJ (2019) Biosensor principles and applications. CRC Press
36. Cruz SMA, Girão AF, Gonçalves G, Marques PAAP (2016) Graphene: the missing piece for cancer diagnosis? Sensors 16:137
37. da Costa GM, Hussain CM (2020) Ethical, legal, social and economics issues of graphene. Compr Anal Chem
38. da Rocha RCF, de Souza FA, Vieira NS, Cestarolli DT, Guerra EM (2020) Synthesis and characterization of V2O5/urease for a biosensor of urea. Biotechnol Appl Biochem
39. Dervisevic M, Çevik E, Şenel M, Nergiz C, Abasiyanik MF (2016) Amperometric cholesterol biosensor based on reconstituted cholesterol oxidase on boronic acid functional conducting polymers. J Electroanal Chem 776:18–24
40. Desai GS, Mathews ST (2014) Saliva as a non-invasive diagnostic tool for inflammation and insulin-resistance. World J Diab 5:730
41. Divies C (1975) Remarques sur l'oxydation de l'éthanol par une électrode microbienne d'Acetobacter xylinum
42. Dong Y-P, Zhou Y, Wang J, Zhu J-J (2016) Electrogenerated chemiluminescence resonance energy transfer between Ru (bpy) 32+ electrogenerated chemiluminescence and gold nanoparticles/graphene oxide nanocomposites with graphene oxide as coreactant and its sensing application. Anal Chem 88:5469–5475
43. Duroux P et al (1991) The ion sensitive field effect transistor (ISFET) pH electrode: a new sensor for long term ambulatory pH monitoring. Gut 32:240–245
44. Dwevedi A (2016) 100 years of enzyme immobilization. In: Enzyme immobilization. Springer, pp 1–20
45. Erickson KA, Wilding P (1993) Evaluation of a novel point-of-care system, the i-STAT portable clinical analyzer. Clin Chem 39:283–287
46. Farzin L, Sadjadi S, Shamsipur M, Sheibani S (2019) An immunosensing device based on inhibition of mediator's faradaic process for early diagnosis of prostate cancer using bifunctional nanoplatform reinforced by carbon nanotube. J Pharm Biomed Anal 172:259–267
47. Gao W et al (2016) Wearable microsensor array for multiplexed heavy metal monitoring of body fluids. Acs Sens 1:866–874
48. Gopalan A, Muthuchamy N, Lee K (2017) A novel bismuth oxychloride-graphene hybrid nanosheets based non-enzymatic photoelectrochemical glucose sensing platform for high performances. Biosens Bioelectron 89:352–360

49. Gowers SA et al (2019) Development of a minimally invasive microneedle-based sensor for continuous monitoring of β-lactam antibiotic concentrations in vivo. ACS Sens 4:1072–1080
50. Guilbault GG, Montalvo JG Jr (1969) Urea-specific enzyme electrode. J Am Chem Soc 91:2164–2165
51. Haghighi N, Hallaj R, Salimi A (2017) Immobilization of glucose oxidase onto a novel platform based on modified TiO_2 and graphene oxide, direct electrochemistry, catalytic and photocatalytic activity. Mater Sci Eng C 73:417–424
52. Hang T et al. (2019) Hierarchical graphene/nanorods-based H_2O_2 electrochemical sensor with self-cleaning and anti-biofouling properties. Sens Actuators B Chem 289:15–23
53. Hatefi-Mehrjardi A (2013) Bienzyme self-assembled monolayer on gold electrode: an amperometric biosensor for carbaryl determination. Electrochim Acta 114:394–402
54. Huang J, Liu Y, Zhang P, Li Y, Ding L (2017) A temperature-triggered fiber optic biosensor based on hydrogel-magnetic immobilized enzyme complex for sequential determination of cholesterol and glucose. Biochem Eng J 125:123–128
55. Huang Y et al (2018) Graphene and Au NPs co-mediated enzymatic silver deposition for the ultrasensitive electrochemical detection of cholesterol. Biosens Bioelectron 102:560–567
56. Ikhsan NI, Rameshkumar P, Huang NM (2016) Electrochemical properties of silver nanoparticle-supported reduced graphene oxide in nitric oxide oxidation and detection. RSC Adv 6:107141–107150
57. Ivanov GR, Marinov YG, Hadjichristov GB, Vlakhov TE (2019) Detection of heavy metal ions by newly designed biosensor with well-formed 3D nano-structure. In: 2019 XXIX international scientific symposium "metrology and metrology assurance" (MMA), 2019. IEEE, pp 1–4
58. Ivanova S, Ivanov Y, Godjevargova T (2013) Urea amperometric biosensors based on nanostructured polypyrrole and poly ortho-phenylenediamine
59. Jafari M et al (2019) Ultrasensitive bioassay of epitope of Mucin-16 protein (CA 125) in human plasma samples using a novel immunoassay based on silver conductive nano-ink: A new platform in early stage diagnosis of ovarian cancer and efficient management. Int J Biol Macromol 126:1255–1265
60. Jagannath B, Lin K-C, Pali M, Sankhala D, Muthukumar S, Prasad S (2020) A sweat-based wearable enabling technology for real-time monitoring of IL-1β and CRP as potential markers for inflammatory bowel disease. Inflamm Bowel Dis 26:1533–1542
61. Jain U, Chauhan N (2017) Glycated hemoglobin detection with electrochemical sensing amplified by gold nanoparticles embedded N-doped graphene nanosheet. Biosens Bioelectron 89:578–584
62. Jang HD, Kim SK, Chang H, Jo EH, Roh KM, Choi J-H, Choi J-W (2015) Synthesis of 3D silver-graphene-titanium dioxide composite via aerosol spray pyrolysis for sensitive glucose biosensor. Aerosol Sci Technol 49:538–546
63. Kailasa S, Rani BG, Reddy MSB, Jayarambabu N, Munindra P, Sharma S, Rao KV (2020) NiO nanoparticles-decorated conductive polyaniline nanosheets for amperometric glucose biosensor. Mater Chem Phys 242:122524
64. Kaur G, Tomar M, Gupta V (2017) Nanostructured NiO-based reagentless biosensor for total cholesterol and low density lipoprotein detection. Anal Bioanal Chem 409:1995–2005
65. Kaur G, Tomar M, Gupta V (2018) Development of a microfluidic electrochemical biosensor: prospect for point-of-care cholesterol monitoring. Sens Actuators B Chem 261:460–466
66. Kawahara J et al (2013) Flexible active matrix addressed displays manufactured by printing and coating techniques. J Polym Sci Part B: Polym Phys 51:265–271
67. Kim J et al (2015) Wearable temporary tattoo sensor for real-time trace metal monitoring in human sweat. Electrochem Commun 51:41–45
68. Kim J, Sung GY, Park M (2020a) Efficient portable urea biosensor based on urease immobilized membrane for monitoring of physiological fluids. Biomedicines 8:596
69. Kim J et al (2014) Non-invasive mouthguard biosensor for continuous salivary monitoring of metabolites. Analyst 139:1632–1636

70. Kim SB et al (2020b) Soft, skin-interfaced microfluidic systems with integrated enzymatic assays for measuring the concentration of ammonia and ethanol in sweat. Lab Chip 20:84–92
71. Kim Y-C, Park J-H, Prausnitz MR (2012) Microneedles for drug and vaccine delivery. Adv Drug Del Rev 64:1547–1568
72. Kulkarni AS, Joshi D, Tagalpallewar G (2014) Biosensors for food and dairy industry. Asian J Dairy Food Res 33:292–296
73. Kumar N, Kumar A, Chaudhry D (2011) A novel approach to use nano-sensor in WSN applications. Int J Comput Appl 975:8887
74. Kumar TV, Sundramoorthy AK (2019) Electrochemical biosensor for methyl parathion based on single-walled carbon nanotube/glutaraldehyde crosslinked acetylcholinesterase-wrapped bovine serum albumin nanocomposites. Anal Chim Acta 1074:131–141
75. Lee P-C, Li N-S, Hsu Y-P, Peng C, Yang H-W (2019) Direct glucose detection in whole blood by colorimetric assay based on glucose oxidase-conjugated graphene oxide/MnO_2 nanozymes. Analyst 144:3038–3044
76. Lee Y-C, Moon J-Y (2020) Bionanotechnology in Medicine. In: Introduction to bionanotechnology. Springer, pp 129–148
77. Lee Y-M, Kwon O-Y, Yoon Y-J, Ryu K (2006) Immobilization of horseradish peroxidase on multi-wall carbon nanotubes and its electrochemical properties. Biotechnol Lett 28:39–43
78. Li H, Liu X, Li L, Mu X, Genov R, Mason AJ (2017) CMOS electrochemical instrumentation for biosensor microsystems: a review. Sensors 17:74
79. Li M, Bo X, Mu Z, Zhang Y, Guo L (2014) Electrodeposition of nickel oxide and platinum nanoparticles on electrochemically reduced graphene oxide film as a nonenzymatic glucose sensor. Sens Actuators B Chem 192:261–268
80. Li R et al (2020) A flexible and physically transient electrochemical sensor for real-time wireless nitric oxide monitoring. Nat Commun 11:1–11
81. Li W et al. (2019) Long-acting reversible contraception by effervescent microneedle patch. Sci Adv 5:eaaw8145
82. Liang B, Cao Q, Mao X, Pan W, Tu T, Fang L, Ye X (2020) An integrated paper-based microfluidic device for real-time sweat potassium monitoring. IEEE Sens J
83. Liedberg B, Nylander C, Lundström I (1995) Biosensing with surface plasmon resonance—how it all started. Biosens Bioelectron 10:i-ix
84. Liu M, Liu R, Chen W (2013) Graphene wrapped Cu_2O nanocubes: non-enzymatic electrochemical sensors for the detection of glucose and hydrogen peroxide with enhanced stability. Biosens Bioelectron 45:206–212
85. Liu Y, Zhang X, He D, Ma F, Fu Q, Hu Y (2016) An amperometric glucose biosensor based on a MnO_2/graphene composite modified electrode. RSC Adv 6:18654–18661
86. Liu Z, Guo Y, Dong C (2015) A high performance nonenzymatic electrochemical glucose sensor based on polyvinylpyrrolidone–graphene nanosheets–nickel nanoparticles–chitosan nanocomposite. Talanta 137:87–93
87. Lu P, Yu J, Lei Y, Lu S, Wang C, Liu D, Guo Q (2015) Synthesis and characterization of nickel oxide hollow spheres–reduced graphene oxide–nafion composite and its biosensing for glucose. Sens Actuators B Chem 208:90–98
88. Luo J, Ma Q, Wei W, Zhu Y, Liu R, Liu X (2016) Synthesis of water-dispersible molecularly imprinted electroactive nanoparticles for the sensitive and selective paracetamol detection. ACS Appl Mater Interfaces 8:21028–21038
89. Maduraiveeran G, Jin W (2020) Functional nanomaterial-derived electrochemical sensor and biosensor platforms for biomedical applications. In: Handbook of nanomaterials in analytical chemistry. Elsevier, pp 297–327
90. Mani V, Devasenathipathy R, Chen S-M, Wang S-F, Devi P, Tai Y (2015) Electrodeposition of copper nanoparticles using pectin scaffold at graphene nanosheets for electrochemical sensing of glucose and hydrogen peroxide. Electrochim Acta 176:804–810
91. Mansourian M, Badiee A, Jalali SA, Shariat S, Yazdani M, Amin M, Jaafari MR (2014) Effective induction of anti-tumor immunity using p5 HER-2/neu derived peptide encapsulated in fusogenic DOTAP cationic liposomes co-administrated with CpG-ODN. Immunol Lett 162:87–93. https://doi.org/10.1016/j.imlet.2014.07.008

92. Mehmeti E, Stanković DM, Chaiyo S, Zavasnik J, Žagar K, Kalcher K (2017) Wiring of glucose oxidase with graphene nanoribbons: an electrochemical third generation glucose biosensor. Microchim Acta 184:1127–1134
93. Meibodi ASE, Haghjoo S (2014) Amperometric urea biosensor based on covalently immobilized urease on an electrochemically polymerized film of polyaniline containing MWCNTs. Synth Met 194:1–6
94. Meier T, Eulenbruch H-P, Wrighton-Smith P, Enders G, Regnath T (2005) Sensitivity of a new commercial enzyme-linked immunospot assay (T SPOT-TB) for diagnosis of tuberculosis in clinical practice. Eur J Clin Microbiol Infect Dis 24:529–536
95. Meng L, Xia Y, Liu W, Zhang L, Zou P, Zhang Y (2015) Hydrogen microexplosion synthesis of platinum nanoparticles/nitrogen doped graphene nanoscrolls as new amperometric glucose biosensor. Electrochim Acta 152:330–337
96. Menon PS, Said FA, Mei GS, Berhanuddin DD, Umar AA, Shaari S, Majlis BY (2018) Urea and creatinine detection on nano-laminated gold thin film using Kretschmann-based surface plasmon resonance biosensor. PLoS One 13:e0201228
97. Meshram B, Agrawal A, Adil S, Ranvir S, Sande K (2018) Biosensor and its application in food and dairy industry: a review. Int J Curr Microbiol Appl Sci 7:3305–3324
98. Metkar SK, Girigoswami K (2019) Diagnostic biosensors in medicine–A review. Biocatal Agric Biotechnol 17:271–283
99. Miodek A, Mejri N, Gomgnimbou M, Sola C, Korri-Youssoufi H (2015) E-DNA sensor of Mycobacterium tuberculosis based on electrochemical assembly of nanomaterials (MWCNTs/PPy/PAMAM). Anal Chem 87:9257–9264
100. Mishra AK (2019) Application of nanotechnology in diagnosis, drug dissolution, drug discovery, and drug carrier. In: Nanobiotechnology in bioformulations. Springer, pp 449–475
101. Moreira FT, Sharma S, Dutra RA, Noronha JP, Cass AE, Sales MGF (2015) Detection of cardiac biomarker proteins using a disposable based on a molecularly imprinted polymer grafted onto graphite. Microchim Acta 182:975–983
102. Narwal V, Deswal R, Batra B, Kalra V, Hooda R, Sharma M, Rana J (2019) Cholesterol biosensors: A review. Steroids 143:6–17
103. Nawaz MAH, Rauf S, Catanante G, Nawaz MH, Nunes G, Marty JL, Hayat A (2016) One step assembly of thin films of carbon nanotubes on screen printed interface for electrochemical aptasensing of breast cancer biomarker. Sensors 16:1651
104. Nayyar A, Puri V, Nguyen NG (2019) Biosenhealth 1.0: A novel internet of medical things (iomt)-based patient health monitoring system. In: International conference on innovative computing and communications. Springer, pp 155–164
105. Nenavathu BP, Dutta RK (2013) Development of nanobiosensor for urea by immobilizing jack bean meal urease on CdS QDs. Adv Mater Lett
106. Nesakumar N, Sethuraman S, Krishnan UM, Rayappan JBB (2016) Electrochemical acetylcholinesterase biosensor based on ZnO nanocuboids modified platinum electrode for the detection of carbosulfan in rice. Biosens Bioelectron 77:1070–1077
107. Nestor U, Frodouard H, Theoneste M (2020) A brief review of how to construct an enzyme-based H_2O_2 sensor involved in nanomaterials. Adv Nanopart 10:1–25
108. Newman JD, Turner AP (2005) Home blood glucose biosensors: a commercial perspective. Biosens Bioelectron 20:2435–2453
109. Nyein HYY et al (2016) A wearable electrochemical platform for noninvasive simultaneous monitoring of Ca2+ and pH ACS nano 10:7216–7224
110. Palchetti I, Mascini M (2010) Biosensor technology: a brief history. In: Sensors and microsystems. Springer, pp 15–23
111. Palladino P, Minunni M, Scarano S (2018) Cardiac Troponin T capture and detection in real-time via epitope-imprinted polymer and optical biosensing. Biosens Bioelectron 106:93–98
112. Pandit S, Dasgupta D, Dewan N, Prince A (2016) Nanotechnology based biosensors and its application. Pharma Innov 5:18
113. Pauliukaite R, Paquim AMC, Brett AMO, Brett CM (2006) Electrochemical, EIS and AFM characterisation of biosensors: trioxysilane sol–gel encapsulated glucose oxidase with two different redox mediators. Electrochimica Acta 52:1–8

114. Pecher O, Carretero-González J, Griffith KJ, Grey CP (2017) Materials' methods: NMR in battery research. Chem Mater 29:213–242
115. Pere CPP et al. (2018) 3D printed microneedles for insulin skin delivery. Int J Pharm 544:425–432
116. Pohanka M (2016) Electrochemical biosensors based on acetylcholinesterase and butyrylcholinesterase. A review. Int J Electrochem Sci 11:7440–7452
117. Pohanka M (2019) Current trends in the biosensors for biological warfare agents assay. Materials 12:2303
118. Raghavan M, Bjorkman PJ (1995) BIAcore: a microchip-based system for analyzing the formation of macromolecular complexes. Structure 3:331–333
119. Rahi A, Sattarahmady N, Heli H (2016) An ultrasensitive electrochemical genosensor for Brucella based on palladium nanoparticles. Anal Biochem 510:11–17
120. Rastogi M, Singh SK (2019) Advances in molecular diagnostic approaches for biothreat agents. In: Defense against biological attacks. Springer, pp 281–310
121. Razumiene J, Gureviciene V, Sakinyte I, Rimsevicius L, Laurinavicius V (2020) The synergy of thermally reduced graphene oxide in amperometric urea biosensor: application for medical technologies. Sensors 20:4496
122. Rezaei B, Shams-Ghahfarokhi L, Havakeshian E, Ensafi AA (2016) An electrochemical biosensor based on nanoporous stainless steel modified by gold and palladium nanoparticles for simultaneous determination of levodopa and uric acid. Talanta 158:42–50
123. Riaz MA et al (2020) Hierarchically porous carbon nanofibers embedded with cobalt nanoparticles for efficient H_2O_2 detection on multiple sensor platforms. Sens Actuators B Chem 319:128243
124. Rizwan M, Mohd-Naim NF, Ahmed MU (2018) Trends and advances in electrochemiluminescence nanobiosensors. Sensors 18:166
125. Rubianes MD, Rivas GA (2005) Enzymatic biosensors based on carbon nanotubes paste electrodes. Electroanal Int J Devoted Fundam Pract Aspects Electroanal 17:73–78
126. Saeedfar K, Heng LY, Ling TL, Rezayi M (2013) Potentiometric urea biosensor based on an immobilised fullerene-urease bio-conjugate. Sensors 13:16851–16866
127. Sagadevan S, Periasamy M (2014) Recent trends in nanobiosensors and their applications-a review. Rev Adv Mater Sci 36:62–69
128. Scognamiglio V, Arduini F (2019) The technology tree in the design of glucose biosensors TrAC. Trends Anal Chem 120:115642
129. Sekine Y et al (2018) A fluorometric skin-interfaced microfluidic device and smartphone imaging module for in situ quantitative analysis of sweat chemistry. Lab Chip 18:2178–2186
130. Seo SE et al (2019) Smartphone with optical, physical, and electrochemical nanobiosensors. J Indus Eng Chem 77:1–11
131. Sezgintürk MK (2020) Introduction to commercial biosensors. In: Commercial biosensors and their applications. Elsevier, pp 1–28
132. Shamagsumova RV, Yu. Efimova O, Gorbatchuk VV, Evtugyn VG, Stoikov II, Evtugyn GA (2019) Electrochemical acetylcholinesterase biosensor based on polylactide–nanosilver composite for the determination of anti-dementia. Drugs Anal Lett 52:1558–1578
133. Shandilya R, Bhargava A, Bunkar N, Tiwari R, Goryacheva IY, Mishra PK (2019) Nanobiosensors: point-of-care approaches for cancer diagnostics. Biosens Bioelectron 130:147–165
134. Sharifi M et al (2019) Cancer diagnosis using nanomaterials based electrochemical nanobiosensors. Biosens Bioelectron 126:773–784
135. Shehab M, Ebrahim S, Soliman M (2017) Graphene quantum dots prepared from glucose as optical sensor for glucose. J Lumin 184:110–116
136. Shen J, Liu C-C (2007) Development of a screen-printed cholesterol biosensor: comparing the performance of gold and platinum as the working electrode material and fabrication using a self-assembly approach. Sens Actuators B Chem 120:417–425
137. Singh S, Kumar V, Dhanjal DS, Datta S, Prasad R, Singh J (2020) Biological biosensors for monitoring and diagnosis. In: Microbial biotechnology: basic research and applications. Springer, pp 317–335

138. Sireesha M, Jagadeesh Babu V, Kranthi Kiran AS, Ramakrishna S (2018) A review on carbon nanotubes in biosensor devices and their applications in medicine. Nanocomposites 4:36–57
139. Solaimuthu A, Vijayan AN, Murali P, Korrapati PS (2020) Nano-biosensors and their relevance in tissue engineering. Curr Opin Biomed Eng 13:84–93
140. Stepurska K, Soldatkin O, Arkhypova V, Soldatkin A, Lagarde F, Jaffrezic-Renault N, Dzyadevych S (2015) Development of novel enzyme potentiometric biosensor based on pH-sensitive field-effect transistors for aflatoxin B1 analysis in real samples. Talanta 144:1079–1084
141. Sun KG, Hur SH (2015) Highly sensitive non-enzymatic glucose sensor based on Pt nanoparticle decorated graphene oxide hydrogel. Sens Actuators B Chem 210:618–623
142. Tai L-C et al (2019) Wearable sweat band for noninvasive levodopa monitoring. Nano Lett 19:6346–6351
143. Taladriz-Blanco P, Pastoriza-Santos V, Pérez-Juste J, Hervés P (2013) Controllable nitric oxide release in the presence of gold nanoparticles. Langmuir 29:8061–8069
144. Tang S, Zhao Q, Tu Y (2016) A sensitive electrochemiluminescent cholesterol biosensor based on Au/hollowed-TiO_2 nano-composite pre-functionalized electrode. Sens Actuators B Chem 237:416–422
145. Tarlani A, Fallah M, Lotfi B, Khazraei A, Golsanamlou S, Muzart J, Mirza-Aghayan M (2015) New ZnO nanostructures as non-enzymatic glucose biosensors. Biosens Bioelectron 67:601–607
146. Taşaltın N, Aydın E, Karakuş S, Kilislioğlu A (2020) K-carrageenan/PVA/nano-eggshell biocomposite-based non-enzymatic electrochemical biosensor for low-level urea detection. Appl Phys A 126:1–9
147. Tiwari A, Aryal S, Pilla S, Gong S (2009) An amperometric urea biosensor based on covalently immobilized urease on an electrode made of hyperbranched polyester functionalized gold nanoparticles. Talanta 78:1401–1407
148. Torrente-Rodríguez RM et al (2020) Investigation of cortisol dynamics in human sweat using a graphene-based wireless mhealth system matter
149. Touhami A (2014) Biosensors and nanobiosensors: design and applications. Nanomedicine 15:374–403
150. Umar A, Ahmad R, Kumar R, Ibrahim AA, Baskoutas S (2016) $Bi_2O_2CO_3$ nanoplates: fabrication and characterization of highly sensitive and selective cholesterol biosensor. J Alloys Compd 683:433–438
151. Wang J (2001) Glucose biosensors: 40 years of advances and challenges. Electroanal Int J Devoted Fundam Pract Aspects Electroanal 13:983–988
152. Wang Z, Dai Z (2015) Carbon nanomaterial-based electrochemical biosensors: an overview. Nanoscale 7:6420–6431
153. Wu G-h, Song X-h, Wu Y-F, Chen X-m, Luo F, Chen X (2013) Non-enzymatic electrochemical glucose sensor based on platinum nanoflowers supported on graphene oxide. Talanta 105:379–385
154. Wu S et al (2017) Development of glucose biosensors based on plasma polymerization-assisted nanocomposites of polyaniline, tin oxide, and three-dimensional reduced graphene oxide. Appl Surf Sci 401:262–270
155. Xu G et al (2019) Smartphone-based battery-free and flexible electrochemical patch for calcium and chloride ions detections in biofluids. Sens Actuators B Chem 297:126743
156. Xu L, Li D, Ramadan S, Li Y, Klein N (2020) Facile biosensors for rapid detection of COVID-19. Biosens Bioelectron:112673
157. Yáñez-Sedeño P, Agüí L, Campuzano S, Pingarrón JM (2019) What electrochemical biosensors can do for forensic science? Unique Features Appl Biosens 9:127
158. Yang Y et al (2020) A laser-engraved wearable sensor for sensitive detection of uric acid and tyrosine in sweat. Nat Biotechnol 38:217–224
159. Ye S, Feng S, Huang L, Bian S (2020) Recent progress in wearable biosensors: from healthcare monitoring to sports analytics. Biosensors 10:205
160. Yin H, Ai S, Xu J, Shi W, Zhu L (2009) Amperometric biosensor based on immobilized acetylcholinesterase on gold nanoparticles and silk fibroin modified platinum electrode for detection of methyl paraoxon, carbofuran and phoxim. J Electroanal Chem 637:21–27

161. Yoo E-H, Lee S-Y (2010) Glucose biosensors: an overview of use in clinical practice. Sensors 10:4558–4576
162. Yuan M, Liu A, Zhao M, Dong W, Zhao T, Wang J, Tang W (2014) Bimetallic PdCu nanoparticle decorated three-dimensional graphene hydrogel for non-enzymatic amperometric glucose sensor. Sens Actuators B Chem 190:707–714
163. Zhang C et al (2016) Dual-function amperometric sensors based on poly (diallydimethylammoniun chloride)-functionalized reduced graphene oxide/manganese dioxide/gold nanoparticles nanocomposite. Sens Actuators B Chem 222:663–673
164. Zhang W, Du Y, Wang ML (2015) Noninvasive glucose monitoring using saliva nanobiosensor. Sens Bio-Sens Res 4:23–29
165. Zhang Z, Pan P, Liu X, Yang Z, Wei J, Wei Z (2017) 3D-copper oxide and copper oxide/few-layer graphene with screen printed nanosheet assembly for ultrasensitive non-enzymatic glucose sensing. Mater Chem Phys 187:28–38
166. Zouari M, Campuzano S, Pingarrón JM, Raouafi N (2020) Femtomolar direct voltammetric determination of circulating miRNAs in sera of cancer patients using an enzymeless biosensor. Anal Chim Acta 1104:188–198

Chapter 10
Biodegradable Materials for Medicinal Applications

R. Deepak Kasai, Devi Radhika, Bhagyavana S. Mudigoudra, Ranvindra B. Chougale, S. Archana, K. Yogesh Kumar, S. B. Benaka Prasad, and Karthik Kannan

1 Introduction

The development of innovative biomaterials has been underway for a number of years and continues to be an area of interest for many scientists. With evolution of human civilization, the field of biomaterials emerged as different materials at multiple length scales from nano to micro and micro to macro level for multipurpose with a simple focus to extend human life and improve the quality of life. Past 1000 years back, silver in several forms was used as an antimicrobial agent to prevent infection also different types of surgical methods has been found during early days of civilization. In this context, probably the most advantageous development occurred in the field of biomaterials over the years 1901–2000 [1]. A biomaterial can be defined as a materials that made with intention to interface with biological systems to treat, augment or replace any tissue, organ or function of the human body [2]. According to the Clemson University Advisory Board for Biomaterials, biomaterial is 'a systematically and pharmacologically stable substance fabricated for implantation within or incorporation with living organisms'. Synthetic or natural biomaterials are used in medical devices or in contact with biological tissues. Nowadays, these biomaterials plays crucial role in human health from all the way, and biomaterial field has seen steady development over its approximately half century of existence and applications ideas from medicine, biology, chemistry, materials science and engineering. A biomaterial

R. D. Kasai · D. Radhika (✉) · S. Archana · K. Y. Kumar · S. B. B. Prasad
Department of Chemistry, Faculty of Engineering and Technology, Jain Deemed-to-be University, Bangalore, Karnataka, India

B. S. Mudigoudra · R. B. Chougale
P. G. Department of Studies in Chemistry, Karnatak University, Dharwad 580003, India

K. Kannan
School of Advanced Materials Science and Engineering, Kumoh National Institute of Technology (KIT), Gumi, Republic of Korea

S. Gopi et al. (eds.), *Nanotechnology for Biomedical Applications*,
Materials Horizons: From Nature to Nanomaterials,
https://doi.org/10.1007/978-981-16-7483-9_11

can be made useful in living systems of to function in intimate contact with living tissue and also used to replace the damaged parts of the living systems. The various applications of biodegradable materials are immense, including medical applications like surgical implants, matrices used for long-term implantable controlled release of drugs, switchers, etc. Also agricultural uses application area like seeding pots, mulch and many traditional packaging applications.

2 Definition of Biomaterial

A biomaterials is a substance which is effectively used in the prosthesis of in a medical devices for contact with the living body for a desired applications and for predefined time period [3]. A biomaterial is defined as a material aimed to interface with biological systems to evaluate, treat, augment and replace tissue, organ or function of the body [2, 4]. Biomaterials are fabricated based on desired applications. First biodegradable suture was approved in the 1960s [5]. Biomaterials must be biocompatible; it means biologically compatible and not to affect the system at the cellular or at the system level. Any biomaterials should be able to perform its intended role without creating adverse reactions in the host body [1]. This emerging paradigm allows to meet multidisciplinary understandings which can effectively integrate different concepts from different fields. It is not only restricted or limited to the chemistry, biology, materials science, mechanical and electrical engineering fields. Collaboration of multidisciplinary ideas can bring effective biomaterials.

Past twentieth century, the fabrication biomaterials for artificial implants used in surgical restorative health care has been majorly dominated by biologically inert and corrosive resistant biomaterial. Such biomaterials either are necessary to be removed through a secondary surgery or sometimes remain fixed in the host body for a lifetime. In such cases, implants often lead to long-term complications like slow but progressive corrosion and deterioration in the biological environment. Therefore, permanent implants have been challenging task over last few decades the fabrication of biodegradable implants biomaterials medical field. Such biodegradable implants can provide the biomechanical support throughout the curing process and eventually get dissolved and can be effectively replaced from the host tissue. In this context, biodegradable polymers are being most effective candidate for biomedical applications due to their ease of fabrication, processing, shaping in desired forms, predictable degradation property and degradation products. Biopolymers are major class of biomaterials because in many implants like bone substitution, bone fixing and dental materials should have long-term stable function in the host body. However, in recent years, progress in the tissue engineering, tissue regenerative medicine, gene therapy and controlled drug release have pose the demand of new properties of novel biodegradable biomaterials. Nowadays, biologically derived and synthetic biodegradable biopolymers have triggered the considerable interest [2]. However, low mechanical properties, acidification of the implantation sites when degradation occur and poor visibility in the body limit the widespread use of biodegradable polymers in medical applications.

Since early 1900s, metal-based plates were used to cure long bone fracture in view of faster and more effective healing. Also by the 1930s, several joint replacement surgeries were being performed and succeeded. The metals and metal alloys have been used in different forms as implants for example joint and dental replacements, bone plates and screws. The metal-based biomaterials are mechanically strong, tough and ductile. For metal-based system, major advantages lies in the ductility. Such materials can be easily prepared and sterilized. But these materials may get corroded in biological media like different body pH, also because of their high densities and strong mechanical properties, these materials are not suitable for hard tissue. Meanwhile, ceramic biomaterials also have been considered advantageous materials due to some highly desirable properties like hip sockets, heart valves joint and dental implants. Aluminium oxide, zirconium oxide, calcium phosphates and bio-glasses are most commonly used biomaterials. Perhaps, these ceramic biomaterials are generally hard, and they exhibit high compressive nature. Also other limitations like brittleness, low tensile strength, impacts strength and are sensitive to notches or microcracks that leads to the fracture. Meanwhile, it is difficult to fabricate ceramic biomaterials as they also corrode in the body, but these can be readily sterilized and make them inert. Furthermore, several metal-based biodegradable biomaterials have been found to have superior mechanical properties which make them promising candidate for vascular, orthopaedic, suturing and other medical applications. But also new biomaterials for biomedical applications still require lot more modifications to fulfil clinical requirements like deployment, efficacy and safety. Therefore, the development of medical technology has set higher expectations for design of biomedical materials and also novel biodegradable materials with specific properties which are in great demand.

3 Properties of Biomaterials

Among different biomaterials properties, the important properties associated with the biomaterials include chemical, physical, mechanical and biological with respect to their surface and bulk properties. Generally chemical properties deal with the chemistry of the selected materials like composition, bonding and atomic structure, while physical properties deal with the microstructures, phases, density and various types of porosity. Mechanical properties deal with the strength, hardness, toughness and failure of the materials. Further, biological properties deal with the behaviour of the biomaterial in a biological environment. All these biomaterial properties are directly related to the chemistry, structure and processing of materials. Therefore, in-depth knowledge about structure and processing of such materials require good understanding these materials. In this context, one must review the advantages and disadvantages of several materials. This allows to selection of materials towards desired applications.

4 Selection of Materials

To fabricate the biodegradable biomaterial, selection of materials is very important and complicated job, because selection depends on several factors like mechanical loading requirements, chemical and structural properties, and biological properties. As these biomaterials are used in tissue engineering and drug delivery applications, various types of biomolecules and drugs can be conducted to desired organ making use of nanostructures and effectively released in a controlled manner. These materials basically focus on issues like biofouling, regenerative medicine, controlled release of therapeutic substances and tissue engineering. Perhaps, these materials have long travel history since the first steel plates were used for fracture curing, but also there is lot more challenges to overcome. In such cases, biological requirements should meet and fulfil the needs, because human biology is so complex and we are just started to the intricate processes and interactions among the materials and host body system. Figure 1 indicates the generic applications of several types of biomaterials [art 17].

Synthetic biodegradable biomaterials offer advantageous properties than natural biomaterials. Several classes and different examples of biomaterials are summarized in Table 1.

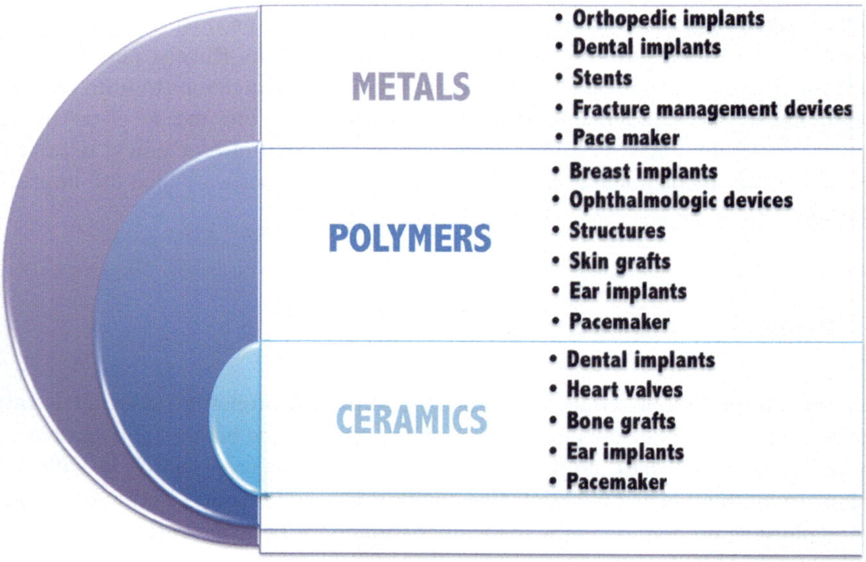

Fig. 1 Applications of biomaterials in various devices

Table 1 Various types of biomaterials

Composites	Carbon- and ceramic-coated materials
Polymers	**Polyurethanes, acrylics, Teflon, silicones and nylon**
Biological materials	Bovine arteries, Porcine skin and heart valves, allographs and autographs
Ceramics	**Hydroxyapatite, carbons, glasses and aluminium oxide**
Metals	Cobalt-based alloys, stainless steels, titanium, titanium-based alloys and gold

5 Classification of Biodegradable Biomaterials

Biodegradable biomaterials are classified as natural and synthetic polymer materials. Several biodegradable synthetic polymer biomaterials like poly(glycolic acid), poly(lactic acid) and their copolymers like poly(p-dioxanone), copolymers of trimethylene carbonate and glycolide have been extensively used in various medical applications. In addition, several natural biodegradable biomaterials like collagen, gelatin and albumins (derived from proteins), polysaccharides such as cellulose, hyaluronate, chitin and alginate [6] (Fig. 2).

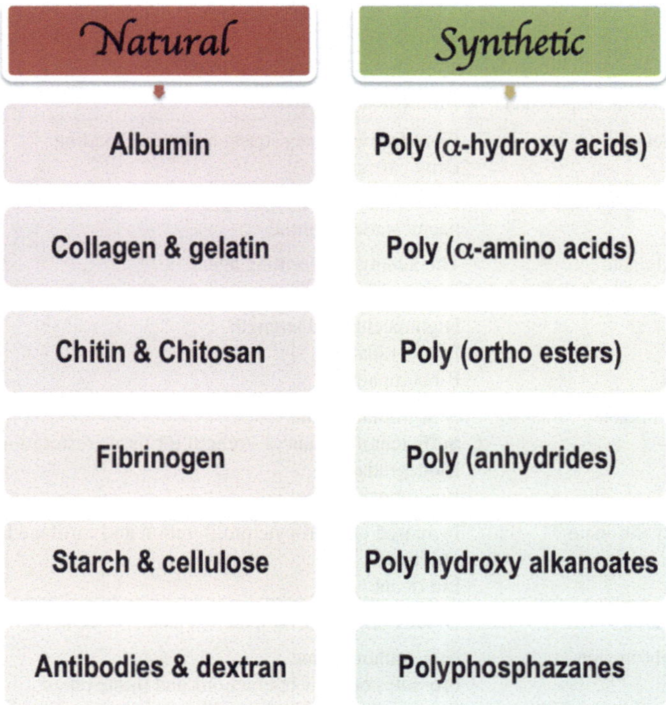

Fig. 2 Classification of biomaterials based on origin

6 Natural Biopolymers

Natural or biological structures are being a sources of inspiration for solving medical challenges in materials science and mechanical engineering. Natural biomaterials are materials which are natural in origin or sometimes through certain chemical modifications used for the fabrication of different biomaterials like scaffolds or implants. Nature has developed functional materials like polymers and minerals with remarkably useful functional properties [7]. Natural biopolymers are biologically degradable materials produced from living organisms. These are generally sustainable and renewable substance and have a surprizing responsibilities in host body. Generally, biopolymers can be fabricated from natural raw materials such as starch, sugar and cellulose [8–10]. The main advantageous of biodegradable and biocompatible biopolymers are utilization of such materials in living organisms would possibly reduce the toxicity, immunological reactions which can be seen with synthetic polymer devices [11]. Furthermore, chemically tailored biopolymers can provide good mechanical and electrical properties for desired applications [12, 13]. Such natural biopolymers have been used in medical filed for centuries [10]. Among several natural biopolymers, most common biopolymers like chitin, chitosan [14], alginate [15–18], collagen, gelatin, silk [19] and fibrin [20, 21] are widely used in biomaterial applications. Table 2 summarizes the biomedical properties of natural

Table 2 Summary the biomedical properties of natural biopolymers

S. No	Natural biopolymer	Biomedical properties
1	Gelatin	Fibre formation via strong hydrogen bonding Drug carrier
2	Chitin/chitosan	Accelerates tissue repair, prevents formation of scar tissue, haemostatic potential
3	Alginate	Thickening/gel-forming agent
4	Silk	Elasticity High mechanical strength Biodegradable Biocompatible
5	Collagen	Non-immunogenic Sufficient mechanical strength for tissue restoration Biodegradable Biocompatible
6	Hyaluronate	Increased chondrocyte proliferation and eartilage ECM production Lubricant
7	Starch	Guides various developmental stages of cells
8	Fibronogen	Non-immunogenic Promotes cellular interactions and tissue repair
9	Cellulose	Stable matrix for tissue engineering applications

biopolymers. The traditional preparation methods have provided ample opportunities for the synthesis of different types of scaffolds and implants. This allows the utilization of natural biomaterials for bone and tissue engineering scaffold production. Natural biopolymers such as bio-ink films can be effectively used to modify the surfaces of wound dressing [22–24]. In addition, bio-printing has brought unique platform to fabricate skin construct which can avoid need for donation of transplants [25–28]. Overall, natural biopolymers have been found very promising for biomedical applications in different forms. This could be due to its low toxicity, biocompatibility, biodegradability, stability and renewable nature.

7 Synthetic Biopolymers

Biodegradable biomaterials with specific properties are required for in vivo applications due to complexity and different vivo environments. In this context, synthetic biopolymers have gained much attention and are alternatives for biomedical applications due to following reasons. (a) They possess good biocompatibility, and some enhances an immune responses in the host body; (b) Ease of chemical modifications when compared to the biodegradable polymers; (c) Chemically modification allows alteration of the bulk properties of the biodegradable material. Therefore, variety of properties can be tuned, and modifications can be done with properly designed synthetic biopolymers without losing the bulk properties [29]. However, various products derived from lactic acid, glycolic acid and other materials such as poly(dioxanone), poly(trimethylene carbonate) copolymers and poly(caprolactone) homopolymers have been widely accepted for use in medical devices. Meanwhile, there is great attention has diverted towards use of polyanhydrides, polyorthoesters, polyphosphazenes and other biodegradable polymers. Among such materials, polyesters such as polyglycolides, polylactides and their copolymers are still considered as class of synthetic biodegradable polymers in medical use [30]. The permanent prosthetic materials have been effectively used as an alternative to biodegradable devices which could help the host body to repair and regenerate the damaged tissue and organs [31].

8 Important Characters of Biomaterials

The most important characteristics to be considered when it is related to biomaterial which is used in the biomedical applications includes toxicity, biocompatibility, functional tissue structure its biology, healing process, dependence on anatomical sites of implantations, mechanical requirements, industrial involvement and ethics. Also biocompatibility testing on both natural and synthetic biomaterials should have most following assessments like cytotoxicity, irritation, sensation, reactivity, subacute and chronic toxicity, genotoxicity, haemocompatibility, biodegradation and developmental toxicity.

9 Toxicology

First of all, biomaterials should be non-toxic, and it is specifically tuned for desired applications like scaffolds and drug delivery systems. The toxicity deals with the material that migrate out of host body, for example, many low molecular weight leachable polymers show some level physiologic activity and cell toxicity. Overall it means that biomaterials should release off anything from its mass as unless it is specifically tuned to do so.

10 Biocompatibility of Biomaterials

Biocompatibility is related to the property of biomaterials in different context. The term elaborates the ability of a substance to function with a suitable host response in a specific condition. Also it refers to the ability of a material when it is in contact with living organism which should not give adverse effect. This allows the understanding and measurement of biocompatibility of biomaterial in medical filed. Meanwhile, the operational definition of biocompatibility allows us little insight in design and tuning of new material.

11 Functional Tissue Structure

A tissue is group of cells which organised to function one or more particular function. Biomaterial integrated into medical devices generally implanted into tissues and organs. So the key principle is to study the structure and function of normal and abnormal tissues, cells and organs. Such fundamental mechanisms of diseases are critically considered and effectively worked in the medical filed.

12 Healing

Injury occurred to tissue will generate well-functioned inflammatory reaction that allows healing process. Also some special process are brought up when biomaterial or device heals in the host body. Generally, in the host body, normal response of the body will be regulated due to the solid implants. However, such reactions will varies in intensity and duration based on the anatomical sites involved. Therefore, for a biomaterials scientist, an understanding of how foreign body changes normal inflammatory reaction is very important.

13 Anatomical Sites of Implantation

Anatomical site of implantation is most common in medical surgery. Each anatomical site implantation has its own challenges in the biomedical device designer. To design an implant, one should concentrate on special requirements for geometry, shape, size compatible bio-responses and mechanical properties. The most common uses of biomaterials are in medical fields like ocular lenses, joint implants, heart valves artery and urinary track replacements. An intraocular lenses may implanted into the anterior chamber of the eye. A hip joints and cardiac muscle will be sutured with articulating joint space and heart valves. Also catheter may be replaced in an artery.

14 Mechanical and Performance Requirements

Mechanical performance of biomaterials is most important property among the all. Such mechanical properties solely depend on the bulk or physical properties of the materials. Generally, requirement for the good biomaterials can be summarized in three categories like mechanical performance, mechanical durability and physical properties for the specific applications, for example, a hip prosthesis should be strong and rigid. Looking into the tendon material, it must be strong as well as flexible but not elastomeric. Looking into its performance, catheter performance is restricted to only for three days, but in bone plate, it extends more than 6 months. Also, leaflet in heart valves must flex 60 times per minute. In dialysis, membrane must have specified permeability for the ions. Therefore, to meet all the requirements, design principles from chemistry, chemical and mechanical engineering, physics and materials sciences are most essential.

15 Applications of Biodegradable Biomaterials in Medical Field

The medical field is not constant it dynamic constantly changing also biomaterials used require lot of adjustments. Generally, biomaterials employed in medical applications must be compatible with tissue and may not be breakdown before its time or function. Therefore, all biomaterials must meet certain criteria, controllable requirements and can be qualified for use in biomedical field. The most common advantageous of biomaterials used is to physically replace damaged or destroyed hard or soft tissues. Also certain tissues and structures of the host body perform for a long period, but sometime they do suffer from several destructive process like fracture, infection and cancer which causes severe pain, loss of function and health issues. Under such conditions, one has to remove damaged tissue or organ and replace it with compatible biomaterial.

16 Orthopaedics

Orthopaedics is one of the most common application place for biomedical biomaterials in orthopaedic implant devices. In human body, osteoarthritis and rheumatoid arthritis generally affects the synovial joints like hip, knee, shoulder, ankle and elbow. The pain in such joints is most common in people. The use of anaesthesia, antisepsis and antibiotics can give only short time relief from the pain. But use of biomaterials devices can significantly reduce the pain as well as improve the patient condition. Bone grafting is performed either alone or combining with arthroplasty, spine, trauma and other orthopaedic methods. Historically, variety of materials has been implanted to restore the host bone anatomy, like wood, coconut shell and different metals [32]. Nowadays, bone graft replacements such as biodegradable and biocompatible polymers, ceramics, metals and composites are most commonly employed in orthopaedics.

17 Cardiovascular Applications

In cardiovascular system, most common problems arises in heart valves and arteries which can successfully treated and replaced by implants. The structural changes occur in heart valves which prevents the valve from fully closing and fully opening, and such diseased valve can be replaced with different biomaterials implants. Also irreversible damage occur from diseases which creates an urgent need to repair the heart. However, treatments used for cardiac disease range from medications to surgical interventions. In such cases, the human donor tissue shortage can be well managed with biomaterials because these materials have become a potential candidate.

18 Ophthalmic

In ophthalmology, surgical implants like ocular and intraocular lenses basically used retinal detachment surgery. Biomaterials are most important materials of the procedure that are employed to enhance and maintain vision. Other major applications of biomaterials in ophthalmology are drainage implants in terminal glaucoma, keratoprosthesis and ophthalmic plastic surgery. In most cases, tissues of eyes suffer from severe diseases which leads to reduced vision and blindness, for example, cataracts which causes cloudiness of the lenses. This can be easily replaced with a synthetic polymer intraocular lenses. According to USA's survey in the year 1991, approximately 35 million people were found active potential contact lens users [33]. A total of 85% soft contact lenses are used as lens wearers, and rest are rigid lenses. Recently, polymer lenses have gained much attention in contact lenses, for example, polymethyl methacrylate (PMMA) have been used since from half century. PMMA exhibits very good optical properties, lightweight, ease of manufacture and well tolerated. PMMA lenses are small in diameter and easily float on the thin layer precorneal tear film. This can be continuously replaced with blinking and lens movement to permit corneal oxygenation. Another effective biomaterial is cellulose acetate butyrate (CAB); it is softer and easily distorted by water absorption when compared to the PMMA. Therefore, fabrication of new materials made the CAB contact lenses which are obsolete.

19 Dental Applications

Recent advances in biomaterials of dentistry would lead to appropriate applications in clinical cases and to improve dental treatment outcomes to better serve to patients [34]. The most important is to select appropriate materials used to generate the tooth structure through processes. In mouth, both tooth and its supporting tissues generally destroyed by bacterially controlled diseases. Teeth and segments of teeth can be exchanged and restored by several biomaterials. Biomaterials and technologies effectively promote the tissue regeneration [35, 36]. Potential material is used for tissue regeneration in dentin, periodontal ligament, dental pulp and even enamel.

20 Wound Healing

In clinical field, one of the oldest uses of implantable biomaterials can be viewed in terms of sutures for wound healing. Tissue repair and wound healing are complex physiological process. In wound healing process, superficial and internal wounds can be treated with variety of dressings which varies from pharmaceutical formulations gauzes and synthetic dressing materials like hydrogels, foams and biomaterials. Generally, type of wound dictates the selection of drug and drug delivery vehicle to assist the tissue repair process. A biopolymers used in the wound dressing have been prepared using variety of materials and different forms such as fibres, mats and hydrogels, and some other may contain additives like silver for antimicrobial properties. Recent progress in the fabrication of advanced wound healing has been the use of biomaterials or incorporation of biologics has ability to stimulate or to promote events in wound dressing [37]. However, conventional wound dressing was prepared from cotton gauze of similar materials. Nowadays, electrospun mats created a lot of attention in wound healing process. Similarly, fibre mats are prepared from polycaprolactone and incorporation of biocompatible materials such as collagen [38] and chitosan to mimics the dermis. Meanwhile, doping of known antimicrobial agents like silver [39] and gentamicin [40] will be added in view of many of these dressings.

Hydrogels are also an important candidate for wound healing process as they have ability to form a barrier from pathogens. Hydrogels generally create hydrated environment to help promote the host body own wound healing responses [41]. Poly(vinyl alcohol) is most commonly used biopolymer in the preparation of hydrogels, and it is effectively used in the wound healing process.

21 Drug Delivery Systems

In medical field, one of the fastest emerging areas for implant applications is controlled and targeted drug delivery. Drug delivery systems carries active agents like growth factors and materials into the targeted location to release the beneficial effects for the diseased cells, and for the treatment disorders [42], osmotic pumps to transfer neurotrophic factors [43] to target the sites. In such cases, drug loading and controlled release are accomplished through interactions of therapeutic drug and delivery systems. In medical field, these biomaterials have significantly contributed to the delivery and efficacy of a range of antibodies, peptides, vaccines, drugs and enzymes [44]. Most commonly, polymer- and lipid-based biomaterials are used in drug delivery systems [45–59]. Silicon rubber was the first biomaterials reported for the controlled drug release [60]. These silicones are biocompatible materials and are used for implantation for a range of biomedical applications like drugs like atropine, histamine, anaesthetics, steroids and antimalarial agents [61–63]. Speaking to hydrogels, three-dimensional network of polymer chains is crosslinked to form matrix exhibiting high water content which are used in delivery and tissue engineering due to their easy modification of physical, chemical and biological properties [64]. PEG is used as a stealth material which enhances the circulation half-life of drugs and reduces drug accumulation in clearance organs [65]. Few authors have reported more comprehensive overviews on hydrogels [66–70].

22 The Future Outlook for Biomedical Biomaterials

There is room for growth and elaboration in several areas of the biomedical biomaterials. Over the past few decades, there is tremendous growth observed in the biomaterials. Several human body tissues like bones, teeth, tendons, arteries, heart valves, ligaments and other load bearing implant replace in the host body successfully with biomaterials. But at present, essential design in the biomedical biomaterials is still to meet lot of requirements with suitable materials. The future demand for long-term implants, tissue engineering and bone replacements is characterized by biocompatibility, bioactivity and mechanical properties of biomaterials, without the immune rejection which has triggered great challenge. In medical field, stainless steel, titanium alloys, zirconia alumina and high molecular weight polymers like polyethylene and polymethyl methacrylate are most common biomaterials which are in use to satisfy the physical, chemical and mechanical properties. Significant research is required to improve the activity of artificial biomaterials. Meanwhile, major efforts has been made to enhance the activity and minimize the wear and lifespan of the biomaterials in the host body.

23 Conclusion

Since from ancient times, biomaterials are being applied in healthcare applications. Also significant revolution in the biomaterials has brought more versatility and has increased its utility. The concept of biomaterials is being effectively used in the treatment of several disease like cardiac failure, fractures, skin and tissue injuries, ocular lenses, orthopaedics, etc. The novel technology and advancements are underway for the fabrication of the sensor chips for regulating the glucose level in the body. Similarly, there is huge development in the area of scaffolds for tissue engineering, and most emerging materials used in the scaffolds are hydrogels and cryogels. Hydrogels have large applications in drug delivery as well as in tissue engineering. In this concept, only one substance cannot fulfil all the requirements of biomaterials. For specific application, one need to have specific materials which can fulfil all biomedical applications. Such requirements can be significantly achieved by the doping different materials in the matrix of materials like composites, polymer-based composites and nanocomposites. Proper selection of matrix system and dopants along with shape, size and volume fraction plays vital role for obtaining significant properties like high strength, biocompatibility and bioactivity. Therefore, with significant development, we expect novel technologies to hit the future market which can immensely bring changes and progress in the healthcare sector.

References

1. Bose S, Bandyopadhyay A, Introduction to biomaterials. Charact Biomater. https://doi.org/10.1016/B978-0-12-415800-9.00001-2
2. Nair LS, Laurencin CT (2007) Biodegradable polymers as biomaterials. Prog Polym Sci 32:762–798
3. Williams DF (2009) Biomaterials 30:5897–5909
4. Levin SN (ed) Materials in biomedical engineering. Annal New York Acad Sci 146 (1968)
5. Middleto JC, Tipton AJ (2000) Synthetic biodegradable polymers as orthopedic devices. Biomaterials 21(23):2335–2346
6. Sabir MI, Xu X, Li L (2009) A review on biodegradable polymeric materials for bone tissue engineering applications. J Mater Sci 44:5713–5724

7. Fratzl P (2007) Biomimetic materials research: what can we really learn from nature's structural materials? J R Soc Interface 4(15):637–642
8. Wohlers T, Gornet T (2014) Wohlers Rep
9. Kruth JP, Leu MC, Nakagawa T (1998) CIRP Ann 4:525–540
10. Ige OO, Umoru LE, Aribo S (2012) ISRN Mater Sci 1–20
11. Yates MR, Barlow CY (2013) Life cycle assessments of biodegradable, commercial biopolymers—a critical review. Resour Conserv Recycl 78:54–66
12. Velde KV, Kiekens P (2002) Biopolymers: overview of several properties and consequences on their applications. Polym Test 21:433–442
13. Kim BS, Baez CE, Atala A (2000) Biomaterials for tissue engineering. World J Urol 18:2–9
14. Daniels A, Chang MKO, Andriano KP, Heller J (2004) J Appl Biomater 1:57–78
15. Athanasiou K, Schmitz J, Agrawal C (1998) Tissue Eng 4:53e63
16. McVicar I, Hatton P, Brook I (1995) Brit J Oral Max Surg 33:220–223
17. Garcia-Gareta E, Coathup MJ, Blunn GW (2015) Osteoinduction of bone grafting materials for bone repair and regeneration. Bone 81:112–121
18. Kurien T, Pearson RG, Scammell BE (2013) Bone graft substitutes currently available in orthopaedic practice: the evidence for their use. Bone Joint J 95-B:583–597
19. Maurus PB, Kaeding CC (2004) Oper Techn Sport Med 12:158–160
20. Lou CW, Yao CH, Chen YS, Hsieh TC, Lin JH, Hsing WH (2008) Text Res J 78:958e965
21. Casey DJ, Rosati L, Jarrett PK, Lehmann LT (1989) Bioabsorbable surgical suture coating, US Patent, No. 4857602
22. Elvira C, Mano JF, San Roman J, Reis RL (2002) Biomaterials 23:1955–1966
23. Kim DH, Viventi J, Amsden JJ, Xiao J, Vigeland L, Kim YS, Blanco JA, Panilaitis B, Frechette ES, Contreras D, Kaplan DL, Omenetto FG, Huang Y, Hwang KC, Zakin MR, Litt B, Rogers JA (2010) Nat Mater 9:511–517
24. Hwang SW, Tao H, Kim D-H, Cheng h, Song J-K, Rill E, Brenckle MA, Panilaitis B, Won SM, Kim YS, Song YM, Yu KJ, Ameen A, Li R, Su Y, Yang M, Kaplan DL, Zakin MR, Slepian MJ
25. Ng WL, Wang S, Yeong WY, Naing MW (2016) Trends Biotechnol 34:689–699
26. He P, Zhao J, Zhang J, Li B, Gou Z, Gou M, Li X (2018) Burns Trauma 6:5
27. Lee W, Debasitis JC, Lee VK, Lee JH, Fischer K, Edminster K, Park JK, Yoo SS (2009) Biomaterials 30:1587–1595
28. Singh M, Jonnalagadda S (2019) Euro J Pharm Sci 105167
29. Tian H, Tang Z, Zhuang X, Chen X, Jing X (2012) Biodegradable synthetic polymers: preparation, functionalization and biomedical application. Prog Polym Sci 37:237–280
30. Gunatillake P, Mayadunne R, Adhikari R (2006) Recent developments in biodegradable synthetic polymers. Biotechnol Annu Rev 12:301–347
31. Soni S, Gupta H, Kumar N, Nishad DK, Mittal G, Bhatnagar A (2010) Biodegradable biomaterials. Recent Patents Biomed Eng 3:30–40
32. Laurencin CT (2003) American academy of orthopaedic surgeons. Bone Graft Substitutes. W. Conshohocken, PA, ASTM International
33. Giordano GG, Refojo MF, Biomaterials in ophthalmology, from: human biomaterials applications
34. Basu B, Katti DS, Kumar A (2009) Advanced biomaterials: fundamentals, processing, and applications. Wiley, Hoboken, NJ, USA
35. Jin QM, Zhao SA, Berry JE, Somerman MJ, Giannobile WV (2003) Cementum engineering with three-dimensional polymer scaffolds. J Biomed Mater Res A 67(1):54–60
36. Bojic S, Volarevic V, Ljujic B, Stojkovic M (2014) Dental stem cells—characteristics and potential. Histol Histopathol 29(6):699–706
37. Farrugia BL, Mi Y, Kim HN, Whitelock JM, Baker S, Wiesmann WP et al (2018) Chitosan-based heparan sulfate mimetics promote epidermal formation in a human organotypic skin model. Adv Funct Mater 28:1802818
38. Ehterami A, Salehi M, Farzamfar S, Vaez A, Samadian H, Sahrapeyma H et al (2018) In vitro and in vivo study of PCL/COLL wound dressing loaded with insulinchitosan nanoparticles on cutaneous wound healing in rats model. Int J Biol Macromol 117:601–609

39. Tra Thanh N, Ho Hieu M, Tran Minh Phuong N, Do Bui Thuan T, Nguyen Thi Thu H, Thai VP, et al (2018) Optimization and characterization of electrospun polycaprolactone coated with gelatin-silver nanoparticles for wound healing application. Mater Sci Eng C 91:318–329
40. Permyakova ES, Polcak J, Slukin PV, Ignatov SG, Gloushankova NA, Zajíčková L et al (2018) Mater Des 153:60–70
41. Kamoun EA, Kenawy ES, Chen X (2017) A review on polymeric hydrogel membranes for wound dressing applications: PVA-based hydrogel dressings. J Adv Res 8:217–233
42. Hosseinkhani H, Hosseinkhani M (2009) Curr Drug Saf 4(1):79–83. https://doi.org/10.2174/157488609787354477
43. Agterberg MJH, Versnel H, Van Dijk LM, De Groot JCMJ, Klis SFL (2009) JARO-J Assoc Res Otolaryngol 10(3):355–367. https://doi.org/10.1007/s10162-009-0170-2
44. Allen TM, Cullis PR (1818) Science 2004:303
45. Langer R (1990) Science 249:1527
46. Ensign LM, Cone R, Hanes J (2012) Adv Drug Delivery Rev 64:557
47. Zhang S, Bellinger AM, Glettig DL, Barman R, Lee YA, Zhu J, Cleveland C, Montgomery VA, Gu L, Nash LD, Maitland DJ, Langer R, Traverso G (2015) Nat Mater 14:1065
48. Lowman AM, Morishita M, Kajita M, Nagai T, Peppas NA (1999) J Pharm Sci 88:933
49. Jeong B, Bae YH, Lee DS, Kim SW (1997) Nature 388:860
50. Guvendiren ML, Lu HD, Burdick JA (2011) Soft Matter 8:260
51. Anselmo AC, Mitragotri S (2014) J Controlled Release 190:15
52. Prausnitz MR, Langer R (2008) Nat Biotechnol 26:1261
53. Patton JS, Byron PR (2007) Nat Rev Drug Discovery 6:67
54. Illum L (2002) Drug Discovery Today 7:1184
55. Gaudana R, Ananthula HK, Parenky A, Mitra AK (2010) AAPS J 12:348
56. Edwards DA, Hanes J, Caponetti G, Hrkach J, Ben-Jebria A, Eskew ML, Mintzes J, Deaver D, Lotan N, Langer R (1868) Science 1997:276
57. Park JH, Allen MG, Prausnitz MR (2005) J Controlled Release 104:51
58. Illum LJ, Jorgensen H, Bisgaard H, Krogsgaard O, Rossing N (1987) Int J Pharm 39:189
59. Bourges JLG, Gautier SE, Delie F, Bejjani RA, Jeanny JC, Gurny R, BenEzra D, Behar-Cohen FF (2003) Invest Ophthalmol Visual Sci 44:3562
60. Folkman J (1990) Biomaterials 11:615
61. Folkman J, Long DM Jr, Rosenbaum R (1966) Science 154:148
62. Folkman J, Reiling W, Williams G (1969) Surgery 66:194
63. Dziuk PJ, Cook B (1966) Endocrinology 78:208; (b) Bass PP, Purdon RA, Wiley JN (1965) Nature 208:591; (c) Powers KG (1965) J Parisitol 51:53
64. Higuchi T (1961) J Pharm Sci 50:874
65. Tibbitt MW, Dahlman JE, Langer R (2016) J Am Chem Soc 138:704
66. Zhang YS, Khademhosseini A (2017) Science 356:3627
67. Annabi N, Tamayol A, Uquillas JA, Akbari M, Bertassoni LE, Cha C, Camci-Unal G, Dokmeci MR, Peppas NA, Khademhosseini A (2014) Adv Mater 26:85
68. Hoare TRK, Kohane DS (1993) Polymer 2008:49
69. Langer RP, Peppas NA (2003) AIChE J 49:2990
70. Hoffman AS (2008) J Controlled Release 132:153

Chapter 11
PLGA-Based Nanomaterials for Cancer Therapy

Yakup Gultekin, Tamer Tekin, Meryem Kocas, Yılmaz Capan, and Adem Sahin

1 Introduction

The field of nanomedicine, which is the branch of nanotechnology dealing with medicine, offers important tools in the diagnosis, treatment, and prophylaxis of diseases. In this context, many materials, especially in nanoscale dimensions, against cancer have been developed and produced. The nanocarriers used to deliver drugs or diagnostic agents consist of biocompatible and biodegradable polymers obtained from natural or synthetic sources. Synthetic polymers are more stable and easier to control than natural polymers. In addition, synthetic polymers are less likely to generate an immune response. Today, the most widely used synthetic polymer is poly (lactic-co-glycolic acid) (PLGA), which has been approved by the US Food and Drug Administration (FDA) and the European pharmaceutical agency (EMA) [1, 2].

PLGA polymer has a wide drug delivery system potential in the field of nanomedicine and has proven itself in many areas since it was used as a biodegradable suture in the early 1970s. Successful results have been achieved with drug delivery systems prepared using PLGA polymer in the treatment of diseases such as cardiovascular diseases, inflammatory diseases, infectious diseases, neurodegenerative diseases, cancer, and genetic diseases. Additionally, significant advances have been made in the diagnosis of diseases using PLGA nanocarriers loaded with contrast agents. Besides the use of nanocarriers prepared with PLGA in the diagnosis and

Adem Sahin—Scopus ID: 57193089016.

Y. Gultekin · T. Tekin · M. Kocas · A. Sahin (✉)
Faculty of Pharmacy, Department of Pharmaceutical Technology, Selcuk University, Konya, Turkey

Y. Capan
Faculty of Pharmacy, Department of Pharmaceutical Technology, Lokman Hekim University, Ankara, Turkey

S. Gopi et al. (eds.), *Nanotechnology for Biomedical Applications*,
Materials Horizons: From Nature to Nanomaterials,
https://doi.org/10.1007/978-981-16-7483-9_13

treatment of diseases, they have also achieved significant success in the prophylaxis of diseases with vaccine studies. In addition to these, with the increase in the medical use of PLGA, a rapid growth has been experienced in many areas such as prostheses, tissue engineering, ophthalmology, medical devices, and implantation of artificial organs, dentistry, and bone repair. In Table 1, PLGA-based approved products developed as a result of these studies up to now are given. Also, various studies are still ongoing for the use of PLGA as nanomedicine [3–7]. In this section, fundamental information about PLGA-based nanomedicines will be given about PLGA-based nanomedicines, after this informations, PLGA nanomedicines in cancer therapy will be evaluated.

Table 1 PLGA-based approved products

Trade name	Active ingredient	Indication	Dosage form
Sublocade	Buprenorphine	Opioid addiction	Microsphere
Zoladex	Goserelin	Prostate cancer	Solid Implant
Lupron Depot	Leuprolide	Prostate cancer	Microparticle
Fensolvi	Leuprolide acetate	Early puberty	Microparticle
Sandostatin	Octreotide	Acromegaly	Microparticle
Atridox	Doxycycline	Periodontitis	In situ gel
Signifor LAR	Pasireotide	Acromegaly	Microparticle
Nutropin	Somatotropin	Growth retardation	Microparticle
Trelstar	Triptorelin	Prostate cancer	Microparticle
Perseris	Risperidone	Schizophrenia	Microparticle
Superfact	Buserelin	Prostate cancer	Solid implant
Somatuline	Lanreotide	Acromegaly	Microparticle
Zilretta	Triamcinolone Acetone	Osteoarthritis	Microsphere
Arrest	Minocycline	Periodontitis	Microparticle
Eligard	Leuprolide	Prostate cancer	In situ gel
Triptodur	Triptorelin	Early puberty	Microparticle
RisperidalConsta	Risperidone	Antipsychotic	Microparticle
Vivitrol	Naltrexone	Opioid addiction	Microparticle
Ozurdex	Dexamethasone	Macular edema	Solid implant
Propel	Mometasone	Chronic sinusitis	Solid implant
Bydureon	Exenatide	Type 2 diabetes	Microparticle
Lupaneta Pack	Leuprolide	Endometrosis	Microparticle
Gliadel	Carmustine	Glioma	Solid implant

2 Fundamental Information About PLGA-Based Nanomedicines

2.1 PLGA Polymer and Nanomedicine Types Prepared with PLGA

It is the most studied and best characterized, biodegradable synthetic copolymer consisting of PLGA, lactic acid (LA) and glycolic acid (GA) monomers or their cyclic diesters lactide and glycolide. Polylactic acid and polyglycolic acid have been the most popular polymers since the 1970s, first as biodegradable sutures and later as auxiliaries for drug delivery systems. LA is 2-hydroxypropanoic acid or glycolic acid containing a methyl group. LA can be produced in two different forms, D and L, by fermentation of corn and other agricultural sources. GA is 2-hydroxyethanoic acid produced by chemical synthesis using chloroacetic acid and sodium hydroxide or by a biochemical enzymatic reaction. Since LA has two enantiomeric isomers, D and L, PLGA also has D-, L- and DL-isomers. Although GA has a very crystalline structure since it does not contain methyl side groups, this feature rapidly disappears when combined with PLGA, and therefore, the PLGA copolymers formed have an amorphous structure. PLGA copolymer can be synthesized by a variety of methods such as the polycondensation process, ring opening polymerization, and segmer assembly polymerization [4, 6, 8–10].

During the polymerization reaction, properties of PLGA such as mechanical strength, crystallinity, glass transition temperature, swelling behavior, and hydrolysis capability can be modified by changing the molar ratio of LA:GA. PLGA is a polymer whose degradation time can vary from a few months to a few years, depending on the ratio of LA and GA copolymer and molecular weight, as it shows the properties of its monomers. Polymers with high LA are more hydrophobic, so they absorb less water and degrade more slowly. Low molecular weight polymers with higher GA are more hydrophilic, so they absorb large amounts of water and therefore have a shorter degradation time. However, as an exception to this rule, PLGA polymer containing LA and GA in the ratio of 50:50 dissolves in a shorter time than the PLGA polymer containing high GA. This feature of PLGA affects the drug release kinetics and providing controlled drug release varying between days and months. The release rate of drugs greatly affects parameters such as pharmacokinetics, pharmacodynamics, and bioavailability. Therefore, it is very important to control the release rate of drugs. PLGA-based nanocarriers release the drugs they carry by mechanisms such as polymer degradation/erosion, diffusion, and solvent penetration/system swelling. Commercially available PLGAs are usually provided with the intrinsic viscosity, and the LA:GA ratio it contains as an auxiliary characterization form for selecting the relevant PLGA in relation to their molecular weight [7, 8, 10–12].

PLGA contains ester or acid groups in its structure. In some cases, both ester and acid groups can coexist. Polymers whose terminal group is ester resist hydrolysis.

Modification of PLGA nanoparticles is usually carried out by attaching the acid group-terminated PLGA polymer to the surfaces of the ligands to be used via carboxyl end groups. This coupling process can be accomplished using covalent conjugation, non-covalent interaction, hydrophobic, or ionic interactions [6, 10].

As a result of the hydrolysis of the ester bonds of PLGA in water, small oligomers are formed first from the polymer chains, followed by two metabolic monomers, lactic acid, and glycolic acid. Since these monomers are natural metabolites of the body, they are eliminated from the lungs and kidneys as carbon dioxide (CO_2) and water (H_2O) through the crepe cycle. Thus, they get away without leaving any traces on the body [3, 5, 7, 9].

In addition to the unique properties of the PLGA, surface modifications play an important role in terms of stability, targeting strategies, release profiles, biocompatibility, and circulation half-life. For example, PLGA acquires various properties when modified with polymers such as polyvinyl alcohol (PVA), polyethylene glycol (PEG), or d-α-tocopheryl PEG 1000 succinate (TPGS). In order to increase the solubility of PLGA in water, various PLGA copolymers were formed with different hydrophilic polymers such as polyethylene oxide (PEO) or polyethylene glycol (PEG), and their circulation half-life and pharmacokinetic performance were increased by preventing their removal by mononuclear phagocyte system. In addition, the hydrophilic property of PLGA reduces random cellular uptake and facilitates targeting in the presence of specific targeting agents. Additionally, heat sensitive triblock copolymers for protein delivery and triblock copolymers have been developed to increase cellular uptake of various cationic vector systems. Apart from this, PLGA has the ability to bind various targeting agents, such as the use of integrin binding peptide as a contrast agent for cancer diagnosis [2, 6, 9, 13].

In recent years, star-shaped PLGA polymers have started to attract attention as they show improved mechanical and rheological properties compared to linear PLGA copolymers. Star-shaped block copolymers are an example of branched polymers with chains and branches extending from a single point. Star-shaped PLGA copolymers can be three-armed or four-armed. Compared to nanocarriers prepared with linear polymers, nanocarriers prepared with star-shaped polymer have advantages such as smaller hydrodynamic diameter, lower solution viscosity, higher encapsulation efficiency, and higher drug loading efficiency [14, 15].

Various nanocarriers such as nanocapsules, nanospheres, polymeric micelles, and nanofibers are prepared using PLGA polymer. Nanoparticles are defined as nano-sized collaidal systems that include nanocapsules and nanospheres. Nanocapsules are defined as a core–shell system with an aqueous or oily inner liquid core surrounded by a polymeric envelope, while nanospheres are defined as a matrix system in the form of a wool ball formed as a result of intertwining of polymers. In nanocapsules the therapeutic agent can be dissolved in the inner core, embedded in the polymeric envelope or attached to the surface, while in nanospheres it can be embedded in the polymeric matrix or attached to the surface. Amphiphilic copolymers with hydrophilic and hydrophobic ends come together spontaneously in the form of aggregates and form micelles. The hydrophobic PLGA polymer can combine with a hydrophilic polymer like PEG to form an amphiphilic copolymer. Micelles are colloidal systems

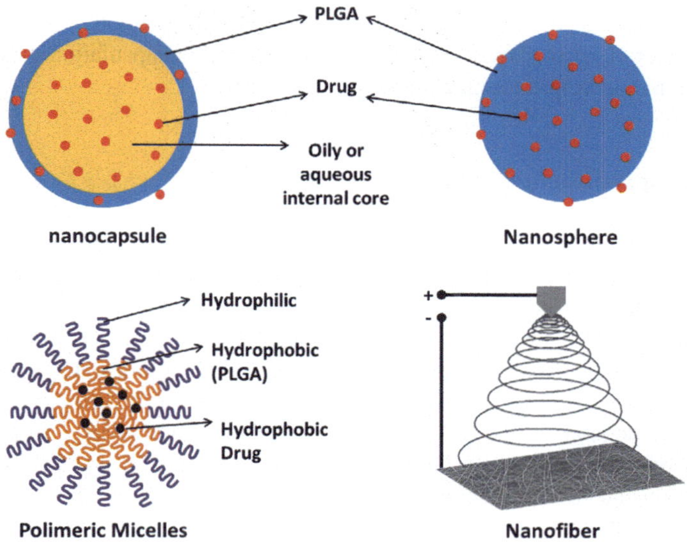

Fig. 1 Nanocarriers prepared using PLGA

in a core–shell structure that can be used in the controlled and targeted transport of anticancer agents with hydrophobic structure and poor bioavailability [16–18]. Nanofibers, these are controlled drug release systems with diameters ranging from 3 to 5000 nm, high surface area/volume ratio, and porosity. Nanofibers are systems that are highly flexible and can be loaded with more than one active substance at the same time [19]. The nanocarriers prepared using PLGA are schematized in Fig. 1.

2.2 *PLGA-Based Nanosphere and Nanocapsule Preparation Methods*

2.2.1 Emulsification-Evaporation Method

Emulsification-evaporation is one of the commonly used methods for preparing PLGA nanoparticles. First, the drug dissolved in a volatile organic solvent immiscible with water (e.g., dichloromethane) is added dropwise to an aqueous phase containing surfactant while stirring. Next, an oil/water emulsion is formed and evaporated to form nanoparticles. To create multiple emulsions, a water/oil type emulsion can be formed and added to an aqueous phase containing surfactant. The encapsulation efficiency of hydrophilic drugs in multiple emulsion techniques is lower than with single emulsion techniques. In this method, the evaporation process and polymer concentration determine the size of the nanoparticles. The higher the polymer concentration, the larger the size of the particles formed. Emulsification method creates nanoparticles of larger size compared to nanoprecipitation and encapsulation efficiency is higher than nanoprecipitation technique [2, 9, 11, 20].

2.2.2 Nanoprecipitation Method

It is a one-step method that is generally used to hold hydrophobic drugs in a polymer matrix. First, the organic phase is prepared by dissolving the polymer, and the hydrophobic drug in an organic solvent that is miscible with water. This organic phase is then added dropwise to the aqueous phase containing an emulsifier or surfactant while mixing. During rapid diffusion of organic solvent into water, nanoparticle formation occurs with precipitation. In this low-energy method, the properties of nanoparticles depend on the molecular weight of the polymer used, the polymer concentration, the properties of the solvents, and the mixing speed. The encapsulation efficiency can be increased with modifications such as salt additives, pH change, and the use of oil solutions [2, 11, 21, 22].

2.2.3 Salting Out Method

In the salting out method, firstly, the polymer and the drug are dissolved in an organic solvent that can be mixed with water. Then, in this solution prepared, an oil/water type emulsion is formed by vigorously mixing with the aqueous phase containing a salt and stabilizing agent that slows the mixing of water with the organic phase. Nanoparticle formation occurs as a result of the diffusion of the organic solvent in the emulsion into the water phase. The added salt is removed using cross-flow filtration or centrifugation. Although salting out method is not suitable for encapsulation of lipophilic agents, it can be used to encapsulate heat sensitive agents such as proteins, DNA, and RNA. In this method, nanoparticle size is not affected by polymer concentration, but is affected by parameters such as polymer molecular weight, solvent type, stabilizing agent type, mixing speed, and duration [2, 6, 11, 23].

2.2.4 Dialysis Method

Dialysis method is generally used to obtain smaller size nanoparticles. The polymer dissolved in a volatile organic solvent is placed in a dialysis tube of the desired pore size. A loss of polymer solubility occurs as a result of the solvent displacement in the dialysis bag. As a result, the polymers gradually come together and form colloidal nanoparticles [11, 24].

2.2.5 Spray Drying Method

This method is an alternative to traditional nanoparticle preparation methods. The advantages of the method can be listed as convenience, speed, and application of less processing parameters. In this method, a water/oil emulsion is sprayed on the hot air stream, which leads to the formation of nanoparticles. However, the adhesion of nanoparticles to the walls of the spray dryer prevents the nanoparticles from collecting effectively, resulting in poor efficiency [11, 25].

2.2.6 Microfluidic Assisted Method

Microfluidic systems are the processing of nanoliter volumes of liquids in micro-scale channels to achieve better results than conventional systems. Depending on the type of reagent flow through the channels, microfluidic systems can be classified into two types as continuous phase flow and droplet phase flow. In the continuous phase flow system, the drug and polymer mixture in an organic solvent are supported by the aqueous phase solution flow through the microchannel, and PLGA nanoparticles are obtained as a result of precipitation in the organic phase. This method has advantages such as nanoparticle synthesis which can be controlled by temperature and reaction time, narrow size distribution, accelerated synthesis, and high reproducibility [2, 9, 26].

2.3 PLGA-Based Polymeric Micelles Preparation Methods

Generally, the method of preparation of polymeric micelles is chosen depending on the water solubility of the drugs and block copolymers. Direct dissolution method can be used for micelles prepared using low molecular weight and relatively hydrophilic polymers. For miscelles prepared using hydrophobic polymers, oil in water emulsification method is used in solvents that do not mix with water; dialysis and film hydration method is used for solvents that can be mixed with water [18, 27, 28].

2.3.1 Direct Dissolution Method

In this method, the active ingredient and the moderately hydrophobic polymer are dissolved together in water at a higher concentration than the critical micelle concentration (CMC), allowing micelle formation spontaneously. This method is the simplest technique that can be used to form polymeric micelles without the need for organic solvents. In addition, the method is easy to scale, but the drug loading capacity is low [28, 29].

2.3.2 Dialysis Method

In this method, primarily, the highly hydrophobic polymer and active substance are dissolved in an organic solvent that is miscible with water. Later, this solvent is slowly removed from the membrane against distilled water by dialysis, resulting in self-assembling polymeric micelles inside the dialysis bag. In addition to the advantage of high drug loading capacity, this method also has disadvantages such as long dialysis time, long time required to remove free drug, and not suitable for scaling up [28, 30, 31].

2.3.3 Oil/Water Emulsion Method

In this method, the polymer and active substance are dissolved in an organic solvent that does not mix with water. Then, the prepared solution is added to the water containing dispersing agent, and emulsification is performed by mixing the two phases. The emulsion droplets formed provide a guide for micelle formation. Active substance-laden micelles are obtained by evaporating the organic phase. With this method, small-sized micelles with narrow size distribution can be obtained [28, 32].

2.3.4 Film Hydration Method

In this method, firstly, the polymer and active substance are dissolved in a volatile organic solvent. The organic solvent is then evaporated, and the film formed is hydrated by adding water. As the hydrophobic block of the polymer interacts, when water is added to the thin film, the hydrophilic blocks dissolve and become hydrated, and micelles are formed as a result of the increase in the interaction between the active substance and hydrophobic blocks. While this method is suitable for scaling up, it requires expensive equipment [28, 33, 34].

2.4 PLGA-Based Nanofiber Preparation Methods

The most known and used method of nanofiber preparation methods is electrospinning. In this method, the polymer or polymer mixture dissolved in a conductive solvent is loaded into a syringe. The positive (+) electrode of the high voltage source is connected to the metal needle at the tip of the syringe, and the negative (−) electrode of the high voltage source is connected to the aluminum plate placed at a certain distance. When an increasing electric field is applied to the needle at the tip of the syringe, the polymer solution first forms a conical structure at the needle tip. After the electric field applied to the polymer solution exceeds a certain threshold value, the polymer leaves the conical structure. Polymer molecules polarized with high voltage follow a chaotic orbit in the direction of the motion axis, creating a random network structure on the collector aluminum plate [35, 36].

3 PLGA-Based Nanomaterials for Cancer Therapy

Cancer is a complex disease that occurs with genetic mutation, uncontrolled division, and proliferation of cells and causes changes in cellular pathology under the influence of genetic and environmental conditions. Despite its rapid progress and increasing cellular pathology, it is very difficult to treat with traditional methods. Nowadays, classical treatments such as chemotherapy, radiotherapy, and surgical methods are used alone or together for cancer treatment. However, these treatments have various disadvantages such as not being selective to cancerous tissues, not reaching sufficient drug concentration in the tumor area, and causing damage to non-cancerous tissues. A wide variety of methods have been developed and tried to eliminate these disadvantages and develop an effective treatment.

In recent years, nanotechnology applications are widely used in cancer research, diagnosis, and treatment. Anticancer-loaded therapeutics produced by nanotechnology have several advantages over conventional preparations such as targeting to the tumor, controlled release, safe transport of higher doses, increased solubility, and bioavailability. In addition, classical anticancer drugs undergo first pass effect are strongly bound to plasma proteins and are rapidly eliminated by the kidneys. Targeted nanoparticular systems are used to overcome these problems and to apply effective chemotherapy. Targeted drug delivery is used to selectively target the drug to tumor tissue and provide higher drug concentration in the tumor tissue through passive or active targeting. These systems have advantages such as a decrease in the frequency of drug delivery compared to conventional systems, a decrease in side effects, and the level of drug in circulation due to being selective to the targeted tissue, thus not affecting healthy tissues.

3.1 Passive Targeted PLGA-Based Nanomedicines for Treatment of Cancer

If biological and pharmacological factors are effective in the accumulation of the drug in a certain area, it is called passive targeting [37]. Passive targeting can also be defined as physical targeting and its principle, it is the transfer of nanocarriers in the body or by passive diffusion. Low molecular weight compounds are transported across the cell membrane by passive diffusion with a concentration differential. The realization of this event in the blood stream of nanocarriers; it depends on the characteristics of circulation, charge, molecule size, and shape [38].

Matsumura and Maeda in 1986 identified two principles that underpin passive nanocarrier targeting. The first principle was the natural accumulation of macromolecular drug carriers applied to solid tumors with leaky vasculature. The second principle was retention of nanocarriers in the tumor with limited lymphatic drainage. These two principles form the basis of the concept of "enhanced permeability and retention" (EPR) effect [39, 40].

In metabolically active cancers, it creates tumors, new blood vessels, and vascular systems to provide nutrient and oxygen supply and help remove metabolic waste products. These tumor vessels are often irregularly branched and have abnormal structures. Therefore, they are structurally different from healthy veins [41]. Dvorak et al. discovered different types of tumor structures with characteristic structures. These different vessels can be listed as follows: (1) feeding artery, (2) main vessel, (3) glomeruloid microvascular proliferation, (4) vascular malformation, (5) capillary, and (6) excretory vein [42]. Main veins, it is characterized by a thin layer of flattened endothelial cells, a disrupted basement membrane, and little or no pericyte cell covers. These properties cause high permeability of the main vessels to plasma proteins and nanocarriers. The spaces between the tumor vessels cause an increase in leakage of the tumor vasculature. The size of these gaps can range from a few nanometers to a few hundred nanometers [40]. These newly formed leaky vessels allow greater than 40 kDa macromolecules and nanosystems to selectively penetrate the tumor stroma [43]. The main vessels show fenestrated endothelial structure that can allow for nanocarrier extravasation. With the difference in concentration between blood and tumor interstitium, nanocarriers are transported passively to the tumor endothelium by extravasation. In order for this transport process to be effective, it may be necessary to increase the blood circulation time of the nanocarriers [40, 41]. Since the discovery of the EPR effect, tumor-targeted drug delivery system applications have been made. Doxil© is the most successful nanocarrier examples on the market, these passive tumor-targeted systems have become important due to their high therapeutic effect [43].

After extensive use of PLGA nanocarriers, various targeting strategies have been used. One of these strategies is passive targeting. Passive targeting takes advantage of the natural size of the nanocarriers, anatomical, and pathophysiological abnormalities of the tumor such as the EPR effect [44]. The shape of PLGA nanocarriers plays an important role in passive targeting. The rod, needle, and cylindrical nanocarriers produced were compared with spherical-shaped PLGA nanocarriers regarding cellular uptake, biodistribution, and half-life. For example, needle-shaped PLGA nanoparticles bind better to endothelial cell membranes and deliver siRNA more efficiently to the cellular cytoplasm than spherical shaped ones. However, needle-shaped PLGA nanoparticles can cause significant cytotoxicity by damaging the lysosome membrane [2].

Modifications of PLGA nanocarriers are important for targeting strategy, biocompatibility, and half-life. To increase the circulation time of PLGA nanocarriers, PEG can be coated with hydrophilic molecules such as polyvinylpyrrolidone, human serum albumin, poloxamers, polysorbate 80, polysorbate 20, vitamin E TPGS, polysaccharides (e.g., dextran), and so on [45]. For example, surface modification by pegylation can be altered to increase blood circulation time by preventing opsonization and uptake by the mononuclear phagocyte system. In another study, curcumin-loaded PLGA nanoparticles were produced with and without pegylation. It was observed that pegylation improved the pharmacokinetic properties of the drug compared to the free drug. It was determined that varying PEG density on the surface of the PLGA nanocarrier (55, 90, and 120 nm) with various dimensions has an effect on the blood circulation times [2].

To provide cancer treatment with PLGA-based nanomedicines by using passive targeting, many studies performed in this part some of them evaluated and listed in Table 2. In one study, passively targeted PEG-PLGA nanopolymerzomes loaded with doxorubicin were prepared by pH-gradient generation technique. Doxorubicin-loaded PEG-PLGA polymeric vesicles showed low inter-batch variation. In an in vitro cytotoxicity study, nanopolymerzomes showed lower cytotoxicity than free doxorubicin. In a study conducted on mice with doxorubicin-loaded nanopolymerzomes in vivo, no toxic effects were observed for a long time in mice, and these results showed that doxorubicin-loaded nanopolymerzomes can increase circulation time and increase the therapeutic efficacy of doxorubicin using passive targeting [46].

Curcumin is known to have the ability to modulate the growth of tumor cells. Curcumin loaded and pegylated, PLGA-based polymeric nanocapsule formulations have been developed. The formulation showed the effects of reaching solid tumors after IV injection, releasing the active drug at the tumor site, inducing apoptosis of proliferating tumor cells. This has shown that passively targeted nanocapsules with PEG can be curcumin carriers and can be effective in cancer treatment [47].

Passively targeted pegylated PLGA nanoparticles of rapamycin were prepared. The nanoparticles prepared had appropriate particle size and escaped uptake by RES, extending the circulation time in the blood. Rapamycin encapsulated in PEGylated PLGA nanoparticles was found to have four times higher AUC values compared to free rapamycin solution. It has been found that the prepared formulation can be an effective chemotherapeutic agent in cancer [48].

With nanosystems, protein peptide therapeutic agents can also be transported. For example, in one study, PLGA nanoparticles loaded with Smac-CPP peptide were prepared. It has been observed that passively targeted nanoparticles can induce apoptosis of tumor cells and inhibit their growth (Priwitaningrum, [49].

In a study with docetaxel active substance; the active ingredient is encapsulated in PEGylated nanoparticles. Drug-loaded nanoparticles, free docetaxel, and non-PEGylated nanoparticles administered intravenously were compared in vivo and in vitro. In the study, it was observed that PEGylated nanoparticles improved the pharmacokinetics and bio-distribution of the active ingredient docetaxel. The optimum size and superficial properties of the nanoparticles produced are seen as the reason for this improvement. PEG modification on the surface of the nanoparticles helped to increase systemic circulation and escape from the removal mechanisms found in the body [50].

Modification of PLGA nanoparticles could be possible with different approaches such as chitosan coating. In these strategies, obtained nanoparticles were not specific

Table 2 Some PLGA-based nanomedicines to provide cancer treatment using passive targeting

Active ingredient(s)	Preparing method	Size (nm)	PDI (Polydispersity index)	Zeta potential (mV)	Nanomedicine	Major finding(s)	References
Doxorubicin	Film rehydration method	134	0.121	–	PEG-b-PLGA nanopolymersomes	Doxorubicin-loaded nanopolymerzomes can increase circulation time and increase the therapeutic efficacy of doxorubicin using passive targeting	[46]
Curcumin	Nanoprecipitation	150.55	0.16	−37.2	Oil-cored PLGA-PEG nanocapsules	The formulated nanocapsule exhibits an extended blood circulation profile compared to the non-PEGylated nanocapsule, and accumulates in the tumors	[47]

(continued)

Table 2 (continued)

Active ingredient(s)	Preparing method	Size (nm)	PDI (Polydispersity index)	Zeta potential (mV)	Nanomedicine	Major finding(s)	References
Rapamycin	Nanoprecipitation	118.8	0.30	−5.9	PEGylated PLGA nanoparticles	PEGylated PLGA nanoparticles was found to have 4 times higher AUC values compared to free rapamycin solution	[48]
Smac peptide and a 14-mer cell penetrating peptide	Double emulsion solvent evaporation method	197	0.1	−16	Polymeric nanoparticles	Passively targeted nanoparticles can induce apoptosis of tumor cells and inhibit their growth	Priwitaningrum, [49]
Docetaxel	Modified emulsification solvent evaporation technique	Non-PEGylated NPs: 123.6 PEGylated NPs: 123.6	Non-PEGylated NPs: 0.245 PEGylated NPs: 0.103	Non-PEGylated NPs: −28.3 PEGylated NPs: −25.9	Non-PEGylated and PEGylated PLGA nanoparticles	PEG modification on the surface of the nanoparticles helped to increase systemic circulation and escape from the removal mechanisms found in the body	[50]

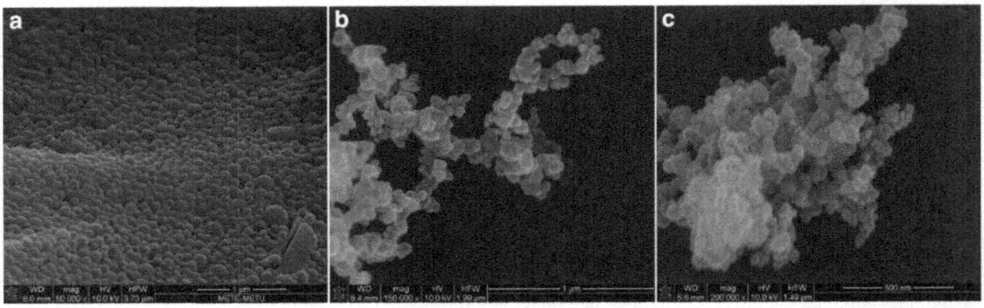

Fig. 2 The scanning electron microscope images of PLGA nanoparticles (**a**), PEGylated-PLGA nanoparticles (**b**), chitosan-coated PLGA-PEG nanoparticles (**c**) [51]

to target. So, these particles should be classified as passive targeted nanoparticles. In recent study, a passive targeted chitosan-coated PLGA-PEG nanoparticles were developed (Fig. 2) for R-flurbiprofen and paclitaxel for the treatment of glioma. In vivo studies which are performed in rat RG2 glioma tumor model showed that chitosan-coated PLGA-PEG nanoparticles loaded with paclitaxel and R-flurbiprofen demonstrated significantly higher therapeutic activity against glioma [51].

3.2 Active Targeted with PLGA-Based Nanomedicines for Treatment of Cancer

Active targeting strategy is a system based on conjugating a ligand specific to these receptors in tumor tissue and ligand-receptor or antigen–antibody interaction [52]. Unlike healthy tissues, tumor tissue has a high percentage of tissue-specific receptors and antigens, and active targeting systems target these receptors. The nanoparticle-bound ligands have the potential for more selective administration and can alter the release and stability of the drug depending on the carrier material used. Additionally, some coating strategy could be evaluated as active targeting strategy because of this coating could provide enhanced nanoparticle uptake to cancer or targeted cells.

Active targeting is based on the uptake of surface-decorated nanoparticles into the cell via receptor-mediated endocytosis by interacting with the target receptor in the cell membrane. In this way, it can deliver the desired amount of therapeutic agent to the tumor site in the target organ in the body, increasing the drug concentration and bioavailability in the tumor tissue.

Targeting selectivity and carrying capacity are two important parameters for the evaluation of active targeting systems. Selectivity is determined by the distribution of the ligand-conjugated nanocarrier in the body and whether it interacts with non-target molecules and cells. The carrying capacity is directly related to the nanoparticle material and structure [53].

High ligand density can be obtained on the surface of nanoparticles with high surface area-to-volume ratio. By using stable peptides or small molecular compounds used for this purpose, the desired level of targeting and stability has been obtained [52]. Studies have reported that the therapeutic efficacy of ligand-receptor-effect nanosystems, targeting and accumulation to the tumor site, cytotoxicity, cellular uptake, and inhibition of tumor growth and survival time increased compared to conventional chemotherapy or passive targeting treatments [54].

To provide cancer treatment with PLGA-based nanomedicines by using active targeting, many studies performed in this part, some of them evaluated and listed in Table 3.

In the treatment of cervical cancer, systems in which PLGA nanoparticles are actively targeted have been designed to reduce the side effects of chemotherapeutics and provide effective treatment. Dana et al., developed an active targeted liposomal formulation containing cisplatin which is a widely used anticancer in chemotherapy treatment of cervical cancer, but its use is limited due to its toxicity and drug resistance. Avastin® which is containing an anti-VEGF antibody bevacizumab was immunolized by chemical conjugation into cisplatin-PLGA liposomes prepared by using double emulsion evaporation method. They tested cell activity by administering liposomal PLGA-cisplatin-Avastin® to mice with cervical tumors by intravenous injection. After all, liposomal PLGA-cisplatin-Avastin® had a significant effect on cell viability and increased the cytotoxicity of SiHa cells. In vitro and in vivo studies of prepared nanoparticles showed that specific transport to cancer cells overexpressing VEGF was increased [55]. In another study conducted for the treatment of cervical cancer, Xu G. et al., prepared PLGA nanoparticles of docetaxel and conjugated with the aptamer to target the uterus. Docetaxel-loaded PLGA nanoparticles were prepared by nanoprecipitation method, after which the AS1411 aptamers were attached to the nanoparticle surface by the Michael addition reaction. Docetaxel aptamer-conjugated PLGA nanoparticles decreased in vitro cytotoxicity and in vivo antitumor analysis when compared with docetaxel-loaded PLGA nanoparticles alone; therapeutic efficacy, survival rate, and cell proliferation are increased [56].

Glioblastoma multiforme (GBM) is the most common primary brain tumor in adults; it is also known as GBM, GBM-IV, and 4th degree brain tumor. The main treatment of GBM consists of surgical resection and chemotherapy. Generally, the prognosis of the disease is poor and the median survival is 12–14 months. Chemotherapeutics used in the treatment of these tumors must cross the blood–brain barrier to be effective. However, anticancer drugs have low solubility and must be given in high doses to cross the blood–brain barrier. Active targeted nanoparticular drug delivery systems are being tested in the treatment of high-dose systemically administered chemotherapeutics to prevent toxicity in healthy tissues before target tissue. Chung et al., designed doxorubicin-loaded PLGA nanoparticles in a study of administration intranasally to target the brain. For active targeting, the nanoparticle surface has been modified with RGD (arginine-glycine-aspartic acid) that binds specifically to integrin $\alpha V\beta 3$ expressed in tumor cells. PLGA nanoparticles were prepared with water in oil in water double emulsion method and NHS/EDC chemical conjugation method was

Table 3 Some PLGA-based nanomedicines to provide cancer treatment using active targeting

Polymer	Preparation method of nanoparticle	Anticancer drug	Ligands	Targeting receptors	Cancer type	References
PLGA	Double emulsion solvent evaporation method	Cisplatin	Bevacizumab Mab	VEGF	Cervical cancer	[55]
M-PLGA-TPGS	Nanoprecepitation	Docetaxel	AS1411 Aptamer	Nucleolin	Cervical cancer	[56]
PLGA	Double emulsion solvent evaporation method	Doxorubicin	RGD (arginine- glycine- aspartic acid)	$\alpha V\beta 3$	Glio blastoma	[57]
PLGA, LMW-Citosan	Nanoprecepitation	Cisplatin	Trastuzumab Mab	HER2	Ovarain Cancer	[58]
PLGA, Poloxamer 407®	Nanoprecepitation	Paclitaxel	Folic acid	Folat receptor	Ovarain Cancer	[59]
PLGA, Citosan	Multiple emulsion solvent evaporation method	Epirubicin	5TR1 Aptamer	MUC1	Breast cancer	[60]
PLGA	Nanoprecepitation	Paclitaxel	S1.5 aptamer	HPA	Breast cancer	[61]
PLGA	Double emulsion solvent evaporation method	Doxorubicin	AS1411 Aptamer	Nucleolin	Lung Cancer	[62]

used for surface modification with RGD. Treatment with RGD-doxurubicin nanoparticles induced reduced tumor growth compared to treatment with free doxorubicin and apoptosis in cancer cells without affecting normal cells [57].

Ovarian cancer is the most common cause of death due to gynecological cancer; the median survival of patients is about six years. Since ovarian cancers are usually diagnosed at an advanced stage, their treatment is based on surgical resection and chemotherapy. The biggest challenge in the treatment of ovarian cancer is that cancer cells develop multiple drug resistance after periodic treatment. Like many other cancers, there are some increased biomarkers in ovarian cancer tumors. For example, HER2 receptor is overexpressed in ovarian cancer. Active targeting systems have been designed with ligands specific to these receptors. Domínguez-Ríos et al., designed HER2 targeted cisplatin-loaded nanoparticles for the treatment of ovarian cancer. Cisplatin nanoparticles were prepared by nanoprecipitation method, and then, trastuzumab, which is a specific antibody to HER2, was added to the surface by carbodiimide conjugation method. The developed nanoparticles were compared with HER2 negative breast cancer cells and analyzed in terms of cytotoxicity, uptake, and drug release kinetics. Consequently, PLGA-trastuzumab nanoparticles showed increased cytotoxicity and tumor uptake compared to free cisplatin [58]. In another study, for the treatment of ovarian cancer, Luiz MT et al., produced folate-modified actively targeted paclitaxel nanoparticles and evaluated their efficacy in treatment in vitro. PLGA nanoparticles containing paclitaxel were prepared by nanoprecipitation technique and polaxamer 407-folic acid conjugate was used as the aqueous phase. Increased cytotoxicity and it has been shown to have 3.6 times greater cellular uptake [59].

Nanoparticular systems targeting anticancer agents for breast cancer treatment have been preferred recently due to their advantages. Epirubicin is an effective antitumor agent widely used in the treatment of breast cancer, but unlike other chemotherapeutics, it is not selective for tumor and its side effects on normal cells limit its use in the treatment. To reduce these limitations, Taghavi et al., designed epirubicin-loaded PLGA nanoparticles targeted to MUC1 receptors overexpressed in breast cancer tumors. The nanoparticles were coated with chitosan to change the surface charge prepared by the multiple emulsion evaporation method. Subsequently, the 5TR1 aptamer ligand that specifically binds to these cells was conjugated by sonication for targeting to MUC1 cells. In vivo and in vitro cytotoxicity studies in mice have shown increased cell delivery and therapeutic efficacy of epirubicin [60]. While estrogen, progesterone, and HER2 receptors were not found in triple negative breast cancer (TNBC) cells, which is a deadly alpine type of breast cancer, it was found that heparanase (HPA) expression increased. For this purpose, Duan T. et al., conducted a study by conjugating S1.5, the anti-HPA aptamer, of paclitaxel which is a widely used anticancer in the treatment, to PLGA nanoparticles. Paclitaxel nanoparticles were prepared by nanoprecipitation method, and S1.5 aptamer was added to the surface using NHS/EDC carbodiimide chemical conjugation method. In the in vitro cytotoxicity and in vivo cell uptake analyzes of the prepared nanoparticles, it was observed that paclitaxel was selective for TNBC cells, and its transmission was effectively increased and accumulated in tumors [61].

There are studies on targeting drug delivery systems that release PLGA-based drug delivery systems to cancer tissue depending on external stimuli such as temperature, pH, and magnetic field. Saravanakumar K., et al., produced a pH-dependent PLGA drug delivery system loaded with doxorubicin in their study. The surface of doxorobicin PLGA nanoparticles, which they produced by double emulsion solvent evaporation method, was conjugated with AS1411 aptamer to target the increased nucleolin receptor in cancer cells. The release profile, in vitro cytotoxicity, and in vivo anti-lung cancer activity of the prepared nanoparticles were evaluated. As a result, PLGA nanoparticles loaded with doxorubicin showed pH-dependent release and increased cytotoxicity and activity in cancer cells compared to treatment with doxorubicin alone. In vivo experiments it was observed that animals given only doksorobusin and doksorobusin-PLGA nanoparticles compared with those given doxorobusin-PLGA-AS1411 nanoparticles showed less animal weight loss, selective to cancer cells, and reduced toxicity in other tissues [62].

4 Conclusion

Cancer is a complex disease that occurs with genetic mutation, uncontrolled division, and proliferation of cells and causes changes in cellular pathology under the influence of genetic and environmental conditions. Despite its rapid progress and increasing cellular pathology, it is very difficult to treat with traditional methods. Nowadays, classical treatments such as chemotherapy, radiotherapy, and surgical methods are used alone or together for cancer treatment.

In recent years, nanotechnology applications are widely used in cancer research, diagnosis, and treatment. Anticancer-loaded therapeutics produced by nanotechnology have several advantages over conventional preparations such as targeting to the tumor, controlled release, safe transport of higher doses, and increased solubility and bioavailability.

Targeted nanoparticular systems are used to overcome these problems and to apply effective chemotherapy. Targeted drug delivery is used to selectively target the drug to tumor tissue and provide higher drug concentration in the tumor tissue through passive or active targeting.

PLGA polymer has a wide drug delivery system potential in the field of nanomedicine. Especially, nanocapsules and nanospheres are frequently used form of PLGA-based nanomedicines against cancer. Easy producing steps and scale up options create a great opportunity to move large scale production. Additionally, successful results have been achieved with passive and active targeted drug PLGA-based nanomedicines in preclinical studies for the treatment of cancer. Hereinafter, it will be important to transfer the experience gained from preclinical studies to the clinical studies.

References

1. Gagliardi A, Paolino D, Costa N, Fresta M, Cosco D (2021) Zein—vs PLGA-based nanoparticles containing rutin: a comparative investigation. Mater Sci Eng C 118:111538
2. Rezvantalab S, Drude NI, Moraveji MK, Guvener N, Koons EK, Shi Y, Lammers T, Kiessling F (2018) PLGA-based nanoparticles in cancer treatment. Front Pharmacol 9:19
3. Danhier F, Ansorena E, Silva JM, Coco R, Le Breton A, Préat V (2012) PLGA-based nanoparticles: an overview of biomedical applications. J Control Release 161(2):505–522
4. Erbetta CD, Alves RJ, Magalh J, de Souza Freitas RF, de Sousa RG (2012) Synthesis and characterization of poly(D,L-Lactide-co-Glycolide) copolymer. J Biomater Nanobiotechnol 3(2):18
5. Kim K-T, Lee J-Y, Kim D-D, Yoon I-S, Cho H-J (2019) Recent progress in the development of poly(lactic-co-glycolic acid)-based nanostructures for cancer imaging and therapy. Pharmaceutics 11(6):280
6. Mir M, Ahmed N, ur Rehman A (2017) (Recent applications of PLGA based nanostructures in drug delivery. Colloids Surf B Biointerfaces 159:217–231
7. Sequeira JAD, Santos AC, Serra J, Veiga F, Ribeiro AJ (2018) Poly(lactic-co-glycolic acid) (PLGA) matrix implants. In: Grumezescu AM (ed) Nanostructures for the engineering of cells, tissues and organs. William Andrew Publishing, pp 375–402
8. Gentile P, Chiono V, Carmagnola I, Hatton PV (2014) An overview of poly(lactic-co-glycolic) acid (PLGA)-based biomaterials for bone tissue engineering. Int J Mol Sci 15(3):3640–3659
9. Kamaly N, Yameen B, Wu J, Farokhzad OC (2016) Degradable controlled-release polymers and polymeric nanoparticles: mechanisms of controlling drug release. Chem Rev 116(4):2602–2663
10. Martins C, Sousa F, Araújo F, Sarmento B (2018) Functionalizing PLGA and PLGA derivatives for drug delivery and tissue regeneration applications. Adv Healthcare Mater 7(1):1701035
11. Sharma S, Parmar A, Kori S, Sandhir R (2016) PLGA-based nanoparticles: a new paradigm in biomedical applications. TrAC, Trends Anal Chem 80:30–40
12. Xu Y, Kim C-S, Saylor DM, Koo D (2017) Polymer degradation and drug delivery in PLGA-based drug–polymer applications: a review of experiments and theories. J Biomed Mater Res B Appl Biomater 105(6):1692–1716
13. Turecek PL, Bossard MJ, Schoetens F, Ivens IA (2016) PEGylation of biopharmaceuticals: a review of chemistry and nonclinical safety information of approved drugs. J Pharm Sci 105(2):460–475
14. Wu B, Liang Y, Tan Y, Xie C, Shen J, Zhang M, Liu X, Yang L, Zhang F, Liu L, Cai S, Huai D, Zheng D, Zhang R, Zhang C, Chen K, Tang X, Sui X (2016) Genistein-loaded nanoparticles of star-shaped diblock copolymer mannitol-core PLGA–TPGS for the treatment of liver cancer. Mater Sci Eng, C 59:792–800
15. Zeng X, Tao W, Mei L, Huang L, Tan C, Feng S-S (2013) Cholic acid-functionalized nanoparticles of star-shaped PLGA-vitamin E TPGS copolymer for docetaxel delivery to cervical cancer. Biomaterials 34(25):6058–6067
16. Gartziandia O, Herran E, Pedraz JL, Igartua M, Hernandez RM (2016) Nanotechnology-based drug-delivery systems releasing growth factors to the CNS: focusing on neurodegenerative disorders. In: Grumezescu AM (ed) Nanobiomaterials in drug delivery. William Andrew Publishing, pp 371–402
17. Gupta M, Sharma V, Chauhan NS (2017) Nanotechnology for oral delivery of anticancer drugs: an insight potential. In: Andronescu E, Grumezescu AM (eds) Nanostructures for oral medicine. Elsevier, pp 467–510
18. Letchford K, Burt H (2007) A review of the formation and classification of amphiphilic block copolymer nanoparticulate structures: micelles, nanospheres, nanocapsules and polymersomes. Eur J Pharm Biopharm 65(3):259–269
19. Mehta PP, Pawar VS (2018) Electrospun nanofiber scaffolds: technology and applications. In: Inamuddin AM, Mohammad A (eds), Applications of nanocomposite materials in drug delivery. Woodhead Publishing, pp 509–573

20. Ramazani F, Chen W, van Nostrum CF, Storm G, Kiessling F, Lammers T, Hennink WE, Kok RJ (2016) Strategies for encapsulation of small hydrophilic and amphiphilic drugs in PLGA microspheres: state-of-the-art and challenges. Int J Pharm 499(1):358–367
21. Fessi H, Puisieux F, Devissaguet JP, Ammoury N, Benita S (1989) Nanocapsule formation by interfacial polymer deposition following solvent displacement. Int J Pharm 55(1):R1–R4
22. Miladi K, Sfar S, Fessi H, Elaissari A (2016) Nanoprecipitation process: from particle preparation to in vivo applications. In: Vauthier C, Ponchel G (eds) Polymer nanoparticles for nanomedicines: a guide for their design, preparation and development. Cham, Springer International Publishing, pp 17–53
23. Lai P, Daear W, Löbenberg R, Prenner EJ (2014) Overview of the preparation of organic polymeric nanoparticles for drug delivery based on gelatine, chitosan, poly(d, l-lactide-co-glycolic acid) and polyalkylcyanoacrylate. Colloids Surf, B 118:154–163
24. Kostag M, Köhler S, Liebert T, Heinze T (2010) Pure cellulose nanoparticles from trimethylsilyl cellulose. Macromol Symp 294(2):96–106
25. Nie H, Lee LY, Tong H, Wang C-H (2008) PLGA/chitosan composites from a combination of spray drying and supercritical fluid foaming techniques: new carriers for DNA delivery. J Control Release 129(3):207–214
26. Li X, Jiang X (2018) Microfluidics for producing poly (lactic-co-glycolic acid)-based pharmaceutical nanoparticles. Adv Drug Deliv Rev 128:101–114
27. Cagel M, Tesan FC, Bernabeu E, Salgueiro MJ, Zubillaga MB, Moretton MA, Chiappetta DA (2017) Polymeric mixed micelles as nanomedicines: achievements and perspectives. Eur J Pharm Biopharm 113:211–228
28. Pham DT, Chokamonsirikun A, Phattaravorakarn V, Tiyaboonchai W (2021) Polymeric micelles for pulmonary drug delivery: a comprehensive review. J Mater Sci 56(3):2016–2036
29. Batrakova EV, Kabanov AV (2008) Pluronic block copolymers: evolution of drug delivery concept from inert nanocarriers to biological response modifiers. J Control Release 130(2):98–106
30. Fournier E, Dufresne M-H, Smith DC, Ranger M, Leroux J-C (2004) A novel one-step drug-loading procedure for water-soluble amphiphilic nanocarriers. Pharm Res 21(6):962–968
31. Kedar U, Phutane P, Shidhaye S, Kadam V (2010) Advances in polymeric micelles for drug delivery and tumor targeting. Nanomed Nanotechnol Biol Med 6(6):714–729
32. Zhang J, Wu M, Yang J, Wu Q, Jin Z (2009) Anionic poly (lactic acid)-polyurethane micelles as potential biodegradable drug delivery carriers. Colloids Surf, A 337(1):200–204
33. Emami J, Maghzi P, Hasanzadeh F, Sadeghi H, Mirian M, Rostami M (2018) PLGA-PEG-RA-based polymeric micelles for tumor targeted delivery of irinotecan. Pharm Dev Technol 23(1):41–54
34. Ozturk N, Kara A, Gulyuz S, Ozkose UU, Tasdelen MA, Bozkir A, Yilmaz O, Vural I (2020) Exploiting ionisable nature of PEtOx-co-PEI to prepare pH sensitive, doxorubicin-loaded micelles. J Microencapsul 37(7):467–480
35. Hu X, Liu S, Zhou G, Huang Y, Xie Z, Jing X (2014) Electrospinning of polymeric nanofibers for drug delivery applications. J Control Release 185:12–21
36. Zhang L, Wang Z, Xiao Y, Liu P, Wang S, Zhao Y, Shen M, Shi X (2018) Electrospun PEGylated PLGA nanofibers for drug encapsulation and release. Mater Sci Eng, C 91:255–262
37. Yadav KS, Mishra DK, Deshpande A, Pethe AM (2019) Levels of drug targeting. Elsevier, Basic fundamentals of drug delivery, pp 269–305
38. Sousa D, Ferreira D, Rodrigues JL, Rodrigues LR (2019) Nanotechnology in targeted drug delivery and therapeutics. In: Mohapatra SS, Ranjan S, Dasgupta N, Mishra RK, Thomas S (eds) Applications of targeted nano drugs and delivery systems. Elsevier, pp 357–409
39. Matsumura Y, Maeda H (1986) A new concept for macromolecular therapeutics in cancer chemotherapy: mechanism of tumoritropic accumulation of proteins and the antitumor agent smancs. Cancer Res 46(12 Pt 1):6387–6392
40. Narum SM, Le T, Le DP, Lee JC, Donahue ND, Yang W, Wilhelm S (2020) Passive targeting in nanomedicine: fundamental concepts, body interactions, and clinical potential. In: Chung EJ, Leon L, Rinaldi C (eds) Nanoparticles for biomedical applications. Elsevier, pp 37–53

41. Pérez-Herrero E, Fernández-Medarde A (2015) Advanced targeted therapies in cancer: drug nanocarriers, the future of chemotherapy. Eur J Pharm Biopharm 93:52–79
42. Dvorak HF, Nagy JA, Dvorak JT, Dvorak AM (1988) Identification and characterization of the blood vessels of solid tumors that are leaky to circulating macromolecules. Am J Pathol 133(1):95–109
43. Attia MF, Anton N, Wallyn J, Omran Z, Vandamme TF (2019) An overview of active and passive targeting strategies to improve the nanocarriers efficiency to tumour sites. J Pharm Pharmacol 71(8):1185–1198
44. Gao C, Pan J, Lu W, Zhang M, Zhou L, Tian J (2009) In-vitro evaluation of paclitaxel-loaded MPEG–PLGA nanoparticles on laryngeal cancer cells. Anticancer Drugs 20(9):807–814
45. Dinarvand R, Sepehri N, Manoochehri S, Rouhani H, Atyabi F (2011) Polylactide-co-glycolide nanoparticles for controlled delivery of anticancer agents. Int J Nanomed 6:877–895
46. Alibolandi M, Sadeghi F, Abnous K, Atyabi F, Ramezani M, Hadizadeh F (2015) The chemotherapeutic potential of doxorubicin-loaded PEG-b-PLGA nanopolymersomes in mouse breast cancer model. Eur J Pharm Biopharm 94:521–531
47. Klippstein R, Wang JT-W, El-Gogary RI, Bai J, Mustafa F, Rubio N, Bansal S, Al-Jamal WT, Al-Jamal KT (2015) Passively targeted curcumin-LOADED PEGylated PLGA Nanocapsules for colon cancer therapy in vivo. Small 11(36):4704–4722
48. Cheraga N, Sun N-C, Huang X-X, Ye Z, Xiao Q-R, Huang N-P (2020) Optimized rapamycin-loaded PEGylated PLGA nanoparticles: preparation, characterization and pharmacokinetics study. J Drug Deliv Sci Technol 102144
49. Priwitaningrum DL, Jentsch J, Bansal R, Rahimian S, Storm G, Hennink WE, Prakash J (2020) Apoptosis-inducing peptide loaded in PLGA nanoparticles induces anti-tumor effects in vivo. Int J Pharmaceut 585:119535
50. Rafiei P, Haddadi A (2017) Docetaxel-loaded PLGA and PLGA-PEG nanoparticles for intravenous application: pharmacokinetics and biodistribution profile. Int J Nanomed 12:935–947
51. Caban-Toktas S, Sahin A, Lule S, Esendagli G, Vural I, Karlı Oguz K, Soylemezoglu F, Mut M, Dalkara T, Khan M, Capan Y (2020) Combination of Paclitaxel and R-flurbiprofen loaded PLGA nanoparticles suppresses glioblastoma growth on systemic administration. Int J Pharmaceut 578:119076
52. Salahpour Anarjan F (2019) Active targeting drug delivery nanocarriers: ligands. Nano-Structures Nano-Objects 19:100370
53. Bertrand N, Wu J, Xu X, Kamaly N, Farokhzad OC (2014) Cancer nanotechnology: the impact of passive and active targeting in the era of modern cancer biology. Adv Drug Deliv Rev 66:2–25
54. Muhamad N, Plengsuriyakarn T, Na-Bangchang K (2018) Application of active targeting nanoparticle delivery system for chemotherapeutic drugs and traditional/herbal medicines in cancer therapy: a systematic review. Int J Nanomed 13:3921–3935
55. Dana P, Bunthot S, Suktham K, Surassmo S, Yata T, Namdee K, Yingmema W, Yimsoo T, Ruktanonchai UR, Sathornsumetee S, Saengkrit N (2020)Active targeting liposome-PLGA composite for cisplatin delivery against cervical cancer. Colloids Surf B Biointerfaces 196:111270
56. Xu G, Yu X, Zhang J, Sheng Y, Liu G, Tao W, Mei L (2016) Robust aptamer-polydopamine-functionalized M-PLGA-TPGS nanoparticles for targeted delivery of docetaxel and enhanced cervical cancer therapy. Int J Nanomed 11:2953–2965
57. Chung K, Ullah I, Kim N, Lim J, Shin J, Lee SC, Jeon S, Kim SH, Kumar P, Lee S-K (2020) Intranasal delivery of cancer-targeting doxorubicin-loaded PLGA nanoparticles arrests glioblastoma growth. J Drug Target 28(6):617–626
58. Domínguez-Ríos R, Sánchez-Ramírez DR, Ruiz-Saray K, Oceguera-Basurto PE, Almada M, Juárez J, Zepeda-Moreno A, del Toro-Arreola A, Topete A, Daneri-Navarro A (2019) Cisplatin-loaded PLGA nanoparticles for HER2 targeted ovarian cancer therapy. Colloids Surf, B 178:199–207
59. Luiz MT, Abriata JP, Raspantini GL, Tofani LB, Fumagalli F, de Melo SMG, Emery FDS, Swiech K, Marcato PD, Lee R, Marchetti JM (2019) In vitro evaluation of folate-modified

PLGA nanoparticles containing paclitaxel for ovarian cancer therapy. Mater Sci Eng C 105:110038
60. Taghavi S, Ramezani M, Alibolandi M, Abnous K, Taghdisi SM (2017) Chitosan-modified PLGA nanoparticles tagged with 5TR1 aptamer for in vivo tumor-targeted drug delivery. Cancer Lett 400:1–8
61. Duan T, Xu Z, Sun F, Wang Y, Zhang J, Luo C, Wang M (2019)HPA aptamer functionalized paclitaxel-loaded PLGA nanoparticles for enhanced anticancer therapy through targeted effects and microenvironment modulation. Biomed Pharmacother 117:109121
62. Saravanakumar K, Hu X, Shanmugam S, Chelliah R, Sekar P, Oh D-H, Vijayakumar S, Kathiresan K, Wang M-H (2019) Enhanced cancer therapy with pH-dependent and aptamer functionalized doxorubicin loaded polymeric (poly D, L-lactic-co-glycolic acid) nanoparticles. Arch Biochem Biophys 671:143–151

Chapter 12
Carbon Nanotubes for Biomedical Applications

Mafalda R. Almeida, João C. F. Nunes, Raquel O. Cristóvão, Joaquim L. Faria, Ana P. M. Tavares, Cláudia G. Silva, and Mara G. Freire

Abbreviations

$^1O_2$	Singlet oxygen
99mTc	Technetium-99m
AF488/647	Alexa-fluor
ALG	Alginate
AT1R	Angiotensin II type 1 receptor
AT2R	Angiotensin II type 2 receptor
AuNPs	Gold nanoparticles
AZ	Azoxystrobin
BA	Betulinic acid
BrdU	Bromodeoxyuridine
BSA	Bovine serum albumin
CDDP	Cisplatin
Ce6	Chlorin e6
CEA	Carcinoembryonic antigen
CHI	Chitosan
CHI-F	Chitosan-folate
CNTs	Carbon nanotubes
CNx-MWCNTs	Nitrogen-doped multi-walled carbon nanotubes
CP	Carboplatin

M. R. Almeida · J. C. F. Nunes · A. P. M. Tavares (✉) · M. G. Freire
CICECO-Aveiro Institute of Materials, Department of Chemistry, University of Aveiro, 3810-193 Aveiro, Portugal
e-mail: aptavares@ua.pt

R. O. Cristóvão · J. L. Faria · C. G. Silva
Laboratory of Separation and Reaction Engineering—Laboratory of Catalysis and Materials (LSRE-LCM), Department of Chemical Engineering, Faculty of Engineering, University of Porto, 4200-465 Porto, Portugal

S. Gopi et al. (eds.), *Nanotechnology for Biomedical Applications*,
Materials Horizons: From Nature to Nanomaterials,
https://doi.org/10.1007/978-981-16-7483-9_14

CPT	Camptothecin
DDS	Drug delivery systems
dendrimerFc	Ferrocene-grafted dendrimer
DEX	Dexamethasone
dMWCNTs	Polyamidoamine dendrimer-modified multi-walled carbon nanotubes
DNA	Deoxyribonucleic acid
Docel™	Commercial DTX injection
DOX	Doxorubicin
DRG	Dorsal root ganglion
DTX	Docetaxel
DWCNTs	Double-walled carbon nanotubes
EAC	Erlich ascites carcinoma
EpCAM	Epithelial cell adhesion molecule
FA	Folic acid
FA-PEG	Folic acid-terminated polyethylene glycol
fCNTs	Functionalized carbon nanotubes
FI	Fluorescein isothiocyanate
fMWCNTs	Functionalized multi-walled carbon nanotubes
FR	Folate receptor
Gd	Gadolinium
GO	Graphene oxide
GOx	Glucose oxidase
GR	Graphene
HA	Hyaluronic acid
hapten AZc6	Carboxylated spacer arm
HER2	Human endothelial receptor 2
HIF-1α	Hypoxia-inducible factor 1 alpha
HMME	Hematoporphyrin monomethyl ether
IGF1R	Insulin-like growth factor 1 receptor
IgY	Immunoglobulin Y
iRGD	Tumor-homing peptide
Lys	Lysine
MDA-MB-231	Human breast carcinoma cells
MDM2	Murine double minute clone 2
MNs	Magnetic nanoparticles
MOF	Metal oxide framework
MSCs	Marrow stromal cells
mTHPC	M-tetrahydroxyphenylchlorin
MTX	Methotrexate
MWCNTs	Multi-walled carbon nanotubes
Nf	Nafion
NGR	Asparagine–glycine–arginine peptide
Ni	Nickel
pAT2	Plasmid AT2

PBS	Phosphate-buffered saline
PCA	Poly citric acid
pDNA	Plasmid deoxyribonucleic acid
PDT	Photodynamic therapy
PLK1	Polo-Like Kinase1
PL-PEG	Phospholipid molecules with polyethylene glycol
PPy	Polypyrrole
Pt	Platinum
PTT	Photothermal therapy
PTX	Paclitaxel
Pyrr	Pyridinium
RAS	Renin-angiotensin system
RES	Reticuloendothelial system
RGD	Cyclic peptide composed of nine amino acids
rhBMP-2	Recombinant human bone morphogenetic protein-2
ROS	Reactive oxidative species
RTA	Recombined ricin A chain protein
SA	Streptavidin
SDS	Sodium dodecyl sulfate
siCas 3	Small-interfering caspase-3
siRNA	Small-interfering ribonucleic acid
siTOX	Cytotoxic small-interfering ribonucleic acid sequence
SS	Cystamine
ssDNA AP	Single-stranded deoxyribonucleic acid aptamer
SWCNTs	Single-walled carbon nanotubes
TAM	Tamoxifen
Tf	Transferrin
TPGS	D-α-Tocopherol polyethylene glycol 1000 succinate
TRAIL	Tumor necrosis factor-related apoptosis-inducing ligand
VA	Vertically aligned
VP-16	Etoposide

1 Introduction

Due to their peculiar properties and nanoscale dimensions, nanomaterials have been largely used as biomaterials in several fields of applications. Specifically, in biomedicine, these miniaturized particles, whose dimensions are in line with those of the living cells, have been opening new and important applications. Among nanoparticles, and particularly due to their chemical and physical properties, carbon-based nanomaterials are becoming highly attractive for biomedical applications, namely for cancer diagnosis and therapy, cell and tissue engineering, targeted drug delivery, and biosensing [1–3]. According to the structural arrangement of their atoms, there are several carbon allotropes, with fullerenes, carbon nanotubes (CNTs) and graphene oxide (GO) being the most used in the field of biomedicine [4, 5].

CNTs were discovered in 1991 by Sumio Iijima [6] using transmission electron microscopy. Since then, they have been one of the most intensely studied nanomaterials, with thousands of reports on a wide variety of applications [7]. CNTs provide improved functionalities due to their specific structural properties [8], strong loading capacity [9], designed biocompatibility [10], high surface area, high strength, and enhanced chemical and thermal stability [11].

CNTs consist entirely of carbon atoms connected through sp^2 bonds structured in several condensed benzene rings rolled up into a cylinder form, presenting lengths in the order of μm and diameters up to 100 nm [12]. The orientation of benzene rings along the cylindrical surface dictates their metallic, semiconducting, or superconducting electron transport properties [13, 14]. Depending on the number of graphitic layers, CNTs are classified into single-walled carbon nanotubes (SWCNTs) or multi-walled carbon nanotubes (MWCNTs). SWCNTs own the simplest morphology, consisting of a single layer of graphene with diameters between 0.4 and 2 nm, generally existing as hexagonal close-packed bundles. MWCNTs comprise two or more concentric cylinders, each made of a single graphene sheet surrounding a hollow core with inner diameters between 1 and 3 nm. The outer diameter fluctuates from 2 to 100 nm [15, 16].

CNTs are generally synthesized by three techniques: chemical vapor deposition using hydrocarbon sources (CO, methane, ethylene, acetylene), electric arc discharge method using arc-vaporization with two carbon rods, or laser ablation method using graphite [17]. To remove CNTs impurities like amorphous carbon, fullerenes, and transition metals (used as catalysts in their synthesis), a purification step is subsequently performed by acid refluxing, surfactant aided sonication, or air oxidation methodologies [18, 19].

Combining CNTs structural dimensions with their exceptional characteristics makes them one of the most potential nanomaterials for biomedical applications [20, 21]. Their unique electrical and optical features make CNTs highly sensitive when exposed to biomolecules, leading to their successful application as sensing components for biosensors to detect proteins, cells, or microorganisms [22]. Photoluminescence is another important characteristic of CNTs [23], which, together with their strong optical absorption in the near-infrared region [24] and Raman scattering [25], allow the CNTs orientation and quantification monitoring in biological media, highly interesting for medical imaging. Their cylindrical graphitic structure and sp^2 character give them excellent mechanical properties, translating into their high strength and stiffness [26]. CNTs also possess an exceptional tensile force, turning them into a robust product with improved thermal conductivity [26]. Notably, the effective possibility of functionalized CNTs to incorporate molecules, namely proteins, and their ability to cross biological barriers, like the cell membrane and penetrate individual cells [27], enable them to act as effective drug carriers [28]. Several authors reported a more effective and safer administration of drugs into cells when bonded to CNTs than by traditional techniques [18, 19]. This function allows overcoming the problems associated with several medications, like low solubility, faster deactivation, and limited bioavailability [29]. The possibility of internalization and release of CNTs from cells, along with their incomparable photothermal response, allows their use as photothermal therapeutic agents to reduce tumor size or even to destroy cancer cells by near-infrared laser irradiation [30, 31].

Despite their relevance in this field, the biomedical application of CNTs raises some questions about safety and potential cytotoxicity issues. Intending a safe CNTs use, high-purity CNTs are preferred to limit potential toxic ions release during the application. However, this can be a real challenge on a large scale, having to reach a compromise between quality and quantity [32]. The CNTs toxicity has been a hot topic of discussion, with many publications but without a clear conclusion.

This book chapter overviews the potential applications of CNTs in the biomedical sector, highlighting their usage on diagnosis and devices, tissue engineering, delivery systems, and target therapies (Fig. 1). Particular attention is also given to the CNTs potential toxicity and attempts to overcome this issue.

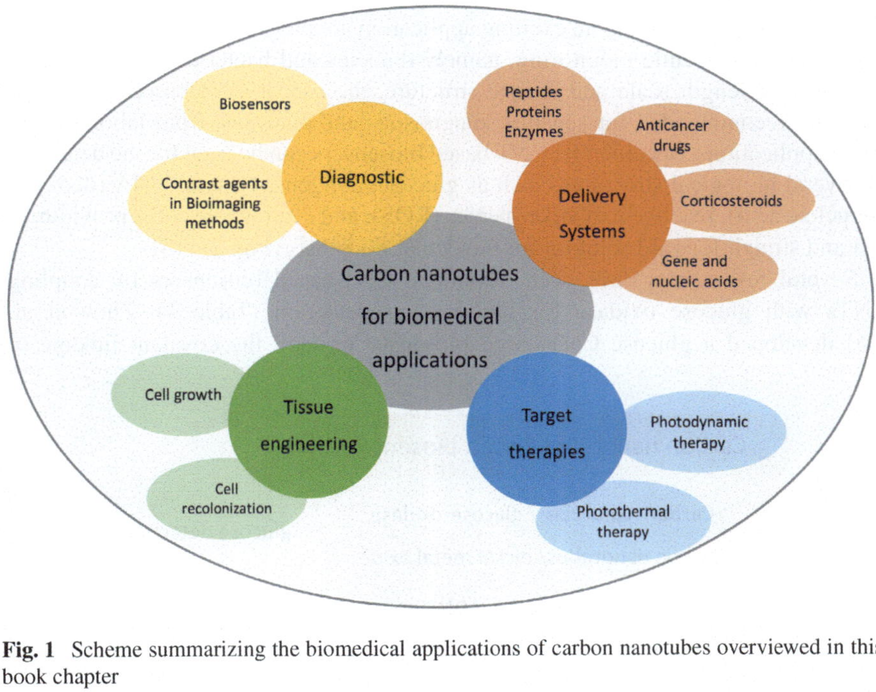

Fig. 1 Scheme summarizing the biomedical applications of carbon nanotubes overviewed in this book chapter

2 Carbon Nanotubes in Biomedical Applications

2.1 *Carbon Nanotubes in Diagnostic*

The success of disease treatment is majorly based on early diagnostic and improved detection methods [32]. In this sense, several technologies based on CNTs have been developed to improve the diagnostic and detection methods. Due to their electronic properties, CNTs have been considered the key element of electrochemical sensors, and different kinds of label-free CNT-based biosensors were developed [32, 33]. Additionally, CNTs can act as contrast agents in other bioimaging methods [32, 34].

2.1.1 Carbon-Nanotube-Based Biosensors

Biosensors are analytical devices that combine a biological component and a physicochemical detector to identify a target analyte [15]. CNT-based biosensors are modern devices, while representing an exciting application area for in vitro and in vivo diagnostics and therapeutic monitoring, namely diabetes and bacterial infections [32]. Due to their length scale and unique structure, the use of CNT-based biosensors is highly recommended for sensitive diagnostics and analyses, from laboratory to clinic applications. Specifically, CNT-based biosensors can be used for the detection of several biological structures, such as glucose, antigens, viruses, DNA (deoxyribonucleic acid), reactive oxidative species (ROS), and even whole cells, providing a fast and straightforward solution for molecular diagnosis (Fig. 2) [15].

Several researchers have been developed CNT-based biosensors by coupling CNTs with glucose oxidase for blood sugar detection (Table 1). Zhou et al. [35] developed a glucose CNT-based biosensor through the covalent linkage of

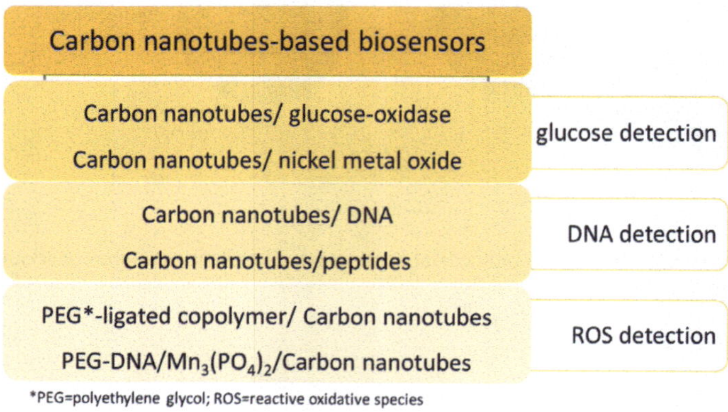

Fig. 2 Carbon-nanotube-based biosensors used in the detection of several biological structures

Table 1 Examples of developed CNT-based biosensors used in the detection of several biological structures and main results

	Biosensor	Main results	References
Glucose detection	**CNTs-CS/dendrimerFc/GOx** Ferrocene-grafted dendrimer covalently linked on the surface of CNTs-chitosan nanocomposite modified electrode	• Fast response time (less than 10 s) • Wide linear range (from 0.02 to 2.91 mM) • Low detection limit (down to 7.5 μM)	[35]
	PPy-Nf-fMWCNTs Bio-nanohybrid film of polypyrrole-Nafion-functionalized MWCNTs nanocomposite prepared on the surface of glassy carbon electrode	• Improved sensitivity (2860.3 μA mM^{-1} cm^{-2}) • Low detection limit (5 μM under a signal/noise ratio of 3) • Long-term stability • Good repeatability and reproducibility • Acceptable measurement of glucose concentration in real serum samples	[36]
	GR-MWCNTs/AuNPs Biopolymer pectin stabilized gold nanoparticles prepared in MWCNTs	• Two linear ranges: 10 μM–2 mM with a limit of detection of 4.1 μM 2 mM–5.2 mM with a limit of detection of 0.95 mM • Appreciable stability, repeatability, reproducibility, and practicality	[37]
	Ni-MOF/CNTs Nickel-based metal organic framework linked to functional CNTs	• Low detection limit (0.82 μM) • High sensitivity (13.85 mA mM^{-1} cm^{-2}) • Linear range (1 μM–1.6 mM) • Good reproducibility and repeatability	[38]
DNA detection	**Gold-supported CNTs–DNA sensors**	• High selectivity and sensitivity	[41]
	CNTs–peptide (Fmoc-RRMEHRMEW)	• Broader sensing range (1.6×10^{-4} to 5 μmol L^{-1}) • Low detection limit (0.88 μg L^{-1})	[42]
ROS detection	**CNTs/polyethylene-DNA/$Mn_3(PO_4)_2$**	• Low detection limit (30×10^{-9} M) • High sensitivity (9.6 and 20.8 μA μM^{-1} cm^{-2})	[44]
	Fluorescent SWCNTs	• Detection of single molecules of H_2O_2	[45]

(continued)

Table 1 (continued)

	Biosensor	Main results	References
	Fluorescent SWCNTs/polyethylene glycol-ligated copolymer	• Selective detection of nitric oxide concentration • Low detection limit (1 μM) • No immune reactivity or other adverse response (for more than 400 days)	[46]

ferrocene-grafted dendrimer on the surface of CNTs chitosan nanocomposite modified electrode able to immobilize a high content of glucose oxidase (CNTs-CHI/dendrimerFc/GOx). Due to the excellent electron transfer acceleration of CNTs and the high loading content of the glucose oxidase and ferrocene mediator on the electrode matrix, this biosensor showed excellent analytical performance, with a fast response time of less than 10 s, a wide linear range from 0.02 to 2.91 mM, and low detection limit down to 7.5 μM [35]. Shrestha et al. [36] developed a highly electroactive bio-nanohybrid film of polypyrrole-Nafion-functionalized MWCNTs nanocomposite prepared on the surface of glassy carbon electrode (PPy-Nf-fMWCNTs) by an easy one-step electrochemical polymerization technique followed by chitosan-glucose oxidase immobilization on its surface. This MWCNT-based biosensor revealed an improved sensitivity (2860.3 μA mM^{-1} cm^{-2}), low detection limit (5 μM under a signal/noise ratio of 3), long-term stability, good repeatability, reproducibility, and acceptable measurement of glucose concentration in real serum samples [36]. Using gold nanoparticles decorated MWCNTs, Devasenathipathy et al. [37] similarly prepared a CNT-based biosensor (GR-MWCNTs/AuNPs) for the determination of glucose. Biopolymer pectin stabilized gold nanoparticles were prepared at graphene and MWCNTs, where glucose oxidase was successfully immobilized. The manufactured glucose biosensor showed two linear ranges: 10 μM–2 mM with a limit of detection of 4.1 μM, and 2 mM–5.2 mM with a limit of detection of 0.95 mM. Moreover, the biosensor exhibited appreciable stability, repeatability, reproducibility, and practicality. Comparing the biosensor performance with other reported sensors revealed a significant improvement in the overall sensor performance [37]. On the other hand, Zhang et al. [38] described a non-enzymatic glucose detector composed of a porous nickel-based metal oxide framework (Ni-MOF/CNTs), where the addition of CNTs enhanced the electrical conductivity. These electrodes exhibited a low detection limit (0.82 μM), a high sensitivity of 13.85 mA mM^{-1} cm^{-2}, a wide linear range from 1 μM to 1.6 mM, and a good reproducibility and repeatability. In this sense, the proposed sensor is found as a relevant alternative for glucose detection through immune complexes immobilization [38].

Other promising CNT-enzyme biosensors, such as CNT-based dehydrogenase biosensors [39] or peroxidase [40] biosensors, have been developed for different therapeutic monitoring and diagnostics. Beyond CNT-based biosensors for blood sugar detection, CNT-based biosensors to detect DNA have been prepared. For example,

He et al. [41] developed a CNT-DNA-based biosensor by modifying CNTs with single-strand DNA chains. A gold-supported aligned CNT array was firstly treated with acetic acid. Then, single-strand DNA chains with an amino group were chemically bonded with the plasma-induced -COOH group on the CNTs via amide formation. Finally, ferrocenecarboxaldehyde was labeled to complementary DNA chains and other non-complementary DNA chains. This CNT-DNA-based biosensor exhibited a high selectivity and sensitivity for probing complementary DNA and target DNA chains of specific sequences [41]. Recently, Li et al. [42] developed a CNT-based biosensor for DNA recognition through a bio-interaction between DNA and biomolecules. A CNT-peptide-based biosensor was prepared using the peptide (with the sequence Fmoc-RRMEHRMEW) discovered by the authors, and which has a special bio-interaction with tiny nucleic acids and causes a significant decrease in zeta potential. Compared to the former reported universal DNA bio-detector and NanoDrop (a spectrometer from Thermo Scientific™), this unique CNT-peptide-based DNA biosensor demonstrated a broader sensing range (from 1.6×10^{-4} to 5 μmol L^{-1}) and a significantly lower detection limit (0.88 μg L^{-1}) [42].

CNTs have been also used in vivo biosensing. Sudibya et al. [43] prepared CNT-based biosensors by the functionalization of CNTs with bioactive monosaccharides. The authors demonstrated that the CNT-monosaccharides-based biosensor could interact with living cells surface and dynamically detect biomolecules [43]. Additionally, Hu et al. [44] reported a CNT-based biosensor through the functionalization of CNT/polyethylene with DNA and $Mn_3(PO_4)_2$ and detected ROS in situ released from cancer cells under drug stimulations. Living human breast carcinoma cells (MDA-MB-231) can grow on the CNT/polyethylene-DNA/$Mn_3(PO_4)_2$ biosensor forming an MDA-MB-231/$Mn_3(PO_4)_2$/DNA-polyethylene/CNT hybrid film, creating a very short diffusion distance between cells and reaction sites, resulting in a highly sensitive method (detection limit of 30×10^{-9} M and sensitivity of 9.6 and 20.8 μA μM^{-1} cm^{-2} depending on the linear range) for detecting ROS. Jin et al. [45] developed a CNT-based biosensor to detect ROS, namely single molecules of H_2O_2, using fluorescent SWCNTs, and found that epidermal growth factor induces 2 nmol H_2O_2 locally for 50 min. For the detection of local nitric oxide, Iverson et al. [46] have modified near-infrared fluorescent SWCNTs with a polyethylene glycol-ligated copolymer, enabling intravenous injection into mice and the selective detection of nitric oxide concentration (detection limit of 1 μM) with no intrinsic immune reactivity or other adverse response for more than 400 days. The half-life for liver retention is 4 h, with sensors clearing the lungs within 2 h after injection, thus avoiding a dominant route of in vivo nanotoxicology. After localization within the liver, it was possible to follow the transient inflammation using nitric oxide as a marker and signaling molecule [46].

2.1.2 Imaging Methods Using Carbon Nanotubes

Standard diagnostic techniques for monitoring and detecting the tissue regeneration process in small animals, namely histology analysis that is the most robust diagnostic technique, lead to significant variation results and difficult the assessments of temporal results in a statistically significant manner. In this sense, and taking advantage of intrinsic properties of CNTs, different CNT-based materials have been developed for applications in a multiplicity of imaging modalities, such as photoluminescence imaging through near-infrared fluorescence, photoacoustic tomography, Raman spectroscopy, magnetic resonance, and radionuclide-based imaging, offering scientists a piece of spatial and temporal information in a faster and more convenient manner (Fig. 3) [47].

In photoluminescence imaging, SWCNTs are excited emitting fluorescence in the near-infrared-I (700–900 nm) and near-infrared-II (1100–1400 nm) ranges, where tissues and water are almost transparent, making it possible to reach deeper penetration depths. In fact, SWCNT fluorescence was emerging as a reliable optical imaging method for vessel observation and used as an alternative optical method to the expensive positron emission tomography-computed tomography for brown fat detection [32]. Welsher and co-authors [48, 49] used this property to optically detect in real time the biodistribution of injected SWCNTs in mice deep tissues and vessels. The authors synthesized phospholipid-polyethylene glycol-coated SWCNTs by sonicating SWCNTs with sodium cholate, followed by surfactant exchange. They used these materials as in vivo near-infrared photoluminescence imaging agents for live mice. The high-resolution intravital image of tumor vessels beneath thick skin was achieved due to deep tissue penetration and low autofluorescence background. Cherukuri et al. [50] also used near-infrared fluorescence microscopy to image SWCNTs in macrophage cells. Macrophage samples were incubated in growth media with SWCNTs at different concentrations to track their ex vivo uptake in

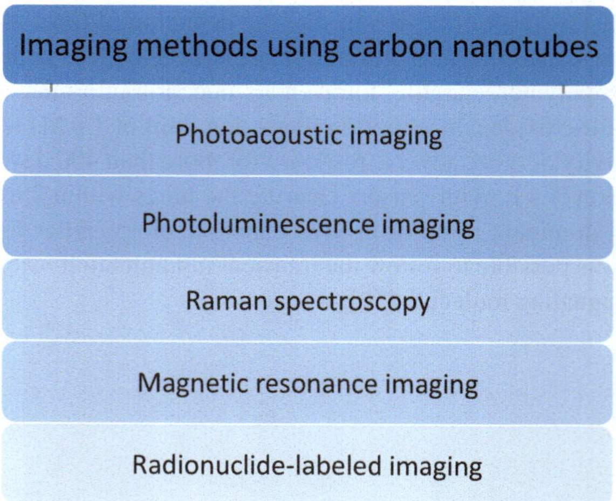

Fig. 3 Imaging modalities where CNTs have been applied

mouse peritoneal macrophage cells. Near-infrared fluorescence imaging revealed no difference in population growth, adhesion, morphology, and confluence between the control and the cultures containing SWCNTs, and the SWCNTs uptake appears to occur through phagocytosis. Additionally, Leeuw et al. [51] have used in vivo near-infrared imaging to assess the biocompatibility of SWCNTs in an intact organism, *Drosophila melanogaster*. This study showed the effectiveness of SWCNTs as near-infrared probes for studying individual nanotubes in tissue specimens or inside living organisms during tissue regeneration. Based on all these studies, photoluminescence imaging seems to be a promising method for tracking SWCNTs in cells and small animals over long durations of time.

Beyond photoluminescence, the photoacoustic effect is another attractive way to use CNT near-infrared absorption. Photoacoustic imaging measures the ultrasounds produced by tissue expansion and heat emitted by CNTs exposed to near-infrared stimulation [32]. Improved contrast on targeted tumor cells, using SWCNTs conjugated with arginine–glycine–aspartate peptides as well as an enhanced sensitivity and detection threshold, was obtained to pristine SWCNT [52, 53]. SWCNT was reported as a technique that differentiates multipotent marrow stromal cells (MSCs) toward osteoblasts [54]. It was demonstrated that a brief (10 min) daily nanoparticle-facilitated exposure of MSCs to nanosecond pulse laser-induced photoacoustic stimulation enhanced differentiation [54].

Raman scattering is also an imaging option due to the vibrational and low-frequency modes of CNTs based on inelastic scattering, which appears at 1590–1600 cm^{-1} (G band, due to stretching along the C–C bond of graphene). Resonance-enhanced Raman bands for SWCNTs appear at 150–300 cm^{-1} (radial breathing modes), which can be related to the diameter of the SWCNTs [32, 55]. Thus, Raman imaging allows for the spatial and chemical imaging of SWCNTs in cells and small animals. Heller et al. [56] applied Raman spectroscopy to monitor cells labeled with CNTs. These nanotube aggregates remained in the cell and displayed scattering even after several cycles of cell division, showing potential as markers in assessing the proliferation and differentiation of cells in culture, long-term labeling of cell populations, and continuous monitoring of the cellular environments [56]. Raman imaging was also applied in study biodistribution and blood circulation of PEGylated SWCNTs in a mouse model [57]. The results showed SWCNT accumulation in the intestine, feces, kidney, and bladder of mice and suggested excretion and clearance of SWCNTs from mice via the biliary and renal pathways [57]. Another study indicated Raman imaging for real-time monitoring of SWCNTs for disease (tumor) targeting [58]. Specifically, disease targeting SWCNTs (RGD–SWCNT, arginine–glycine–aspartic acid peptides) were examined for localization in diseased (tumor) mice as against plain SWCNT [58]. The images showed a higher accumulation of RGD–SWCNTs in the tumor than the non-functionalized SWCNT mouse group, with a lower accumulation in the liver and spleen [58].

CNTs have also been shown to increase the contrast enhancement efficacy of metal ions, such as gadolinium, commonly applied in magnetic resonance imaging as a contrast agent [59]. Magnetic resonance imaging is a non-invasive imaging modality regularly used to diagnose several diseases and physiochemical states of tissue and blood flow since it provides anatomical details [60]. In magnetic resonance imaging, it is applied an external magnetic field and radiofrequency pulses, exciting proton spin states. From the excited state, spin to equilibrium is required a period, the relaxation time. This relaxation causes a change in the flux of the receiving coil of the magnetic resonance imaging scanner, inducing a magnetic resonance signal. The relaxation time affects the quality of the image, especially the contrast [60]. Contrast agents, such as gadolinium, are used to improve the contrast since they can modulate the relaxation time. Sitharaman et al. [61] reported the use of SWCNTs as carriers of gadolinium. The authors described the encapsulation of Gd^{3+} ions with SWCNTs to sequester the toxic effect of Gd^{3+} ions and amplify the contrast-enhancing efficacy. The results showed an increase in proton relaxation time, increasing the contrast and enhancing the efficacy [61]. Therefore, GB-SWCNT-based magnetic resonance imaging contrast agents could be dispersed throughout an engineered tissue, and for instance, to monitor the process of tissue regeneration [61].

Efficient contrast agents were also developed by functionalizing MWCNTs with gadolinium chelating agent, diethylene triamine penta-acetic acid, enabling tight attachment of gadolinium (Gd) atoms onto the nanotube surface [59]. The resulting Gd-MWNTs contrast agents were found to be three times better than the clinically approved T1 contrast agent Magnevist. Moreover, the Gd-MWNTs contrast agents were stable over two weeks in water and mouse serum [59]. The functionalization method used by servant and co-workers [59] was previously applied by radio probes composed of metal halides ($Na^{125}I$), encapsulated within SWCNTs, which were developed for single-photon emission computed tomography imaging, another application of CNTs in bioimaging [62]. In brief, metal halides were sealed inside SWCNTs to create high-density radio-emitting crystals, while SWCNTs surfaces were then covalently sealed with biantennary carbohydrates, improving dispersibility and biocompatibility. The intravenous administration of these radio probes in mice was tracked in vivo using single-photon emission computed tomography imaging, resulting in ultrasensitive imaging and delivery of unprecedented radio dose density [62].

Beyond the examples of CNT-based contrast agents on studies that harness the intrinsic properties of these nanomaterials, different investigations have shown that only functionalization of the external carbon sheath of CNTs (SWCNTs or MWCNTs) with appropriate imaging agents allows molecular imaging [63, 64].

2.2 Carbon Nanotubes in Tissue Engineering

Tissue engineering combines biological and material sciences in which the fast development of nanotechnology and nanomaterials is an opportunity to enhance biomedical engineering. CNTs present several and very distinct properties which can be assembled into different morphologies, such as fibers, films, and three-dimensional (3D) structures, and have a positive influence on living cells. All these properties make of these nanomaterials suitable for tissue engineering [32, 65]. Several authors have described promising results in this field with Lovat et al. [66] reporting CNTs as a good surface for cellular growth and as potential devices able to improve neural signal transfer while supporting dendrite elongation and cell adhesion. It was demonstrated the growth of neuronal circuits on a CNT grid accompanied by a significant increase in the network activity, which may be related to the specific properties of CNT materials, such as the high electrical conductivity [66].

Mazzatenta and co-authors [67] developed an integrated SWCNT–neuron system to investigate whether electrical stimulation delivered via SWCNTs can induce neuronal signaling. To this end, hippocampal cells were grown on pure SWCNTs substrates and patch clamped, and neuronal responses to voltage steps delivered either via conductive SWCNTs substrates or via the patch pipette were compared. The results, also supported by mathematical models, revealed that SWCNTs could directly stimulate brain circuit activity, highlighting SWCNTs as a promising material [67]. Similarly, it was proved that neurons would preferentially grow on thin layers of double-walled carbon nanotubes (DWCNTs) compared to SiO_2 surface, possibly explained by a favorable surface texture and better adsorption of culture medium proteins, allowing patterning of neurons networks [68]. Additionally, cell differentiation was also enhanced when the growth occurred on DWCNTs [68]. The success of pristine MWCNTS was reported in the growth and proliferation of pancreatic cancer cells, exhibiting a new approach to studying this kind of cancer [69]. Due to the ability of CNTs to form 3D architectures, MWCNT-based 3D networks as scaffolds for cell growth were developed [70]. The 3D interlocking resistive network of MWCNTs was prepared by treating a vertically aligned MWCNT array with exerting chemically induced capillary forces [70]. The proposed MWCNT-based 3D networks could be shaped to fit the best natural tissue morphology but also remained stable in vivo. Seven days later, fibroblasts had covered entirely almost the whole surface of the MWCNT-based 3D scaffolds [70].

For regenerative purposes, drugs or biofunctional molecules have been delivered by CNT-based 3D porous scaffolds to control their release profile and specific interactions with the attached cells [71]. The common strategy is to bind proteins, such as growth factors or enzymes onto functionalized CNTs [71]. MWCNT/chitosan composite scaffolds were prepared to support cell recolonization [72]. Scaffolds were composed of MWCNTs (up to 89%) and chitosan. These MWCNT/chitosan composite scaffolds showed a remarkable biocompatible character in both in vitro and in vivo experiments. The authors also investigated the evolution of a cell line toward an osteoblastic lineage in the presence of the recombinant human bone morphogenetic protein-2 (rhBMP-2) in vivo [72]. The results obtained are based on rhBMP-2-adhered MWCNT/chitosan scaffolds used into subcutaneous pockets in the mouse back muscle tissue for three weeks [72]. After this time, most of the MWCNT/chitosan scaffold structure was degraded and replaced by cells, and bone tissue regeneration was clearly observed [72]. A fibronectin/CNT hybrid was also produced, and it was found that fibronectins maintained their native structures upon CNT binding [73]. The hybrid material displayed a stronger affinity to human mesenchymal stem cells than the conventional fibronectin coated glass.

Selective and accelerated growth of the cells was also demonstrated on the patterned fibronectin/CNT hybrid nanostructures [73].

CNTs have been applied in nerve tissue engineering, namely by investigating the growth and organization of neural networks to treat neural diseases. Keefer et al. [74] prepared CNTs coated with conventional tungsten and stainless steel wire electrodes using electrochemical techniques under ambient conditions. The CNTs coating enhanced both recording and electrical stimulation of neurons in culture, rats, and monkeys by decreasing the electrode impedance and increasing charge transfer. Similarly, platinum/tungsten microelectrodes coated with a polypyrrole-CNT composite were reported as responsible for significantly reducing microelectrode impedance at all neuronal signal frequencies and an improvement of the signal-to-noise ratio [75]. Lee et al. [76] pre-treated rats with amine-modified SWCNTs and found that the pre-treatment can protect neurons and enhance the recovery of behavioral functions in rats with induced stroke. Less tissue damage was observed in SWCNTs-treated rats than in controls. To understand the mechanisms in which CNTs might affect the activity of cultured neuronal networks, single-cell electrophysiology techniques, electron microscopy analysis, and theoretical modeling were applied [77]. It was showed that CNTs improved the responsiveness of neurons by forming tight contacts with the cell membranes that might favor electrical shortcuts between the proximal and distal compartments of the neuron [77].

2.3 Carbon Nanotubes in Delivery Systems

All types of active molecules can be potentially linked to functionalized CNTs (fCNTs), allowing their application in drug, gene, and vaccine delivery [78, 79]. In fact, fCNTs are broadly applied as a delivery tool of therapeutic agents, namely anticancer drugs, antihypertensive drugs and corticosteroids, genes and nucleic acids, proteins, enzymes, peptides and ligands to their specific sites [80, 81] (Fig. 4).

2.3.1 Anticancer Drugs, Antihypertensive Drugs, and Corticosteroids

Drug delivery systems (DDS) are tools that allow the introduction of therapeutic agents in the body through a targeted and controlled release, improving their efficacy and safety [82]. Regarding DDS with anticancer drugs linked to CNTs (Table 2), Tan et al. [83] developed a betulinic acid-oxidized MWCNTs (BA-MWCNTs) nanocomposite through a 14.5%–14.8% (w/w) loading of BA (low water-soluble drug) in MWCNT, while keeping its tubular structures. Along 22 h, 98% of BA was released using phosphate-buffered saline (PBS) solution at pH 7.4. Following 72 h, BA-MWCNTs (<50 $\mu g\ mL^{-1}$) showed no cytotoxicity for mouse embryo fibroblast cells. Nevertheless, BA-MWCNTs displayed anticancer activity against human lung and liver cancer cell lines, revealing a half-maximal inhibitory concentration of 2.7 and 11.0 $\mu g\ mL^{-1}$, respectively [83].

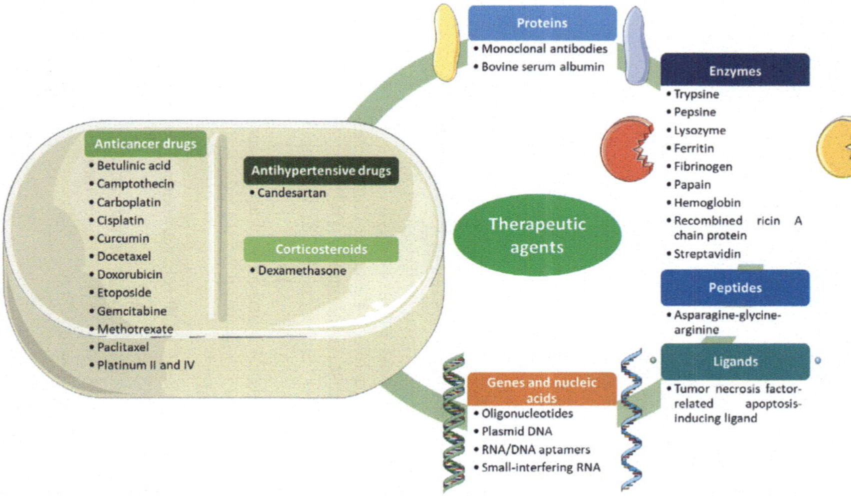

Fig. 4 Therapeutic agents applied in delivery systems using functionalized carbon nanotubes

Additionally, Tian et al. [84] designed camptothecin-functionalized MWCNTs (CPT-fMWCNTs) supramolecular complexes via the attachment of a non-water-soluble anticancer drug (CPT) onto MWCNTs previously coated with Pluronic P123, a tri-block copolymer, through non-covalent π-stacking interaction to enhance its solubility. While fMWCNT only presented 10% of antitumor activity, CPT-fMWCNTs showed around 37% of antitumor activity for HeLa cell lines [84].

The platinum (Pt)-based anticancer drug carboplatin (CP) was successfully loaded onto CNTs, as demonstrated by Arlt et al. [85], reaching a loading yield of 0.20 mg per mg of material. CP-CNT continuously released up to 68% of its Pt load within 14 days, allowing its use for drug storage. CP-CNT impaired up to 76%–94% of the clonogenic survival of prostate carcinoma (DU145, PC-3), bladder carcinoma (EJ28), and renal cell carcinoma (A498) cells. However, free CP simply induced 12% and 30% of apoptosis against DU145 cells, CP-CNT induced nearly 24% and 58% using 2.5 $\mu g\ mL^{-1}$ and 5.0 $\mu g\ mL^{-1}$, respectively [85]. Moreover, Guven et al. [86] encapsulated cisplatin (CDDP), a broadly used anticancer drug, into ultra-short SWCNTs, whose coating with the surfactant, Pluronic-F108, allowed a slow, controlled CDDP release upon dialysis in PBS at 37 °C and a higher cytotoxicity against breast cancer (MCF-7, MDA-MB-231) cell lines than free CDDP following 24 h [86].

Singh et al. [87] developed curcumin-CHI-ALG-SWCNTs through the loading of the hydrophobic anticancer drug curcumin, within SWCNTs, previously functionalized with polysaccharides, namely alginate (ALG) and chitosan (CHI), enabling a good drug loading performance and constant drug release. Treatment of human lung cancer (A549) cells with this DDS displayed enhanced cytotoxicity than free curcumin at analogous dosage, while presenting dose-related apoptotic induction [87].

Table 2 Anticancer drugs, antihypertensive drugs, and corticosteroids used in drug delivery systems with carbon nanotubes and corresponding main results

	Therapeutic agents	Drug delivery systems	Main results	Reference
Anticancer drugs	**Betulinic acid (BA)**	**BA-MWCNTs nanocomposite**	• 14.5%–14.8% (w/w) BA loading • 98% of BA was released at pH 7.4 (along 22 h) • No cytotoxicity for mouse embryo fibroblast cells • Anticancer activity against human lung and liver cancer cell lines	[83]
	Camptothecin (CPT)	**CPT-fMWCNTs supramolecular complexes** Designed via the attachment of CPT onto MWCNTs previously coated with Pluronic P123, a tri-block copolymer, through non-covalent π-stacking interaction	• Around 37% of antitumor activity for HeLa cell lines	[84]
	Carboplatin (CP)	**CP-CNT**	• 0.20 mg per mg of material of loading yield • Continuously released up to 68% of its platinum (Pt) load within 14 days • Impaired up to 76%-94% of the clonogenic survival of prostate carcinoma (DU145, PC-3), bladder carcinoma (EJ28), and renal cell carcinoma (A498) cells • Induced nearly 24% and 58% of apoptosis against DU145 cells using 2.5 $\mu g\ mL^{-1}$ and 5.0 $\mu g\ mL^{-1}$	[85]
	Cisplatin (CDDP)	**CDDP-SWCNTs** CDDP encapsulation into ultrashort SWCNTs coated with the surfactant Pluronic-F108	• Slow, controlled CDDP release upon dialysis in PBS at 37 °C • Higher cytotoxicity against breast cancer (MCF-7, MDA-MB-231) cell lines than free CDDP following 24 h	[86]

(continued)

Table 2 (continued)

	Therapeutic agents	Drug delivery systems	Main results	Reference
	Curcumin	**Curcumin-CHI-ALG-SWCNTs** Loading of curcumin, within SWCNTs, previously functionalized with polysaccharides, namely alginate (ALG) and chitosan (CHI)	• Excellent drug loading performance and constant drug release • Treatment of human lung cancer (A549) cells displayed enhanced cytotoxicity than free curcumin at analogous dosage while presenting dose-related apoptotic induction	[87]
	Docetaxel (DTX)	**DTX-TPGS-MWCNT** Carboxylated and acylated MWCNTs coated or covalently attached with D-α-Tocopherol polyethylene glycol 1000 succinate (TPGS), loaded with DTX and fluorescent coumarin-6 via the nano-extraction technique	• Drug encapsulation efficiency up to 76% • IC_{50} value reduction of 80-fold when comparing with commercial DTX injection (Docel™), following 24 h incubation with A549 cells • Following 72 h in PBS, 45 μg to 102 μg of DTX was released, which could generate anticancer impact since those concentrations are superior to their IC_{50} • Substantially superior cytotoxicity, early or late apoptosis, apoptotic cell death percentage, minimal ROS generation, and toxicity	[88]

(continued)

Table 2 (continued)

	Therapeutic agents	Drug delivery systems	Main results	Reference
		DTX-MWCNTs-Tf Carboxylated MWCNTs loaded with TPGS and DTX, later covalently linked with transferrin (Tf)	• Drug encapsulation efficiency up to 74.9% • DTX continuous release throughout 72 h • Enhanced cytotoxicity, an IC_{50} value reduction up to 136-fold than Docel™, following 24 h incubation against A549 cells • Augmented apoptotic cell death percentage, with nearly all cells in early or late apoptotic stage using a 100-fold lower DTX concentration (0.25 $\mu g\ ml^{-1}$) than Docel™ (25 $\mu g\ ml^{-1}$), while exhibiting low ROS levels	[89]
		DTX-CHI-F-MWCNTs DTX/ coumarin-6 loaded chitosan-folate (CHI-F) conjugated MWCNTs	• Drug encapsulation efficiency up to 79% • IC_{50} value reduction up to 89-fold than Docel™, after 24 h incubation with A549 cell lines • Internalization into cancer cells via a folate receptor-mediated endocytic route substantially increased cytotoxicity and reduced toxicity	[90]
	Doxorubicin (DOX)	**DOX-HA-MWCNT** Hyaluronic acid (HA)-MWCNT conjugate loaded with DOX through π-π stacking interaction	• 3.2 times superior cytotoxicity, plus improved apoptotic activity than free DOX in similar concentrations • Five times higher tumor-growth inhibitory impact than free DOX in similar concentrations while lacking any detectable cardiotoxicity, hepatotoxicity, and nephrotoxicity in mice	[91]

(continued)

Table 2 (continued)

	Therapeutic agents	Drug delivery systems	Main results	Reference
		DOX-BSA-CEA antibody-SWCNTs Multiple functionalization of oxidized SWCNTs with DOX, fluorescein-labeled bovine serum albumin (BSA), and a monoclonal carcinoembryonic antigen (CEA) antibody at non-competing binding sites	• Cellular uptake with consequent DOX translocation to the nucleus, whereas the SWCNTs remained in the cytoplasm of WiDr colon cancer cells	[92]
		DOX-SWCNTs SWCNTs coated with DOX via hydrazone linkage	• Successfully taken up by liver cancer (HepG2) cells leading to enhanced cytotoxicity	[93]
		DOX-FA-CHI-ALG-SWCNTs SWCNTs functionalized with the polysaccharides alginate (ALG) and chitosan (CHI), along with the targeting promoter FA and DOX	• pH-triggered DOX release, consequent nucleus penetration, and induction of HeLa cells death	[94]
		DOX-FI-FA-fMWCNTs DOX loaded onto poly(amidoamine) dendrimers altered with fluorescein isothiocyanate (FI) and FA covalently coupled to fMWCNTs	• Drug encapsulation efficiency up to 97.8% • pH-triggered DOX release and enhanced cytotoxicity against human epithelial carcinoma (KB) cells	[95]
		DOX-MWCNTs	• Increased cytotoxicity against MCF-7 cells	[96]

(continued)

Table 2 (continued)

	Therapeutic agents	Drug delivery systems	Main results	Reference
		DOX-FA-MN-MWCNT DOX loading of MWCNTs functionalized with poly(acrylic acid) conjugated with iron oxide magnetic nanoparticles (MNs) coupled with FA-enabled double-targeted DOX delivery via a magnetic field, plus ligand-receptor interactions	• High DOX loading efficiency • Increased cytotoxicity toward human glioblastoma (U87) cells • Internalized by U87 cells with consequent intracellular DOX release and transport into the nucleus, whereas the nanoparticles remained in the cytoplasm	[97]
		DOX-FA-PEG-SWCNTs DOX loaded onto the surface of SWCNTs, previously coated by FA-terminated polyethylene glycol (FA-PEG) via adsorption	• Extremely high loading efficiency (149.3 ± 4.1%) • Negligible cytotoxicity against normal (3T3) cells, whereas selectively entered HeLa cells via clathrin-mediated endocytosis, where DOX was released, subsequently entered the nucleus, thus causing cell death	[98]
		DOX-loaded anti-CD105 conjugated magnetic SWCNTs	• Improved targeting toward lung metastatic sites in a breast cancer animal model • Potential in cancer treatment efficiency at early, plus late phases of cancer progression	[99]
		DOX-DEX-fMWCNTs DOX loaded onto dexamethasone (DEX) conjugated fMWCNTs	• High loading efficiency (92.6 ± 0.5%) • Low hemolyticity • Higher cytotoxicity against A549 cells	[100]

(continued)

Table 2 (continued)

	Therapeutic agents	Drug delivery systems	Main results	Reference
		Nanocomplex DOX-PEI-B-MWCNTs-siRNA Polyethyleneimine (PEI) covalently coupled with betaine (PEI-B) and oMWCNTs and combined with DOX and survivin siRNA	• Able to co-deliver DOX and siRNA into A549 cells, leading to the increase of the apoptotic index • Able to reduce the tumor volume in nude mice with A549 tumor cells without causing any damages to normal tissues	[101]
	Etoposide (VP-16)	**PEG-grafted MWCNTs (PEG-oMWCNTs) hybrid nanosystem** Carrier of the lung cancer drug VP-16 and Bcl-2 phosphorothioate antisense deoxyoligonucleotides	• Higher cytostatic efficacy • Enhancement of the apoptotic cells against the small cell (SCLC; DMS53) and non-small cell lung cancer (NSCLC; NCIH2135) cell lines	[102]
	Gemcitabine (GEM)	**GEM-SWCNTs′; GEM-PEG-SWCNTs″** GEM-SWCNTs and GEM-PEG-SWCNTs via carboxylation, acylation, amination, and PEGylation of pristine SWCNTs followed by gemcitabine (GEM) coupling	• GEM loading capacity of 43.14% (w/w)′ and 37.32%″ • More cytotoxic against A549 and human pancreatic carcinoma (MIA PaCa-2) cell lines • Superior efficacy in tumor growth suppression in B6 nude mice"	[103]

(continued)

Table 2 (continued)

	Therapeutic agents	Drug delivery systems	Main results	Reference
	Methotrexate (MTX)	**MTX-FA-AF-fMWCNTs** Theranostic prodrug, based on fMWCNTs co-tethered with MTX, a tumor targeting module (folic acid (FA)), a fluorochrome (Alexa-fluor (AF488/647)), and a radionuclide (Technetium-99 m (99mTc))	• Selectively internalized by folate receptor (FR) positive A549 and MCF-7 cancer cells via FR-mediated endocytosis, consequently presenting superior anticancer activity • The tumor-specific accumulation of MTX-FA-AF-fMWCNTs in Ehlrich Ascites tumor-bearing mice was nearly 19 and 8.6-fold superior to free MTX and MWCNTs without FA, respectively • MTX-FA-AF-fMWCNTs displayed superior tumor growth arrest while not showing any hepatotoxicity, cardiotoxicity, and nephrotoxicity in mice	[104]
		MTX-CHI-MWCNT nanohybrid	• Negligible toxicity against healthy (MRC-5) cells • Selective and pH-responsive MTX delivery to human lung cancer (H1299) cells	[105]
	Paclitaxel (PTX)	**PTX-fSWCNT/graphene oxide flakes**	• Superior cell death against NCI-H460 cells and mitogen-associated protein kinase activation	[106]
		PTX-PCA-MWCNTs MWCNTs initially functionalized with hyperbranched poly citric acid (PCA), whose carboxyl functional groups were afterward coupled with PTX	• Drug encapsulation efficiency: 38% (w/w) • pH-responsive PTX release into tumor tissues and cells • Negligible toxicity against A549 and ovary cancer (SKOV3) cells • Higher cytotoxicity than free PTX over an inferior incubation time, indicating enhanced cell penetration	[107]

(continued)

Table 2 (continued)

	Therapeutic agents	Drug delivery systems	Main results	Reference
		PTX-PEG-SWCNT[a]; PTX-PEG-MWCNT[b] PTX physically loaded via immersion of PEG-SWCNT, plus PEG-MWCNT	• Loading capability: 26% (w/w)[a] and 36% (w/w)[b] • Sustainable PTX release in buffer solution • Superior efficacy against HeLa and MCF-7 cells	[108]
		PTX-SWCNTs Co-exposure of SWCNTs and PTX	• Increased cell death and apoptosis induction of human ovarian cancer (OVCAR3) cells	[109]
		PTX/HA-CHI-SWCNTs Chitosan non-covalently linked to SWCNTs, combined with HA connected to the outer CHI layer, later attached to PTX	• Negligible toxicity toward normal fibroblast cells, besides enhanced toxicity and apoptosis induction to A549 cells	[110]
	Pt(IV)	**Pt(IV)-fSWCNT conjugate**	• Successfully delivered a lethal dosage of cis-[$Pt(NH_3)_2Cl_2$] upon intracellular reduction • Substantially enhanced cytotoxicity profile against testicular carcinoma (NTera-2) cells	[111]
		Pt(IV)-fMWCNTs Properly load an inert Pt(IV) complex into fMWCNTs	• Administration in mice considerably augmented the Pt content in certain tissues (especially lung, liver, and spleen) while not causing inflammatory reaction or necrosis	[112]
Antihypertensive drugs	**Candesartan (CD)**	**iRGD-PEI-MWCNT-SS-CD/pAT2 complexes** Tumor-homing peptide (iRGD) attached to PEI connected to MWCNT skeleton, combined with CD linked via cystamine (SS) to the MWCNT, all of which assembled by plasmid AT2 (pAT2)	• Significant neovascularization and tumor growth suppression in A549 xenograft nude mice	[113]

(continued)

Table 2 (continued)

	Therapeutic agents	Drug delivery systems	Main results	Reference
Corticosteroids	**Glucocorticoid**	**VA-fMWCNTs** Vertically aligned (VA)-MWCNTs functionalized and PEGylated, before the loading of glucocorticoid	• Enhancement of drug release, displaying 55%, 65%, and 95% release at pH 7.4, 6.5, and 5.5, respectively • Shallow cytotoxicity effect in PC-12 cells	[114]

Drug delivery systems: ′GEM-SWCNTs; ″GEM-PEG-SWCNTs; [a] PTX-PEG-SWCNT; [b] PTX-PEG-MWCNT

Regarding DDS with docetaxel (DTX), carboxylated and acylated MWCNTs coated or covalently attached with D-α-Tocopherol polyethylene glycol 1000 succinate (TPGS), plus loaded with DTX and fluorescent coumarin-6 via the nano-extraction technique were reported [88]. A drug encapsulation efficiency of up to 76%, with an IC_{50} value reduction of 80-fold compared with commercial DTX injection (Docel™), was obtained with A549 cells after 24 h of incubation [88]. Following 72 h in PBS (pH 7.4), 45 μg to 102 μg of DTX were released, which could generate an anticancer impact since those concentrations are superior to their IC_{50}. DTX-TPGS-MWCNT presented substantially superior cytotoxicity, early or late apoptosis, apoptotic cell death percentage, besides minimal ROS generation and toxicity [88]. Singh et al. [89] showed carboxylated MWCNTs also loaded with TPGS and DTX, later covalently linked with transferrin (Tf), reached a similar drug encapsulation efficiency (up to 74.9%) and a DTX continuous release throughout 72 h. However, DTX-MWCNTs-Tf displayed enhanced cytotoxicity, with an IC_{50} value reduction up to 136-fold than Docel™, following 24 h incubation against A549 cells. DTX-MWCNTs-Tf presented augmented apoptotic cell death percentage, with nearly all cells in early or late apoptotic stage using a 100-fold lower DTX concentration (0.25 $\mu g\ ml^{-1}$) than Docel™ (25 $\mu g\ ml^{-1}$), while exhibiting low ROS levels, thus showing its efficacy and safety [89]. In the same line, Singh et al. [90] showed that DTX/coumarin-6 loaded chitosan-folate (CHI-F) conjugated MWCNTs reached a slightly higher drug encapsulation efficiency (up to 79%), as well as attained an IC_{50} value reduction up to 89-fold than Docel™, after 24 h incubation with A549 cell lines. Furthermore, DTX-CHI-F-MWCNTs exhibited internalization into cancer cells via a folate receptor-mediated endocytic route, substantially increased cytotoxicity and reduced toxicity, confirming its potential for targeted lung cancer treatment [90].

Concerning the use of the anticancer drug doxorubicin (DOX) in DDS, a hyaluronic acid-MWCNT (HA-MWCNT) conjugate loaded with DOX through π-π stacking interaction was developed. DOX-HA-MWCNT showed a 3.2 times superior cytotoxicity, plus improved apoptotic activity than free DOX in similar concentrations [91]. The conjugate displayed a five times higher tumor-growth inhibitory impact than free DOX in similar concentrations, while lacking detectable cardiotoxicity, hepatotoxicity, and nephrotoxicity in mice [91]. Multiple functionalization of oxidized SWCNTs with DOX, fluorescein-labeled bovine serum albumin (BSA), and a monoclonal carcinoembryonic antigen (CEA) antibody was also displayed at non-competing binding sites [92]. The cellular uptake of DOX-BSA-CEA antibody-SWCNTs with consequent DOX translocation to the nucleus was enabled, whereas the SWCNTs remained in the cytoplasm of WiDr colon cancer cells [92]. SWCNTs coated with DOX via hydrazone linkage were also used for DDS, which were successfully taken up by liver cancer (HepG2) cells leading to enhanced cytotoxicity [93]. Polysaccharides ALG and CHI DDS were used to functionalize SWCNTs, along with the targeting promoter FA and DOX, which enabled pH-triggered DOX release, consequent nucleus penetration, and induction of HeLa cells death [94]. DDS consisting of DOX loaded onto poly(amidoamine) dendrimers altered with fluorescein isothiocyanate (FI) and FA covalently coupled to fMWCNTs attained a drug encapsulation efficiency up to 97.8%, besides allowing a pH-triggered DOX release and enhanced cytotoxicity against human epithelial carcinoma (KB) cells [95]. DOX-MWCNTs also led to increased cytotoxicity against MCF-7 cells [96].

DDS based on DOX loading of MWCNTs functionalized with poly(acrylic acid) conjugated with iron oxide magnetic nanoparticles (MNs) and coupled with FA enabled double-targeted DOX delivery via a magnetic field, plus ligand-receptor interactions [97]. DOX-FA-MN-MWCNT displayed high DOX loading efficiency, as well as increased cytotoxicity toward human glioblastoma (U87) cells [97]. DOX-FA-MN-MWCNT was internalized by U87 cells with consequent intracellular DOX release and transport into the nucleus, whereas the nanoparticles remained in the cytoplasm [97]. DDS, including DOX, loaded onto the surface of SWCNTs, previously coated by FA-terminated polyethylene glycol (FA-PEG) via adsorption, reached an extremely high loading efficiency (149.3 ± 4.1%) [98]. DOX-FA-PEG-SWCNTs showed negligible cytotoxicity against normal (3T3) cells, whereas selectively entered HeLa cells via clathrin-mediated endocytosis, where DOX was released due to the acidic pH, subsequently entered the nucleus, thus causing cell death via transcription inhibition [98]. The association between non-invasive magnetic resonance imaging plus specific magnets positioning was reported [99]. Improved DOX-loaded anti-CD105 conjugated magnetic SWCNTs targeting lung metastatic sites in a breast cancer animal model were obtained, showing its potential in cancer treatment efficiency at early plus late phases of cancer progression [99]. Besides, it was shown that DOX loaded onto dexamethasone (DEX) conjugated fMWCNTs (DEX-fMWCNTs) attained high loading efficiency (92.6 ± 0.5%), while displaying low hemolyticity, plus higher cytotoxicity against A549 cells, suggesting DOX-DEX-fMWCNTs as a promising carrier for future cancer therapy [100]. The nanocomplex DOX-PEI-B-MWCNTs-siRNA (polyethyleneimine (PEI) covalently coupled with betaine (PEI-B) and MWCNTs and combined with DOX and survivin siRNA) were able to co-deliver DOX and siRNA into A549 cells, leading to the increase of the apoptotic index [101]. DOX-PEI-B-MWCNTs-siRNA reduced the tumor volume in nude mice with A549 tumor cells without causing any damages to normal tissues, confirming its potential for co-delivery of anticancer drugs, as well as genes [101].

PEG-grafted MWCNTs (PEG-MWCNTs) hybrid nanosystem was used as a carrier of the lung cancer drug etoposide (VP-16) and Bcl-2 phosphorothioate antisense deoxyoligonucleotides leading to higher cytostatic efficacy, besides an enhancement of the apoptotic cells against the small cell (SCLC; DMS53) and non-small cell lung cancer (NSCLC; NCIH2135) cell lines [102].

Novel anticancer DDS, e.g., GEM-SWCNTs and GEM-PEG-SWCNTs via carboxylation, acylation, amination, and PEGylation of pristine SWCNTs followed by gemcitabine (GEM) coupling, were developed, achieving a GEM loading capacity of 43.14% (w/w) and 37.32%, respectively [103]. While GEM-SWCNTs were more cytotoxic against A549 and human pancreatic carcinoma (MIA PaCa-2) cell lines, GEM-PEG-SWCNTs presented superior efficacy in tumor growth suppression in B6 nude mice [103].

Regarding the use of methotrexate (MTX), a new theranostic prodrug, based on fMWCNTs co-tethered with MTX, a tumor targeting module (folic acid (FA)), a fluorochrome (Alexa-fluor (AF488/647)), and a radionuclide (Technetium-99 m (99mTc)) was developed [104]. MTX-FA-AF-fMWCNTs were selectively internalized by folate receptor (FR) positive A549 and MCF-7 cancer cells via FR-mediated endocytosis, consequently presenting superior anticancer activity due to lysosomal trafficking [104]. The tumor-specific accumulation of MTX-FA-AF-fMWCNTs in Ehrlich Ascites tumor-bearing mice was nearly 19 and 8.6-fold superior to free MTX and MWCNTs without FA, respectively [104]. MTX-FA-AF-fMWCNTs displayed superior tumor growth arrest without showing any hepatotoxicity, cardiotoxicity, and nephrotoxicity in mice [104]. Overall, MTX-FA-AF-fMWCNTs allowed enhanced MTX therapeutic efficacy against FR positive cancer cells, besides real-time monitoring of the treatment reaction via multimodal imaging [104]. A new MTX-CHI-MWCNT nanohybrid was developed, showing negligible toxicity against healthy (MRC-5) cells, in addition to a selective and pH-responsive MTX delivery to human lung cancer (H1299) cells [105].

A novel synergic lung cancer treatment based on the ROS-mediated cooperation between fSWCNT/graphene oxide flakes and paclitaxel (PTX) was shown to display superior cell death against NCI-H460 cells and mitogen associated protein kinase activation, as proven by the bromodeoxyuridine (BrdU) incorporation against A549 cells [106]. In addition, Sobhani et al. [107] reported MWCNTs initially functionalized with hyperbranched poly citric acid (PCA), whose carboxyl functional groups were afterward coupled with PTX shaping PTX-PCA-MWCNTs, reaching a drug encapsulation efficiency of 38% (w/w), pH-responsive PTX release into tumor tissues and cells, negligible toxicity against A549 and ovary cancer (SKOV3) cells, higher cytotoxicity than free PTX over an inferior incubation time, and enhanced cell penetration. Furthermore, PTX physically loaded via immersion of PEG-SWCNT and PEG-MWCNT exhibited a saturated PTX solution in methanol, achieved a loading capability of 26% (w/w) and 36% (w/w), respectively [108]. Both PEG-SWCNT and PEG-MWCNT displayed sustainable PTX release in buffer solution and superior efficacy against HeLa and MCF-7 cells, as shown by the IC_{50} value inferior to free PTX, suggesting its potential for cancer therapeutics [108]. The co-exposure of SWCNTs and PTX led to increased cell death and apoptosis induction of human ovarian cancer (OVCAR3) cells, implying its potential for cancer treatment progress [109]. A novel DDS PTX/HA-CHI-SWCNTs comprising CHI non-covalently linked to SWCNTs was shown to enhance biocompatibility, combined with HA connected to the outer CHI layer to provide targeting features later attached to PTX [110]. PTX/HA-CHI-SWCNTs displayed negligible toxicity toward normal fibroblast cells and enhanced toxicity and apoptosis induction to A549 cells, showing potential for cancer therapy [110].

Feazell et al. [111] developed a Pt(IV)-fSWCNT conjugate, which successfully delivered a lethal dosage of cis-[$Pt(NH_3)_2Cl_2$] upon intracellular reduction. However, cis-[$Pt(NH_3)_2Cl_2$] was almost non-toxic, Pt(IV)-fSWCNT exhibited a substantially enhanced cytotoxicity profile against testicular carcinoma (NTera-2) cells [111]. Besides that, the properly load of CDDP or of an inert Pt(IV) complex into fMWCNTs was managed, whose administration in mice considerably augmented the Pt content in certain tissues (especially lung, liver, and spleen) while not causing inflammatory reaction or necrosis [112].

In the field of CNTs, antihypertensive drugs have also been applied in DDS (Table 2). The only work found showed a dual-targeting, plus co-delivery method, applying iRGD-PEI-MWCNT-SS-CD/pAT2 complexes formed by tumor-homing peptide (iRGD) attached to PEI connected to MWCNT skeleton, combined with CD linked via cystamine (SS) to the MWCNT, all of which were assembled by plasmid AT2 (pAT2) [113]. iRGD-PEI-MWCNT-SS-CD/pAT2 complexes were designed for anti-angiogenesis therapy in lung cancer, which targeted the renin-angiotensin system (RAS) dysregulation through the synergistic regulation of angiotensin II type 1 receptor (AT1R), plus type 2 receptor (AT2R) pathway. iRGD-PEI-MWCNT-SS-CD/pAT2 complexes displayed significant neovascularization and tumor growth suppression in A549 xenograft nude mice [113].

In addition to the previously described drugs, corticosteroids have also been used (Table 2). Komane et al. [114] showed vertically aligned (VA)-MWCNTs functionalized and PEGylated, before the loading of the anti-inflammatory glucocorticoid, led to the enhancement of its release, displaying 55%, 65%, and 95% release at pH 7.4, 6.5, and 5.5, respectively. VA-fMWCNTs (20 to 20,000 $\mu g\ mL^{-1}$) exhibited a shallow cytotoxicity effect in PC-12 cells, revealing their potential in ischemic stroke treatment [114].

2.3.2 Genes and Nucleic Acids

Several DDS containing genes and nucleic acids, such as oligodeoxynucleotides, oligonucleotides, plasmid DNA (pDNA), aptamers, and small-interfering RNA (siRNA) linked to CNTs, were studied aiming their intracellular transporting and controlled release [27, 115–118]. In this field, Jia et al. [115] developed a double functionalization of MWCNT delivery method comprising therapeutic gene antisense oligodeoxynucleotides and fluorescent labeling probes mercaptoacetic acid-capped CdTe quantum dots, through electrostatically layer-by-layer assembling. These fMWCNTs enabled a successful intracellular transporting, cell nucleus localization, delivery efficiency, and anticancer activity against HeLa cells, revealing the potential for biological delivery and gene therapy [115]. Polyamidoamine dendrimer-modified MWCNTs (dMWCNTs) coupled with the FI-labeled antisense c-myc oligonucleotides displayed high transfection efficiencies and successfully hindered cell growth in time- and dose-dependent ways in MCF-7, MDA-MB-435, and HepG2 cells, showing its potential for cancer therapy and molecular imaging [117].

A novel pDNA-fSWCNT-based gene delivery vector method was developed [27] based on fSWCNTs complexed with pDNA. These were prepared via electrostatic interactions, which could bind and enter HeLa cells, while enhancing DNA uptake and gene expression in Chinese hamster ovary (CHO) cells [27]. Ammonium-functionalized SWCNTs with pDNA (pDNA-NH_3^+-SWCNT) complexes, as well as established lysine (Lys)-functionalized SWCNTs (pDNA-NH_3^+-Lys-SWCNT) and NH_3^+-functionalized MWCNTs with pDNA (pDNA-NH_3^+-MWCNT) complexes, were able to transfect A549 cells with distinct efficiency levels, displaying their potential as a successful gene delivery vector method [119].

EpCAM RNA aptamer-piperazine-PEI-SWCNT based on RNA aptamer against epithelial cell adhesion molecule (EpCAM) physically attached to SWCNT conjugated to piperazine-PEI derivative was reported to specifically induce apoptosis in MCF-7 cells, besides leading to BCL9l (related to breast and colorectal cancers) protein level decrease, further showing its targeted silencing activity [120]. In a similar work, anti MUC1-aptamer-cMWCNT hybrid based on anti MUC1 (membrane-associated glycoform overexpressed on a wide range of epithelial cancer cells)-aptamers covalently grafted onto cMWCNTs were able to translocate into MCF-7 cytoplasm via a receptor-independent manner [116].

Regarding DDS with siRNA, Varkouhi et al. [121] showed the potential of MWCNTs covalently functionalized with PEI (PEI-MWCNTs) and MWCNTs non-covalently functionalized with cationic pyridinium (Pyrr-MWCNTs) complexed with anti-luciferase siRNA in displaying a silencing activity between 10 and 30% and cytotoxicity between 10 and 60%. These results were similar to the controls (PEI and reference transfectants (Lipofectamine 2000, pDMAEMA)) [121]. Polo-Like Kinase1 (PLK1) was confirmed as a cancer gene therapy target through the association between PLK1 siRNA and fMWCNTs as delivery vectors [118]. This association allowed the enhancement of siRNA retention in a solid tumor and increased tumor cellular uptake in vivo, which subsequently prolonged animal survival of human lung carcinoma (Calu6) xenografts bearing mice [118]. Furthermore, siTOX-NH_3^+-MWCNT complexes based on proprietary cytotoxic siRNA sequence (siTOX) complexed with NH_3^+-functionalized MWCNTs (NH_3^+-MWCNT) were able to delay tumor growth [122]. siTOX-NH_3^+-MWCNT complexes enhanced the survival of xenograft-bearing animals following its intratumoral administration [122]. Al-Jamal et al. [123] revealed the perilesional stereotactic administration of Caspase-3 siRNA (siCas 3) delivered by NH_3^+-MWCNT enabled neuroprotection, following focal ischemic damage caused by Endothelin-1 in the motor cortex of both mice and rats, leading to functional recovery as displayed by the "skilled reaching" test in living rats. In the same line, Ladeira et al. [124] showed that siRNA-fSWCNTs could be applied as delivery systems to different cell types, such as non-metastatic human hepatocellular carcinoma (SKHep1), cardiomyocytes, and rat dorsal root ganglion (DRG) neuron cells, while displaying a successful target gene silencing with high specificity and reduced toxicity. Kam et al. [125] also developed siRNA-SS-PL-PEG-SWCNTs based on the strong binding of phospholipid molecules with PEG (PL-PEG) to SWCNTs, followed by the cleavable disulfide linkage between PL-PEG-SWCNTs and siRNA, which attained high siRNA efficiency delivery in HeLa cells. Chen et al. [126] reported SWCNTs functionalization with DSPE-PEG-amine allowed siRNA binding via disulfide bonds, reaching a siRNA loading efficiency of 83.55%. In fact, siRNA specifically targeting the oncogene Murine Double Minute Clone 2 (MDM2 siRNA) coupled with fSWCNTs managed to enter breast cancer (B-Cap-37) cells, with MDM2 siRNA being subsequently released due to disulfide bonds digestion by intracellular enzymes, thus inhibiting cell proliferation and promoting apoptosis [126]. Bartholomeusz et al. [127] demonstrated that siRNA-SWCNTs complexes with hypoxia-inducible factor 1 alpha (HIF-1α)-targeted siRNA (HIF-1α siRNA) were able to inhibit cellular HIF-1α activity in a pancreatic cell line with HIF-1α/luciferase reporter (MiaPaCa2-HRE) cells. HIF-1α siRNA-SWCNTs complexes intratumorally administrated in mice bearing MiaPaCa-2/HRE tumors led to HIF-1α activity inhibition, showing its potential in human cancer cases with resistance to therapy [127].

2.3.3 Proteins, Peptides, and Ligands

Numerous proteins, peptides, and ligands, namely BSA, trypsin, pepsin, lysozyme, ferritin, fibrinogen, papain, hemoglobin, recombined ricin A chain protein (RTA), Streptavidin (SA), asparagine–glycine–arginine (NGR), and tumor necrosis factor-related apoptosis-inducing ligand (TRAIL), linked to CNTs were applied in DDS, envisioning their uptake by cells and increased cytotoxicity [128–131]. Parra et al. [132] developed AZc6-BSA-CNTs conjugates based on azoxystrobin (AZ), a strobilurin fungicide, and a derivative bearing a carboxylated spacer arm (hapten AZc6) covalently linked to BSA, afterward covalently associated to fCNTs with distinct shapes and sizes. Immunization of New Zealand rabbits and BALB/c mice with AZc6-BSA-CNTs conjugates triggered significant IgG responses with and without an adjuvant, thus showing CNTs' self-adjuvant ability. From all tested conditions, the best antibody response was produced by short and thick AZc6-BSA-CNTs conjugates. In fact, low dosages (0.05 μg) of AZc6-BSA-CNTs conjugates had strong anti-AZ immune responses in rabbits [132]. CHI NPs-MWCNT hybrids were synthesized via an ionotropic gelation method enabled high (76.8%) BSA immobilization efficiency and biocompatibility with HeLa cells since cell viability is 81%, following 24 h incubation with 100 μg mL^{-1} [133]. SWCNTs were functionalized with cationic-designed amphiphilic polypeptides and subsequent protein binding, e.g., BSA, trypsin, pepsin, lysozyme, ferritin, fibrinogen, papain, and hemoglobin or DNA via a bilayer approach [128]. Their delivery efficiency into HeLa cells was affected by the complexes charge, showing its potential as delivery systems [128]. This model that predicts natural proteins' adsorption onto SWCNT, based on its structure and composition, was reported for the first time [128]. RTA-MWCNTs conjugates induced target destruction of tumor cells, particularly HeLa cells, whose cell mortality reached around 75% [134]. Actually, HER2-RTA-MWCNTs obtained by attaching HER2 onto RTA-MWCNTs conjugates led to selective destruction of HER2 overexpressing breast cancer cells [134]. SA-biotin-SWCNTs conjugates based on fluoresceinated SA, a protein with anticancer applications complexed with the biotin-SWCNTs transporter, were uptake through endocytosis into human promyelocytic leukemia (HL60) and human T (Jurkat) cells, displaying dose-dependent cytotoxicity [129].

Concerning the use of peptides in DDS, TAM-NGR-SWCNTs conjugates based on SWCNTs functionalized with NGR, subsequently loaded with tamoxifen (TAM), an anticancer drug, attained high cellular uptake, cytotoxicity, and cell apoptosis in vitro due to its receptor-mediated tumor-targeting potential [130]. In another work with SWCNTs, TRAIL-based SWCNTs nanovectors shown to be highly efficient in triggering cancer cell killing, increasing apoptosis in vitro due to their capacity to enhance caspase-8 activation [131].

2.4 Carbon Nanotubes in Targeted Therapies

CNTs have intrinsic physical features, which allied with suitable functionalization, enable their use in photothermal therapy (PTT), photodynamic therapy (PDT), and the combination of both [30, 135, 136].

2.4.1 Photothermal Therapy

PTT is a non-invasive, harmless, and efficient therapeutic technology that enables tumor cells' death via locally heated tumor tissues by near-infrared region absorption [30, 137, 138]. Therefore, CNTs have been applied in PTT due to their ability to absorb light in the near-infrared region and successfully transform absorbed energy into released heat [30, 139–142].

Regarding PTTs using SWCNTs, Shao et al. [143] prepared anti-IGF1-anti-HER2-SWCNTs hybrids based on SWCNTs functionalized with insulin-like growth factor 1 receptor (IGF1R), as well as human endothelial receptor 2 (HER2) specific antibodies. These hybrids targeted their corresponding receptors in breast cancer cells, allowing SWCNTs' internalization, and by a near-infrared region irradiation, they were able to destroy all treated cells, whereas non-specific-antibody–SWCNT hybrids only killed 20% of the cells [143]. Specific binding of SWCNTs dispersions were prepared with human anti-CD22 antibodies, followed by exposure to near-infrared region leading to highly specific ablation of human Burkitt's lymphoma cells in vitro [31]. FA-CoMoCAT® SWCNTs conjugates exhibited an effective increased photothermal destruction on tumor cells while sparing non-targeted normal cells in both in vitro and in vivo experiments [144]. Xiao et al. [145] demonstrated the selective photothermal ablation after near-infrared region irradiation of HER2 immunoglobulin Y (IgY)-SWCNT complexes based on the covalent conjugation between cSWCNTs and anti-HER2 chicken IgY antibodies, specifically targeted HER2-expressing SK-BR-3 cells. HER2 IgY-SWCNT complexes allowed the detection, and selective photothermal ablation of receptor-positive breast cancer cells without cells' internalization [145]. A combination between intratumoral injection of PEG-SWCNTs and near-infrared irradiation allowed the complete destruction of solid malignant tumors devoid of side effects or tumor recurrence over six months [146]. The majority of injected SWCNTs were excreted from mice bodies in two months via biliary or urinary pathways [146]. The combination between the injection of SWCNTs and ten minutes of near-infrared region irradiation eradicated squamous cell carcinomas, and SWCNTs remained exclusively in the tumor even three months following injection [147]. Mitochondria targeting SWCNTs, after near-infrared region irradiation, were able to destroy the targeted mitochondria selectively, thus inducing apoptosis, and tumor growth suppression was achieved in a breast cancer model [148]. Zhou et al. [149] displayed glycated CHI-SWCNTs, after near-infrared region irradiation, induced photothermal death of mouse mammary tumor (EMT6) cells in vitro and in vivo, enabling complete tumor regression, as well as long-term survival in several cases. NGR-siRNA-PEI-SWCNTs, based on the functionalization of SWCNTs with PEI, were further conjugated with the tumor-targeting NGR peptide and loaded with hTERT siRNA, which enabled in vitro apoptosis induction in PC-3 cells, besides high antitumor activity in vivo [150]. The combination of DOX-loaded SWCNTs (DOX-SWCNTs) and near-infrared region irradiation enabled an efficient MDA-MB-231 cell death via mitochondrial disruption, as well as ROS generation [151]. The combination of water-soluble fSWCNTs, followed by two minutes of radiofrequency field treatment, led to the destruction of almost all hepatocellular cancer (HepG2, Hep3B), plus pancreatic adenocarcinoma (Panc-1) cells in vitro [140]. In fact, fSWCNTs injection into hepatic (VX2) tumors in rabbits, immediately followed by radiofrequency field, led to total necrosis of all treated tumors after 48 h [140].

Concerning PTTs using MWCNTs, Torti et al. [152] reported that nitrogen-doped MWCNTs (CNx-MWCNTs) led to the destruction of kidney cancer cells following near-infrared region irradiation. Anti-GD2 monoclonal antibody conjugated to MWCNTs highly internalized by neuroblastoma (stNB-V1) cells, through GD2-mediated endocytosis and irradiated with near-infrared region treatment, led to necrosis of all stNB-V1 cells, whereas all non-targeted normal cells remained viable [153]. Complete photothermal kidney tumor ablation was achieved using MWCNTs; 80% of mice were alive, plus tumor-free, following 3 months treatment with a proper amount of MWCNTs [154]. Intratumoral injection of DNA-encased MWCNTs, followed by near-infrared region irradiation, enabled a total eradication of PC3 xenograft tumors in all treated mice [142]. In fact, DNA-encasement allowed a threefold decrease in the concentration of MWCNTs needed for a 10 °C temperature rise in bulk solution temperature [142]. Also, Burlaka et al. [155] demonstrated MWCNTs led to the destruction of 95.2% of Ehrlich Ascites Carcinoma (EAC) cells, following 1.5 min of near-infrared region irradiation [155] and the conjugation between P-Glycoprotein-targeted MWCNTs and near-infrared region irradiation caused high cytotoxicity in multidrug resistance cancer cells [156].

2.4.2 Photodynamic Therapy

PDT is a local non-invasive cancer treatment, which comprises the administration of a non-toxic photosensitizer, afterward activated by a suitable harmless light source [135, 157, 158]. The photosensitizer generates highly reactive singlet oxygen ($^1O_2$) via light energy transfer to the tissue oxygen, an aggressive chemical species that mediates cellular toxicity leading to cell damage and eventually cell death [135, 157, 158]. For instance, Zhu et al. [135] reported, as a proof of concept for PDT, that single-stranded DNA aptamer (ssDNA AP)-photosensitizer-SWCNTs based on the attachment of ssDNA AP and photosensitizer on SWCNTs, allowed controllable $^1O_2$ generation via target binding [159]. The non-covalent interaction between pyrenyl-functionalized distyryl-dodipy sensitizer and SWCNTs generated $^1O_2$ after irradiation at 660 nm [159]. Additionally, Xiao et al. [160] prepared CHI–Ce6–SWCNTs based on the non-covalent attachment of the photosensitizer chlorin e6 (Ce6) and SWCNTs, afterward wrapped by CHI to improve biocompatibility and aqueous solubility, and demonstrated their low dark toxicity and efficient PDT efficacy toward HeLa cells. PEI-SWCNTs based on SWCNTs covalent functionalized with PEI revealed a strong photodynamic in vitro and in vivo effect on mice melanoma B16F10 cells and tumor-bearing mice, showing their potential for cancer PDT [161].

2.4.3 Combination of Photothermal and Photodynamic Therapies

The combination of PTD and PDT has received significant interest in cancer treatment due to its potential synergistic effect [162]. Shi et al. [163] reported the HMME-HA-SWCNTs based on HA-derivatized SWCNTs, followed by the adsorption of hematoporphyrin monomethyl ether (HMME), enabled in vivo melanoma B16F10-bearing mice growth inhibition via a synergistic effect of local specific PDT, as well as external near-infrared region PTT. Ru-SWCNTs were loaded with two-photon luminescent Ru(II) complexes onto SWCNTs [162]. After irradiation at 808 nm, Ru(II) complexes are released, producing $^1O_2$ upon two-photon laser irradiation. High anticancer efficacies were obtained in HeLa cancer cells, in addition to multicellular tumor spheroids, as well as in vivo tumor ablation [162]. mTHPC-MWCNTs performance comprising the photosensitizer m-tetrahydroxyphenylchlorin (mTHPC) and MWCNTs depend on the irradiation conditions, leading to the synergistic apoptosis of the whole SKOV3 cell population [136].

3 Toxicity of Carbon Nanotubes

The toxicity of CNTs is a subject of some controversy in the literature. A significant amount of works dealing with this issue can be found, particularly to address whether or not CNTs are potentially harmful to organisms and if they have potential application in biomedicine. CNTs toxicity was determined by both in vivo and in vitro tests. In vivo toxicity analysis of CNTs is commonly performed by evaluating specific organs and biodistribution in mice and rats, while in vitro toxicity tests use animal cell lines and microbes (bacteria, yeast). In particular, the CNTs toxicity was evaluated through cell proliferation/viability tests, apoptosis detection, ROS generation, and superoxide dismutase quantification [164, 165].

Injection, ingestion, and inhalation are the three possible ways for CNTs to enter the human body. Once inside the organism, CNTs can interact with cells in different ways or even cross them, reaching the bloodstream and spread to numerous parts of the body [166]. The main effects behind CNTs toxicity are oxidative stress, inflammatory responses, DNA damage or mutation, malignant transformation, interstitial fibrosis, and granuloma [19, 167–169], thus affecting several organs, like liver and kidney [170], and pulmonary [171], reproductive [172, 173] and nervous [174] systems. For instance, Czarny et al. [175] reported a high CNTs bio-persistence, remaining in the organism for 12 months. CNTs organ biodistribution and bio-persistence depend on diverse physicochemical properties, such as CNTs degree and type of surface functionalization, dispersibility, length, and metal impurities [176].

Several authors have been reporting the reduction of CNTs toxicity by functionalization. For example, Ali-Boucetta et al. [176] showed that the higher the functionalization degree of the CNT surface, the less the accumulation in reticuloendothelial system (RES) organs and the higher the CNTs urinary excretion. Jain et al. [177] also observed a rapid elimination from the organism of highly oxidized MWCNTs through renal excretion without causing nephrotoxicity. In contrast, pristine and less oxidized MWCNTs revealed greater retention in RES organs. Another work showed reduced toxic effects on human umbilical vein endothelial cells when exposed to hydroxylated or carboxylated MWCNTs [178]. The conjugation with biocompatible materials, like PEG, has also proved to reduce CNTs toxicity [179]. No toxic effects of PEGylated SWCNTs was reported after three months of in vivo tests in mice [57]. Conversely, De Marchi et al. [180] demonstrated that carboxylated MWCNTs induced toxicity in suspension-feeding polychaetes *Ficopomatus. enigmaticus*, negatively affecting their biochemical and physiological status.

The degree of dispersibility is also an important parameter to take into account regarding CNTs toxicity. More dispersed CNTs cause less toxic effects since agglomerated particles tend to accumulate more quickly in vital organs, leading to inflammatory responses or possible cell-damaging [181]. Yang et al. [182] detected long-term cytotoxic effects promoted by the aggregation of SWCNTs and their accumulation in the lungs of mice. However, the toxicity of the dispersant and its dosage must be taken into account. An increase in the SWCNTs toxicity was demonstrated with the solubilizing agent toxicity [183]. Additionally, the SWCNTs dispersion in water by sodium dodecyl sulfate (SDS) was studied, revealing an inhibition of the growth of kidney epithelial cells (NRK-52E) in rats at low concentrations and an increase in the number of cells in the apoptotic sub-G(1) phase with increasing dispersant concentration [184].

The particle length affects the CNT toxicity essentially regarding their cellular internalization. Raffa et al. [185] demonstrated that the smaller the CNTs, the easier the penetration to the cell membrane. In turn, Murphy et al. [186] studied the direct instillation of long and short CNTs into the pleural cavity. They observed the clearance of short CNTs, whereas the retention of the long ones caused acute inflammation on the parietal pleura. It has to be noted that during the functionalization process, most of the CNTs are broken into smaller particles, making it difficult to confirm which parameter most influences their toxicity, whether functionalization or length.

Nevertheless, more recently, in a study about the influence of MWCNTs diameters on the toxicity to human endothelial cells, a decrease in cytotoxicity with the increase of MWCNTs diameters was reported [187]. The smallest MWCNTs induced cytokine release, THP-1 monocyte adhesion, and ROS generation. Also, Yamashita et al. [188] revealed that the CNTs toxicity depends not only on their length but also on their thickness. These authors evaluated SWCNTs and different sized MWCNTs, detecting DNA damage and severe inflammatory effects caused by long and thick MWCNTs, but not by short and thin ones or by SWCNTs.

Another important factor in CNTs toxicity is the amount and type of metal impurities introduced during the synthesis step. It was observed that commercial non-purified CNTs caused oxidative stress in cells, while acid-purified ones, with almost no metal content, revealed slight or no cytotoxic effects [189]. Inhibition of cell proliferation by MWCNTs impurities was also reported [190]. Nevertheless, Cheng et al. [191] did not detect differences in human macrophage cell death after exposure to non-purified or heat-purified MWCNTs. In fact, the CNTs impurities contribution toward toxicity should consider the impurities dosage, the administration method, and the method used to detect that contribution [19].

The potential CNTs toxicity is still a topic of debate, which is even more exacerbated due to the CNTs panoply, i.e., with diverse morphologies and purities, as well as several biological media with different concentrations (in vitro, in vivo, primary cells, cancer cell lines, several animal models) and numerous techniques of dispersion, exposure, and characterization of CNTs. Moreover, the CNTs exposure to real circumstances is difficult to measure, and the dosages used are far from reality. Further toxicity studies are crucially needed to identify the CNTs safety limit value and to elucidate their toxicological mechanism.

4 Conclusions and Future Perspectives

CNTs are promising materials and have been extensively investigated for biomedical applications due to their unique structure and excellent properties, such as strong loading capacity, biocompatibility, high surface area, high strength, and enhanced chemical and thermal stability. CNTs have been applied in the pharmacy and medicine fields, namely in diagnostics and diseases monitoring, due to their high conductivity; tissue engineering; drug and gene delivery to cells or organs owing to their large surface area; and targeted therapies, on account of their nanoscale diameters that enable them to penetrate the cell membrane. However, their use in biomedical applications raises some questions in what concerns safety and potential toxicity. However, up to date it is clear that CNTs toxicity depends on their structure, morphology, and surface functional groups. To decrease their toxicity and improve biocompatibility, surface functionalization of CNTs has been widely investigated.

Although numerous reports described CNTs for biosensing, bioimaging, drug delivery, and therapy, additional investigations are still needed to spread the CNTs use in the biomedical field in the future and under safe conditions. In specific, more efforts on the evolution of the in vitro and in vivo CNTs toxicity are required to clarify their long-term effects so that their application could become more frequent and widespread.

Acknowledgements This work was developed within the scope of the project CICECO-Aveiro Institute of Materials, UIDB/50011/2020 & UIDP/50011/2020, financed by national funds through the Portuguese Foundation for Science and Technology/MCTES. This work was financially supported by Base Funding—UIDB/EQU/50020/2020 of the Associate Laboratory LSRE-LCM—funded by national funds through FCT/MCTES (PIDDAC), and POCI-01-0145-FEDER-031268—funded by FEDER, through COMPETE2020—Programa Operacional Competitividade e Internacionalização (POCI), and by national funds (OE), through FCT/MCTES. João C. F. Nunes acknowledges SPQ and FCT for the PhD fellowship (SFRH/BD/150671/2020). Raquel O. Cristóvão acknowledges FCT funding under DL57/2016 Transitory Norm Programme. Ana P. M. Tavares acknowledges the FCT for the research contract CEECIND/2020/01867.

References

1. Tejendra Kumar G, Pattabhi Ramaiah B, Sivakumar Reddy C, Sudhir Sastry YB, Marco P, Stephane PB (2019) Advances in carbon based nanomaterials for bio-medical applications. Curr Med Chem 26:6851–6877. https://doi.org/10.2174/0929867326666181126113605
2. Hong G, Diao S, Antaris AL, Dai H (2015) Carbon nanomaterials for biological imaging and nanomedicinal therapy. Chem Rev 115:10816–10906. https://doi.org/10.1021/acs.chemrev.5b00008
3. Rajakumar G, Zhang X-H, Gomathi T, Wang S-F, Ansari MA, Mydhili G, Nirmala G, Alzohairy MA, Chung I-M (2020) Current use of carbon-based materials for biomedical applications—a prospective and review. Processes 8:355
4. Maiti D, Tong X, Mou X, Yang K (2019) Carbon-based nanomaterials for biomedical applications: a recent study. Front Pharmacol 9:1401. https://doi.org/10.3389/fphar.2018.01401
5. Bhattacharya K, Mukherjee SP, Gallud A, Burkert SC, Bistarelli S, Bellucci S, Bottini M, Star A, Fadeel B (2016) Biological interactions of carbon-based nanomaterials: From coronation to degradation. Nanomed Nanotechnol, Biol Med 12:333–351. https://doi.org/10.1016/j.nano.2015.11.011
6. Iijima S (1991) Helical microtubules of graphitic carbon. Nature 354:56–58. https://doi.org/10.1038/354056a0
7. Balasubramanian K, Burghard M (2005) Chemically functionalized carbon nanotubes. Small 1:180–192. https://doi.org/10.1002/smll.200400118
8. Dettlaff A, Sawczak M, Klugmann-Radziemska E, Czylkowski D, Miotk R, Wilamowska-Zawłocka M (2017) High-performance method of carbon nanotubes modification by microwave plasma for thin composite films preparation. RSC Adv 7:31940–31949. https://doi.org/10.1039/C7RA04707J
9. Dong X, Wei C, Liang J, Liu T, Kong D, Lv F (2017) Thermosensitive hydrogel loaded with chitosan-carbon nanotubes for near infrared light triggered drug delivery. Colloids Surf B Biointerfaces 154:253–262. https://doi.org/10.1016/j.colsurfb.2017.03.036
10. Verma SK, Modi A, Bellare J (2019) Polyethersulfone-carbon nanotubes composite hollow fiber membranes with improved biocompatibility for bioartificial liver. Colloids Surf B Biointerfaces 181:890–895. https://doi.org/10.1016/j.colsurfb.2019.06.051
11. Raphey VR, Henna TK, Nivitha KP, Mufeedha P, Sabu C, Pramod K (2019) Advanced biomedical applications of carbon nanotube. Mater Sci Eng C 100:616–630. https://doi.org/10.1016/j.msec.2019.03.043

12. Tasis D, Tagmatarchis N, Bianco A, Prato M (2006) Chemistry of carbon nanotubes. Chem Rev 106:1105–1136. https://doi.org/10.1021/cr050569o
13. Wu H-C, Chang X, Liu L, Zhao F, Zhao Y (2010) Chemistry of carbon nanotubes in biomedical applications. J Mater Chem 20:1036–1052. https://doi.org/10.1039/B911099M
14. Wang J (2005) Carbon-nanotube based electrochemical biosensors: a review. Electroanalysis 17:7–14. https://doi.org/10.1002/elan.200403113
15. He H, Pham-Huy LA, Dramou P, Xiao D, Zuo P, Pham-Huy C (2013) Carbon nanotubes: applications in pharmacy and medicine. Biomed Res Int 2013:578290. https://doi.org/10.1155/2013/578290
16. Anzar N, Hasan R, Tyagi M, Yadav N, Narang J (2020) Carbon nanotube—a review on synthesis, properties and plethora of applications in the field of biomedical science. Sensors Int 1:100003. https://doi.org/10.1016/j.sintl.2020.100003
17. Monthioux M, Serp P, Flahaut E, Razafinimanana M, Laurent C, Peigney A, Bacsa W, Broto J-M (2010) Introduction to carbon nanotubes. In: Bhushan B (ed) Springer handbook of nanotechnology. Springer, Berlin Heidelberg, Berlin, Heidelberg, pp 47–118
18. Singh BGP, Rao CH, Pispati V, Pathipati H, Muthy N, Prassana SRV, Rathode BG (2012) Carbon nanotubes. a novel drug delivery system. Int J Res Pharm Chem 2:523–532
19. Usui Y, Haniu H, Tsuruoka S, Saito N (2012) Carbon nanotubes innovate on medical technology. Med Chem (Los Angeles) 02. https://doi.org/10.4172/2161-0444.1000105
20. Alshehri R, Ilyas AM, Hasan A, Arnaout A, Ahmed F, Memic A (2016) Carbon nanotubes in biomedical applications: factors, mechanisms, and remedies of toxicity. J Med Chem 59:8149–8167. https://doi.org/10.1021/acs.jmedchem.5b01770
21. Elhissi AMA, Ahmed W, Hassan IU, Dhanak VR, D'Emanuele A (2012) Carbon nanotubes in cancer therapy and drug delivery. J Drug Deliv 2012:837327. https://doi.org/10.1155/2012/837327
22. Yang N, Chen X, Ren T, Zhang P, Yang D (2015) Carbon nanotube based biosensors. Sens Actuators B Chem 207:690–715. https://doi.org/10.1016/j.snb.2014.10.040
23. Lefebvre J, Austing DG, Bond J, Finnie P (2006) Photoluminescence imaging of suspended single-walled carbon nanotubes. Nano Lett 6:1603–1608. https://doi.org/10.1021/nl060530e
24. Murakami Y, Einarsson E, Edamura T, Maruyama S (2005) Polarization dependent optical absorption properties of single-walled carbon nanotubes and methodology for the evaluation of their morphology. Carbon N Y 43:2664–2676. https://doi.org/10.1016/j.carbon.2005.05.036
25. Dresselhaus MS, Dresselhaus G, Saito R, Jorio A (2005) Raman spectroscopy of carbon nanotubes. Phys Rep 409:47–99. https://doi.org/10.1016/j.physrep.2004.10.006
26. Yu M-F, Lourie O, Dyer MJ, Moloni K, Kelly TF, Ruoff RS (2000) Strength and breaking mechanism of multiwalled carbon nanotubes under tensile load. Science (80-) 287:637. https://doi.org/10.1126/science.287.5453.637
27. Pantarotto D, Singh R, McCarthy D, Erhardt M, Briand JP, Prato M, Kostarelos K, Bianco A (2004) Functionalized carbon nanotubes for plasmid DNA gene delivery. Angew Chemie—Int Ed 43:5242–5246. https://doi.org/10.1002/anie.200460437
28. Pantarotto D, Partidos CD, Graff R, Hoebeke J, Briand J-P, Prato M, Bianco A (2003) Synthesis, structural characterization, and immunological properties of carbon nanotubes functionalized with peptides. J Am Chem Soc 125:6160–6164. https://doi.org/10.1021/ja034342r
29. Zhang W, Zhang Z, Zhang Y (2011) The application of carbon nanotubes in target drug delivery systems for cancer therapies. Nanoscale Res Lett 6:555. https://doi.org/10.1186/1556-276x-6-555
30. Kam NW, O'Connell M, Wisdom JA, Dai H (2005) Carbon nanotubes as multifunctional biological transporters and near-infrared agents for selective cancer cell destruction. Proc Natl Acad Sci USA 102:11600–11605. https://doi.org/10.1073/pnas.0502680102
31. Chakravarty P, Marches R, Zimmerman NS, Swafford ADE, Bajaj P, Musselman IH, Pantano P, Draper RK, Vitetta ES (2008) Thermal ablation of tumor cells with antibody-functionalized single-walled carbon nanotubes. Proc Natl Acad Sci USA 105:8697–8702. https://doi.org/10.1073/pnas.0803557105

32. Simon J, Flahaut E, Golzio M (2019) Overview of carbon nanotubes for biomedical applications. Mater (Basel, Switzerland) 12:624. https://doi.org/10.3390/ma12040624
33. Yang W, Ratinac KR, Ringer SP, Thordarson P, Gooding JJ, Braet F (2010) Carbon nanomaterials in biosensors: should you use nanotubes or graphene? Angew Chemie Int Ed 49:2114–2138. https://doi.org/10.1002/anie.200903463
34. Punbusayakul N, Talapatra S, Ajayan PM, Surareungchai W (2013) Label-free as-grown double wall carbon nanotubes bundles for Salmonella typhimuriumimmunoassay. Chem Cent J 7:102. https://doi.org/10.1186/1752-153X-7-102
35. Zhou J, Li H, Yang H, Cheng H, Lai G (2017) Immobilization of glucose oxidase on a carbon nanotubes/dendrimer-ferrocene modified electrode for reagentless glucose biosensing. J Nanosci Nanotechnol 17:212–216. https://doi.org/10.1166/jnn.2017.12388
36. Shrestha BK, Ahmad R, Mousa HM, Kim I-G, Kim JI, Neupane MP, Park CH, Kim CS (2016) High-performance glucose biosensor based on chitosan-glucose oxidase immobilized polypyrrole/Nafion/functionalized multi-walled carbon nanotubes bio-nanohybrid film. J Colloid Interface Sci 482:39–47. https://doi.org/10.1016/j.jcis.2016.07.067
37. Devasenathipathy R, Mani V, Chen S-M, Huang S-T, Huang T-T, Lin C-M, Hwa K-Y, Chen T-Y, Chen B-J (2015) Glucose biosensor based on glucose oxidase immobilized at gold nanoparticles decorated graphene-carbon nanotubes. Enzyme Microb Technol 78:40–45. https://doi.org/10.1016/j.enzmictec.2015.06.006
38. Zhang X, Xu Y, Ye B (2018) An efficient electrochemical glucose sensor based on porous nickel-based metal organic framework/carbon nanotubes composite (Ni-MOF/CNTs). J Alloys Compd 767:651–656. https://doi.org/10.1016/j.jallcom.2018.07.175
39. Li Y-W, Chen Y, Ma Y-H, Shi J-G, Wang Y-X, Qi C-H, Li Q-S (2014) Recent advances in the dehydrogenase biosensors based on carbon nanotube modified electrodes. Chinese J Anal Chem 42:759–765. https://doi.org/10.1016/S1872-2040(13)60733-1
40. Kafi AKM, Naqshabandi M, Yusoff MM, Crossley MJ (2018) Improved peroxide biosensor based on horseradish peroxidase/carbon nanotube on a thiol-modified gold electrode. Enzyme Microb Technol 113:67–74. https://doi.org/10.1016/j.enzmictec.2017.11.006
41. He P, Dai L (2004) Aligned carbon nanotube–DNA electrochemical sensors. Chem Commun 348–349. https://doi.org/10.1039/B313030B
42. Li W, Gao Y, Zhang J, Wang X, Yin F, Li Z, Zhang M (2020) Universal DNA detection realized by peptide based carbon nanotube biosensors. Nanoscale Adv 2:717–723. https://doi.org/10.1039/C9NA00625G
43. Sudibya HG, Ma J, Dong X, Ng S, Li L-J, Liu X-W, Chen P (2009) Interfacing glycosylated carbon-nanotube-network devices with living cells to detect dynamic secretion of biomolecules. Angew Chemie Int Ed 48:2723–2726. https://doi.org/10.1002/anie.200805514
44. Hu FX, Kang YJ, Du F, Zhu L, Xue YH, Chen T, Dai LM, Li CM (2015) Living cells directly growing on a DNA/Mn3(PO4)2-immobilized and vertically aligned CNT array as a free-standing hybrid film for highly sensitive in situ detection of released superoxide anions. Adv Funct Mater 25:5924–5932. https://doi.org/10.1002/adfm.201502341
45. Jin H, Heller DA, Kalbacova M, Kim J-H, Zhang J, Boghossian AA, Maheshri N, Strano MS (2010) Detection of single-molecule H_2O_2 signalling from epidermal growth factor receptor using fluorescent single-walled carbon nanotubes. Nat Nanotechnol 5:302–309. https://doi.org/10.1038/nnano.2010.24
46. Iverson NM, Barone PW, Shandell M, Trudel LJ, Sen S, Sen F, Ivanov V, Atolia E, Farias E, McNicholas TP, Reuel N, Parry NMA, Wogan GN, Strano MS (2013) In vivo biosensing via tissue-localizable near-infrared-fluorescent single-walled carbon nanotubes. Nat Nanotechnol 8:873–880. https://doi.org/10.1038/nnano.2013.222
47. Hong H, Gao T, Cai W (2009) Molecular imaging with single-walled carbon nanotubes. Nano Today 4:252–261. https://doi.org/10.1016/j.nantod.2009.04.002
48. Welsher K, Liu Z, Sherlock SP, Robinson JT, Chen Z, Daranciang D, Dai H (2009) A route to brightly fluorescent carbon nanotubes for near-infrared imaging in mice. Nat Nanotechnol 4:773–780. https://doi.org/10.1038/nnano.2009.294

49. Welsher K, Sherlock SP, Dai H (2011) Deep-tissue anatomical imaging of mice using carbon nanotube fluorophores in the second near-infrared window. Proc Natl Acad Sci 108:8943 LP–8948. https://doi.org/10.1073/pnas.1014501108
50. Cherukuri P, Bachilo SM, Litovsky SH, Weisman RB (2004) Near-infrared fluorescence microscopy of single-walled carbon nanotubes in phagocytic cells. J Am Chem Soc 126:15638–15639. https://doi.org/10.1021/ja0466311
51. Leeuw TK, Reith RM, Simonette RA, Harden ME, Cherukuri P, Tsyboulski DA, Beckingham KM, Weisman RB (2007) Single-walled carbon nanotubes in the intact organism: near-IR imaging and biocompatibility studies in drosophila. Nano Lett 7:2650–2654. https://doi.org/10.1021/nl0710452
52. De La Zerda A, Zavaleta C, Keren S, Vaithilingam S, Bodapati S, Liu Z, Levi J, Smith BR, Ma T-J, Oralkan O, Cheng Z, Chen X, Dai H, Khuri-Yakub BT, Gambhir SS (2008) Carbon nanotubes as photoacoustic molecular imaging agents in living mice. Nat Nanotechnol 3:557–562. https://doi.org/10.1038/nnano.2008.231
53. de la Zerda A, Liu Z, Bodapati S, Teed R, Vaithilingam S, Khuri-Yakub BT, Chen X, Dai H, Gambhir SS (2010) Ultrahigh sensitivity carbon nanotube agents for photoacoustic molecular imaging in living mice. Nano Lett 10:2168–2172. https://doi.org/10.1021/nl100890d
54. Green DE, Longtin JP, Sitharaman B (2009) The effect of nanoparticle-enhanced photoacoustic stimulation on multipotent marrow stromal cells. ACS Nano 3:2065–2072. https://doi.org/10.1021/nn900434p
55. Paratala BS, Sitharaman B (2011) Carbon nanotubes in regenerative medicine. In: Carbon nanostructures. pp 27–39
56. Heller DA, Baik S, Eurell TE, Strano MS (2005) Single-walled carbon nanotube spectroscopy in live cells: towards long-term labels and optical sensors. Adv Mater 17:2793–2799. https://doi.org/10.1002/adma.200500477
57. Liu Z, Davis C, Cai W, He L, Chen X, Dai H (2008) Circulation and long-term fate of functionalized, biocompatible single-walled carbon nanotubes in mice probed by Raman spectroscopy. Proc Natl Acad Sci 105:1410. https://doi.org/10.1073/pnas.0707654105
58. Zavaleta C, de la Zerda A, Liu Z, Keren S, Cheng Z, Schipper M, Chen X, Dai H, Gambhir SS (2008) Noninvasive Raman spectroscopy in living mice for evaluation of tumor targeting with carbon nanotubes. Nano Lett 8:2800–2805. https://doi.org/10.1021/nl801362a
59. Servant A, Jacobs I, Bussy C, Fabbro C, da Ros T, Pach E, Ballesteros B, Prato M, Nicolay K, Kostarelos K (2016) Gadolinium-functionalised multi-walled carbon nanotubes as a T1 contrast agent for MRI cell labelling and tracking. Carbon N Y 97:126–133. https://doi.org/10.1016/j.carbon.2015.08.051
60. Mansfield P (2004) Snapshot magnetic resonance imaging (Nobel Lecture). Angew Chemie Int Ed 43:5456–5464. https://doi.org/10.1002/anie.200460078
61. Sitharaman B, Kissell KR, Hartman KB, Tran LA, Baikalov A, Rusakova I, Sun Y, Khant HA, Ludtke SJ, Chiu W, Laus S, Tóth É, Helm L, Merbach AE, Wilson LJ (2005) Superparamagnetic gadonanotubes are high-performance MRI contrast agents. Chem Commun 3915–3917. https://doi.org/10.1039/B504435A
62. Hong SY, Tobias G, Al-Jamal KT, Ballesteros B, Ali-Boucetta H, Lozano-Perez S, Nellist PD, Sim RB, Finucane C, Mather SJ, Green MLH, Kostarelos K, Davis BG (2010) Filled and glycosylated carbon nanotubes for in vivo radioemitter localization and imaging. Nat Mater 9:485–490. https://doi.org/10.1038/nmat2766
63. Liu Z, Cai W, He L, Nakayama N, Chen K, Sun X, Chen X, Dai H (2007) In vivo biodistribution and highly efficient tumour targeting of carbon nanotubes in mice. Nat Nanotechnol 2:47–52. https://doi.org/10.1038/nnano.2006.170
64. McDevitt MR, Chattopadhyay D, Kappel BJ, Jaggi JS, Schiffman SR, Antczak C, Njardarson JT, Brentjens R, Scheinberg DA (2007) Tumor targeting with antibody-functionalized, radiolabeled carbon nanotubes. J Nucl Med 48:1180–1189. https://doi.org/10.2967/jnumed.106.039131
65. Harrison BS, Atala A (2007) Carbon nanotube applications for tissue engineering. Biomaterials 28:344–353. https://doi.org/10.1016/j.biomaterials.2006.07.044

66. Lovat V, Pantarotto D, Lagostena L, Cacciari B, Grandolfo M, Righi M, Spalluto G, Prato M, Ballerini L (2005) Carbon nanotube substrates boost neuronal electrical signaling. Nano Lett 5:1107–1110. https://doi.org/10.1021/nl050637m
67. Mazzatenta A, Giugliano M, Campidelli S, Gambazzi L, Businaro L, Markram H, Prato M, Ballerini L (2007) Interfacing neurons with carbon nanotubes: electrical signal transfer and synaptic stimulation in cultured brain circuits. J Neurosci 27:6931 LP–6936. https://doi.org/10.1523/JNEUROSCI.1051-07.2007
68. Béduer A, Seichepine F, Flahaut E, Loubinoux I, Vaysse L, Vieu C (2012) Elucidation of the role of carbon nanotube patterns on the development of cultured neuronal cells. Langmuir 28:17363–17371. https://doi.org/10.1021/la304278n
69. Matta-Domjan B, King A, Totti S, Matta C, Dover G, Martinez P, Zakhidov A, La Ragione R, Macedo H, Jurewicz I, Dalton A, Velliou EG (2018) Biophysical interactions between pancreatic cancer cells and pristine carbon nanotube substrates: potential application for pancreatic cancer tissue engineering. J Biomed Mater Res B Appl Biomater 106:1637–1644. https://doi.org/10.1002/jbm.b.34012
70. Correa-Duarte MA, Wagner N, Rojas-Chapana J, Morsczeck C, Thie M, Giersig M (2004) Fabrication and biocompatibility of carbon nanotube-based 3D networks as scaffolds for cell seeding and growth. Nano Lett 4:2233–2236. https://doi.org/10.1021/nl048574f
71. Jun Han Z, Rider AE, Ishaq M, Kumar S, Kondyurin A, Bilek MMM, Levchenko I, Ostrikov KK (2013) Carbon nanostructures for hard tissue engineering. RSC Adv 3:11058–11072. https://doi.org/10.1039/C2RA23306A
72. Abarrategi A, Gutiérrez MC, Moreno-Vicente C, Hortigüela MJ, Ramos V, López-Lacomba JL, Ferrer ML, del Monte F (2008) Multiwall carbon nanotube scaffolds for tissue engineering purposes. Biomaterials 29:94–102. https://doi.org/10.1016/j.biomaterials.2007.09.021
73. Namgung S, Kim T, Baik KY, Lee M, Nam J-M, Hong S (2011) Fibronectin–carbon-nanotube hybrid nanostructures for controlled cell growth. Small 7:56–61. https://doi.org/10.1002/smll.201001513
74. Keefer EW, Botterman BR, Romero MI, Rossi AF, Gross GW (2008) Carbon nanotube coating improves neuronal recordings. Nat Nanotechnol 3:434–439. https://doi.org/10.1038/nnano.2008.174
75. Baranauskas G, Maggiolini E, Castagnola E, Ansaldo A, Mazzoni A, Angotzi GN, Vato A, Ricci D, Panzeri S, Fadiga L (2011) Carbon nanotube composite coating of neural microelectrodes preferentially improves the multiunit signal-to-noise ratio. J Neural Eng 8:66013. https://doi.org/10.1088/1741-2560/8/6/066013
76. Lee HJ, Park J, Yoon OJ, Kim HW, Lee DY, Kim DH, Lee WB, Lee N-E, Bonventre JV, Kim SS (2011) Amine-modified single-walled carbon nanotubes protect neurons from injury in a rat stroke model. Nat Nanotechnol 6:121–125. https://doi.org/10.1038/nnano.2010.281
77. Cellot G, Cilia E, Cipollone S, Rancic V, Sucapane A, Giordani S, Gambazzi L, Markram H, Grandolfo M, Scaini D, Gelain F, Casalis L, Prato M, Giugliano M, Ballerini L (2009) Carbon nanotubes might improve neuronal performance by favouring electrical shortcuts. Nat Nanotechnol 4:126–133. https://doi.org/10.1038/nnano.2008.374
78. Chen Z, Zhang A, Wang X, Zhu J, Fan Y, Yu H, Yang Z (2017) The advances of carbon nanotubes in cancer diagnostics and therapeutics. J Nanomater 2017:1–13. https://doi.org/10.1155/2017/3418932
79. Bianco A, Kostarelos K, Partidos CD, Prato M (2005) Biomedical applications of functionalised carbon nanotubes. Chem Commun 571–577. https://doi.org/10.1039/b410943k
80. Hasnain MS, Ahmad SA, Hoda MN, Rishishwar S, Rishishwar P, Nayak AK (2019) Stimuli-responsive carbon nanotubes for targeted drug delivery. In: Stimuli responsive polymeric nanocarriers for drug delivery applications. Elsevier, pp 321–344
81. Sheikhpour M, Naghinejad M, Kasaeian A, Lohrasbi A, Shahraeini SS, Zomorodbakhsh S (2020) The applications of carbon nanotubes in the diagnosis and treatment of lung cancer: a critical review. Int J Nanomedicine 15:7063–7078. https://doi.org/10.2147/IJN.S263238
82. Jain KK (2008) Drug delivery systems—an overview. Methods Mol Biol 437:1–50. https://doi.org/10.1007/978-1-59745-210-6_1

83. Tan JM, Karthivashan G, Arulselvan P, Fakurazi S, Hussein MZ (2014) Characterization and in vitro studies of the anticancer effect of oxidized carbon nanotubes functionalized with betulinic acid. Drug Des Devel Ther 8:2333–2343. https://doi.org/10.2147/DDDT.S70650
84. Tian Z, Yin M, Ma H, Zhu L, Shen H, Jia N (2011) Supramolecular assembly and antitumor activity of multiwalled carbon nanotube-camptothecin complexes. J Nanosci Nanotechnol 11:953–958. https://doi.org/10.1166/jnn.2011.3100
85. Arlt M, Haase D, Hampel S, Oswald S, Bachmatiuk A, Klingeler R, Schulze R, Ritschel M, Leonhardt A, Fuessel S, Büchner B, Kraemer K, Wirth MP (2010) Delivery of carboplatin by carbon-based nanocontainers mediates increased cancer cell death. Nanotechnology 21. https://doi.org/10.1088/0957-4484/21/33/335101
86. Guven A, Rusakova IA, Lewis MT, Wilson LJ (2012) Cisplatin@US-tube carbon nanocapsules for enhanced chemotherapeutic delivery. Biomaterials 33:1455–1461. https://doi.org/10.1016/j.biomaterials.2011.10.060
87. Singh N, Sachdev A, Gopinath P (2017) Polysaccharide functionalized single walled carbon nanotubes as nanocarriers for delivery of curcumin in lung cancer cells. J Nanosci Nanotechnol 18:1534–1541. https://doi.org/10.1166/jnn.2018.14222
88. Singh RP, Sharma G, Sonali SS, Kumar M, Pandey BL, Koch B, Muthu MS (2016) Vitamin E TPGS conjugated carbon nanotubes improved efficacy of docetaxel with safety for lung cancer treatment. Colloids Surfaces B Biointerfaces 141:429–442. https://doi.org/10.1016/j.colsurfb.2016.02.011
89. Singh RP, Sharma G, Sonali SS, Patne SCU, Pandey BL, Koch B, Muthu MS (2016) Effects of transferrin conjugated multi-walled carbon nanotubes in lung cancer delivery. Mater Sci Eng C 67:313–325. https://doi.org/10.1016/j.msec.2016.05.013
90. Singh RP, Sharma G, Sonali SS, Bharti S, Pandey BL, Koch B, Muthu MS (2017) Chitosan-folate decorated carbon nanotubes for site specific lung cancer delivery. Mater Sci Eng C 77:446–458. https://doi.org/10.1016/j.msec.2017.03.225
91. Datir SR, Das M, Singh RP, Jain S (2012) Hyaluronate tethered, "smart" multiwalled carbon nanotubes for tumor-targeted delivery of doxorubicin. Bioconjug Chem 23:2201–2213. https://doi.org/10.1021/bc300248t
92. Heister E, Neves V, Tîlmaciu C, Lipert K, Beltrán VS, Coley HM, Silva SRP, McFadden J (2009) Triple functionalisation of single-walled carbon nanotubes with doxorubicin, a monoclonal antibody, and a fluorescent marker for targeted cancer therapy. Carbon N Y 47:2152–2160. https://doi.org/10.1016/j.carbon.2009.03.057
93. Gu YJ, Cheng J, Jin J, Cheng SH, Wong WT (2011) Development and evaluation of pH-responsive single-walled carbon nanotube-doxorubicin complexes in cancer cells. Int J Nanomedicine 6:2889–2898. https://doi.org/10.2147/ijn.s25162
94. Zhang X, Meng L, Lu Q, Fei Z, Dyson PJ (2009) Targeted delivery and controlled release of doxorubicin to cancer cells using modified single wall carbon nanotubes. Biomaterials 30:6041–6047. https://doi.org/10.1016/j.biomaterials.2009.07.025
95. Wen S, Liu H, Cai H, Shen M, Shi X (2013) Targeted and pH-responsive delivery of doxorubicin to cancer cells using multifunctional dendrimer-modified multi-walled carbon nanotubes. Adv Healthc Mater 2:1267–1276. https://doi.org/10.1002/adhm.201200389
96. Ali-Boucetta H, Al-Jamal KT, McCarthy D, Prato M, Bianco A, Kostarelos K (2002) Multi-walled carbon nanotube-doxorubicin supramolecular complexes for cancer therapeutics. Chem Commun 8:459–461. https://doi.org/10.1039/b712350g
97. Lu YJ, Wei KC, Ma CCM, Yang SY, Chen JP (2012) Dual targeted delivery of doxorubicin to cancer cells using folate-conjugated magnetic multi-walled carbon nanotubes. Colloids Surf B Biointerfaces 89:1–9. https://doi.org/10.1016/j.colsurfb.2011.08.001
98. Niu L, Meng L, Lu Q (2013) Folate-conjugated PEG on single walled carbon nanotubes for targeting delivery of doxorubicin to cancer cells. Macromol Biosci 13:735–744. https://doi.org/10.1002/mabi.201200475
99. Al Faraj A, Shaik AS, Halwani R, Alfuraih A (2016) Magnetic targeting and delivery of drug-loaded SWCNTs theranostic nanoprobes to lung metastasis in breast cancer animal model: noninvasive monitoring using magnetic resonance imaging. Mol Imaging Biol 18:315–324. https://doi.org/10.1007/s11307-015-0902-0

100. Lodhi N, Mehra NK, Jain NK (2013) Development and characterization of dexamethasone mesylate anchored on multi walled carbon nanotubes. J Drug Target 21:67–76. https://doi.org/10.3109/1061186X.2012.729213
101. Cao Y, Huang H-Y, Chen L-Q, Du H-H, Cui J-H, Zhang LW, Lee B-J, Cao Q-R (2019) Enhanced lysosomal escape of pH-responsive polyethylenimine-betaine functionalized carbon nanotube for the codelivery of survivin small interfering RNA and doxorubicin. ACS Appl Mater Interfaces 11:9763–9776. https://doi.org/10.1021/acsami.8b20810
102. Heger Z, Polanska H, Krizkova S, Balvan J, Raudenska M, Dostalova S, Moulick A, Masarik M, Adam V (2017) Co-delivery of VP-16 and Bcl-2-targeted antisense on PEG-grafted oMWCNTs for synergistic in vitro anti-cancer effects in non-small and small cell lung cancer. Colloids Surf B Biointerfaces 150:131–140. https://doi.org/10.1016/j.colsurfb.2016.11.023
103. Razzazan A, Atyabi F, Kazemi B, Dinarvand R (2016) In vivo drug delivery of gemcitabine with PEGylated single-walled carbon nanotubes. Mater Sci Eng C 62:614–625. https://doi.org/10.1016/j.msec.2016.01.076
104. Das M, Datir SR, Singh RP, Jain S (2013) Augmented anticancer activity of a targeted, intracellularly activatable, theranostic nanomedicine based on fluorescent and radiolabeled, methotrexate-folic acid-multiwalled carbon nanotube conjugate. Mol Pharm 10:2543–2557. https://doi.org/10.1021/mp300701e
105. Cirillo G, Vittorio O, Kunhardt D, Valli E, Voli F, Farfalla A, Curcio M, Spizzirri UG, Hampel S (2019) Combining carbon nanotubes and chitosan for the vectorization of methotrexate to lung cancer cells. Materials (Basel) 12:2889. https://doi.org/10.3390/ma12182889
106. Arya N, Arora A, Vasu KS, Sood AK, Katti DS (2013) Combination of single walled carbon nanotubes/graphene oxide with paclitaxel: a reactive oxygen species mediated synergism for treatment of lung cancer. Nanoscale 5:2818–2829. https://doi.org/10.1039/c3nr33190c
107. Sobhani Z, Dinarvand R, Atyabi F, Ghahremani M, Adeli M (2011) Increased paclitaxel cytotoxicity against cancer cell lines using a novel functionalized carbon nanotube. Int J Nanomedicine 705. https://doi.org/10.2147/IJN.S17336
108. Lay CL, Liu HQ, Tan HR, Liu Y (2010) Delivery of paclitaxel by physically loading onto poly(ethylene glycol) (PEG)-graftcarbon nanotubes for potent cancer therapeutics. Nanotechnology 21. https://doi.org/10.1088/0957-4484/21/6/065101
109. Zhang W, Zhang D, Tan J, Cong H (2012) Carbon nanotube exposure sensitize human ovarian cancer cells to paclitaxel. J Nanosci Nanotechnol 12:7211–7214. https://doi.org/10.1166/jnn.2012.6506
110. Yu B, Tan L, Zheng R, Tan H, Zheng L (2016) Targeted delivery and controlled release of Paclitaxel for the treatment of lung cancer using single-walled carbon nanotubes. Mater Sci Eng C 68:579–584. https://doi.org/10.1016/j.msec.2016.06.025
111. Feazell RP, Nakayama-Ratchford N, Dai H, Lippard SJ (2007) Soluble single-walled carbon nanotubes as longboat delivery systems for platinum(IV) anticancer drug design. J Am Chem Soc 129:8438–8439. https://doi.org/10.1021/ja073231f
112. Li J, Pant A, Chin CF, Ang WH, Ménard-Moyon C, Nayak TR, Gibson D, Ramaprabhu S, Panczyk T, Bianco A, Pastorin G (2014) In vivo biodistribution of platinum-based drugs encapsulated into multi-walled carbon nanotubes. Nanomed Nanotechnol, Biol Med 10:1465–1475. https://doi.org/10.1016/j.nano.2014.01.004
113. Su Y, Hu Y, Wang Y, Xu X, Yuan Y, Li Y, Wang Z, Chen K, Zhang F, Ding X, Li M, Zhou J, Liu Y, Wang W (2017) A precision-guided MWNT mediated reawakening the sunk synergy in RAS for anti-angiogenesis lung cancer therapy. Biomaterials 139:75–90. https://doi.org/10.1016/j.biomaterials.2017.05.046
114. Komane PP, Kumar P, Marimuthu T, du Toit LC, Kondiah PPD, Choonara YE, Pillay V (2018) Dexamethasone-loaded, pegylated, vertically aligned, multiwalled carbon nanotubes for potential ischemic stroke intervention. Molecules 23. https://doi.org/10.3390/molecules23061406
115. Jia N, Lian Q, Shen H, Wang C, Li X, Yang Z (2007) Intracellular delivery of quantum dots tagged antisense oligodeoxynucleotides by functionalized multiwalled carbon nanotubes. Nano Lett 7:2976–2980. https://doi.org/10.1021/nl071114c

116. Van Den Bossche J, Al-Jamal WT, Tian B, Nunes A, Fabbro C, Bianco A, Prato M, Kostarelos K (2010) Efficient receptor-independent intracellular translocation of aptamers mediated by conjugation to carbon nanotubes. Chem Commun 46:7379–7381. https://doi.org/10.1039/c0cc02092c
117. Pan B, Cui D, Xu P, Ozkan C, Feng G, Ozkan M, Huang T, Chu B, Li Q, He R, Hu G (2009) Synthesis and characterization of polyamidoamine dendrimer-coated multi-walled carbon nanotubes and their application in gene delivery systems. Nanotechnology 20. https://doi.org/10.1088/0957-4484/20/12/125101
118. Guo C, Al-Jamal WT, Toma FM, Bianco A, Prato M, Al-Jamal KT, Kostarelos K (2015) Design of cationic multiwalled carbon nanotubes as efficient siRNA vectors for lung cancer xenograft eradication. Bioconjug Chem 26:1370–1379. https://doi.org/10.1021/acs.bioconjchem.5b00249
119. Singh R, Pantarotto D, McCarthy D, Chaloin O, Hoebeke J, Partidos CD, Briand JP, Prato M, Bianco A, Kostarelos K (2005) Binding and condensation of plasmid DNA onto functionalized carbon nanotubes: toward the construction of nanotube-based gene delivery vectors. J Am Chem Soc 127:4388–4396. https://doi.org/10.1021/ja0441561
120. Mohammadi M, Salmasi Z, Hashemi M, Mosaffa F, Abnous K, Ramezani M (2015) Single-walled carbon nanotubes functionalized with aptamer and piperazine-polyethylenimine derivative for targeted siRNA delivery into breast cancer cells. Int J Pharm 485:50–60. https://doi.org/10.1016/j.ijpharm.2015.02.031
121. Varkouhi AK, Foillard S, Lammers T, Schiffelers RM, Doris E, Hennink WE, Storm G (2011) SiRNA delivery with functionalized carbon nanotubes. Int J Pharm 416:419–425. https://doi.org/10.1016/j.ijpharm.2011.02.009
122. Podesta JE, Al-Jamal KT, Herrero MA, Tian B, Ali-Boucetta H, Hegde V, Bianco A, Prato M, Kostarelos K (2009) Antitumor activity and prolonged survival by carbon-nanotube-mediated therapeutic siRNA silencing in a human lung xenograft model. Small 5:1176–1185. https://doi.org/10.1002/smll.200801572
123. Al-Jamal KT, Gherardini L, Bardi G, Nunes A, Guo C, Bussy C, Herrero MA, Bianco A, Prato M, Kostarelos K, Pizzorusso T (2011) Functional motor recovery from brain ischemic insult by carbon nanotube-mediated siRNA silencing. Proc Natl Acad Sci USA 108:10952–10957. https://doi.org/10.1073/pnas.1100930108
124. Ladeira MS, Andrade VA, Gomes ERM, Aguiar CJ, Moraes ER, Soares JS, Silva EE, Lacerda RG, Ladeira LO, Jorio A, Lima P, Fatima Leite M, Resende RR, Guatimosim S (2010) Highly efficient siRNA delivery system into human and murine cells using single-wall carbon nanotubes. Nanotechnology 21. https://doi.org/10.1088/0957-4484/21/38/385101
125. Kam NWS, Liu Z, Dai H (2005) Functionalization of carbon nanotubes via cleavable disulfide bonds for efficient intracellular delivery of siRNA and potent gene silencing. J Am Chem Soc 127:12492–12493. https://doi.org/10.1021/ja053962k
126. Chen H, Ma X, Li Z, Shi Q, Zheng W, Liu Y, Wang P (2012) Functionalization of single-walled carbon nanotubes enables efficient intracellular delivery of siRNA targeting MDM2 to inhibit breast cancer cells growth. Biomed Pharmacother 66:334–338. https://doi.org/10.1016/j.biopha.2011.12.005
127. Bartholomeusz G, Cherukuri P, Kingston J, Cognet L, Lemos R, Leeuw TK, Gumbiner-Russo L, Weisman RB, Powis G (2009) In vivo therapeutic silencing of hypoxia-inducible factor 1 alpha (HIF-1α) using single-walled carbon nanotubes noncovalently coated with siRNA. Nano Res 2:279–291. https://doi.org/10.1007/s12274-009-9026-7
128. Sanz V, Coley HM, Silva SRP, McFadden J (2012) Modeling the binding of peptides on carbon nanotubes and their use as protein and DNA carriers. J Nanoparticle Res 14. https://doi.org/10.1007/s11051-011-0695-2
129. Kam NWS, Jessop TC, Wender PA, Dai H (2004) Nanotube molecular transporters: internalization of carbon nanotube-protein conjugates into mammalian cells. J Am Chem Soc 126:6850–6851. https://doi.org/10.1021/ja0486059
130. Chen C, Hou L, Zhang H, Zhu L, Zhang H, Zhang C, Shi J, Wang L, Jia X, Zhang Z (2013) Single-walled carbon nanotubes mediated targeted tamoxifen delivery system using

aspargine-glycine-arginine peptide. J Drug Target 21:809–821. https://doi.org/10.3109/1061186X.2013.829071
131. Zakaria AB, Picaud F, Rattier T, Pudlo M, Saviot L, Chassagnon R, Lherminier J, Gharbi T, Micheau O, Herlem G (2015) Nanovectorization of TRAIL with single wall carbon nanotubes enhances tumor cell killing. Nano Lett 15:891–895. https://doi.org/10.1021/nl503565t
132. Parra J, Abad-Somovilla A, Mercader JV, Taton TA, Abad-Fuentes A (2013) Carbon nanotube-protein carriers enhance size-dependent self-adjuvant antibody response to haptens. J Control Release 170:242–251. https://doi.org/10.1016/j.jconrel.2013.05.019
133. Li C, Yang K, Zhang Y, Tang H, Yan F, Tan L, Xie Q, Yao S (2011) Highly biocompatible multi-walled carbon nanotube-chitosan nanoparticle hybrids as protein carriers. Acta Biomater 7:3070–3077. https://doi.org/10.1016/j.actbio.2011.05.005
134. Weng X, Wang M, Ge J, Yu S, Liu B, Zhong J, Kong J (2009) Carbon nanotubes as a protein toxin transporter for selective HER2-positive breast cancer cell destruction. Mol Biosyst 5:1224–1231. https://doi.org/10.1039/b906948h
135. Zhu Z, Tang Z, Phillips JA, Yang R, Wang H, Tan W (2008) Regulation of singlet oxygen generation using single-walled carbon nanotubes. J Am Chem Soc 130:10856–10857. https://doi.org/10.1021/ja802913f
136. Marangon I, Ménard-Moyon C, Silva AKA, Bianco A, Luciani N, Gazeau F (2016) Synergic mechanisms of photothermal and photodynamic therapies mediated by photosensitizer/carbon nanotube complexes. Carbon N Y 97:110–123. https://doi.org/10.1016/j.carbon.2015.08.023
137. Gobin AM, Lee MH, Halas NJ, James WD, Drezek RA, West JL (2007) Near-infrared resonant nanoshells for combined optical imaging and photothermal cancer therapy. Nano Lett 7:1929–1934. https://doi.org/10.1021/nl070610y
138. Nomura S, Morimoto Y, Tsujimoto H, Arake M, Harada M, Saitoh D, Hara I, Ozeki E, Satoh A, Takayama E, Hase K, Kishi Y, Ueno H (2020) Highly reliable, targeted photothermal cancer therapy combined with thermal dosimetry using a near-infrared absorbent. Sci Rep 10:1–7. https://doi.org/10.1038/s41598-020-66646-x
139. Brennan ME, Coleman JN, Drury A, Lahr B, Kobayashi T, Blau WJ (2003) Nonlinear photoluminescence from van Hove singularities in multiwalled carbon nanotubes. Opt Lett 28:266. https://doi.org/10.1364/ol.28.000266
140. Gannon CJ, Cherukuri P, Yakobson BI, Cognet L, Kanzius JS, Kittrell C, Weisman RB, Pasquali M, Schmidt HK, Smalley RE, Curley SA (2007) Carbon nanotube-enhanced thermal destruction of cancer cells in a noninvasive radiofrequency field. Cancer 110:2654–2665. https://doi.org/10.1002/cncr.23155
141. O'Connell MJ, Bachilo SH, Huffman CB, Moore VC, Strano MS, Haroz EH, Rialon KL, Boul PJ, Noon WH, Kittrell C, Ma J, Hauge RH, Weisman RB, Smalley RE (2002) Band gap fluorescence from individual single-walled carbon nanotubes. Science (80-) 297:593–596. https://doi.org/10.1126/science.1072631
142. Ghosh S, Dutta S, Gomes E, Carroll D, D'Agostino R, Olson J, Guthold M, Gmeiner WH (2009) Increased heating efficiency and selective thermal ablation of malignant tissue with DNA-encased multiwalled carbon nanotubes. ACS Nano 3:2667–2673. https://doi.org/10.1021/nn900368b
143. Shao N, Lu S, Wickstrom E, Panchapakesan B (2007) Integrated molecular targeting of IGF1R and HER2 surface receptors and destruction of breast cancer cells using single wall carbon nanotubes. Nanotechnology 18. https://doi.org/10.1088/0957-4484/18/31/315101
144. Zhou F, Xing D, Ou Z, Wu B, Resasco DE, Chen WR (2009) Cancer photothermal therapy in the near-infrared region by using single-walled carbon nanotubes. J Biomed Opt 14:021009. https://doi.org/10.1117/1.3078803
145. Xiao Y, Gao X, Taratula O, Treado S, Urbas A, Holbrook RD, Cavicchi RE, Avedisian CT, Mitra S, Savla R, Wagner PD, Srivastava S, He H (2009) Anti-HER2 IgY antibody-functionalized single-walled carbon nanotubes for detection and selective destruction of breast cancer cells. BMC Cancer 9:351. https://doi.org/10.1186/1471-2407-9-351
146. Moon HK, Lee SH, Choi HC (2009) In vivo near-infrared mediated tumor destruction by photothermal effect of carbon nanotubes. ACS Nano 3:3707–3713. https://doi.org/10.1021/nn900904h

147. Huang N, Wang H, Zhao J, Lui H, Korbelik M, Zeng H (2010) Single-wall carbon nanotubes assisted photothermal cancer therapy: animal study with a murine model of squamous cell carcinoma. Lasers Surg Med 42:798–808. https://doi.org/10.1002/lsm.20968
148. Zhou F, Wu S, Wu B, Chen WR, Xing D (2011) Mitochondria-targeting single-walled carbon nanotubes for cancer photothermal therapy. Small 7:2727–2735. https://doi.org/10.1002/smll.201100669
149. Zhou F, Wu S, Song S, Chen WR, Resasco DE, Xing D (2012) Antitumor immunologically modified carbon nanotubes for photothermal therapy. Biomaterials 33:3235–3242. https://doi.org/10.1016/j.biomaterials.2011.12.029
150. Wang L, Shi J, Zhang H, Li H, Gao Y, Wang Z, Wang H, Li L, Zhang C, Chen C, Zhang Z, Zhang Y (2013) Synergistic anticancer effect of RNAi and photothermal therapy mediated by functionalized single-walled carbon nanotubes. Biomaterials 34:262–274. https://doi.org/10.1016/j.biomaterials.2012.09.037
151. Oh Y, Jin JO, Oh J (2017) Photothermal-triggered control of sub-cellular drug accumulation using doxorubicin-loaded single-walled carbon nanotubes for the effective killing of human breast cancer cells. Nanotechnology 28. https://doi.org/10.1088/1361-6528/aa5d7d
152. Torti SV, Byrne F, Whelan O, Levi N, Ucer B, Schmid M, Torti FM, Akman S, Liu J, Ajayan PM, Nalamasu O, Carroll DL (2007) Thermal ablation therapeutics based on CNx multi-walled nanotubes. Int J Nanomed 2:707–714
153. Wang CH, Huang YJ, Chang CW, Hsu WM, Peng CA (2009) In vitro photothermal destruction of neuroblastoma cells using carbon nanotubes conjugated with GD2 monoclonal antibody. Nanotechnology 20. https://doi.org/10.1088/0957-4484/20/31/315101
154. Burke A, Ding X, Singh R, Kraft RA, Levi-Polyachenko N, Rylander MN, Szot C, Buchanan C, Whitney J, Fisher J, Hatcher HC, D'Agostino R, Kock ND, Ajayan PM, Carroll DL, Akman S, Torti FM, Torti SV (2009) Long-term survival following a single treatment of kidney tumors with multiwalled carbon nanotubes and near-infrared radiation. Proc Natl Acad Sci USA 106:12897–12902. https://doi.org/10.1073/pnas.0905195106
155. Burlaka A, Lukin S, Prylutska S, Remeniak O, Prylutskyy Y, Shuba M, Maksimenko S, Ritter U, Scharff P (2010) Hyperthermic effect of multi-walled carbon nanotubes stimulated with near infrared irradiation for anticancer therapy: in vitro studies. Exp Oncol 32:48–50
156. Suo X, Eldridge BN, Zhang H, Mao C, Min Y, Sun Y, Singh R, Ming X (2018) P-Glycoprotein-targeted photothermal therapy of drug-resistant cancer cells using antibody-conjugated carbon nanotubes. ACS Appl Mater Interfaces 10:33464–33473. https://doi.org/10.1021/acsami.8b11974
157. Triesscheijn M, Baas P, Schellens JHM, Stewart FA (2006) Photodynamic therapy in oncology. Oncologist 11:1034–1044. https://doi.org/10.1634/theoncologist.11-9-1034
158. Zhu S, Gu Z (2018) Lanthanide-doped materials as dual imaging and therapeutic agents. Elsevier Inc.
159. Erbas S, Gorgulu A, Kocakusakogullari M, Akkaya EU (2009) Non-covalent functionalized SWNTs as delivery agents for novel Bodipy-based potential PDT sensitizers. Chem Commun 4956–4958. https://doi.org/10.1039/b908485a
160. Xiao H, Zhu B, Wang D, Pang Y, He L, Ma X, Wang R, Jin C, Chen Y, Zhu X (2012) Photodynamic effects of chlorin e6 attached to single wall carbon nanotubes through noncovalent interactions. Carbon N Y 50:1681–1689. https://doi.org/10.1016/j.carbon.2011.12.013
161. Wang L, Shi J, Liu R, Liu Y, Zhang J, Yu X, Gao J, Zhang C, Zhang Z (2014) Photodynamic effect of functionalized single-walled carbon nanotubes: a potential sensitizer for photodynamic therapy. Nanoscale 6:4642–4651. https://doi.org/10.1039/c3nr06835h
162. Zhang P, Huang H, Huang J, Chen H, Wang J, Qiu K, Zhao D, Ji L, Chao H (2015) Noncovalent ruthenium(II) complexes-single-walled carbon nanotube composites for bimodal photothermal and photodynamic therapy with near-infrared irradiation. ACS Appl Mater Interfaces 7:23278–23290. https://doi.org/10.1021/acsami.5b07510
163. Shi J, Ma R, Wang L, Zhang J, Liu R, Li L, Liu Y, Hou L, Yu X, Gao J, Zhang Z (2013) The application of hyaluronic acid-derivatized carbon nanotubes in hematoporphyrin monomethyl ether-based photodynamic therapy for in vivo and in vitro cancer treatment. Int J Nanomed 8:2361–2373. https://doi.org/10.2147/IJN.S45407

164. Girardello R, Baranzini N, Tettamanti G, de Eguileor M, Grimaldi A (2017) Cellular responses induced by multi-walled carbon nanotubes: in vivo and in vitro studies on the medicinal leech macrophages. Sci Rep 7:8871. https://doi.org/10.1038/s41598-017-09011-9
165. Madani SY, Mandel A, Seifalian AM (2013) A concise review of carbon nanotube's toxicology. Nano Rev 4:21521. https://doi.org/10.3402/nano.v4i0.21521
166. Wang JTW, Fabbro C, Venturelli E, Ménard-Moyon C, Chaloin O, Da Ros T, Methven L, Nunes A, Sosabowski JK, Mather SJ, Robinson MK, Amadou J, Prato M, Bianco A, Kostarelos K, Al-Jamal KT (2014) The relationship between the diameter of chemically-functionalized multi-walled carbon nanotubes and their organ biodistribution profiles in vivo. Biomaterials 35:9517–9528. https://doi.org/10.1016/j.biomaterials.2014.07.054
167. Lindberg HK, Falck GC, Suhonen S, Vippola M, Vanhala E, Catalán J, Savolainen K, Norppa H (2009) Genotoxicity of nanomaterials: DNA damage and micronuclei induced by carbon nanotubes and graphite nanofibres in human bronchial epithelial cells in vitro. Toxicol Lett 186:166–173. https://doi.org/10.1016/j.toxlet.2008.11.019
168. Liu Y, Zhao Y, Sun B, Chen C (2013) Understanding the toxicity of carbon nanotubes. Acc Chem Res 46:702–713. https://doi.org/10.1021/ar300028m
169. Kobayashi N, Izumi H, Morimoto Y (2017) Review of toxicity studies of carbon nanotubes. J Occup Health 59:394–407. https://doi.org/10.1539/joh.17-0089-RA
170. Yu S, Su X, Du J, Wang J, Gao Y, Zhang L, Chen L, Yang Y, Liu X (2018) The cytotoxicity of water-soluble carbon nanotubes on human embryonic kidney and liver cancer cells. New Carbon Mater 33:36–45. https://doi.org/10.1016/S1872-5805(18)60325-7
171. Knudsen KB, Berthing T, Jackson P, Poulsen SS, Mortensen A, Jacobsen NR, Skaug V, Szarek J, Hougaard KS, Wolff H, Wallin H, Vogel U (2019) Physicochemical predictors of multi-walled carbon nanotube-induced pulmonary histopathology and toxicity one year after pulmonary deposition of 11 different multi-walled carbon nanotubes in mice. Basic Clin Pharmacol Toxicol 124:211–227. https://doi.org/10.1111/bcpt.13119
172. Johansson HKL, Hansen JS, Elfving B, Lund SP, Kyjovska ZO, Loft S, Barfod KK, Jackson P, Vogel U, Hougaard KS (2017) Airway exposure to multi-walled carbon nanotubes disrupts the female reproductive cycle without affecting pregnancy outcomes in mice. Part Fibre Toxicol 14:17. https://doi.org/10.1186/s12989-017-0197-1
173. Xu C, Liu Q, Liu H, Zhang C, Shao W, Gu A (2016) Toxicological assessment of multi-walled carbon nanotubes in vitro: potential mitochondria effects on male reproductive cells. Oncotarget 7:39270–39278. https://doi.org/10.18632/oncotarget.9689
174. Facciolà A, Visalli G, La Maestra S, Ceccarelli M, D'Aleo F, Nunnari G, Pellicanò GF, Di Pietro A (2019) Carbon nanotubes and central nervous system: environmental risks, toxicological aspects and future perspectives. Environ Toxicol Pharmacol 65:23–30. https://doi.org/10.1016/j.etap.2018.11.006
175. Czarny B, Georgin D, Berthon F, Plastow G, Pinault M, Patriarche G, Thuleau A, L'Hermite MM, Taran F, Dive V (2014) Carbon nanotube translocation to distant organs after pulmonary exposure: insights from in situ 14C-radiolabeling and tissue radioimaging. ACS Nano 8:5715–5724. https://doi.org/10.1021/nn500475u
176. Ali-Boucetta H, Kostarelos K (2013) Pharmacology of carbon nanotubes: toxicokinetics, excretion and tissue accumulation. Adv Drug Deliv Rev 65:2111–2119. https://doi.org/10.1016/j.addr.2013.10.004
177. Jain S, Thakare VS, Das M, Godugu C, Jain AK, Mathur R, Chuttani K, Mishra AK (2011) Toxicity of multiwalled carbon nanotubes with end defects critically depends on their functionalization density. Chem Res Toxicol 24:2028–2039. https://doi.org/10.1021/tx2003728
178. Li Z, Liu T, Long J, Wu Y, Yan B, Ma P, Cao Y (2019) The toxicity of hydroxylated and carboxylated multi-walled carbon nanotubes to human endothelial cells was not exacerbated by ER stress inducer. Chinese Chem Lett 30:582–586. https://doi.org/10.1016/j.cclet.2018.12.011
179. Hindumathi R, Jagannatham M, Haridoss P, Sharma CP (2018) Novel nano-cocoon like structures of polyethylene glycol multiwalled carbon nanotubes for biomedical applications. Nano-Struct Nano-Objects 13:30–35

180. De Marchi L, Oliva M, Freitas R, Neto V, Figueira E, Chiellini F, Morelli A, Soares AMVM, Pretti C (2019) Toxicity evaluation of carboxylated carbon nanotubes to the reef-forming tubeworm Ficopomatus enigmaticus (Fauvel, 1923). Mar Environ Res 143:1–9. https://doi.org/10.1016/j.marenvres.2018.10.015
181. Qu G, Bai Y, Zhang Y, Jia Q, Zhang W, Yan B (2009) The effect of multiwalled carbon nanotube agglomeration on their accumulation in and damage to organs in mice. Carbon N Y 47:2060–2069. https://doi.org/10.1016/j.carbon.2009.03.056
182. Yang S-T, Wang X, Jia G, Gu Y, Wang T, Nie H, Ge C, Wang H, Liu Y (2008) Long-term accumulation and low toxicity of single-walled carbon nanotubes in intravenously exposed mice. Toxicol Lett 181:182–189. https://doi.org/10.1016/j.toxlet.2008.07.020
183. Alpatova AL, Shan W, Babica P, Upham BL, Rogensues AR, Masten SJ, Drown E, Mohanty AK, Alocilja EC, Tarabara VV (2010) Single-walled carbon nanotubes dispersed in aqueous media via non-covalent functionalization: effect of dispersant on the stability, cytotoxicity, and epigenetic toxicity of nanotube suspensions. Water Res 44:505–520. https://doi.org/10.1016/j.watres.2009.09.042
184. Nam C-W, Kang S-J, Kang YK, Kwak M-K (2011) Cell growth inhibition and apoptosis by SDS-solubilized single-walled carbon nanotubes in normal rat kidney epithelial cells. Arch Pharm Res 34:661–669. https://doi.org/10.1007/s12272-011-0417-4
185. Raffa V, Ciofani G, Nitodas S, Karachalios T, D'Alessandro D, Masini M, Cuschieri A (2008) Can the properties of carbon nanotubes influence their internalization by living cells? Carbon N Y 46:1600–1610. https://doi.org/10.1016/j.carbon.2008.06.053
186. Murphy FA, Poland CA, Duffin R, Al-Jamal KT, Ali-Boucetta H, Nunes A, Byrne F, Prina-Mello A, Volkov Y, Li S, Mather SJ, Bianco A, Prato M, Macnee W, Wallace WA, Kostarelos K, Donaldson K (2011) Length-dependent retention of carbon nanotubes in the pleural space of mice initiates sustained inflammation and progressive fibrosis on the parietal pleura. Am J Pathol 178:2587–2600. https://doi.org/10.1016/j.ajpath.2011.02.040
187. Zhao X, Chang S, Long J, Li J, Li X, Cao Y (2019) The toxicity of multi-walled carbon nanotubes (MWCNTs) to human endothelial cells: the influence of diameters of MWCNTs. Food Chem Toxicol 126:169–177. https://doi.org/10.1016/j.fct.2019.02.026
188. Yamashita K, Yoshioka Y, Higashisaka K, Morishita Y, Yoshida T, Fujimura M, Kayamuro H, Nabeshi H, Yamashita T, Nagano K, Abe Y, Kamada H, Kawai Y, Mayumi T, Yoshikawa T, Itoh N, Tsunoda S, Tsutsumi Y (2010) Carbon nanotubes elicit DNA damage and inflammatory response relative to their size and shape. Inflammation 33:276–280. https://doi.org/10.1007/s10753-010-9182-7
189. Pulskamp K, Diabaté S, Krug HF (2007) Carbon nanotubes show no sign of acute toxicity but induce intracellular reactive oxygen species in dependence on contaminants. Toxicol Lett 168:58–74. https://doi.org/10.1016/j.toxlet.2006.11.001
190. Haniu H, Matsuda Y, Takeuchi K, Kim YA, Hayashi T, Endo M (2010) Proteomics-based safety evaluation of multi-walled carbon nanotubes. Toxicol Appl Pharmacol 242:256–262. https://doi.org/10.1016/j.taap.2009.10.015
191. Cheng C, Müller KH, Koziol KKK, Skepper JN, Midgley PA, Welland ME, Porter AE (2009) Toxicity and imaging of multi-walled carbon nanotubes in human macrophage cells. Biomaterials 30:4152–4160. https://doi.org/10.1016/j.biomaterials.2009.04.019

Chapter 13
"Biomedical Applications of Porphyrin Nanohybrids"

K. Simi Pushpan and Ajalesh B. Nair

1 Introduction

In nature, porphyrinoids are ubiquitous pigments of life that perform a variety of functions ranging from photosynthesis, electron transfer, oxygen transfer, and storage to the destruction of cells. Porphyrins are heteroaromatic tetrapyrrolic macrocycles linked to each other at a position by a methine bridge. It has 22 π electrons of which 18 π electrons are involved in providing its extended aromaticity. Figure 1 represents the basic unit of porphin scaffold.

Core and peripheral modification of porphyrins and related macrocycles like phthalocyanine, calixarene rings have gradually developed as a stimulating subfield of organic chemistry, which allows alteration of the molecular properties of these macrocycles. Such a structural alteration of these macrocycles opened pathways for studying novel reactivities and intriguing π-conjugation phenomena in these macrocyclic systems.

Porphyrinoids act as an attractive ligand for coordination chemistry for a variety of transition metals. The protonated form of these macrocycles can be tuned as specific anion binding agents depending on their cavity size. The strong absorption in near-infrared owing to extended conjugation and planarity allow these macrocycles to exhibit exciting optoelectronic properties and are considered as the most powerful building blocks for supramolecular assembly. They also exhibit interesting fluorescence behaviors and are clinically proven photosensitizers apart from their ability to

K. Simi Pushpan (✉) · A. B. Nair (✉)
Post Graduate and Research Department of Chemistry, Union Christian College, Aluva, Kerala 683 102, India
e-mail: simi.pushpan@uccollege.edu.in

A. B. Nair
e-mail: ajaleshbnair@uccollege.edu.in

S. Gopi et al. (eds.), *Nanotechnology for Biomedical Applications*,
Materials Horizons: From Nature to Nanomaterials,
https://doi.org/10.1007/978-981-16-7483-9_15

Fig. 1 Porphyrin unit

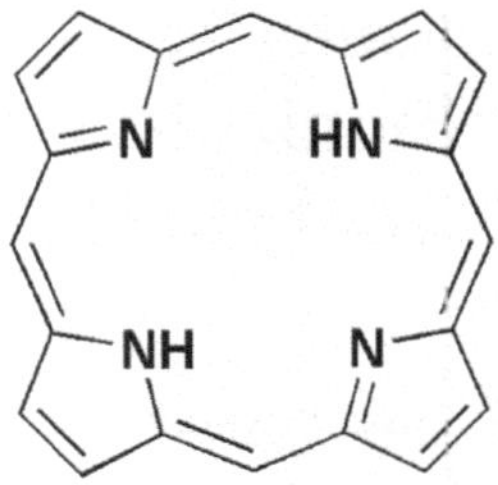

chelate metals in a neutral state and functioning as anion binders in the protonated stage [1, 2].

Herein, we highlight recent research on porphyrin-derived nanohybrids utilized in multimodal imaging techniques, bio-sensing, photodynamic therapy, drug delivery, and probe design, illustrating their promising future for biomedical applications. This chapter is envisioned to provide a comprehensive summary of the biomedical applications of porphyrin-based nanohybrids and their anticipated utility as the next-generation diagnostic and therapeutic class of biomaterials.

2 Photophysics and Photochemistry of Porphyrins

Investigation of photophysics, photochemistry, and spectroscopic properties of porphyrins and metalloporphyrins has been a hot research topic for many decades. The emission properties of porphyrins and their derivatives are based on the relaxation of excited states to ground states through photon emission. The porphyrin macrocycle displays two types of fluorescence, namely, strong singlet I fluorescence between 550–800 nm, and weak singlet II peak is observed between 400-and 550 nm. The simplified Jablonski diagram which explains excitation and relaxation processes is shown in Fig. 2. The singlet excited state of porphyrin has known to undergo vibrational relaxation and radiative decay like fluorescence and phosphorescence [3]. On excitation, irradiation energy lost due to heat is very negligible (quantum yield of intersystem crossing and fluorescence is ~ 95%). Hence porphyrin and their chelate complexes are found to be very useful as photosensitizers. Tuning of photophysical properties is possible by superficially modifying the structure of these macrocycles.

3 Porphyrin-Based Nanomaterials

Nanomaterials find extensive utility in medicine and pharmaceuticals owing to their specific mechanical, electrical, magnetic, and optical behavior which also renders increased bioavailability, lower toxicity, better efficacy, and miniaturization of diagnostics [4]. Judiciously designed porphyrin-based nanomaterials have

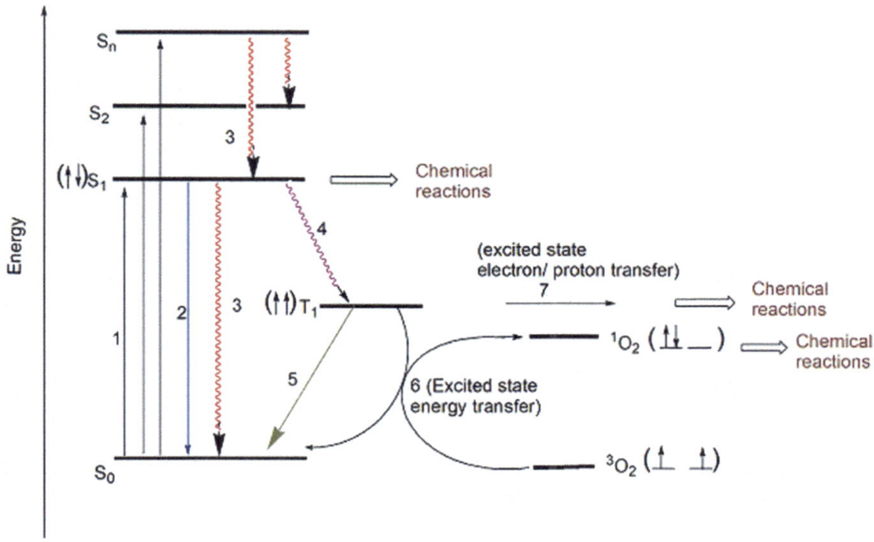

Fig. 2 Modified Jablonski: 1. Absorption of light; 2. Fluorescence; 3. Internal conversion; 4. Inter-systemcrossing 5. Phosphorescencec; 6. Singlet oxygen production (Type II photoprocess) 7. Hydrogen or electron transfer (Type 1 photoprocess)

been found to play a pivotal role in the varied treatment and diagnostic modalities like photodynamic and sonodynamic therapy, MRI, fluorescence, and photoacoustic imaging wherein, such macrocycles have found real clinical applications as shown in Fig. 3. Apart from biomedical applications, these nanostructures were found to have additional applications in the area of catalysis, light-harvesting, and electron

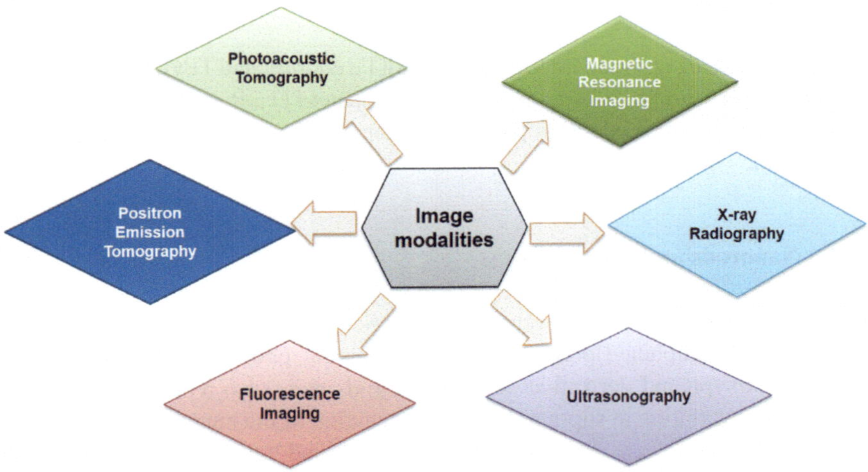

Fig. 3 Diagnostic Image modalities of porphyrin-based nanoparticles [28]

transfer devices, wherein the functionality can be decoupled from the morphology of nanostructures [5].

Porphyrin nanocomposites can be designed to selectively target diseased cells and can be tuned to incorporate longer tissue residence time, immune tolerance apart from making it pharmacodynamically favorable. Hydrophobicity of these macrocycles elevates its aggregation tendency, hence encapsulation of these nanoparticles into drug delivery vehicles improves the efficiency to work in aqueous media. Cancer theranostics, healing, and restoring damaged organs are other medical applications of these macrocyclic nanocomposites [6].

3.1 Classifications

Porphyrin-based nanomaterials are classified into the following categories according to their structural and electronic characteristics [7].

3.1.1 Nanoparticles

This is the first reported class of nanomaterials based on porphyrins which consist of colloidal suspension bearing dimensions of typical nanomaterials. Potential applications listed in literature for these nanomaterials are drug delivery and catalysis. Recent literature reports suggest surfactant mediated solvent mixture [8] and reprecipitation technique [9, 10] to prepare porphyrin nanoparticle-based supramolecular systems.

3.1.2 Nanosheets

In 2002, Milic et al., reported large nanosheets of Pd coordinated TPyP units as square planar nonameric porphyrin columnar stack aggregates [11]. Porphyrinic nanosheets of SnPyTriPP complex were prepared by Wang et al. in 2007 by reprecipitation method [12].

3.1.3 Nanorods and Nanorings

Initial report of porphyrin nanorods in 2003 by Schwab et al., based on 5,10,15,20-tetrakis(4-sulfonatophenyl)porphyrin with fascinating photoconductivity [13a-b]. Later a carboxyphenyl derivative derived nanorods were also reported that exhibited structural variation in supramolecular aggregation of the nanomaterial obtained wherein, nanorods were obtained in nitric acid and nanoring formation was induced with hydrochloric acid [14]. Few more of these nanorods were reported in following years by utilizing surface dewetting process [15] and sonication method

[16]. 5,10,15,20- tetrakis(4-pyridyl)porphyrin Zn (ZnTPyP) was encapsulated in fullerene and existed as hexagonal shaped nanorods and the supramolecular assembly exhibited photoinduced electron transfer between fullerene and porphyrin. These chromophoric nanorods used in photoelectrochemical cells displayed significant improvement in solar energy conversion [17].

3.1.4 Nanowires

The assembly of Sn-porphyrin complex forms dispersed nanowires whose dimensions and morphology of nanostructures can be dictated by the addition of amphiphilic surfactant Pluronic F127. The structures obtained depends mainly on the synthetic protocol and presence of surfactant which can give rise to nanowires to nanoplates apart from tuning its hydrophilicity and precipitation from aqueous medium [18].

3.1.5 Nanotubes and Fibers

Wang et al. reported that porphyrin units of opposite charges aggregate to form nanotubes. J-aggregates formed by ionic self assembly of two oppositely charged porphyrin units gave rise to nanotubular materials [19]. Further they assessed how the variation of the porphyrin structures influenced the morphology of nanostructures obtained by self assembly. Self assembly of oxoantimony complex of tetraphenylporphyrin with tetrasulfonatophenylporphyrin, produced nanofibers while that of Mo(V) complex of the dodecaphenylporphyrin gave nanotubes [20]. Photocatalytic behavior of these supramolecular nano structures were studied [21].

3.1.6 Nanosponges

Porphyrin-based nanosponges are nanoporous solids bearing molecular size cavities formed by a supramolecular network assembled by weak interactions stabilized by solvate molecules [22]. TPyP metal complexes with high thermal stability act as a 3D network of organic zeolite analogs having profound selectivity towards guest molecules. This work inspired researchers to come up with scaffolds like H_2TCPP and resorcinol-substituted porphyrins giving rise to self-assembling porphyrin-based ordered solids held together by peripheral hydrogen bonds and inner cavities being utilized for molecular recognition [23]. Later, Kojima et al. published the utility of porphyrins of bis- and tetraresorcinol substituted porphyrins used as selective sensing layer in Quartz Crystal Microbalance [24].

3.1.7 Nanochannels

Saddle distortion of porphyrin leads to the formation of nanochannel. Such compounds were first reported by Harada and Kojima [23]. Dodecaphenylporphyrin dihydrochloride salt were utilized then to selectively for guest molecules. Porphyrin dictation acts as electron-deficient host wherein, hydroquinone derivatives, act as guests [25].

3.2 Porphyrin-Based MOFs, COFs, APOPs, and HOFs

Porphyrin-based porous nanomaterials which can mimic photosynthesis have received great attention in photocatalysis owing to greater surface area, synthetically tunable scaffold, and promising photocatalytic features. Porous porphyrin-based nanomaterials are classified into four main classes, [26] mainly;

(a) Metal–organic frameworks (MOFs)
(b) Covalent-organic frameworks (COFs)
(c) Amorphous porous organic polymers (APOPs) and
(d) Hydrogen bonding-organic frameworks (HOFs).

3.3 Miscellaneous Porphyrin-Based Nanomaterials

Tetraphenylporphyrin (TPP) based supramolecular amphiphilic aggregates wherein TPP has cationic attachments in the periphery of the macrocycle and cation-π non-covalent interactions help them to be utilized as chemical sensors, especially for amines or olefins [27].

4 Porphyrin-Based Bioactive Nanomaterials

Porphyrin-based bioactive nanomaterials find utility in the following fields:

4.1 Nanotheranostics

Nanoparticles with therapeutic and diagnostic competencies are known as nanotheranostics. Such multifunctional nanoparticles show promising dual characteristics of being a therapeutic agent that can also track the efficacy after the drug is deployed.

Porphyrin-based macrocycles are used in advanced imaging techniques like functional Magnetic Resonance Imaging as MRI contrast agents and the systemic administration of these agents evokes adverse side effects, toxicity, and solubility issues, making it less bioavailable. Nanoparticles bear several advantages like tailoring the nanoparticles for preferential uptake in organs and their utility for diagnostics which allows targeted and controlled drug delivery to diseased tissues [5].

4.2 Diagnostic Imaging

Porphyrin-based nanoparticles find its applications in clinical imaging modalities like x-ray imaging and computed tomography (CT), ultrasound imaging, magnetic resonance imaging, positron emission tomography (PET), and single photon emission computed tomography (SPECT) as shown in Fig. 3. These materials are highly sought after due to inherent properties like lower cytotoxicity in the dark, tunable photophysical properties, better tumor selectivity and intake, possible variety of structural modifications.

MRI was first introduced in 1977, and it remains even now as one of the most reliable non-invasive diagnostic tools. Porphyrins in blood, heme serves as the bright red contrast agent indicates injury sites apart from being the sole oxygen carrier in our body. Porphyrins serve as the first exogenous class of compound utilized as intrinsic agents utilized in multimodal theranostics owing to its ability to act as a photosensitizer and its near IR fluorescence emission. This macrocyclic ligand and its chelates with a wide range of metals are being used as diagnostic fluorescence agents since 1924, PET ligand since 1951, and MRI contrast agent since 1987. Multimodality and better biocompatibility make these class of compounds much sought after as theranostic agents [29].

4.3 Photothermal Therapy (PTT)

PTT involves photothermal sensitizers and laser light to thermally ablate tumor cells. Porphysomes are self-assembled organic phospholipid-porphyrin conjugate with various biophotonic activities [30]. Different metals like Gd, Mn^{3+} ions incorporated into porphysomes that imparts MRI sensitivity while enhancing unprecedented photostability apart from bearing high photothermal efficiency and low toxicity. Researchers have also prepared expanded porphyrin-based MRI contrast agents, especially texaphyrins and their metal complexes have garnered enough attention as efficacious agents [31, 32]. Dual diagnostic functionality, namely dual imaging diagnostic tools like introducing nuclear medicine ^{64}Cu with chelated porphyrin-phospholipid liposomes was introduced by Lou et al. in 2017, wherein PET imaging was complemented by fluorescence imaging to diagnose and treat the disease using one drug [33]. Another popular concept developed during the last decade is that porphyrin-conjugates play a major role in imparting dual therapeutic modalities, namely PTT and PDT by synergistically functioning to bring in better selectivity, biocompatibility, and efficacy to the drug molecule [34].

5 Porphyrin-Based Nanohybrids

Porphyrin-based nanohybrids are classified under the category organic molecule-coated nanohybrids (OMCNHs). Literature suggests ionic self assembly, reprecipitation, Microwave irradiation, coordination polymerization, continuous flow methods to synthesize nanohybrids [35].

5.1 Design/Synthesis/Preparation/Self Assembly

Room-temperature nanoscale precipitation, microwave-assisted synthesis, solvothermal method, and continuous flow synthesis are being utilized along with methodologies to prepare porphyrin core and a variety of substitutions to incorporate biomolecules, inorganic compounds, etc. Ionic self assembly is employed to make nanotubes containing J-aggregates, while phase transfer mediated self assembly gave rise to nanofibers [5]. ZnT(4-Py)P was prepared by reprecipitation reaction which resulted in hollow hexagonal nanoprisms. SnPyTriPP was formed as nanosheets while SnDiPyP which is a dipyridyl porphyrin exhibited nanorod-like morphology. The effect of temperature and microwave was also studied in detail. Coordination polymerization was also observed in specific cases which resulted in the formation of nanospheres. Solvothermal synthesis can give rise to nanomaterials with controlled particle size and shape. Microwave-assisted synthesis and continuous flow method offer fewer disadvantages, shorter reaction time and mild reaction conditions, and better control on the size of the nanoparticle [5]. Self-assembled nanohybrids can be designed in different morphology, fabricated to different sizes which can give rise to a variety of applications. Metal–organic frameworks (MOFs) along with porous coordination polymers (POPs) can be tuned into multifunctional nanoplatforms with potential utility in multiple therapeutic and diagnostic fields [36]. Porphyrin -hydrogel with four-arm PEG-PCL copolymer(II) is utilized by Dong et al. as a dual fluorescent drug delivery system [37]. Cyclodextrin conjugated with Zn–porphyrin was utilized for combining chemotherapy and immunotherapy at cellular level [38] while Gardella et al. synthesized conjugated graphite and poly(L-lactide) for drug delivery [39]. Silica coated liposomes consisting of alkylchains, lipid molecules with triethoxysilane, and a connector molecule are known as cerasomes. Porphyrin-containing cerasomes prepared by sol–gel methodology by Liang et al. were found to have enhanced biodistribution, stability, and tumor selectivity [40].

5.2 *Characterization of Nanohybrids*

Spectroscopic and microscopic Characterization of nanohybrids can be performed for porphyrin-based nanohybrids to determine the structure and size of the nanoparticle formed which heavily depends on the reaction conditions. UV–vis spectra, Atomic force microscopy (AFM) data, Transmission electron microscopy (TEM), fluorescence spectra, and resonance light scattering, and if the final product is chiral, then circular dichroism is also employed to determine the structure and morphology. Raman and SERS spectral data can also be recorded [5].

5.3 *Importance of Porphyrin-Based Nanohybrids in Biomedical Applications*

Functionalized nanohybrids find a variety of applications in the field of medicine (treatment and diagnostics) and catalysis. Inherent photochemical and photophysical properties exhibited by the porphyrins can be controlled and tuned according to the requirements making it a worthy class of compounds that can be developed further into a clinical candidate. The structure of porphyrin can be peripherally and core modified to incorporate biomolecules which can impart better biocompatibility and toxicity profile. It can also be expanded and thus improved upon to modify and incorporate a variety of metals in its core and can also be utilized as drug carriers and contrast agents which can impart di/tri modality to the treatment.

6 Applications Porphyrin-Based Nanohybrids

6.1 *Bio- Imaging*

Non-invasive medical imaging technologies are crucial in today's world to accelerate the progression of drug design and development. Porphyrin-based biomaterials exhibit preferential accumulation in tumor cells. Radiolabeled metalloporphyrins contain positron emitting nuclei like^{13}N, ^{18}F, ^{11}C, ^{64}Cu, ^{52} Fe, ^{111}In, and ^{67} Ga [41]. In 2002, Murugesan et al. demonstrated the ability of a specific porphyrin like macrocyclic ligand, Chlorin, to detect and monitor the progression of cancer [42].

6.1.1 Diagnostic Imaging Agents

^{68} Ga-fluorinated porphyrin complex as a PET ligand was reported by Fazaeli et al. [43]. and Gadolinium motexafin was synthesized by Thomas et al. and used for treatment of brain cancer [44]. Tetranuclear gadolinium complexed with tetraphenylporphyrin was used as theranostic agent wherein, it does a dual job of a photosensitizer for PDT and MRI contrast agent [45]. A water soluble porphyrin was reported to act as a dual agent for MRI and fluorescence zinc sensing [46].

Zackrisson et al. has given an exhaustive account of photoacoustic imaging (PAI) and molecules utilized in such translational diagnostic imaging without ionizing radiation involving real time non-invasive imaging technique which extracts functional and molecular information of deep-seated tissues with high spatial resolution [47, 48]. Exogenous imaging or contrast agents like porphyrins find utility as good photoacoustic agents. In 2017, Banala et al. [49] utilized porphyrins bearing quinones as contrast agents and its Zn complex for PAI. Wu et al. [50] utilized carbon nanodots implanted on porphyrin (PNDS) for PAI, PDT, and In vivo breast cancer ablation. He utilized tetrakis (4-aminophenyl)porphyrin (TAPP) and citric acid to form carbon nanodots [48].

6.1.2 Photodynamic Therapy (PDT)

Porphyrins have the unique ability to accumulate or selectively prefer tumor cells bearing lower pH values arising due to their hyperproliferating ability. This characteristic nature of the macrocycle makes them excellent photosensitizers, which can also act as transporting agents for other anti-cancer drugs into tumor tissues creating synergistic treatment by PDT and chemotherapy [51]. Selective detection and treatment modality offered by the engineered porphyrin-based nanoparticles which can efficiently destroy tumor cells, and microbes in the body has been acquiring attention and momentum in the area of tumor therapy and antimicrobial chemotherapy [52]. Representative examples of some tetraphenyl-based porphyrins are depicted in Fig. 4. Clinical challenges posed by porphyrin-derived macrocycles in photodynamic therapy were addressed by supramolecular porphyrin structures [53]. Scientists have synthesized three generations of porphyrin-based photosensitizers (PBP). Each generation addressed the drawbacks of the previous generation like longer half-life, skin accumulation and subsequent sensitivity, poor pharmacokinetic features and current set of sensitizers have increased tumor cell specificity by the help of structural modifications including moieties like targeting biomolecules, nanomaterials (Quantum Dots such CuInS/ZnS and Au), and other moieties like SPIONS [54–58].

Photodynamic therapy (PDT) utilized to kill microorganisms is referred to as photodynamic antimicrobial chemotherapy (PACT). The singlet oxygen generated by porphyrin-based photosensitizers on irradiation gives rise to its antimicrobial activity. Inherent flexibility of hydrogels to be molded to any required shapes find utility around wounds. Tissue engineering, drug delivery, PACT, and PDT are some of the applications for which porphyrin-based hydrogels [2, 29]. In 2009, cross linked

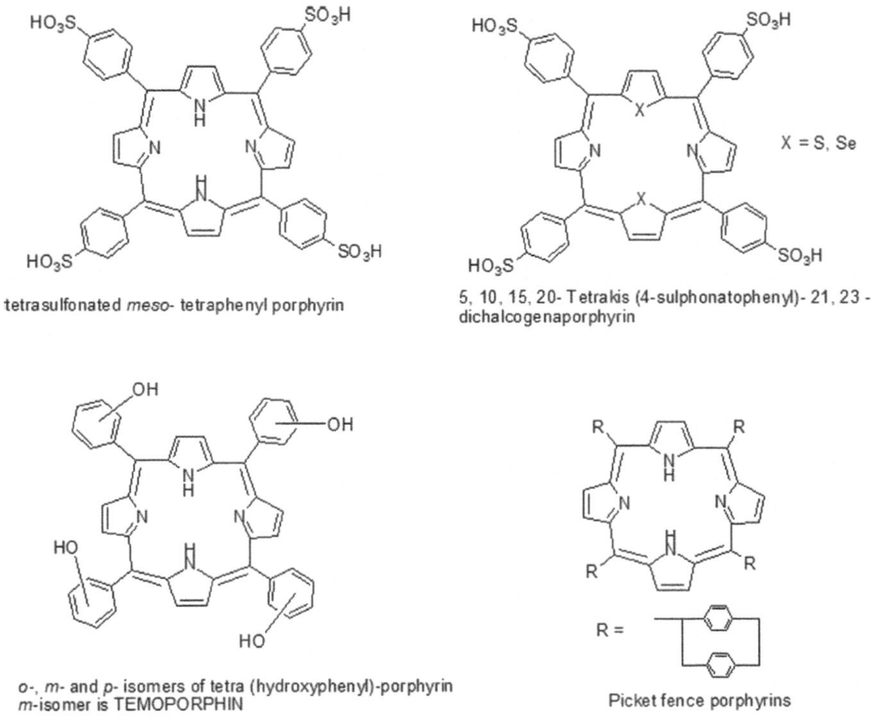

Fig. 4 Some tetraphenylporphyrin based photosensitizers

PVA hydrogel loaded with meso-tetra (N-methyl-4-pyridyl) porphine tetratosylate was utilized as antimicrobial agent, and another report of porphyrin-based hydrogel developed for the treatment of intraocular lens-associated infectious endophthalmitis. Another cationic porphyrin tetrakis-meso(4-N-methylpyridyl) porphyrin in conjugation with 2-(hydroxyethyl)methacrylate and methacrylic acid showed antibacterial activity against S. epidermidis [59, 60]. Table 1 gives us the current list of photosensitizers that are clinically approved.

6.1.3 Drug Delivery

Drug delivery is an important area in disease treatment and codelivery of drug and bioactive molecules is an important selective treatment of cancer that helps in target-specific release of medicine. Porphyrin-based metal–organic frameworks with porous surfaces are endowed with tunable functionality, better toxicity profile, better biodegradability, biocompatibility and can release the drug in a controlled fashion [36, 51]. Nanomaterials are held in the respective order by the hydrophobic interaction, hydrogen bonding, Vander waal's forces allows sustained release as a stimuli response.

Table 1 List of porphyrin-based clinical agents

Photosensitizer	Country Approved For Use	Disease Condition/ Modality	Trade Name
Hematoporphyrin	Worldwide	Lung, esophageal, bile duct, bladder, brain, ovarian, and cervical cancer PDT, myopic maculopathy	Photofrin
5 -ALA	US & EU	Actinic keratosis, basal-cell carcinoma, head and neck, and gynecological cancer PDT. Brain, head and neck, bladder cancer imaging	Levulan
5-ALA methyl ester (MAL)	Worldwide	Actinic keratosis, Bowen's disease, basal-cell carcinoma PDT. Bladder cancer diagnosis	Metric
Verteporfin	US, Canada, EU	Age-related macular degeneration, pathologic myopia, histoplasmosis	Visudyne
Talaporfin	Japan	Lung cancer PDT	NPe6
Temoporfin	EU	head and neck, prostate, pancreatic cancer PDT	Foscan

6.1.4 Sonodynamic Therapy and Radiotherapy of Cancer:

Porphysome research yielded enhanced drug delivery using sonoactive porphyrin-based liposomes conjugated to doxorubicin (Dox) for anti-cancer treatments first reported by Wang et al. [61] The release mechanism of sonoactivatable doxorubicin (Dox) loaded porphyrin-phospholipid-liposome for anti-cancer treatment is shown in Fig. 5. Manganese(III) complexed TPPS was conjugated with PMMA nanoparticles with anionic surface was utilized as dual active sonodynamic and MRI contrast agent [62].

Porphyrin-based dual functional (radiotherapy and radiodynamic therapy) therapeutic agents for cancer [63, 64]. Porphyrin-based nanoparticles and MOFs are utilized as radiosensitizers have been reviewed exhaustively in various reports [65, 66].

6.1.5 Nano Biosensors

The planar aromatic macrocyclic scaffold of porphyrins is an adaptable platform for peripheral adornment with functional groups that can offer non-covalent interactions

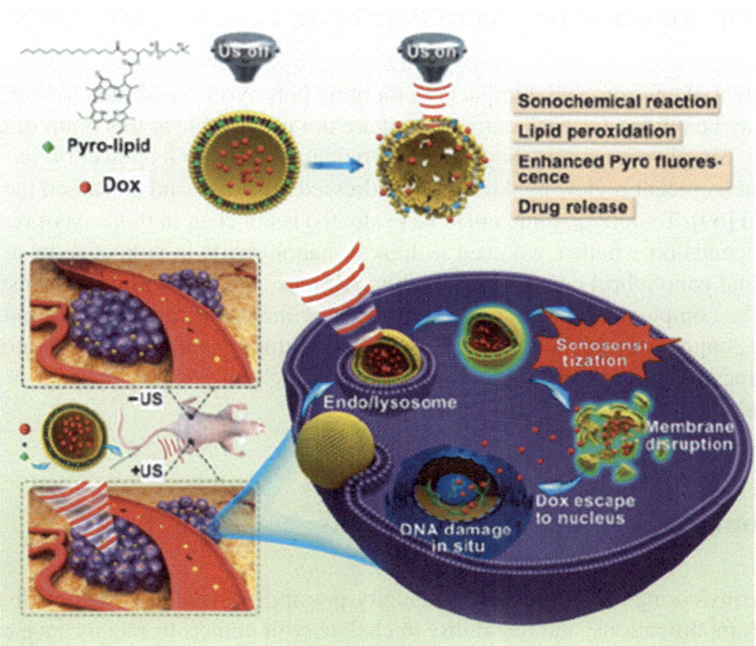

Fig. 5 Star-shaped polymer conjugated porphyrin material for cellular drug delivery applications [61]

like hydrogen bond, π–π stacking interactions, electrostatic and coordinative interactions which are beneficial in the formation of supramolecular assemblies. The chemical information stored in such well-defined porphyrin assemblies can be transferred to solid systems with tailored properties that can be exploited for making thin films of novel materials which find their utility in the area of solid-state sensors and molecular recognition. Chiral appendages on the meso and β positions of the porphyrins resulted in supramolecular chirality which in turn gave rise to novel chiroptical properties which elevated its molecular recognition property to the next level. The flexibility in synthesizing modified porphyrins (core, expanded, β-substituted, and isomers) which can potentially create ordered supramolecular structures coupled with its ability to incorporate chirality into the system in multiple levels makes it a wonderful framework for preparing rationally designed nanostructures. Such characteristic properties have been found to open up better sensitivity and selectivity for these nanomaterials which can potentially offer superior supramolecular assemblies for chiral discrimination, functional devices like chemical sensors [68, 69].

7 Environmental Impact of Porphyrin-Based Nanohybrids

The potential environmental impact of emerging porphyrin-based nanohybrid materials may be unique for each complex and are not predictable at this point of development. Simple parameters may not be sufficient to evaluate its power on our environment. A recent review by Aich et al. addressed this topic and discussed the issue in detail [69]. Toxicity parameters to be evaluated is not clear in this area of research as it depends on whether we need to look at nanohybrids in collective or as their individual nanohybrid component stability. All these questions remain unanswered while the complexity of nanohybrid structures is increasing as we are introducing infinite conjugation and functionalization of two or more nanomaterials to introduce multifunctionality into these materials. A systematic risk assessment study of these materials is the need of the hour.

8 Conclusions and Future Prospects

Porphyrins, owing to its extended aromaticity, possibility to create structural diversity through modifications, and the ability to chelate with numerous metals have earned a unique place in the area of biomedicine. Porphyrin-based nanomaterials exhibit fascinating photochemical, photoredox, and photophysical properties that make them highly researched multimodal theranostic agents with enormous possibilities in the area of biomedicine. It is used in the multimodal treatment and multidiagnostic area of cancer, inflammation, microbes apart from being used in other material science avenues. This chapter relies mainly on how these macrocycles and their hybrids with advanced properties can be advantageous for fabricating innovative, synergistic, and functional materials for clinical applications. Efforts around the globe are focusing on making them a more effective and selective therapeutic and diagnostic agent. We need improvement on selectivity, biocompatibility, safety profile, stability against demetallation, off-target phototoxicity, high singlet quantum yield, enhancement of relaxivity values, better absorption in the visible-NIR region of spectra for developing porphyrin-based nanohybrids into the superior therapeutic and diagnostic agent. Apart from that, we need more scientific efforts to bring the aforementioned characteristics into these macrocycles so that it can be useful as theranostic agents in multiple disease areas. Only a few or negligible efforts are concentrated on the environmental compatibility of newer nanohybrid materials, which needs more attention from the research community along with the improvement of such nanohybrids for extensive use as theranostic agents.

References

1. Fathi P, Pan D (2020) Current trends in pyrrole and porphyrin-derived nanoscale materials for biomedical applications. Nanomedicine 15(25):2493–2515. https://doi.org/10.2217/nnm-2020-0125
2. Zhang Q, He J, Yu W, Li Y, Liu Z, Zhou B, Liu Y (2020) A promising anticancer drug: a photosensitizer based on the porphyrin skeleton. RSC Med Chem. 11(4):427–437. https://doi.org/10.1039/c9md00558g
3. Valiscek Z, Kiss MP, Fodor MA, Imran M, Horvath O (2017) Formation, Photophysics, photochemistry and quantum chemistry of out of plane metalloporphyrins. Hung J Ind Chem 45(1):29–36. https://doi.org/10.1515/hjic-2017-0006
4. Damodharan J (2020) Nanomaterials in medicine—an overview. Mater Today: Proc. In press https://doi.org/10.1016/j.matpr.2020.05.380
5. Medforth CJ, Wang Z, Martin KE, Song Y, Jacobsen JL, Shelnutt JA (2009) Self-assembled porphyrin nanostructures. Chem Commun 7261–7277. https://doi.org/10.1039/b914432c
6. Rabiee N, Tavakkoli Yaraki MT, Garakani SM, Garakani SM, Ahmadi S, Lajevardi A, Bagherzadeh M, Rabiee M, Tayebi L, Tahriri M, Hamblin MR (2020) Recent advances in porphyrin-based nanocomposites for effective targeted imaging and therapy. Biomaterials 232:119707. https://doi.org/10.1016/j.biomaterials.2019.119707
7. Monti D, Nardis S, Stefanelli M, Paolesse R, Natale CI, D'Amico A (2009) Porphyrin-based nanostructures for sensing applications. J Sens 10. https://doi.org/10.1155/2009/856053
8. Drain CM, Smeureanu G, Patel S, Gong X, Garno J, Arijeloye J (2006) Porphyrin nanoparticles as supramolecular systems. New J Chem 30(12):1834–1843. https://doi.org/10.1039/b607289e
9. Sandanayaka ASD, Araki Y, Wada T, Hasobe T (2008) Structural and photophysical properties of self-assembled porphyrin nanoassemblies organized by ethylene glycol derivatives. J Phys Chem C 112(49):19209–19216. https://doi.org/10.1021/jp805202y
10. Perepogu AK, Bangal PR (2008) Preparation and characterization of free-standing pure porphyrin nanoparticles. J Chem Sci 120(5):485–491. https://doi.org/10.1007/s12039-008-0076-2
11. Milic TN, Chi N, Yablon DG, Flynn GW, Batteas JD, Drain CM (2002) Controlled hierarchical self-assembly and deposition of nanoscale photonic materials. Angew Chem Int Ed 41(12):2117–2119 PMID: 19746614
12. Wang Z, Li Z, Medforth CJ, Shelnutt JA (2007) Self assembly and self-metallization of porphyrin nanosheets. J Am Chem Soc 129(9):2440–2441. https://doi.org/10.1021/ja068250o
13. (a) Schwab AD, Smith DE, Rich CS, Young ER, Smith WF, de Paula JC (2003) Porphyrin nanorods. J Phys Chem B 107(41):11339–11345. https://doi.org/10.1021/jp035569b. (b) Schwab AD, Smith DE, Bond-Watts B, et al. (2004) Photoconductivity of self-assembled porphyrin nanorods. Nano Lett 4(7):1261–1265. https://doi.org/10.1021/nl049421v
14. Doan SC, Shanmugham S, Aston DE, McHale JL (2005) Counterion dependent dye aggregates: nanorods and nanorings of tetra(p-carboxyphenyl)porphyrin. J Am Chem Soc 127(16):5885–5892. https://doi.org/10.1021/ja0430651
15. Lensen MC, Takazawa K, Elemans JAAW, et al. (2004) Aided self-assembly of porphyrin nanoaggregates into ring-shaped architectures. Chemistry 10(4):831–839. https://doi.org/10.1002/chem.200305436
16. Hasobe T, Oki H, Sandanayaka ASD, Murata H (2008) Sonication-assisted supramolecular nanorods of mesodiaryl-substituted porphyrins. Chem Commun 6:724–726. https://doi.org/10.1039/B713971C
17. Hasobe T, Sandanayaka ASD, Wada T, Araki Y (2008) Fullerene-encapsulated porphyrin hexagonal nanorods. An anisotropic donor-acceptor composite for efficient photoinduced electron transfer and light energy conversion. Chem Commun 29:3372–3374. https://doi.org/10.1039/B806748A
18. Lee SJ, Hupp JT, Nguyen ST (2008) Growth of narrowly dispersed porphyrin nanowires and their hierarchical assembly into macroscopic columns. J Am Chem Soc 130(30):9632–9633. https://doi.org/10.1021/ja801733t

19. Wang Z, Medforth CJ, Shelnutt JA (2004) Porphyrin nanotubes by ionic self-assembly. J Am Chem Soc 126(49):15954–15955. https://doi.org/10.1021/ja045068j
20. Wang Z, Medforth CJ, Shelnutt JA (2006) Porphyrin nanofiber bundles from phase-transfer ionic self-assembly and their photocatalytic self-metallization. Advan Mater 18(19):2557–2560. https://doi.org/10.1002/adma.200600539
21. Kojima T, Harada R, Nakanishi T, Kaneko K, Fukuzumi S (2007) Porphyrin nanotubes based on self-assembly of Mo(V)-dodecaphenylporphyrin complexes and inclusion of Mo-oxo clusters: synthesis and characterization by Xray crystallography and transmission electron microscopy. Chem Mater 19(1):51–58. https://doi.org/10.1021/cm062031k
22. Chou J-H, Kosal ME, Nalwa HS, Rakow NA, Suslick KS (2000) Applications of porphyrins and metalloporphyrins to materials chemistry. In: Kadish KM, Smith KM, Guilard R (eds) The porphyrin handbook, vol 4. Academic Press, San Diego, California, USA, pp 43–131
23. Ariga K, Endo K, Aoyama Y, Okahata Y (2000) QCM analyses on adsorption of gaseous guests to cast films of porphyrinresorcinol derivatives. Colloids Surf A 169(1–3):177–186. https://doi.org/10.1016/S0927-7757(00)00434-9
24. Kojima T, Nakanishi T, Honda T, Fkuzumi S (2009) Photoinduced electron transfer in supramolecular assemblies involving saddle-distorted porphyrins and phthalocyanines. J Porphyrins Phthalocyanines 13:14–21. https://doi.org/10.1142/S1088424609000164
25. Harada R, Kojima T (2005) A porphyrin nanochannel: formation of cationic channels by a protonated saddle-distorted porphyrin and its inclusion behaviour. Chem Commun 716–718. https://doi.org/10.1039/B414483J
26. Monti D, Venanzi M, Russo M et al (2004) Spontaneous deposition of amphiphilised porphyrin ordered film on glass. New J Chem 28:1123–1128. https://doi.org/10.1142/9789812701770_0006
27. Jin L, Lv S, Miao Y, Liu D, Song F (2021) Recent development of porous porphyrin-based nanomaterials for photocatalysis. ChemCatChem 13:140–152. https://doi.org/10.1002/cctc.202001179
28. Dini F, Martinelli E, Pomarico G et al (2009) Chemical sensitivity of self-assembled porphyrin nano-aggregates. Nanotechnology 20(5):8. https://doi.org/10.1088/0957-4484/20/5/055502
29. Tsolekile N, Nelana S, Oluwafemi OS (2019) Porphyrin as diagnostic and therapeutic agent. Molecules 24:2669. https://doi.org/10.3390/molecules24142669
30. MacDonald TD, Zheng G (2014) Porphysome nanoparticles: tailoring treatments with nature's pigments. Photon Lasers Med 3(3):183–191. https://doi.org/10.1515/plm-2013-0059
31. Lovell JF, Jin CS, Huynh E, Jin H, Kim C, Rubinstein JL, Chan WC, Cao W, Wang LV, Zheng G (2011) Porphysome nanovesicles generated by porphyrin bilayers for use as multimodal biophotonic contrast agents. Nat Mater 10(4):324–332. https://doi.org/10.1038/nmat2986
32. Lin Y, Zhou T, Bai R, Xie Y (2020) Chemical approaches for the enhancement of porphyrin skeleton-based photodynamic therapy. J Enzym Inhib Med Chem 35(1):1080–1099. https://doi.org/10.1080/14756366.2020.1755669
33. Luo D, Goel S, Liu H, Carter KA, Jiang D, Geng J, Kutyreff CJ, Engle JW, Huang W, Shao S et al (2017) Intrabilayer ^{64}Cu labeling of photoactivatable doxorubicin-loaded stealth liposomes. ACS Nano 11:12482–12491. https://doi.org/10.1021/acsnano.7b06578
34. Zeng J, Yang W, Shi D, Li X, Zhang H, Chen M (2018) Porphyrin derivative conjugated with gold nanoparticles for dual-modality photodynamic and photothermal therapies in vitro. ACS Biomater Sci Eng 4:963–972
35. Zhang K, Zhao N, Xu F-J (2020) Self-assembled organic/metal ion nanohybrids for theranostics. VIEW 1:e17. https://doi.org/10.1002/viw2.17
36. Huxford RC, Della Rocca J, Lin W (2010) Metal–organic frameworks as potential drug carriers. Curr Opin Chem Biol 14:262–268. https://doi.org/10.1016/j.cbpa.2009.12.012
37. Dong X, Chen H, Qin J, Wei C, Liang J, Liu T, Lv F (2017) Thermosensitive porphyrin-incorporated hydrogel with four-arm PEG-PCL copolymer(II): doxorubicin loaded hydrogel as a dual fluorescent drug delivery system for simultaneous imaging tracking in vivo. Drug Deliv 24:641–650. https://doi.org/10.1080/10717544.2017.1289570

38. Kejík Z, Bříza T, Králová J, Poučková P, Král A, Martásek P, Král V (2011) Coordination conjugates of therapeutic proteins with drug carriers: a new approach for versatile advanced drug delivery. Bioorgan Med Chem Lett 21:5514–5520. https://doi.org/10.1016/j.bmcl.2011.06.101
39. Gardella L, Colonna S, Fina A, Monticelli O (2016) A novel electrostimulated drug delivery system based on PLLA composites exploiting the multiple functions of graphite nanoplatelets. ACS Appl Mater Interfaces 8:24909–24917. https://doi.org/10.1021/acsami.6b08808
40. Liang X, Li X, Yue X, Dai Z (2011) Conjugation of porphyrin to nanohybrid cerasomes for photodynamic diagnosis and therapy of cancer. Angew Chem Int Ed 50(49):11622–11627. https://doi.org/10.1002/anie.201103557
41. Holland JP, Williamson MJ, Lewis JS (2010) Unconventional nuclides for radiopharmaceuticals. Mol Imaging 9:1–20
42. Murugesan S, Shetty SJ, Srivastava TS, Samuel AM, Noronha OPD (2002) Preparation and biological evaluation of the new chlorin photosensitizer T3, 4BCPC for detection and treatment of tumors. J Photochem Photobiol B Biol 68:33–38. https://doi.org/10.1016/S1011-1344(02)00329-9
43. Fazaeli Y, Jalilian AR, Amini MM, Ardaneh K, Rahiminejad A, Bolourinovin F, Majdabadi A (2012) Development of a ^{68}Ga-fluorinated porphyrin complex as a possible PET imaging agent. Nucl Med Mol Imaging 46:20–26. https://doi.org/10.1007/s13139-011-0109-510.2147/nano.2007.2.1.79
44. Thomas SR, Khuntia D (2007) Motexafin gadolinium injection for the treatment of brain metastases in patients with non-small cell lung cancer. Int J Nanomed 2:79. https://doi.org/10.2147/nano.2007.2.1.79
45. Luo J, Chen LF, Hu P, Chen ZN (2014) Tetranuclear gadolinium(III) porphyrin complex as a theranostic agent for multimodal imaging and photodynamic therapy. Inorg Chem 53:4184–4191. https://doi.org/10.1021/ic500238s
46. Zhang XA, Lovejoy KS, Jasanoff A, Lippard SJ (2007) Water-soluble porphyrins as a dual-function molecular imaging platform for MRI and fluorescence zinc sensing. Proc Natl Acad Sci USA 104:10780–10785. https://doi.org/10.1073/pnas.0702393104
47. Zackrisson S, van de Ven SMWY, Gambhir SS (2014) Light in and sound out: emerging translational strategies for photoacoustic imaging. Cancer Res 74(4):979–1004. https://doi.org/10.1158/0008-5472.CAN-13-2387
48. Steinberg I, Huland DM, Vermesh O, Frostig HE, Tummers WS, Gambhir SS (2019) Photoacoustic clinical imaging. Photoacoustics 14:77–98. https://doi.org/10.1016/j.pacs.2019.05.001
49. Banala S, Fokong S, Brand C, Andreou C, Kräutler B, Rueping M, Kiessling F (2017) Quinone-fused porphyrins as contrast agents for photoacoustic imaging. Chem Sci 8:6176–6181. https://doi.org/10.1039/C7SC01369H
50. Wu F, Su H, Cai Y, Wong W, Jiang W, Zhu X (2018) Porphyrin-Implanted Carbon Nanodots for Photoacoustic Imaging and in Vivo Breast Cancer Ablation. ACS Appl Bio Mater 1:110–117. https://doi.org/10.1021/acsabm.8b00029
51. Huang H, Song W, Rieffel J, Lovell JF (2015) Emerging applications of porphyrins in photomedicine. Front Phys 3:23. https://doi.org/10.3389/fphy.2015.00023
52. Cheng Q, Li ZH, Sun YX (2019) Controlled synthesis of a core-shell nanohybrid for effective multimodal image-guided combined photothermal/photodynamic therapy of tumors. NPG Asia Mater 11:63. https://doi.org/10.1038/s41427-019-0164-4
53. Rajora MA, Lou JWH, Zheng G (2017) Advancing porphyrin's biomedical utility via supramolecular chemistry. Chem Soc Rev 46:6433–6469. https://doi.org/10.1039/C7CS00525C
54. Feng Y, Liu L, Hu S, Liu Y, Ren Y, Zhang X (2016) Forster resonance energy transfer properties of a new type of near-infrared excitation PDT photosensitizer: $CuInS_2$/ZnS quantum dots-5-aminolevulinic acid conjugates. RSC Adv 6:55568–55576. https://doi.org/10.1039/C6RA06937A

55. Theodossiou TA, GonÅalves AR, Yannakopoulou K, Skarpen E, Berg K (2015) Photochemical Internalization of tamoxifens transported by a "Trojan-Horse" nanoconjugate into breast-cancer cell lines. Angew Chem Int Ed 54:1–6. https://doi.org/10.1002/anie.201500183
56. Bera K, Maiti S, Maity M, Mandal C, Maiti NC (2018) Porphyrin−gold nanomaterial for efficient drug delivery to cancerous cells. ACS Omega 3:4602–4619. https://doi.org/10.1021/acsomega.8b00419
57. Arja K, Elgland M, Appelqvist H, Konradsson P, Lindgren M, Nilsson KPR (2018) Synthesis and characterization of novel fluoroglycosylated porphyrins that can be utilized as theranostic agents. ChemistryOpen 7:495–503. https://doi.org/10.1002/open.201800020
58. Fakayode OJ, Tsolekile N, Songca SP, Oluwafemi OS (2018) Applications of functionalized nanomaterials in photodynamic therapy. Biophys Rev 10:49–67. https://doi.org/10.1007/s12551-017-0383-2
59. Bonnett R, Buckley DG, Burrow T, Galia ABB, Saville B, Songca SP (1993) Photobactericidal materials based on porphyrins and phthalocyanines. J Mater Chem 3:323–324. https://doi.org/10.1039/JM9930300323
60. Parsons C, McCoy CP, Gorman SP, Jones DS, Bell SEJ, Brady C et al (2009) Anti-infective photodynamic biomaterials for the prevention of intraocular lensassociated infectious endophthalmitis. Biomaterials 30:597–602. https://doi.org/10.1016/j.biomaterials.2008.10.015
61. Wang X, Yan F, Liu X, Wang P, Shao S, Sun Y, Zheng H (2018) Enhanced drug delivery using sonoactivatable liposomes with membrane-embedded porphyrins. J Control Release 286:358–368. https://doi.org/10.1016/j.jconrel.2018.07.048
62. Schaue D, McBride WH (2015) Opportunities and challenges of radiotherapy for treating cancer. Nat Rev Clin Oncol 12(9):527–540. https://doi.org/10.1038/nrclinonc.2015.120
63. Thariat J, Hannoun-Levi J-M, Sun Myint A, Vuong T, Gérard J-P (2013) Past, present, and future of radiotherapy for the benefit of patients. Nat Rev Clin Oncol 10(1):52–60. https://doi.org/10.1038/nrclinonc.2012.203
64. Chen W, Zhang J (2006) Using nanoparticles to enable simultaneous radiation and photodynamic therapies for cancer treatment. J Nanosci Nanotechnol 6(4):1159–1166. https://doi.org/10.1166/jnn.2006.327
65. Luksiene Z, Juzenas P, Moan J (2006) Radiosensitization of tumours by porphyrins. Cancer Lett 235(1):40–47 [PubMed: 15946797]
66. Varchi G, Foglietta F, Canaparo R, Ballestri M, Arena F, Sotgiu G, Guerrini A, Nanni C, Cicoria G, Cravotto G (2015) Engineered porphyrin loaded core-shell nanoparticles for selective sonodynamic anticancer treatment. Nanomedicine 10(23):3483–3494 [PubMed: 26607137]
67. Balaban TS (2005) Tailoring porphyrins and chlorins for self-assembly in biomimetic artificial antenna systems. Acc Chem Res 38(8):612–623. https://doi.org/10.1021/ar040211z
68. Monti D, Nardis S, Stefanelli M, Paolesse R, Natale CI, D'Amico A (2009) Porphyrin-based nanostructures for sensing applications. J Sens 10 https://doi.org/10.1155/2009/856053
69. Aich N, Plazas-Tuttle J, Lead JR, Saleh NB (2014) A critical review of nanohybrids: synthesis, applications and environmental implications. Environ Chem 11:609–623. https://doi.org/10.1071/EN14127

Chapter 14
Revealing Glycobiology by Quantum Dots Conjugated to Lectins or "Borono-Lectins"

Weslley F. Oliveira, Camila A. P. Monteiro, Cássia R. A. Cunha, Carinna N. Lima, Mariana P. Cabrera, Beate S. Santos, Luana C. B. B. Coelho, Maria T. S. Correia, Paulo E. Cabral Filho, and Adriana Fontes

1 Introduction

Nanostructures have been increasingly used for biomedical applications, including drug delivery, biosensing, and image-guided technologies [1]. Indeed, nanoparticles (NPs) have been also applied to assist the elucidation of many biological processes, revealing the interaction of biomolecules at the cellular level or even in whole organisms [2]. Overall, many advances in the biomedical area have been supported by

W. F. Oliveira · L. C. B. B. Coelho · M. T. S. Correia
Departamento de Bioquímica, Centro de Biociências, Universidade Federal de Pernambuco—UFPE, Recife, PE, Brazil

C. A. P. Monteiro · C. N. Lima · P. E. Cabral Filho · A. Fontes (✉)
Departamento de Biofísica e Radiobiologia, Centro de Biociências, Universidade Federal de Pernambuco—UFPE, Recife, PE, Brazil
e-mail: adriana.fontes@ufpe.br

P. E. Cabral Filho
e-mail: paulo.euzebio@ufpe.br

C. R. A. Cunha
Laboratório Federal de Defesa Agropecuária em Pernambuco, Recife, PE, Brazil

M. P. Cabrera
Departamento de Química Fundamental, Centro de Ciências Exatas e da Natureza, Universidade Federal de Pernambuco—UFPE, Recife, PE, Brazil

B. S. Santos
Departamento de Ciências Farmacêuticas, Centro de Ciências da Saúde, Universidade Federal de Pernambuco—UFPE, Recife, PE, Brazil

W. F. Oliveira · C. A. P. Monteiro · C. R. A. Cunha · C. N. Lima · M. P. Cabrera · B. S. Santos · L. C. B. B. Coelho · M. T. S. Correia · P. E. Cabral Filho · A. Fontes
Biomedical Nanotechnology Research Group, UFPE, Recife, Brazil

S. Gopi et al. (eds.), *Nanotechnology for Biomedical Applications*,
Materials Horizons: From Nature to Nanomaterials,
https://doi.org/10.1007/978-981-16-7483-9_16

fluorescence-based nanotools, especially due to their high sensitivity, and among them stand out the quantum dots (QDs) [3].

The Russian physicist A. Ekimov pioneered the field of QDs, synthesizing in 1981 those nanostructures in a glass matrix. Years later, in 1985, the researcher L. Brus prepared, for the first time, QDs in a colloidal medium. Nevertheless, it was only in 1988 that the scientist M. Reed coined the term "quantum dot" [4]. After about ten years, the first biological applications of colloidal QDs were reported [5, 6].

QDs are nanocrystals formed by semiconductor materials with few nanometers in size (typically from 1.5 to 10 nm). Since the report of the first biological applications, QDs have been growingly used in a variety of biological assays [7]. The unique properties of QDs consolidated them as powerful fluorescent probes for biomedical applications. QDs have high brightness, great photostability, and chemically active surface, which make these nanocrystals very specials to promote sensitive and specific labeling of various biological structures [8], such as carbohydrates.

Glycans are usually present on molecules exposed on the external biological surfaces, such as in glycoproteins of viral capsids, peptidoglycan of bacteria, mannoproteins of fungi, as well as in glycoproteins/glycolipids on mammalian cell membranes [9]. Glycosylation is a common post-translational modification, and glycans are essential for different biological processes, participating in various interactions within a biological system, such as protein–ligand, cell–cell, cell–matrix, and host–pathogen relationship [10]. Alterations in the glycosylation profile can be associated with several diseases and the study of the glycocode can lead to the discovery of relevant glycan biomarkers and targets for novel diagnostic or therapeutic technologies [11].

Among the approaches used to investigate glycosylation, the use of lectins has been stood out. Lectins are proteins able to reversibly bind to different types of carbohydrates [9]. Various lectins, mainly from plants, have been used in diverse biotechnological applications, from (cyto) histochemistry analyses to the development of biosensors [12]. Besides lectins, phenylboronic acids (PBAs) are synthetic molecules that have also drawn attention from the glycobiology area due to their capability of recognizing carbohydrates, then PBAs have been called "borono-lectins" [13]. Hence, when lectins or "borono-lectins" are conjugated to QDs, the full potential of those molecules can be explored, yielding versatile and sensitive fluorescence-based nanotools for glycobiological studies.

Therefore, the present chapter reports biological applications of glyco-specific fluorescent nanoprobes consisted of QDs conjugated with lectins or "borono-lectins", in the last five years. This review also presents an overview on QDs, lectins, "borono-lectins", and QD-based conjugation, in the context of the reported applications. We hope that this chapter can offer a background for future researches in areas such as glycobiology and nanobiotechnology.

2 Fundamentals on Quantum Dots

QDs are semiconductor nanocrystals that proved to be powerful nanotools for several applications, such as electroluminescent devices, optical switching, photovoltaic cells, and biomedical applications [14]. In particular, the unique physicochemical properties that made colloidal QDs widely employed in diverse biological applications include (i) fluorescence emission from visible to near-infrared region, tunable according to the size or composition, (ii) high quantum yields providing bright signals; (iii) narrow fluorescence emission and broad absorption spectra, enabling multicolor analyses using a single-excitation light source; (iv) high resistance to photodegradation, allowing long-term studies; and (v) active surface with chemical functional groups for conjugation with molecules, NPs, and surfaces of other materials [15, 16].

QDs, as a class of semiconductor materials, are characterized by a valence band (VB) and a conduction band (CB) separated by an energy band gap (Eg) <4 eV at room temperature [17, 18]. QDs, nevertheless, are under a three-dimensional quantum confinement regime, which results in a size-tunable Eg accompanied by a discretization of electronic energy levels [19], as exemplified in Fig. 1a, b.

In a semiconductor material, electrons from VB can reach CB after receiving energy—for example, in the form of photons—leaving "holes" (h^+) in VB. With

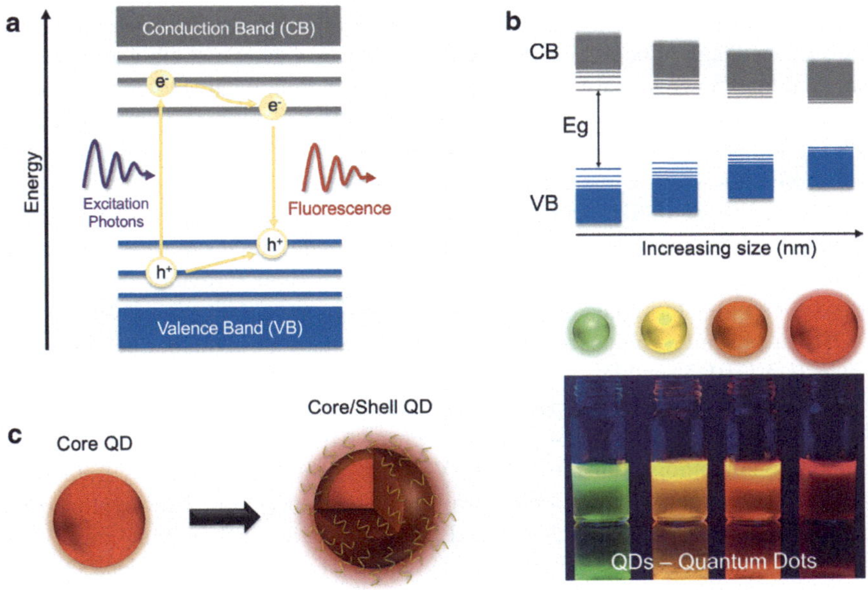

Fig. 1 Schematic representations of the QD fluorescence emission (**a**), the discretization of electronic energy levels and the size-tunable fluorescence emission as a consequence of the tridimensional quantum confinement effect in QDs (**b**), and the final QD structure after passivation and functionalization (**c**). Image from Biomedical Nanotechnology Research Group—UFPE

the return of electrons to VB and the electron–hole recombination, fluorescence photons are emitted at a specific wavelength corresponding to Eg. The electron–hole pair is called "exciton" and the electron–hole distance is known as exciton Bohr radius (a_B), which is typical for each semiconductor material type [20]. For instance, the a_B of CdTe is about 7.3 nm [21]. In this context, the term "quantum dot" is then employed when the NP radius becomes smaller than a_B [20]. Moreover, as previously mentioned, there are changes in the photophysical properties, and Eg gets inversely proportional to the QD size (Fig. 1b). Thus, larger QDs will emit toward the infrared spectral region, and smaller QDs the other way around [22, 23].

The bottom-up strategy, applying colloidal synthetic routes, is the most widely used method to produce QDs for biological applications. Typically, the synthesis process involves at least a metal, a chalcogenide, and a stabilizing agent as precursors, dispersed in an aqueous solvent. The synthesis usually occurs under controlled atmosphere, temperature, and/or pH conditions. Those parameters, the molar ratio of precursors, the reaction velocity, and/or the synthesis time can be managed, for example, to tune the size and modulate the optical properties of QDs [24]. Besides, when a hydrophobic route is adopted, it is usually required a post-synthesis ligand exchange employing alkyl thiol molecules, or surface silanization, to render QDs hydrophilic for biological applications [25, 26]. Those additional procedures, however, can decrease the emission intensity of the nanocrystals [27].

QDs have a core/shell structure (Fig. 1c) at the end of the synthetic procedure. The inorganic crystalline core—associated with the composition, shape, and size—is responsible for the fluorescence spectral profile [20]. The shell passivates the core surface dangling bonds, ensuring effective electron–hole recombination and an intense fluorescence signal [28, 29]. QDs applied in biological studies are usually type-I, presenting a shell composed of a thin layer from other semiconductor material with Eg higher than the core. The QD external layer is commonly formed by alkyl thiol molecules, such as mercaptosuccinic acid (MSA), 3-mercaptopropionic acid (MPA), glutathione (GSH), and cysteamine (CYS), which promote the QD colloidal stabilization by electrostatic repulsion. Besides, in hydrophilic synthetic routes, the stabilizing agent thiol group also contributes to the shell formation, whereas its carboxyl or amine terminal confers functional groups for conjugation [27].

QDs are usually composed of binary combinations of elements belonging to groups II–VI, III–V, and IV–VI of the periodic table, leading, for example, to CdTe, InP, and PbSe nanocrystals, respectively [22]. Furthermore, more recently, the assembling of ternary QDs have also grown, such as $CuInS_2$ nanocrystals, formed by compounds containing atoms from I–III–VI groups [30]. In particular, hydrothermal synthesis, in which pressure is also involved, has been explored for synthesizing ternary QDs [31]. Moreover, QDs doped with metal ions, such as Mn^{2+}, also have been synthesized to provide new optical and physical chemical properties [32].

Tables 1 and 2 summarize the QD core types mostly employed for conjugation with lectins or "borono-lectins", respectively, molecules focused on this chapter. As observed, CdTe QDs have been widely applied in glycobiology studies with lectins. There was only one report with lectins employing $CuInS_2$ QDs, which were

Table 1 Characteristics and applications of conjugates based on QDs and lectins

QD core	Stabilizing/Functionalizing agent	Lectin	Biological application of the conjugate	References
CdTe	MSA	Con A	Labeling of yeast cells and biofilm of *Candida albicans*	[33]
CdTe	MSA	Cramoll	Evaluation of the best conjugation strategy using *C. albicans* labeling	[34]
CdTe	MSA	Cramoll	Studying glucose/mannose expression in different species of *Candida* (*C. albicans*, *C. glabrata*, and *C. parapsilosis* sensu stricto)	[35]
CdTe	GSH	WSMoL	Monitoring larvicidal action of the lectin in *Aedes aegypti*	[36]
CdTe	MSA	UEA I	Evaluation of the expression of the L-fucose in red blood cells	[37]
CdTe	TGA	DSA and LCA	Analyzing the expression of carbohydrates in liver cancer cells and normal human cells (from the liver and endothelium)	[38]
CdS	Pectin	Jacalin	Labeling of chronic myeloid leukemia cells (K562)	[39]
CdTe	MSA	Cramoll	Evaluation of glucose/mannose profile in normal, fibroadenoma, and invasive ductal carcinoma tissues	[40]
CdTe	Not informed	TBL	Evaluation of the best conjugation strategy for cancer cell recognition	[41]
CdTe	MSA	BmoLL	Labeling of human red blood cells with specific detection of galactose in type B	[42]

(continued)

Table 1 (continued)

QD core	Stabilizing/Functionalizing agent	Lectin	Biological application of the conjugate	References
CdTe	TGA	Con A	Biosensor for detection of free lipopolysaccharide from *Serratia marcescens* and whole cells	[43]
$CuInS_2$	MPA	Con A	Biosensor for detection of transferrin and glucose oxidase	[44]
CdSe	Mercaptoethylamine	C-type lectin from eggshell	Formation of a hybrid system with anisotropic silver nanoparticles also sensitized with the lectin to detect molecules containing carbohydrates on *Staphylococcus aureus*	[45]
CdS	Methacryloylamido cysteine and methacryloylamido tyrosine	Con A	Biosensor for detection of immunoglobulin M in an aqueous environment	[46]

MSA mercaptosuccinic acid; *GSH* glutathione; *TGA* thioglycolic acid; *MPA* 3-mercaptopropionic acid; *Con A* Concanavalin A; *UEA I Ulex europaeus* I; *DSA Datura stramonium* agglutinin; *LCA Lens culinaris* agglutinin; *TBL Tepary bean* lectin

combined with Concanavalin A (Con A). In studies involving "borono-lectins", a greater diversity of applications employing binary and ternary QDs was observed.

The following sections will present fundamentals on lectins and "borono-lectins", and an overview on QD-based conjugation.

3 Lectins

Lectins form a heterogeneous group of proteins capable of recognizing and reversibly binding to monosaccharides, oligosaccharides, or glycoconjugates. Those sugars can be free in a medium or present on biological structures. These proteins, which can be found in the form of a monomer, dimer, or tetramer, have at least one non-catalytic site, named carbohydrate recognition domain (CRD), that binds to specific carbohydrates, being able to agglutinate cells or cause precipitation of glycoconjugates [55, 56]. Thus, based on those characteristics, lectins are distinguished from sensor/transport proteins that can detect only free mono- and oligosaccharides,

Table 2 Characteristics and applications of conjugates based on QDs and "borono-lectins"

QD core	Stabilizing/Functionalizing agent	"Borono-Lectin"	Biological application of the conjugate	References
CdSeTe	GSH	APBA	Evaluation of sialic acid levels in K562 leukemia cells treated and non-treated with AZT	[47]
Gd-doped $CuInS_2$	GSH	APBA	Preparation of a bimodal nanosystem for recognition of tumor folate receptor and sialic acid	[48]
CdSeTe	TGA	VPBA	Preparation of theragnostic nanoprobe based on QDs and doxorubicin for pH-responsive antitumor drug release associated with photodynamic therapy	[49]
CdTe	MSA	MPBA[1]	Development of a site-specific nanotool using red blood cells as biological models and evaluation of sialic acid levels in KG-1 and K562 leukemia cells	[50]
Mn-doped CdS	MPA	APBA	Reusable photoelectrochemical biosensor to detect K562 cells	[51]
Mn-doped ZnS	The "Borono-Lectin"	MPBA[2]	Biosensing platform to detect transferrin biomolecules	[52]
CdTe	TGA	APBA	Evaluation of glucose in photosynthetic organisms	[53]
CdSe	MPA	APBA	Detection of intracellular glucose based on QD fluorescence lifetime	[54]

GSH glutathione; *TGA* thioglycolic acid; *MSA* mercaptosuccinic acid; *MPA* 3-mercaptopropionic acid; *APBA* 3-aminophenylboronic acid; *VPBA* 4-vinylbenzeneboronic acid; [1]*MPBA* 3-mercaptophenylboronic acid; [2]*MPBA* 4-mercaptophenylboronic acid; *AZT* 3'-azido-3'-deoxythymidine carbohydrate-recognizing immunoglobulins, and enzymes that act on glycans used as substrates [57].

Interactions between lectins and their target sugars are governed mainly by hydrogen bonds with the contribution of hydrophobic interactions and Van der Waals forces [12]. Moreover, lectins can interact with their ligands through other further mechanisms, (i) stacking pyranose ring with aromatic radical from lectins that possess amino acids such as tryptophan, phenylalanine, and tyrosine; (ii) electrostatic interactions, for example, between cationic arginines and negatively charged sialic acids (SAs); (iii) metal ion coordination that can stabilize the CRD and increase the protein–sugar interaction; and (iv) solvent-mediated interaction, such as, desolvation may be important for the interaction of lectin with their target polysaccharide [58].

The affinity of lectins for monosaccharides is relatively weak and their dissociation constant (K_d) is in the range of 10^{-3} M due to the shallow nature of their binding pockets that can compete with solvent interactions. On the opposite, the binding of lectins with complex oligosaccharides (or glycans) can occur through multivalent interactions since those proteins can be assembled in oligomeric structures with several binding sites for different arms of a branched oligosaccharide (or different glycans) in the same glycoconjugate. This behavior results in an improvement in the lectin affinity, leading to K_d values in the range of 10^{-6} to 10^{-8} M, and is named cluster effect [59, 60]. Moreover, the spatial orientation of the glycidic residues can also influence the binding of lectin to these structures [61].

Lectins are ubiquitous and can be isolated from viruses to higher organisms. Their biological roles are variable and depend on the organism synthesizing them. Thus, (i) lectins from viruses, bacteria, fungi, and parasites, sometimes also named adhesins, can act in adhesion and colonization processes; (ii) animal lectins can mediate processes of cell migration and adhesion, opsonization, phagocytosis, glycoprotein production, and activation of complement pathway; and (iii) plant lectins can play important functions in defense mechanisms against the attack of pathogens and insects, participating also in fostering modulation [56, 62].

Due to the broad distribution and easier purification strategies, lectins from plants have been the most studied, with about hundreds already isolated and characterized. These lectins are even the most widely used to determine glycan profiles in different biological systems due to their stability, availability, and varied specificity [56, 63–65]. A large part of these lectins has been isolated from seeds [56], like most of the lectins conjugated to QDs presented in this review, such as Cramoll from *Cratyllia mollis* and Con A from *Canavalia ensiformis* seeds. Besides, lectins can also be found in other plant parts, such as roots and flowers [66] or leaves, for example, BmoLL lectin from *Bauhinia monandra.*

Lectins, mainly those extracted from plants, have attracted the attention of researchers for more than 130 years, being used as valuable tools in glycobiology [66]. Since the first discovery of the hemagglutinating capacity of ricin [67], many studies have been focused on the potential applications of those proteins to assist lectinology, revealing that the lectin importance goes beyond the hemagglutination of cells [66]. Many of these applications are in the field of biomedical sciences, in which these proteins help to unveil the information encoded by carbohydrates [68, 69].

Many approaches in this chapter have explored legume lectins such as Con A, BmoLL, Tepary bean (TBL), Cramoll, *Lens culinaris* agglutinin (LCA), and *Ulex europaeus* I (UEA I), which belong to the group of plant lectins used in glycobiology studies. Besides of composing the largest group of plant lectins, these proteins also have a broad range of specificity when compared to other families [56], being also one of the groups that have attracted increasing attention to studies related to their biochemical and pharmacological activities [12].

Plant lectins can also be purified with high yields and can show specificity to different carbohydrates [55]. Thus, those lectins can be valuable tools for studying changes in the expression of sugars on cells and tissues, which can be associated with pathological processes, as well as for detecting glycan levels that may be altered in biological fluids.

Many studies have revealed interesting applications of lectin-glycan interactions when those proteins are combined with QDs, as compiled in Table 1. All these applications highlight the relevance of using fluorescent probes to elucidate the functions of lectins and their interaction with various biological systems. Besides lectins, other important molecules for glycobiology applications are the PBAs, which have also been employed to functionalize QDs for developing versatile glyco-specific conjugates.

4 "Borono-Lectins"

Due to the relevance of carbohydrates in many biological processes, besides lectins, other types of sugar-binding molecules have attracted a growing interest in the glycobiology area. In this way, PBAs are promising synthetic molecules to develop glyco-specific nanoprobes. PBAs have not only shown a chemical affinity for carbohydrates but can contain functional groups for conjugation, such as thiol, amine, vinyl, and carboxyl [13]. Additionally, PBAs are small-sized molecules and stable in an aqueous medium. Their chemical structure can be tailored and they do not induce cell adhesion to substrates. Therefore, these features led to PBAs being widely explored [70, 71].

Over the years, several researchers have been studying the interaction of PBAs with sugars. One of the pioneered publications in this area was the study of J. P. Lorand and J. O. Edwards in 1959. The authors systematically analyzed the binding between PBAs and diol groups of different carbohydrates—by changing the pH—to determine the binding constant [72]. In the mid-80s, the term "artificial lectin" began to be used after that Burnett et al. [73] evaluated the ability of PBA to bind to the surface of *Bacillus subtilis* bacterium, presumably through carbohydrate residues. More recently, researchers have employed the term "borono-lectin" since PBAs recognize carbohydrates [13, 74]. The chemical principle of forming stable boronate compounds between PBA and carbohydrates is based on an acid–base reaction that is pH-dependent. PBAs act like Lewis acids, while carbohydrates operate as Lewis bases. It is dated in the literature that PBAs can form bonds with carbohydrates containing 1,2- or 1,3-diol structures under mild alkaline aqueous conditions (e.g., reactional pH close or higher than pKa value of PBA). Based on the ionization state of the boron atom, two kinds of compounds with diol-groups could be formed: (i) boronic esters or (ii) boronate esters, when the boron atom is found as neutral trigonal or anionic tetrahedral, respectively. Nevertheless, in an aqueous medium, the formation of anionic tetrahedral PBAs is favored. On the other hand, PBAs have shown a low affinity for most carbohydrates at weakly acidic conditions; except for SA, for which PBAs can have a high affinity near physiological pH [70, 75–78].

Indeed, the high affinity and selectivity of "borono-lectin" for sugars (e.g., galactose, glucose, mannose, and SA) have been reported in the literature [76, 79, 80]. Furthermore, the function of PBAs as Lewis acid has also been explored to detect Lewis bases such as fluoride, cyanate, pyrophosphate, and amines species, expanding their potential for application in biosensors [81].

Therefore, the association of these synthetic molecules with QDs have yielded interesting nanotools, opening up new ways of elucidating the complex glycosylation machinery and the cellular functions related to glycans. Table 2 illustrates the diversity of biological studies applying PBAs combined with QDs. Additionally, it is possible to observe that PBAs functionalized with amine groups were the most employed in the reported applications.

5 Overview on Conjugation

An efficient conjugation between QDs and biological molecules, such as lectins and "borono-lectins", is a challenge and a fundamental step for a successful application. Over the years, different covalent and non-covalent strategies to conjugate QDs with biomolecules have been proposed [22, 82]. Regarding this chapter, the adsorption interaction and the carboxyl-amine covalent binding, mediated by *N*-ethyl-3-(3-dimethylaminopropyl) carbodiimide hydrochloride (EDC) or EDC/*N*-hydroxysuccinimide (NHS) chemistry, were the main approaches applied to conjugate QDs and lectins. The compound *N*-hydroxysulfosuccinimide sodium salt (Sulfo-NHS) can also be used in the place of NHS, especially because it has a higher hydrophilic character [83]. The adsorption approach was indeed widely explored in the applications with lectins. This strategy can involve, for example, electrostatic attraction, hydrophobic interactions, and/or hydrogen bond, depending on the QDs and biomolecules employed [27, 84]. Although the adsorption interaction is usually considered weaker than covalent binding, QD-lectins conjugates obtained by the non-covalent strategy presented in this chapter have shown quite efficient overtime functional stability [34]. Moreover, while the conjugation via EDC and NHS esters is favored in pH around 5.0–6.0 [82], the best pH for adsorption interaction, however, depends on the biomolecules involved. Indeed, the overall behavior of biomolecules determines the best conjugation strategy to be adopted. It is also important to mention that carboxylic groups activated by coupling agents used in the covalent conjugation can react non-specifically with the biological system, so a final step is required to block those unreacted moieties. This could be performed by incubating the conjugate, for example, with glycine [85] or Tris-Base—tris(hydroxymethyl)aminomethane [37].

Regarding "borono-lectins", the carboxyl-amine covalent binding conjugation procedure, employing the coupling agents, was also widely used in various applications described in this chapter. Besides, the direct attachment via thiol-metal affinity was as well as explored to conjugate thiolated "borono-lectins" with QDs, as reported in the following sections.

An appropriated conjugation procedure should, at least, (i) yield fluorescent and colloidally stable conjugates, (ii) maintain the activity of biomolecules, and (iii) avoid non-specific interactions between conjugates and the biological system [22]. Thus, the evaluation of the efficiency and specificity of conjugates, as well as their optical and functional properties, is a required step before applications.

The optical properties of QDs after conjugation are frequently monitored by UV–Vis absorption spectroscopy and fluorimetry. Additionally, as lectins are able to aggregate red blood cells (RBCs), a conventional approach to inspect their functional activity after conjugation with QDs has been the hemagglutination assay.

Considering the studies described in this chapter, techniques, such as UV–Vis spectroscopy, Fluorescence Correlation Spectroscopy (FCS), Fourier-transform infrared (FTIR) spectroscopy, Dynamic Light Scattering (DLS), electrophoresis, and zeta potential (ζ) analyses have been applied to confirm the conjugation. While FCS and DLS compare the hydrodynamic radius of conjugates and bare QDs to suggest the conjugation, ζ analyses explore changes on surface charges of NPs. The electrophoresis, on the other hand, uses the differential migration of QDs and conjugates based on their size and charge to evaluate the conjugation. FTIR, however, gives information on chemical binding between QD and molecule [22, 86].

In respect of the conjugate specificity evaluation, additional tests, especially inhibition assays, have been widely employed with lectins and "borono-lectins". In inhibition assays, before having contact with the biological systems, the conjugates are incubated with a carbohydrate by which the lectin or the "borono-lectin" has an affinity. Since the molecule binding sites become occupied before the interaction of the conjugates with the biological system target, a decrease or absence of the labeling is used to indicate that the nanoprobe has specificity.

Therefore, diverse techniques, usually combined, have been used to develop efficient and specific QD-based conjugates, yielding interesting glycobiological applications, as depicted as follows.

6 Applications of QD-Lectin Conjugates

6.1 Pathogens

The abnormal expression of carbohydrates on cell surfaces can be associated with the development of many diseases, as well as with the pathogenicity and the resistance of microorganisms [87–90]. *Candida albicans* is an opportunistic fungal pathogen responsible for high rates of morbidity and mortality. Nevertheless, besides its importance to public health, *C. albicans* yeast cells can also be a versatile biological model to assess the effectiveness and the specificity of QD-lectin conjugates. *C. albicans* yeast cell wall is rich in carbohydrate residues, mainly mannose and glucose, which can be recognized by various lectins [91].

Tenório et al. [33] used *C. albicans* yeast cells as a biological model to develop a conjugate consisted of carboxyl-coated CdTe QDs and Con A lectin. The authors also applied the conjugate to detect carbohydrates in *C. albicans* biofilms. Con A binds specifically to mannose/glucose residues [92], and QDs were conjugated to Con A using the adsorption strategy, at pH 8.0. According to circular dichroism, the secondary structure of Con A lectin was conserved after the conjugation. Furthermore, the QD-Con A conjugate displayed a hemagglutination activity equivalent to the lectin alone, reinforced the preservation of biological functions after the conjugation. Subsequently, confocal fluorescence microscopy and flow cytometry were used to evaluate the efficiency of the conjugate in labeling *C. albicans* yeast cells. An inhibition assay was also performed by incubating the conjugate with methyl-α-D-mannopyranoside before the interaction with yeast cells. As a result, confocal fluorescence microscopy images showed bright labeling for *C. albicans* yeasts after incubation with the conjugate. Besides, flow cytometry analyses confirmed that about 93% of yeast cells were labeled by the QD-Con A nanosystem. The specificity of the conjugate was confirmed since a considerable labeling decrease was observed after incubation with the inhibited nanoprobe. Fluorescence microscopy also revealed the conjugate potential to study glucose/mannose profiles on cells and the extracellular matrix of biofilms. Therefore, the authors concluded that the QD-Con A conjugate can be a practical tool for studying carbohydrate profiles related to pathogens [33].

Cunha et al. [34] also used *C. albicans* yeast cells as a biological model to develop a conjugate based on carboxyl-coated CdTe QDs and Cramoll lectin. This protein is extracted from the seeds of *C. mollis*, a leguminous from the Brazilian Caatinga biome, and belongs to the same family of Con A, which also has an affinity for glucose and mannose residues [93]. Cramoll is an interesting lectin that has several biological potentialities, such as immunomodulatory effect [94], antiparasitic action [95], and antitumor activity [96]. Cunha et al. [34] evaluated two conjugation strategies, adsorption interaction and covalent binding, applying different pH values, to develop the QD-Cramoll nanoprobe. The covalent approach was carried out using EDC and Sulfo-NHS. To evaluate the best conjugation strategy, all conjugates were incubated with *C. albicans* yeast cells and the labeling was analyzed by fluorescence microscopy and flow cytometry. According to the authors, the best result

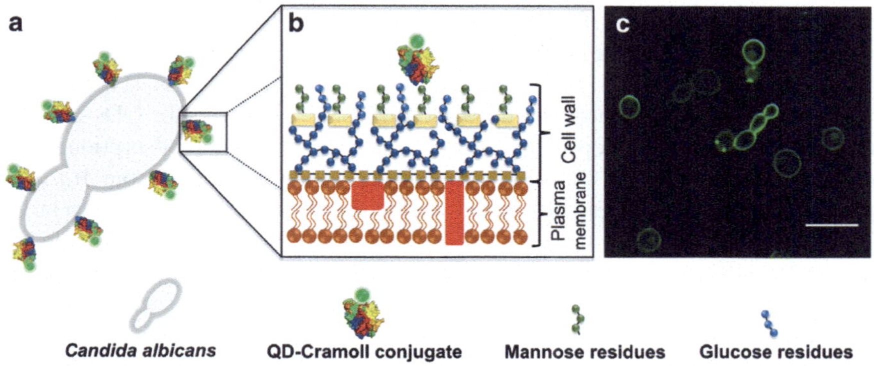

Fig. 2 **a** Representation of the interaction between the QD-Cramoll conjugate and glucose/mannose residues on the yeast cell surface. **b** Schematic overview of the fungal cell wall and the interaction of QD-Cramoll conjugate with mannose residues on the yeast cell surface. **c** Fluorescence confocal microscopy image of *Candida albicans* labeled with QD-Cramoll conjugates, scale bar—10 μm. Image from the Biomedical Nanotechnology Research Group—UFPE related to the study of Oliveira et al. [35]

was achieved by the conjugate prepared by adsorption at pH 7.0, which provided intense labeling of *ca.* 92% of the cells. Besides, an inhibition assay, incubating the conjugate with methyl-α-D-mannopyranoside before interaction with cells was also performed. Those results revealed that the labeling decreased to only 3%, confirming the conjugate specificity. Therefore, the authors were very successful in assembling a nanoprobe able to track carbohydrate profiles on biological systems. Furthermore, the study reinforces that *C. albicans* yeast cell is a compelling model to evaluate the best strategy for obtaining efficient QD-lectins conjugates [34].

Oliveira et al. [35] incubated strains of *C. albicans*, *C. glabrata*, and *C. parapsilosis* sensu stricto with the QD-Cramoll conjugate, to further investigate and compare glucose and mannose profiles among *Candida spp*, as represented in Fig. 2. Different from the previous study, herein, the authors employed the conjugation procedure established by Cunha et al. [34] for green-emitting carboxyl-coated CdTe QDs. Besides evaluating the conjugate specificity by performing inhibition assays with methyl-α-D-mannopyranoside, the authors also carried out a saturation experiment, incubating cells with an excess of lectin. The labeling efficiency of *Candida* cells was analyzed by confocal fluorescence microscopy and flow cytometry; and more than 90% of yeasts from the three species were labeled by the conjugate. Nevertheless, considerable labeling reductions, after interaction with the inhibited conjugate, were reached only for *C. albicans* and *C. glabrata*, not for *C. parapsilosis* sensu stricto. The labeling displayed an effective decrease for this species only after the saturation assay. The authors suggested that the higher lectin affinity for multiple sugar residues (due to the cluster effect) than for single monosaccharides [97] could justify the results obtained for *C. parapsilosis* sensu stricto after the inhibition assay. Then, inhibition and saturation experiments indicated that *C. parapsilosis* sensu stricto had a higher content of glucose/mannose on its surface when compared with the other two species. Therefore, concluded that the QD-Cramoll nanoprobe is promising to distinguish the glucose/mannose content in different species of *Candida* [35].

Another QD-lectin nanoprobe composed of GSH capped CdTe QDs and the lectin WSMoL was developed by Arruda et al. [36], also using adsorption interaction. WSMoL is known as a chitin-binding lectin and is purified from *Moringa oleifera* seeds, a plant widely distributed worldwide [98]. WSMoL has shown larvicidal activity against *Aedes aegypti* larvae [99]. This mosquito is responsible for the transmission of viruses that lead to serious diseases, like dengue, yellow fever, chikungunya, and zika [100]. Thus, the QD-WSMoL conjugate was applied to study the interaction of this lectin with *A. aegypti* larvae. The conjugation was confirmed by electrophoresis and isothermal titration calorimetry (ITC) experiments, suggesting that the interaction between GSH-CdTe QDs and WSMoL occurred through hydrogen bonds. Larvae collected at the third stage (L3) were incubated with the nanoprobe, and the labeling (already at the L4 stage) was analyzed by conventional and confocal fluorescence microscopy. A previous report suggested that the affinity of WSMoL by the larvae peritrophic matrix (a chitinous membrane presented in the gut of insects that protect against pathogens) may be involved in the mechanism of larvicidal action of this lectin [99]. The results showed an accumulation of the conjugate along with the whole digestive tract, mainly in chitin-rich regions, where the peritrophic matrix is present, reinforcing the mechanism of action previously proposed. Lastly, the authors also carried out a conjugate inhibition assay with glucose, a carbohydrate for which WSMoL can also present affinity [101]. Larvae incubation with the inhibited nanosystem showed less intense labeling, confirming the nanoprobe specificity [36].

6.2 Mammalian Cells and Tissues

The application of QD-lectin conjugates can also bring valuable information on the glycocode of mammalian cells and tissues. Aiming to evaluate A, B, and H antigens expressed on human RBCs using flow cytometry, green, and orange emitting carboxyl-coated CdTe QDs were synthesized and conjugated, respectively, to UEA I lectin and anti-A or anti-B antibodies. A covalent conjugation approach using EDC and Sulfo-NHS was chosen by the authors. UEA I lectin can identify H antigens on RBCs by L-fucose affinity. The efficiency of the conjugations was proven by using FCS. For the QD-UEA I nanosystem, the conjugation was also confirmed by incubating the conjugate with L-fucose to inhibit carbohydrate-domain lectin sites before the interaction with RBCs. As O RBC membranes are rich in H antigens, a high labeling percentage (about 85%) was observed for those cells after incubation with the QD-UEA I conjugate. On the other hand, since B, A_1B, and A_1 RBCs present few exposed L-fucose on their membranes, labeling smaller than 8% was found after interaction with the QD-UEA I nanosystem. Besides, after carbohydrate inhibition assay with L-fucose, the QD-UEA I conjugate was not able to label O RBCs, further confirming the specificity of the nanoprobe. The authors also compared the expression of H antigens in A_1 and A_{weak} (A_x and A_{el}) blood groups. They observed that H antigen expression decreased in the order $A_X = A_{el} > A_1$. Overall, the conjugates were able to distinguish some A phenotypes and quantify A, B, and H antigens in RBCs of different blood groups. The conjugates were stable and active for at least 6 months showing to be complementary tools for immuno hematological investigations [37].

The carbohydrates also play key roles in malignancy, progression, and tumor metastasis [38, 102]. Thus, the investigation of carbohydrates as tumor biomarkers has attracted increasing attention, not only aiming at early diagnosis but also to understand biological processes related to cancer [38, 103]. For this purpose, He et al. [38] developed fluorescent QD-lectin conjugates to evaluate the expression of carbohydrates in liver cancer cells (HepG2) compared with liver (LO2) normal cells. Carboxyl-coated CdTe QDs were covalently conjugated to *Datura stramonium* agglutinin (DSA) or LCA lectins, using EDC as the coupling agent. DSA binds to *N*-acetylgalactosamine (GalNAc) and (β-1,4) triantennary *N*-linked glycans, while LCA recognizes *N*-acetylglucosamine (GlcNAc) and mannose [74]. To confirm the conjugation effectiveness, agarose gel electrophoresis, hemagglutination activity test, and carbohydrate inhibition assays were performed. The results suggested that QDs were efficiently conjugated to lectins, which remained active after the conjugation process. Fluorescence microscopy imaging and flow cytometry analyses revealed intense and specific labeling for HepG2 cells using QD-DSA and QD-LCA nanoprobes. The labeling was 2.8-fold and 2.0-fold higher than those obtained for LO2 cells, for the respective conjugates. The difference in the carbohydrate expression in normal and cancer cells, revealed by QD-DSA and QD-LCA conjugates, suggested that this glycidic profile could be used as a cancer biomarker. Besides, results also indicated that those fluorescent nanotools can provide simple and fast carbohydrate detection [38].

It is still challenging to preserve the optical properties of QDs and the specificity of the biomolecule after the conjugation process [34]. Based on that, Ayaz Ahmed et al. [39] investigated the interaction between pectin-coated CdS QDs and jacalin lectin by adsorption strategy, also reporting the potential of this conjugate to label cancer cells. Thermodynamic analysis indicated that binding was spontaneously driven by enthalpy and that polar interactions were involved in the conjugation process. Jacalin lectin has a high affinity for galactose residues [104]. The conjugation efficiency was proved by hemagglutination and galactose-inhibition assays. Thomsen-Friedenreich disaccharide antigen (T-antigen), an important biomarker for cancer, is rich in galactose-residues. Thus, the authors used the QD-jacalin conjugate to analyze T-antigen expression in chronic myeloid leukemia cells (K562). Fluorescence images showed that the conjugate promoted intense labeling of K562 cells. Besides, no significant labeling was observed when the nanosystem was incubated with healthy human peripheral lymphocytes (PBMC). Due to the affinity for T-antigen (one of the main antigens expressed in primary cancers), the QD-jacalin conjugate proved to be a potential fluorescent nanoprobe for studying cancer cell biology [39].

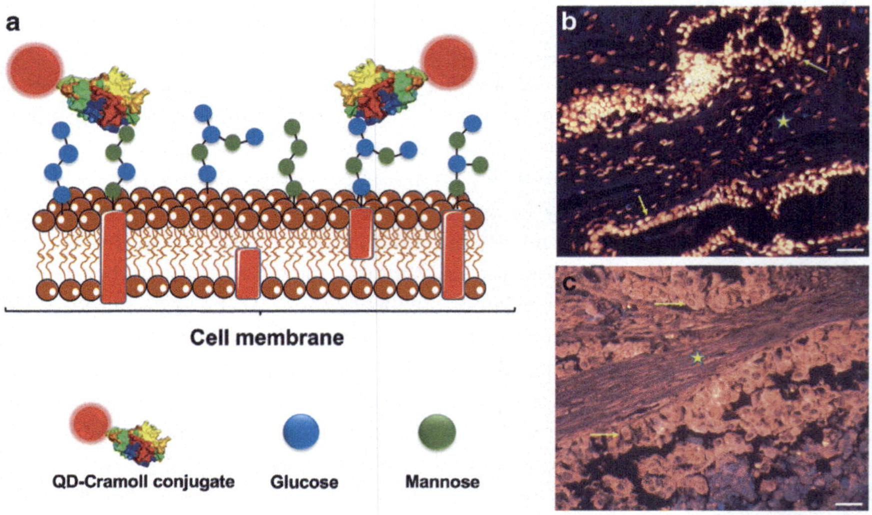

Fig. 3 **a** Representative image of the interaction between the QD-Cramoll conjugate and glucose/mannose residues on the cell surface, **b** and **c** shows, respectively, microscopy images of fibroadenoma and invasive ductal carcinoma tissues labeled by the QD-Cramoll conjugate (in orange). The stars indicate stromal areas, yellow arrows indicate ductal cells, and in blue is the tissue autofluorescence. Scale bar 50 μm; **b** and **c** images from Biomedical Nanotechnology Research Group—UFPE related to the study of Carvalho et al. [40]

The use of lectins in histopathological analyses to identify changes in carbohydrate expressions on cell surfaces, induced by the malignancy process such as breast cancer, is one of the focuses of glycobiology [40, 105]. Carvalho et al. [40] employed the same procedure proposed by Cunha et al. [34] to conjugate Cramoll lectin with carboxyl-coated CdTe QDs. Carvalho et al. [40] applied the conjugate to investigate the glucose/mannose profiles in normal, fibroadenoma (FB), and invasive ductal carcinoma (IDC) tissues (as represented in Fig. 3a), using fluorescence microscopy and an innovative quantitative method based on a fluorescence microplate assay (FMA). Following Cunha et al. [34], the conjugation efficiency was confirmed by labeling *C. albicans* yeast cells through flow cytometry, before and after an inhibition assay performed with methyl-α-D-mannopyranoside. Saturation experiments, offering an excess of lectin before incubation with the conjugate, were also performed. The QD-Cramoll conjugate promoted intense and specific labeling of *C. albicans* cells. Furthermore, fluorescence images revealed that ductal cells from normal and FB tissues were labeled by the conjugate, while in IDC tissues, such labeling occurred both in cells and the stroma (as shown in Fig. 3b, c). These results reinforce the notorious role of the stroma and other matrix components in several stages of the cell transformation process [106]. After the inhibition assay using methyl-α-D-mannopyranoside, the labeling in FB tissues was significantly reduced, while in IDC tissue a small decrease was observed. The authors attributed this result to the cluster effect. Considerable IDC labeling reduction was reached only after the saturation assay. Besides, using FMA, it was possible to quantify and note that there is a correlation between the level of glucose/mannose residues and the degree of tissue transformation, being those sugars more expressed in tissues with a higher degree of malignancy, as in IDC [40].

Cervantes-Jiménez et al. [41] conjugated carboxyl-coated CdTe QDs to TBL using a microfluidic approach, which allowed the use of fewer sample quantities for conjugation. To assess the efficiency of the proposed method, the authors also conjugated the same dilutions of lectins and QDs by covalent strategy, using glutaraldehyde as the coupling agent. By transmission electron microscopy (TEM) images, the conjugate produced by the microfluidic method presented a more homogenous profile than the covalent one. Besides, about 75% of the lectin hemagglutination activity was preserved after conjugation by the new method. Cytotoxicity analysis in colon cancer cells (HT-29) revealed that there was no significant difference between free or conjugated lectins. Therefore, the microfluidic method showed to be an efficient and fast conjugation procedure. Although the authors had not mentioned the lectin specificity, they reported that this protein has an anti-tumorigenic effect on colon cancer, concluding that the conjugate has the potential to be applied in studies related to tumor recognition [41].

Oliveira et al. [42] proposed the conjugation of the lectin extracted from the leaf of BmoLL with carboxyl-coated CdTe QDs, also using the adsorption interaction. Since BmoLL is a galactose-binding lectin [107], Oliveira et al. [42] used RBCs from the ABO blood group system to evaluate the specificity and efficiency of this new conjugate, which reported that the conjugation did not affect the hemagglutinating activity of the lectin. Moreover, the QD-BmoLL conjugate was able to intensely label about 71% of B RBCs, whose membranes are rich in galactose residues, by flow cytometry. Low labeling was also observed, not only in A_1 and O RBCs but also for B type after the conjugate inhibition with galactose, confirming the specificity of the nanoprobe. Moreover, the conjugate was stable and able to promote efficient labeling, at least for 5 months. Besides, there were no signs of hemolysis during the studies performed. Thus, through a simple and effective conjugation procedure, a specific and stable fluorescent nanotool for galactose detection was developed, to assist in several glycobiology studies [42].

6.3 Biosensors

QD-based fluorescent biosensors have been excelled due to their sensitivity and the ability to perform simultaneous multiplex detection of different analytes [108]. Thus, QD-lectin conjugates can be applied to detect glycoconjugates associated with diseases and infectious processes. This detection can be, followed by monitoring the decrease in the conjugate fluorescence intensity after interaction with the analyte, as schematized in Fig. 4. In this context, Ebrahim et al. [43] conjugated the lectin Con A to carboxyl-coated CdTe QDs using glutaraldehyde adsorbed on the NP surface.

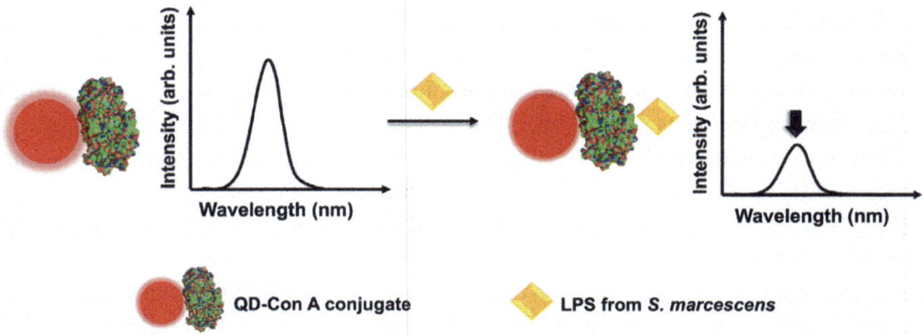

Fig. 4 Schematic diagram of detection from the LPS (lipopolysaccharide) containing glucose-L-rhamnose disaccharide from *Serratia marcescens* bacteria by monitoring QD-Con A conjugate fluorescence intensity

The conjugate obtained was used to detect a lipopolysaccharide (LPS), containing D-glucose-L-rhamnose disaccharide, from *Serratia marcescens* bacteria. The authors verified a linear reduction in the conjugate fluorescence intensity according to the increase in LPS concentration (from 10 to 90 fg mL^{-1}). Besides, they applied the QD-Con A nanosystem to capture *S. marcescens* bacteria and the fluorescence intensity of conjugates also decreased with the rise in bacterium concentration (from 10 to 1 $\times 10^6$ CFU mL^{-1}).

Förster resonance energy transfer (FRET) based biosensing applying QD-lectin conjugates is another approach for detecting glycobiological molecules. FRET is based on the energy transfer between a fluorophore donor to an acceptor that occurs by charge-charge interaction of the donor–acceptor pair dipoles. To this process occurs, the donor and the acceptor need to be at a distance of approximately 1–10 nm, and it is requested an overlap between the donor emission fluorescence spectrum and the acceptor absorption band [109]. The photophysical properties of QDs make them advantageous as FRET donors or acceptors. The use of QDs in FRET can be benefited from their large stokes shift, high extinction coefficient, and narrow and tunable emission. Those properties can be used to optimize the donor–acceptor spectral overlap, avoid direct acceptor excitation, and increase the brightness of the FRET system. This configuration can provide sensitive detection of molecules in low concentrations [110]. Thus, Gao et al. [44] proposed a FRET-based biosensor using QD-lectin conjugates to detect two mannose-containing glycoproteins, transferrin (Tf), and glucose oxidase. The system used carboxyl-coated $CuInS_2$ QDs conjugate to Con A as the acceptor and glucosamine-coated rhodamine as the donor. The QD-lectin conjugation was performed by EDC/NHS as coupling agents. In this optical sensor, a competitive reaction was explored, when the glycan moiety of glycoproteins interacted with the lectin from the conjugate, inhibition of donor–acceptor energy transfer was observed. Then, a correlation between glycoprotein concentrations and the fluorescence intensity ratio at the wavelengths 654 and 577 nm (from the acceptor and the donor) was established. In this way, the results showed a linear behavior for glucose oxidase from 3.35×10^{-10} to 6.70×10^{-8} mol L^{-1} and for Tf from 6.90 $\times$

10^{-10} to 3.45×10^{-8} mol L^{-1}. The detection limit for glucose oxidase and Tf was 1.5×10^{-10} mol L^{-1} and 2.5×10^{-10} mol L^{-1}, respectively [44].

Bacterial cell surfaces have molecules containing carbohydrates, such as peptidoglycans and lipopolysaccharides, and then, these microorganisms can be recognized by lectins, such as the C-type lectin from the avian eggshell. In this way, lectins with this binding capacity can be used to develop biosensors for bacteria [111]. In this context, nanosystems, containing mercaptoethylamine modified CdSe QDs and silver NPs (spherical or anisotropic), were evaluated to detect *Staphylococcus aureus* bacteria. The most promising system has consisted of a QD-anisotropic silver NPs hybrid system. To develop the best biosensor, both NPs were sensitized with C-type lectin from hen eggshell to recognize carbohydrate-containing structures on the bacterial surface. The extinction spectrum of silver NPs overlapped with the QD fluorescence spectrum, and the aggregation of the nanostructures enabled resonance energy transfer between these nanostructures. As a result, a quenching of the QD fluorescence, directly proportional to the increase in the *S. aureus* concentration occurred, allowing the detection of bacteria in the range from 6×10^3 to 2×10^7 cells mL^{-1} [45].

Con A lectin has also been conjugated to CdS QDs to detect immunoglobulin M (IgM) antibody, which contains mannose residues in their structure. The conjugate was prepared using amino acid decorated QDs functionalized with methacryloylamido cysteine (MACys) and methacryloylamido tyrosine (MATyr) monomers. Then, the covalent crosslinking immobilization of Con A on the nanocrystals occurred by incubating the lectin in the presence of photosensitive ruthenium-based amino acid oligomer and ammonium persulfate. After the functionalization steps with the two monomers, there was an increase in the fluorescence intensity of the QDs. However, the conjugation of Con A in the presence of the photosensitive ruthenium-based monomer reduced the QD intensity due to a quenching assigned to the FRET phenomenon. After the interaction with IgM, the fluorescence intensity of the conjugate decreased when compared with the previous intensity indicating the binding of the antibody to the nanosystem. Nevertheless, among the concentrations of IgM tested (0.01, 0.5, and 1 ppm), there was a discreet rise in the fluorescence intensity with the increase of the concentration [46].

7 Applications of QD-"Borono-Lectin" Conjugates

7.1 Mammalian Cells and Tissues

PBAs have attracted the interest of researchers since these molecules can interact with different carbohydrates present on the surfaces of viruses, bacteria, fungi, or mammalian cells. Besides, "borono-lectins" have the potential to give contributions related to therapy, diagnosis, drug delivery, and comprehension of biological processes [13, 78, 79, 112, 113].

Cao et al. [47] proposed a microfluidic platform to detect SAs at the level of a single cell using conjugates consisted of QDs and 3-aminophenylboronic acid (APBA). GSH capped CdSeTe@ZnS-SiO_2 QDs were covalently bound to APBA via EDC. The conjugation was confirmed by FTIR analyses. Then, the nanoprobe was evaluated through the recognition of SAs in K562 cells, non-treated, and treated with the drug AZT (3'-azido-3'-deoxythymidine) at two different concentrations (20 and 40 μmol L^{-1}). The cell labeling was analyzed by fluorescence microscopy. To evaluate the specificity for SA, cells were incubated with QDs alone, and as a result, only weak labeling was observed. On the other hand, when incubated with the QD-APBA conjugate, cells presented a high brightness. The results showed that the fluorescence intensity increased after the treatment with 20 μmol L^{-1} and 40 μmol L^{-1}, indicating that SA levels increased after the treatments. The strategy presented by the authors provided an alternative, sensitive, and effective tool based on QD-APBA conjugates for glycan analysis [47].

Gedda et al. [48] prepared a bimodal nanosystem combining abilities for fluorescence analyses and magnetic resonance imaging (MRI). Thus, GSH capped $CuInS_2$/ZnS QDs doped with Gd^{3+} (CIGS/ZnS QDs) were synthesized and conjugated to folic acid (FA) and APBA, aiming at targeting both folate receptors and SA simultaneously. The conjugation was performed using EDC and Sulfo-NHS as coupling agents and confirmed by ζ measurements as well as FTIR analyses. Then, under confocal fluorescence microscopy, HeLa cells (human epithelial cervical carcinoma) incubated with the QD-FA|APBA conjugate showed labeling in the cytoplasm, indicating a successful internalization. Besides, HeLa cells were also incubated with QDs conjugated only to FA or only to APBA, and just a slightly labeling was visualized when compared to the result obtained with the QD-FA|APBA nanosystem. The specificity of the nanoplatform was verified by pre-treating the cells with the enzyme sialidase, and later saturating them with FA, and results showed low labeling. Due to its bimodal potentialities, the QD-FA|APBA conjugate was also evaluated as a contrast agent for T_1-weighted MRI. As a result, significant contrast in cell pellets incubated with the QD-FA|APBA nanosystem was observed, when compared to the control (without this conjugate); and the nanosystem showed low cytotoxicity. The results of the analyses of embryonic mortality and the hatching rate of zebrafish eggs also showed that the nanoplatform did not induce any detectable abnormality. Furthermore, zebrafish excreted the QD-FA|APBA conjugate without a significant effect on the growth. Therefore, the authors concluded that the developed nanoplatform holds promising for site-specific bimodal imaging in clinical applications [48].

Qi et al. [49] prepared polymer nanomicelles composed also of diethylenetriaminepentaacetic acid (DTPA) chelating agent, carboxyl-coated CdSeTe QDs, and 4-vinyl benzene boronic acid (VPBA). The nanosystem CdSeTe@PC7A-PEG-DTPA-VI-VPBA was loaded with doxorubicin (DOX—a drug used for cancer chemotherapy). VPBA was select because of its affinity to SA. QDs were also chosen due to their potential as photosensitizers for photodynamic therapy. A set of analyses was used to characterize the nanomicelles, such as DLS and ζ measurements. Moreover, the authors found that DOX was released from nanomicelles faster in a weakly acidic medium (pH 5.0) than in physiological pH (7.4), which may be interesting considering the pH of the tumor environment. For evaluating the potential of the multimodal nanomicelle to treat cancer, Qi et al. incubated DOX alone and DOX@CdSeTe@PC7A-PEG-DTPA-VI-VPBA for 4 h and 10 h with B16F10 cells (mouse melanoma) and analyzed them by fluorescence microscopy. Low fluorescence was observed in the first 4 h after incubation with the nanosystem when compared to DOX alone. Nevertheless, after 10 h, the fluorescence in cells incubated with nanomicelles increased significantly, indicating DOX releasing from nanostructures. Afterward, the authors inoculated B16F10 cells in C57BL mice to analyze the nanosystem efficacy for tumor suppression and its biocompatibility in vivo. DOX@CdSeTe@PC7A-PEG-DTPA-VI-VPBA was injected for 14 days and irradiated with LED (40 W) for 20 min every day (free DOX and saline were the controls). DOX@CdSeTe@PC7A-PEG-DTPA-VI-VPBA inhibited tumor growth more significantly than DOX alone. Besides, using DTPA chelating ^{99m}Tc and single-photon emission tomography (SPECT) imaging, the nanosystem had an efficient accumulation in the tumor site in vivo. In addition, comparing the treatment with and without irradiation, it was possible to observe a change in the tumor volume because CdSeTe QDs generate reactive oxygen species (ROS) under light conditions. Finally, the authors also suggested that this new nanotool can be a promising antitumoral drug carrier for cancer [49].

Monteiro et al. [50] developed a QD-based nanoplatform using mercaptophenylboronic acid (MPBA) for SA evaluation in different biological systems. Thus, carboxyl-coated CdTe QDs were functionalized with MPBA exploring the affinity of the thiolate molecules to the nanocrystal surface, in a simple and fast procedure. The conjugation was confirmed by FCS and inductively coupled plasma spectrometry (ICP). To assess the nanoplatform applicability, the authors used RBCs as models since their membranes are rich in SA, as depicted in Fig. 5. It was reached labeling of about 91%, through flow cytometry, after incubating RBCs with the QD-MPBA

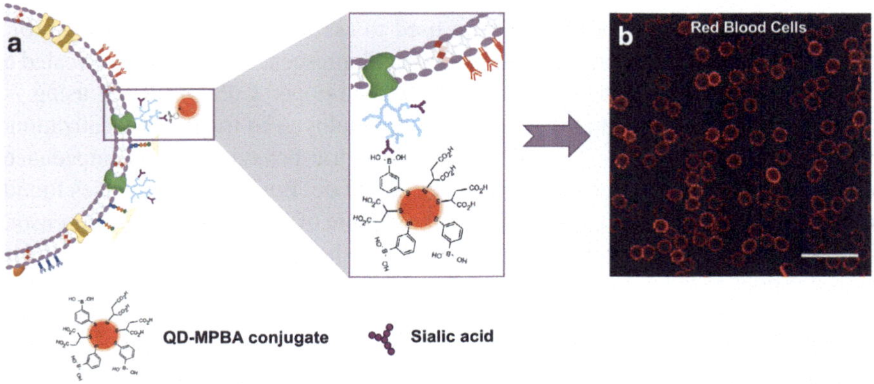

Fig. 5 **a** Schematic representation of the interaction between SAs and QD-MPBA conjugates. **b** Fluorescence confocal microscopy image depicturing the labeling of RBCs by QD-MPBA conjugates. Scale bar—25 μm. Image from Biomedical Nanotechnology Research Group—UFPE, related to the study of Monteiro et al. [50]

conjugate. On the other hand, reduced labeling (17%) was observed after the conjugate interacts with RBCs pretreated with neuraminidase (an enzyme that cleaves SA). Low labeling of about 5% was also achieved after RBC incubation with bare QDs only. Those assays together with other labeling analyses, applying RBCs and conjugates pre-incubated with different carbohydrates (glucose, galactose, and mannose), confirmed the nanoprobe specificity under the pH conditions used. After, the QD-MPBA conjugate was incubated with chronic (K562) and acute (KG-1) myeloid leukemia cells. The results revealed that KG-1 had a higher level of SA in their membrane when compared with K562 cells. Flow cytometry labeling for KG-1 cells was 100%, while for K562 was about 94%. Besides, the median of fluorescence intensity (MFI) for K562 cells was *ca.* 2.5-fold lower than KG-1 cells. Therefore, the study concludes that the QD-MPBA conjugate holds a high potential for SA evaluation in a variety of biological systems [50].

7.2 Biosensors

Nowadays, many researchers are looking for the development of improved and innovative biosensors, and the approaches based on QDs and PBAs molecules have great promises.

Wu et al. [51] reported an attractive and sensitive biosensor based on the photoelectrochemical (PEC) process, exploring photo-excitation and electrochemical detection in a single system. Then, the authors employed carboxyl-coated CdS QDs doped with Mn^{2+} to improve the PEC performance and used APBA to recognize carbohydrate moieties on K562 cells [114]. The PEC biosensing platform was assembled by immobilizing Mn-doped CdS QDs onto a glassy carbon electrode (GCE). Then, MPA was added to Mn-doped CdS QDs/GCE, and a covalent binding between QDs and APBA was promoted using EDC/NHS. Fluorescence microscopy and electrochemical impedance spectroscopy (EIS) were used to assess the PEC biosensor performance (i.e., capture and release of K562 cells). Fluorescence microscopy revealed a satisfactory capture of cancer cells onto APBA/Mn-doped CdS QDs/GCE using 7-amino-4-chloromethylcoumarin dye (CMAC), employed to track cell localization. EIS also displayed a typical profile of cell dielectric behavior and an impedance increase upon cell capturing on the modified electrode. Besides, lower values found for photocurrent measurements reinforced the capture of K562 cells by the biosensor. The authors showed a linear relationship between the photocurrent and the cellular concentration, in the range of 3.5×10^2 to 7.0×10^6 cells mL^{-1}, with a detection limit of 200 leukemia cells mL^{-1}. Furthermore, a photocurrent recovery with the release of cancer cells after adding fructose (50 mmol L^{-1}) suggested that the PEC platform could be reused. Fructose was applied since APBA can have an affinity for this monosaccharide at moderately alkaline pH [114]. Lastly, the authors demonstrated that the PEC biosensor could switch between cell capture and release for five cycles, showing potential as a reusable platform [51].

Differently from Wu et al. [51] that investigated biosensing with PBA in whole cells, Chang et al. [52] employed a platform consisted of Mn-doped ZnS QDs stabilized/functionalized with MPBA to detect a glycoprotein, the Tf, exploring the MPBA affinity for Tf glycosylated sites. An improvement of the fluorescence occurred according to the rise of the Tf concentration (binding constant equal to 2.75×10^6 mol L^{-1}). Besides, the Tf detection showed a linear range from 0.10 to 10.0 μmol L^{-1}, and the detection limit was 5.69×10^{-9} mol L^{-1}. The authors also studied the conjugate behavior in the presence of other molecules, such as glucose, fructose, and mannose. Non-glycosylated proteins, like cytochrome C, lysozyme, and albumin, were used. The interaction with horseradish peroxidase (HRP—containing glycosylation sites) was also observed. No evident changes occurred in the presence of sugar molecules and non-glycosylated proteins. Nevertheless, with HRP, there was quenching of the fluorescence, and the authors attributed this distinct behavior to differences in the isoelectric point and glycan positions between HRP and Tf. Moreover, the fluorescence-based biosensor also allowed obtaining satisfactory results for the Tf detection in more complex medium, as fetal bovine serum [52].

Li et al. [53] proposed glucose detection in photosynthetic organisms, algae cells from *Chara zeylanica,* and plant leaf tissues from *Arabidopsis thaliana*, using fluorescence biosensing. The optical biosensing method used two probes, one composed of bare carboxyl-coated CdTe/CdS QDs without response to glucose (control system), and a QD-APBA conjugate responsive to glucose. The conjugation was performed using EDC/NHS and characterized by FTIR, DLS, and UV–Vis absorption analyses. Afterward, fluorescence intensities came from bare QDs and the conjugate were analyzed in vitro in the presence of glucose and other sugars (sucrose, fructose, lactose, maltose, and galactose). Assays revealed that a notable decrease of fluorescence intensity occurred for the QDs-APBA conjugate (not for bare QDs) only in the presence of glucose. The authors attributed this result to the conjugate aggregation caused by the affinity between PBA on the QD surface and two pairs of cis-diols on glucose. The other sugars have only one pair of cis-diols, and thus they did not lead to aggregation. In vivo analyses with the photosynthetic organisms showed similar results. Glucose detection was possible at concentrations above 500 and 100 μmol L^{-1} for in vivo and in vitro assays, respectively. Confocal fluorescence microscopy allowed observing the localization of QD-based probes and their response to glucose in the photosynthetic organisms. The probes were found in the cytosol of algae cells and on the leaf mesophyll cells of the *Arabidopsis* plant. Therefore, the authors demonstrated an interesting method to detect glucose, even in vivo in photosynthetic organisms [53].

Ripoll et al. [54] addressed their research by applying time-resolved spectroscopy and fluorescence lifetime imaging microscopy (FLIM), aiming at improving the glucose biosensing performance. It was mentioned that this approach could display benefits for intracellular glucose detection when compared to conventional detections based only on the fluorescence intensity. For instance, fluorescence lifetime does not depend on the nanoprobe concentration, variations in the excitation light power, or cell autofluorescence. Ripoll et al. prepared the fluorescence nanosensor for glucose detection using carboxyl-coated CdSe/ZnS QDs conjugated to APBA via

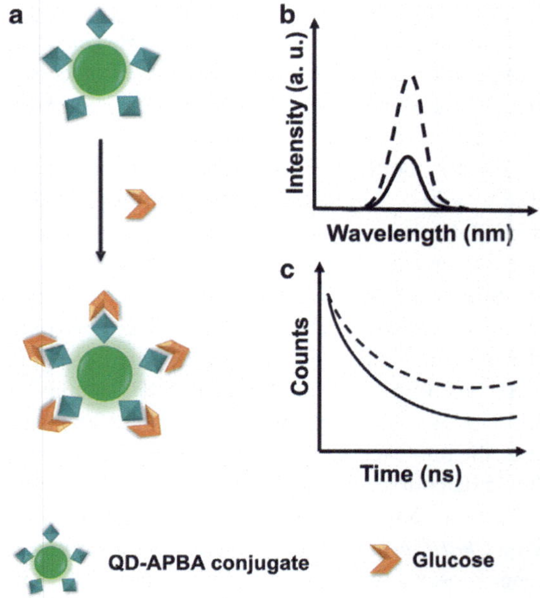

Fig. 6 **a** Schematic representation of the QD-APBA conjugate detecting glucose in alkaline medium. An improvement for **b** the fluorescence intensity along with **c** longer fluorescence lifetime was observed for the QD-APBA conjugate after binding with the glucose (dashed line) when compared with QD-APBA conjugate (filled line)

EDC/NHS covalent coupling. Results showed that the interaction of APBA molecules with QDs led to a decreasing in the fluorescence intensity and lifetime, which also confirmed the conjugation. Nevertheless, upon glucose recognition by the QD-APBA conjugate, a fluorescence intensity recovery and longer lifetimes were observed, as schematized in Fig. 6. Furthermore, the selectivity for glucose was studied by incubating the nanosensor with samples containing ions usually found in the intracellular environment (Na^+, K^+, Ca^{2+}, or Mg^{2+}) or bovine serum albumin, and almost none interference in fluorescence lifetimes occurred. The glucose detection limit based on fluorescence lifetimes was 0.1 mmol L^{-1}. Moreover, the fluorescence nanosensor was also used to detect glucose in MDA-MB-231 breast cancer cells by FLIM, and satisfactory outcomes were achieved. The authors conclude that the proposed method could be potentially applied to cancer studies and diagnosis [54].

8 Conclusions

QDs have been conjugated mainly to lectins purified from plants, usually exploring either adsorption or covalent strategies. Regarding "borono-lectins", both the thiol-metal affinity and the covalent linkage have been used to develop the conjugates. Inhibition assays with carbohydrates were broadly applied to confirm the efficiency and specificity of the conjugation.

A diversity of applications was found in the studies using the conjugates with purposes of diagnosis and therapeutic approaches, as well as comprehension of biological processes. Techniques such as fluorescence microscopy and flow cytometry were employed in the analyses, demonstrating the potential of QD-lectin or QD-"borono-lectin" conjugates to unravel the glycocode in mammalian cells and tissues, microorganisms, and insect larva. Moreover, the nanoprobes showed great performance in detecting free sugars and glycoproteins through emission intensities, fluorescence lifetimes, FRET phenomenon, or PEC assays.

Therefore, QD-lectin and QD-"borono-lectin" conjugates have been providing many advances to glycobiology, encouraging not only the preparation of novel nanosystems but also innovative applications with conjugates already developed.

Acknowledgements The authors are thankful to the Conselho Nacional de Desenvolvimento Científico e Tecnológico (CNPq), Coordenação de Aperfeiçoamento de Pessoal de Nível Superior (CAPES), and Fundação de Amparo à Ciência e Tecnologia do Estado de Pernambuco (FACEPE). This work is also linked to the National Institute of Photonics (INCT-INFo) and LARNANO/UFPE.

References

1. Kaushik A (2019) Biomedical nanotechnology related grand challenges and perspectives. Front Nanotechnol 1:1–4
2. Auría-Soro C, Nesma T, Juanes-Velasco P et al (2019) Interactions of nanoparticles and biosystems: microenvironment of nanoparticles and biomolecules in nanomedicine. Nanomaterials 9(10):1365
3. Qamar SA, Asgher M, Khalid N, Sadaf M (2019) Nanobiotechnology in health sciences: current applications and future perspectives. Biocatal Agr Biotechnol 22:101388
4. Xu J, Zheng J (2018) Quantum dots and nanoclusters. In: Nano-inspired biosensors for protein assay with clinical applications. Elsevier, pp 67–90
5. Bruchez M Jr, Moronne M, Gin P et al (1998) Semiconductor nanocrystals as fluorescent biological labels. Science 281:2013–2016
6. Chan WCW, Nie S (1998) Quantum dot bioconjugates for ultrasensitive nonisotopic detection. Science 281:2016–2018
7. Keçili R, Büyüktiryaki S, Hussain CM (2019) Advancement in bioanalytical science through nanotechnology: past, present and future. TrAC – Trends Anal Chem 110:259–276
8. Badıllı U, Mollarasouli F, Bakirhan NK et al (2020) Role of quantum dots in pharmaceutical and biomedical analysis, and its application in drug delivery. TrAC - Trends Anal Chem 131:116013
9. Oswald DM, Cobb BA (2018) Emerging glycobiology tools: a renaissance in accessibility. Cell Immunol 333:2–8
10. Wu H, Kohler J (2019) Photocrosslinking probes for capture of carbohydrate interactions. Curr Opin Chem Biol 53:173–182
11. Reily C, Stewart TJ, Renfrow MB, Novak J (2019) Glycosylation in health and disease. Nat Rev Nephrol 15:346–366
12. Gautam AK, Sharma D, Sharma J, Saini KC (2020) Legume lectins: Potential use as a diagnostics and therapeutics against the cancer. Int J Biol Macromol 142:474–483
13. Matsumoto A, Miyahara Y (2018) 'Borono-lectin' based engineering as a versatile platform for biomedical applications. Sci Technol Adv Mater 19:18–30
14. Vokhmintcev KV, Samokhvalov PS, Nabiev I (2016) Charge transfer and separation in photoexcited quantum dot-based systems. Nano Today 11:189–211
15. Chan WCW, Maxwell DJ, Gao X et al (2002) Luminescent quantum dots for multiplexed biological detection and imaging. Curr Opin Biotechnol 13:40–46

16. Michalet X, Pinaud FF, Bentolila LA et al (2005) Quantum dots for live cells, in vivo imaging, and diagnostics. Science 307:538–544
17. Klingshirn CF (2007) Semiconductor optics, 3rd edn. Springer, Berlin
18. Smith AM, Nie S (2010) Semiconductor nanocrystals: structure, properties, and band gap engineering. Acc Chem Res 43:190–200
19. Ranjbar-Navazi Z, Omidi Y, Eskandani M, Davaran S (2019) Cadmium-free quantum dot-based theranostics. TrAC - Trends Anal Chem 118:386–400
20. Reshma VG, Mohanan PV (2019) Quantum dots: applications and safety consequences. J Lumin 205:287–298
21. Duan J, Song L, Zhan J (2009) One-pot synthesis of highly luminescent CdTe quantum dots by microwave irradiation reduction and their Hg^{2+}-sensitive properties. Nano Res 2:61–68
22. Pereira G, Monteiro CAP, Albuquerque GM et al (2019) (Bio)conjugation strategies applied to fluorescent semiconductor quantum dots. J Braz Chem Soc 30:2536–2560
23. Wegner KD, Hildebrandt N (2015) Quantum dots: bright and versatile *in vitro* and *in vivo* fluorescence imaging biosensors. Chem Soc Rev 44:4792–4834
24. Farzin MA, Abdoos H (2021) A critical review on quantum dots: From synthesis toward applications in electrochemical biosensors for determination of disease-related biomolecules. Talanta 224:121828
25. Lim SJ, Ma L, Schleife A, Smith AM (2016) Quantum dot surface engineering: toward inert fluorophores with compact size and bright, stable emission. Coord Chem Rev 320–321:216–237
26. Zhang Y, Clapp A (2011) Overview of stabilizing ligands for biocompatible quantum dot nanocrystals. Sensors 11:11036–11055
27. Karakoti AS, Shukla R, Shanker R, Singh S (2015) Surface functionalization of quantum dots for biological applications. Adv Coll Interface Sci 215:28–45
28. Hunsur Ravikumar C, Schiffman JD, Balakrishna RG (2018) Quantum dots as fluorescent probes: synthesis, surface chemistry, energy transfer mechanisms, and applications. Sens Actuators B Chem 258:1191–1214
29. Jha S, Mathur P, Ramteke S, Jain NK (2018) Pharmaceutical potential of quantum dots. Artif Cells Nanomed Biotechnol 46:57–65
30. Liu S, Su X (2014) The synthesis and application of I-III-VI type quantum dots. RSC Adv 4:43415–43428
31. Liu S, Zhang H, Qiao Y, Su X (2012) One-pot synthesis of ternary $CuInS_2$ quantum dots with near-infrared fluorescence in aqueous solution. RSC Adv 2:819–825
32. Patel J, Jain B, Singh AK et al (2020) Mn-doped ZnS quantum dots–An effective nanoscale sensor. Microchem J 155:104755
33. Tenório DPLA, Andrade CG, Cabral Filho PE et al (2015) CdTe quantum dots conjugated to concanavalin a as potential fluorescent molecular probes for saccharides detection in *Candida albicans*. J Photochem Photobiol B 142:237–243
34. Cunha CRA, Andrade CG, Pereira MIA et al (2018) Quantum dot–Cramoll lectin as novel conjugates to glycobiology. J Photochem Photobiol B 178:85–91
35. Oliveira WF, Cabrera MP, Santos NRM et al (2020) Evaluating glucose and mannose profiles in *Candida* species using quantum dots conjugated with Cramoll lectin as fluorescent nanoprobes. Microbiol Res 230:126330
36. Arruda CCL, Freitas DV, Seabra MABL et al (2020) CdTe-GSH as luminescent biomarker for labeling the larvicidal action of WSMoL lectin in *Aedes aegypti* larvae. Colloids Surf B Biointerfaces 187:110672
37. Cabral Filho PE, Pereira MIA, Fernandes HP et al (2015) Blood group antigen studies using CdTe quantum dots and flow cytometry. Int J Nanomed 10:4393–4404
38. He D, Wang D, Shi X et al (2017) Simultaneous fluorescence analysis of the different carbohydrates expressed on living cell surfaces using functionalized quantum dots. RSC Adv 7:12374–12381
39. Ayaz Ahmed KBA, Raja MRC, Mahapatra SK, Anbazhagan V (2017) Interaction of cadmium sulfide quantum dots with jacalin for specific recognition of cancer cells. J Lumin 182:283–288

40. Carvalho MET, Oliveira WF, Cunha CRA et al (2019) Evaluating the glycophenotype on breast cancer tissues with quantum dots-Cramoll lectin conjugates. Int J Biol Macromol 138:302–308
41. Cervantes-Jiménez R, Sánchez-Segura L, Estrada-Martínez LE et al (2020) Quantum dot labelling of tepary bean (*Phaseolus acutifolius*) lectins by microfluidics. Molecules 25:1041
42. Oliveira WF, Santos NRM, Cabrera MP et al (2020) *Bauhinia monandra* leaf lectin (BmoLL) conjugated with quantum dots as fluorescent nanoprobes for biological studies: application to red blood cells. Methods Appl Fluoresc 8:035009
43. Ebrahim S, Reda M, Hussien A, Zayed D (2015) CdTe quantum dots as a novel biosensor for *Serratia marcescens* and lipopolysaccharide. Spectrochim Acta Part A Mol Biomol Spectrosc 150:212–219
44. Gao X, Li D, Tong Y et al (2015) Highly sensitive fluorescence detection of glycoprotein based on energy transfer between $CuInS_2$ QDs and rhodamine B. Luminescence 30:1389–1394
45. Hovhannisyan VA, Bazukyan IL, Gasparyan VK (2017) Application of silver nanoparticles and CdSe quantum dots sensitized with of C-like lectin for detection of *St. aureus*. Comparison of various approaches. Talanta 175:366–369
46. Büyüktiryaki S, Yılmaz F, Say R, Ersöz A (2019) Proteinous polymeric shell decorated nanocrystals for the recognition of immunoglobulin M. J Fluoresc 29:609–617
47. Cao JT, Zhang PH, Liu YM et al (2015) Versatile microfluidic platform for the assessment of sialic acid expression on cancer cells using quantum dots with phenylboronic acid tags. ACS Appl Mater Interfaces 7:14878–14884
48. Gedda G, Chen GR, Yao YY et al (2017) Aqueous synthesis of dual-targeting Gd-doped $CuInS_2$/ZnS quantum dots for cancer-specific bimodal imaging. New J Chem 41:14161–14170
49. Qi Q, Zeng X, Peng L et al (2020) Tumor-targeting and imaging micelles for pH-triggered anticancer drug release and combined photodynamic therapy. J Biomater Sci Polym Ed 31:1385–1404
50. Monteiro CAP, Silva RC, Assis LG et al (2020) Quantum dots functionalized with 3-mercaptophenylboronic acids as novel nanoplatforms to evaluate sialic acid content on cell membranes. Colloids Surf B 193:111142
51. Wu P, Pan J-B, Li X-L et al (2015) Long-lived charge carriers in Mn-doped CdS quantum dots for photoelectrochemical cytosensing. Chem Eur J 21:5129–5135
52. Chang L, Wu H, He X et al (2017) A highly sensitive fluorescent turn-on biosensor for glycoproteins based on boronic acid functional polymer capped Mn-doped ZnS quantum dots. Anal Chim Acta 995:91–98
53. Li J, Wu H, Santana I et al (2018) Standoff optical glucose sensing in photosynthetic organisms by a quantum dot fluorescent probe. ACS Appl Mater Interfaces 10:28279–28289
54. Ripoll C, Orte A, Paniza L, Ruedas-Rama MJ (2019) A quantum dot-based FLIM glucose nanosensor. Sensors 19:4992
55. De Oliveira DR, Dos Santos ML, Migliolo L, Franco OL (2015) Insights into animal and plant lectins with antimicrobial activities. Molecules 20:519–541
56. Mishra A, Behura A, Mawatwal S et al (2019) Structure-function and application of plant lectins in disease biology and immunity. Food Chem Toxicol 134:110827
57. Ludwig AK, Kaltner H, Kopitz J, Gabius HJ (2019) Lectinology 4.0: altering modular (ga)lectin display for functional analysis and biomedical applications. Biochim Biophys Acta Gen Subj 1863:935–940
58. Johnson QR, Lindsay RJ, Petridis L, Shen T (2015) Investigation of carbohydrate recognition via computer simulation. Molecules 20:7700–7718
59. Bhutia SK, Panda PK, Sinha N et al (2019) Plant lectins in cancer therapeutics: targeting apoptosis and autophagy-dependent cell death. Pharmacol Res 144:8–18
60. Tommasone S, Allabush F, Tagger YK et al (2019) The challenges of glycan recognition with natural and artificial receptors. Chem Soc Rev 48:5488–5505
61. Müller C, Despras G, Lindhorst TK (2016) Organizing multivalency in carbohydrate recognition. Chem Soc Rev 45:3275–3302

62. Martínez-Alarcón D, Blanco-Labra A, García-Gasca T (2018) Expression of lectins in heterologous systems. Int J Mol Sci 19:616
63. Cagliari R, Kremer FS, Pinto L da S (2018) *Bauhinia* lectins: Biochemical properties and biotechnological applications. Int J Biol Macromol 119:811–820
64. Dang K, Zhang W, Jiang S et al (2020) Application of lectin microarrays for biomarker discovery. ChemistryOpen 9:285–300
65. Vandenborre G, Smagghe G, Van Damme EJM (2011) Plant lectins as defense proteins against phytophagous insects. Phytochemistry 72:1538–1550
66. Tsaneva M, Van Damme EJM (2020) 130 years of plant lectin research. Glycoconj J 37:533–551
67. Stillmark H (1888) Ueber Ricin, ein giftiges Fragment aus den Samen von Ricinus comm. L und einigen anderen Euphorbiaceen; Ph.D. Thesis, Kaiserliche Universität zu Dorpat: Tartu, Estonia
68. André S, Kaltner H, Manning JC et al (2015) Lectins: getting familiar with translators of the sugar code. Molecules 20:1788–1823
69. Singh S, Wong JH, Ng TB et al (2019) Biomedical applications of lectins from traditional Chinese medicine. Curr Protein Pept Sci 20(3):220–230
70. Otsuka H, Uchimura E, Koshino H et al (2003) Anomalous binding profile of phenylboronic acid with N-acetylneuraminic acid (Neu5Ac) in aqueous solution with varying pH. J Am Chem Soc 125:3493–3502
71. Uchimura E, Otsuka H, Okano T et al (2001) Totally synthetic polymer with lectin-like function: induction of killer cells by the copolymer of 3-acrylamidophenylboronic acid with N,N-dimethylacrylamide. Biotechnol Bioeng 72(3):307–314
72. Lorand JP, Edwards JO (1959) Polyol complexes and structure of the benzeneboronate ion. J Org Chem 24(6):769–774
73. Burnett TJ, Peebles HC, Hageman JH (1980) Synthesis of a fluorescent boronic acid which reversibly binds to cell walls and a diboronic acid which agglutinates erythrocytes. Biochem Biophys Res Commun 96(1):157–162
74. Jin S, Cheng Y, Reid S et al (2010) Carbohydrate recognition by boronolectins, small molecules, and lectins. Med Res Rev 30:171–257
75. Dowlut M, Hall DG (2006) An improved class of sugar-binding boronic acids, soluble and capable of complexing glycosides in neutral water. J Am Chem Soc 128:4226–4227
76. Han E, Ding L, Ju H (2011) Highly sensitive fluorescent analysis of dynamic glycan expression on living cells using glyconanoparticles and functionalized quantum dots. Anal Chem 83:7006–7012
77. Matsumoto A, Stephenson-Brown AJ, Khan T et al (2017) Heterocyclic boronic acids display sialic acid selective binding in a hypoxic tumor relevant acidic environment. Chem Sci 8:6165–6170
78. Nakagawa Y, Yukishige I (2012) Molecular architecture and therapeutic potential of lectin mimics. Adv Carbohydr Chem Biochem 68:1–58
79. Bie Z, Chen Y, Ye J et al (2015) Boronate-affinity glycan-oriented surface imprinting: a new strategy to mimic lectins for the recognition of an intact glycoprotein and its characteristic fragments. Angew Chem 127:10349–10353
80. Liu A, Peng S, Soo JC et al (2011) Quantum dots with phenylboronic acid tags for specific labeling of sialic acids on living cells. Anal Chem 83:1124–1130
81. Wu X, Chen XX, Jiang YB (2017) Recent advances in boronic acid-based optical chemosensors. Analyst 142:1403–1414
82. Hermanson GT (2013) Bioconjugate techniques. Academic press, London, p 2013
83. Klykov O, Weller MG (2015) Quantification of N-hydroxysuccinimide and N-hydroxysulfosuccinimide by hydrophilic interaction chromatography (HILIC). Anal Methods 7:6443–6448
84. Foubert A, Beloglazova NV, Rajkovic A et al (2016) Bioconjugation of quantum dots: review & impact on future application. TrAC - Trends Anal Chem 83:31–48

85. Shan Y, Wang L, Shi Y et al (2008) NHS-mediated QDs-peptide/protein conjugation and its application for cell labeling. Talanta 75:1008–1014
86. Kulkarni SK (2015) Analysis Techniques. Nanotechnology: principles and practices (3rd Edition). Springer, NY, USA, pp 135–198
87. Brandley BK, Schnaar RL (1986) Cell-surface carbohydrates in cell recognition and response. J Leukoc Biol 40:97–111
88. Gorelik E, Galili U, Raz A (2001) On the role of cell surface carbohydrates and their binding proteins (lectins) in tumor metastasis. Cancer Metastasis Rev 20:245–277
89. Maldonado RF, Sá-Correia I, Valvano MA (2016) Lipopolysaccharide modification in gram-negative bacteria during chronic infection. FEMS Microbiol Rev 40:480–493
90. Moran AP, Gupta A, Joshi L (2011) Sweet-talk: Role of host glycosylation in bacterial pathogenesis of the gastrointestinal tract. Gut 60:1412–1425
91. Ruiz-Herrera J, Victoria Elorza M, Valentín E, Sentandreu R (2006) Molecular organization of the cell wall of *Candida albicans* and its relation to pathogenicity. FEMS Yeast Res 6:14–29
92. Fontaniella B, Millanes A-M, Vicente C et al (2004) Concanavalin A binds to a mannose-containing ligand in the cell wall of some lichen phycobionts. Plant Physiol Biochem 42:773–779
93. Correia MTS, Coelho LCBB (1995) Purification of a glucose/mannose specific lectin, isoform 1, from seeds of *Cratylia mollis* Mart. (Camaratu Bean). Appl Biochem Biotechnol 55:261–273
94. Oliveira PSS de, Rêgo MJB de M, Da Silva RR et al (2013) *Cratylia mollis* 1, 4 lectin: a new biotechnological tool in IL-6, IL-17A, IL-22, and IL-23 induction and generation of immunological memory. BioMed Res Int 2013:263968
95. Fernandes MP, Inada NM, Chiaratti MR et al (2010) Mechanism of *Trypanosoma cruzi* death induced by *Cratylia mollis* seed lectin. J Bioenerg Biomembr 42:69–78
96. Andrade CAS, Correia MTS, Coelho LCBB et al (2004) Antitumor activity of *Cratylia mollis* lectin encapsulated into liposomes. Int J Pharm 278:435–445
97. Lee RT, Lee YC (2000) Affinity enhancement by multivalent lectin-carbohydrate interaction. Glycoconj J 17:543–551
98. Santos AFS, Argolo ACC, Coelho LCBB, Paiva PMG (2005) Detection of water soluble lectin and antioxidant component from *Moringa oleifera* seeds. Water Res 39:975–980
99. Coelho JS, Santos NDL, Napoleão TH et al (2009) Effect of *Moringa oleifera* lectin on development and mortality of *Aedes aegypti* larvae. Chemosphere 77:934–938
100. Higuera A, Ramírez JD (2019) Molecular epidemiology of dengue, yellow fever, Zika and Chikungunya arboviruses: an update. Acta Trop 190:99–111
101. Moura KS, Silva HRC, Dornelles LP et al (2016) Coagulant activity of water-soluble *Moringa oleifera* lectin is linked to lowering of electrical resistance and inhibited by monosaccharides and magnesium ions. Appl Biochem Biotechnol 180:1361–1371
102. Gallegos IB, Pérez-Campos E, Martinez M et al (2012) Expression of antigen tf and galectin-3 in fibroadenoma. BMC Res Notes 5:694
103. An HJ, Kronewitter SR, de Leoz MLA, Lebrilla CB (2009) Glycomics and disease markers. Curr Opin Chem Biol 13:601–607
104. Sankaranarayanan R, Sekar K, Banerjee R et al (1996) A novel mode of carbohydrate recognition in jacalin, a *Moraceae* plant lectin with a β-prism fold. Nat Struct Biol 3:596–603
105. Rêgo MJBDM, Vieira De Mello GS, Da Silva Santos CA et al (2013) Implications on glycobiological aspects of tumor hypoxia in breast ductal carcinoma in situ. Med Mol Morphol 46:92–96
106. Friedl P, Alexander S (2011) Cancer invasion and the microenvironment: plasticity and reciprocity. Cell 147:992–1009
107. Coelho LCBB, Silva MBR (2000) Simple method to purify milligram quantities of the galactose-specific lectin from the leaves of *Bauhinia monandra*. Phytochem Anal 11:295–300
108. Ma F, Li C-C, Zhang C-Y (2018) Development of quantum dot-based biosensors: principles and applications. J Mater Chem B 6:6173–6190

109. Zhang X, Hu Y, Yang X et al (2019) FÖrster resonance energy transfer (FRET)-based biosensors for biological applications. Biosens Bioelectron 138:111314
110. Cardoso Dos Santos M, Algar WR, Medintz IL, Hildebrandt N (2020) Quantum dots for Förster resonance energy transfer (FRET). TrAC - Trends Anal Chem 125:115819
111. Wellman-Labadie O, Lakshminarayanan R, Hincke MT (2008) Antimicrobial properties of avian eggshell-specific C-type lectin-like proteins. FEBS Lett 582:699–704
112. Adamczyk-Woźniak A, Sporzyński A (2020) The influence of ortho-substituents on the properties of phenylboronic acids. J Organomet Chem 913:121202
113. Schauer R, Kamerling JP (2018) Exploration of the Sialic acid world. In: Advances in carbohydrate chemistry and biochemistry. Academic Press Inc., pp 1–213
114. Zhong X, Bai H-J, Xu J-J et al (2010) A reusable interface constructed by 3-aminophenylboronic acid-functionalized multiwalled carbon nanotubes for cell capture, release, and cytosensing. Adv Func Mater 20:992–999

Chapter 15
Nanotechnology in Venom Research: Recent Trends and Its Application

Pushpendra Singh, Manish Kumar Tripathi, and Dhruv Kumar

1 Introduction

Nanotechnology is well-thought-out to be the leading engineering novel innovation in the different fields of sciences such as computer science, electrical science, medical science, food science, and agricultural science. A nanoparticle is generally distinct as a particle of matter that is the diameter in between 1 and 100 nm (nm). The principal parameters of nanoparticles are their size, shape, quantum mechanics effects, solvent affinity surface, interfacial layer, characteristics, and inner structure. Nanoparticles habitually have exclusive physical–chemical and biological properties. Many different types of nanosized structures are developed to improve nanotechnology strategies in healthcare systems including nanoparticles, nanorods, nanotubes, nanowires, nanoribbons, nanostars, nanosheets, nanosheets, nanobelts, nanofibers, quantum dots, and hollow spheres [1]. Nowadays application of nanotechnology is more exciting in the biological science areas, where developments are being made in both the diagnostics and treatment areas at a molecular level. Nanoparticles are capable vehicles for drug delivery by easy handling to formulate carriers with the objective of distributing the drugs to specific target sites; such an advantage

P. Singh
ICAR- National Institute of High Security Animal Diseases Bhopal, Bhopal, Madhya Pradesh 462021, India

M. K. Tripathi
Department of Pharmaceutical Engineering & Technology, Indian Institute of Technology (Banaras Hindu University), Varanasi 221005, India

D. Kumar (✉)
Amity Institute of Molecular Medicine & Stem Cell Research, Amity University Uttar Pradesh Noida (UP), Noida 201313, India
e-mail: dkumar13@amity.edu

S. Gopi et al. (eds.), *Nanotechnology for Biomedical Applications*,
Materials Horizons: From Nature to Nanomaterials,
https://doi.org/10.1007/978-981-16-7483-9_17

improves drug safety. Nanoparticles efficiently carry drugs, vaccine, proteins, and DNA to target sites.

Nanotechnology at present has bifurcated in a huge number of diverse areas of sciences for of its excellent benefits in terms of improving the performance in many fields such as biomedical, energy and environment, manufacturing. As the precise use of nanotechnology in the health system, nanomedicine has appeared to offer novel and innovative elucidations for the medicine's unsolved complications in diagnostics and treatments. Nanomedicine is generally the technology or area of nanotechnologies in healthcare systems and the significant benefits that have already been seen implicate the use of nanoparticles to enhance the performance of drug ingredients.

This chapter presents a brief appraisal of the recent trends to nanotechnology in nanomedicine used for venom research. It also covers the recent developments in the fabrication of nanomedicine based on two types of application in venom research; first is nanoformulation of anti-snake venom (antidote) against snake venom and second is venom components-based nanomedicine against various diseases such as cancer and arthritis.

2 Nanomaterials

Overall, there are three main types of nanomaterials-based components of nanoparticles that are, natural and synthetic (engineered). Natural nanomaterials are formed either by through natural processes or through anthropogenic activities. Naturally synthesized nanomaterials can be produced by the bodies of organisms, insects, plants, animals, and human bodies along with their components such as proteins, toxin, enzymes, and secondary metabolites. Synthetic (engineered) nanomaterials are complex in shape and require specific processes for their fabrication such as engine exhaust, mechanical grinding, and smoke, or are synthesized by physical, chemical, hybrid methods [2]. Nanomaterials have their own characteristics such as chemical bioactivity, biological reactivity as well as tissue and organ diffusion ability, and countless bioavailability. Nanomaterials have made a remarkable impact in the management of various types of diseases, as demonstrated by the numerous nanoparticle-based nanomedicine and distribution systems that are in clinical use.

3 Venom Research

Snake venoms are mixtures of different protein families, and each of these families contains many various toxins or toxin isoforms. Snake venoms contain different compounds typically proteins, polypeptides, enzymes, and numerous other substances with venomous and fatal properties that vary extensively between and within snake species. Venoms are rich and complex natural sources of biologically active molecules such as enzymes (phospholipases, serine proteases, metalloproteinase, lectins, l-amino-acid oxidases proteases, oxidases, etc.), non-enzymatic components (bradykinin potentiating factors, natriuretic factors, and integrin) and different peptide toxins with specific pathophysiological functions. Snake venom is a biologically significant functional attribute in venomous snakes, and its composition and activity regularly coevolve with the physiology of the target animals it is deployed against. Peptide or enzymes toxins isolated from venoms target are mainly membrane receptors, ion channels, and components of the hemostatic system with high selectivity and affinity. The pharmacological usage of snake venom and their components as a natural product resource of constituents for biological applications has received much attention, presently, from pharmaceutical companies and experts in the fields of nanotechnology research. Venom research originally began with the interest of thoughtful snakebites and associated medical treatments. Consequently, additional reasons that make snake venoms attractive to scientist around the world are richness, specialization, and efficiency of their components. Presently, snake venom and their components are used as valuable and prevailing pharmacologically research tools. Venom derived medicine is produced by the pharmaceutical industry as Captopril, Aggrastat, and Eptifibatide, all designed based on snake venom components [3]. The recent trends in venom research merge with the advancement of nanotechnology. Linking nanomaterials with venom and their components could act as an enhanced interface for an antidote, targeted therapy, drug delivery, and superior molecular level interaction thereby increasing the efficacy of the venom molecule having therapeutics value (Fig. 1).

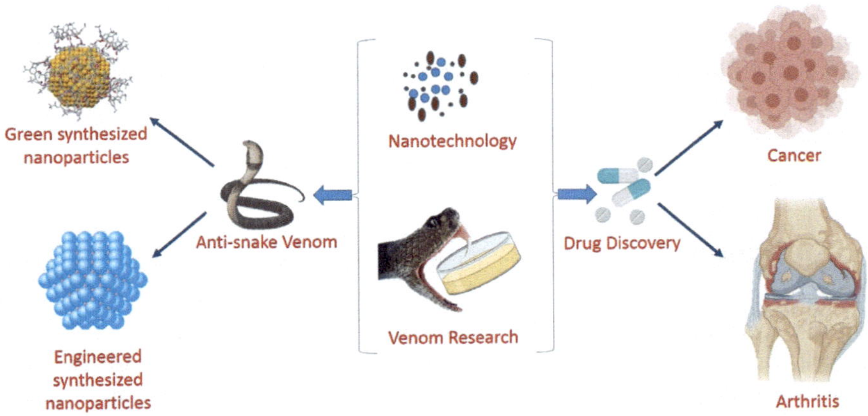

Fig. 1 Recent trends and developments in "Venom Nanotechnology"

3.1 *Nanomaterials Synthesized from Venom and Their Components in Drug Discovery*

Snake venoms are rich sources of different types of toxin such as enzymes, proteolytic proteins, peptides, neurotransmitters, other non-toxic compounds [4]. As a consequence of development and natural selection, toxins from snake venoms show wide-ranging pharmacological activities. Since the venom toxin targets are related to biological functions, with many of them have significant roles in human diseases, some of the venom and their components could use in the design of new therapeutic agents. Development of novel drugs signifies one of the greatest interesting activities in the biotechnology and pharmaceutical industry. Nanotechnology methods to develop new drugs have involved much consideration in many areas to design nanoparticle conjugates with venom for improved efficacy of the potential venom components (Table 1).

4 Anti-Arthritic

Gold nanoparticle (GNP) is conjugated with purified *Naja kaouthia* protein toxin (NKCT1), a cytotoxic protein toxin from Indian cobra venom synthesized by NaBH4 reduction method for the evaluation of anti-arthritic activity and toxicity in male albino rats. This study established that GNP-NKCT1 conjugation improved the anti-arthritic activity and reduced toxicity profile of NKCT1 [5]. And another study revealed that GNP-NKCT1 conjugate down-regulated PI3K/Akt and mTOR expression which induces an alternate cell death pathway other than apoptosis in the form of autophagy in the leukemic cell [6]. Snake venom protein toxin *Naja kaouthia* cytotoxin 1 (NKCT1) is conjugated with gold nanoparticle; this functionalized conjugated has potential to the restoration of the arthritic markers and bone degradation in treatment of osteoarthritic [7].

5 Anticancer

The efficacy of GNP-NKCT1 conjugate on cancer cells was analyzed in vitro fibroblast cell and peripheral blood mononuclear lymphocytes assay along with in vivo hepato-carcinoma mice, and this study suggested that GNP-NKCT1 conjugate might be an effective agent against anti-carcinoma or anti-sarcoma targets [8]. Another gold nanoparticles are conjugated with the purified NN-32 toxin from *Naja naja* venom and its evaluated anticancer potential against human breast cancer cell MCF-7, and MDA-MB-231 lines result in show arrested cell division and enhance its biomedical function [9]. Iron oxide, multifunctional super-paramagnetic nanoparticles, conjugated with chlorotoxin (CTX), an essential toxin of venom that binds specifically to

Table 1 Nanomaterials conjugated with venom and their components in drug discovery

S. No	Nanoparticles functionalized with	Venom source	Activity	References
1	Gold nanoparticle is conjugate with venom peptide	*Vespa orientalis* wasp venom	Antimicrobial activity against multidrug-resistant pathogens	[11]
2	Gold nanoparticle is conjugated with NKCT1	*Naja kaouthia* snake venom	Anti-arthritic activity and toxicity in male albino rats	[5]
3	Gold nanoparticle is conjugated with NKCT1	*Naja kaouthia* snake venom	Regulated PI3K/Akt and mTOR expression in apoptosis and autophagy in leukemic cell	[6]
4	Gold nanoparticle is conjugated with NKCT1	*Naja kaouthia* snake venom	Anti-carcinoma or anti-sarcoma activity	[8]
5	Gold nanoparticle is conjugated with purified NN-32 toxin	*Naja naja* snake venom	Anticancer potential against human breast cancer cell MCF-7 and MDA-MB-231 lines	[9]
6	Iron oxide nanoparticle is conjugated with chlorotoxin	scorpion venom	Effective therapeutic agents in treatment of glioma	[10]
7	Silica nanoparticles combined with extracted snake venom	*Walterinnesia aegyptia snake venom*	Inhibited tumor growth in human MM cell line U266 and MM-bearing mouse model. Treatment of breast cancer- and prostate cancer-bearing experimental mouse models	[12, 13]

the glioma cell surface as a specific chloride channel and matrix metalloproteinase-2 blocker which show effective therapeutic agents in the treatment of glioma [10].

5.1 Anti-Snake Venom (Antidote) Made from Nanoparticles in Treat Snakebites

Envenoming by snakebites found a severe and profound health issue worldwide that affects approximately 2.5 million people and causes more than 100,000 deaths annually [14]. This condition is mainly acute in tribal and rural settings of sub-Saharan Africa and Asian countries. The support therapy in the snakebite envenoming is the only parenteral administration of animal-derived antivenoms, constituted by antibodies (purified IgG or IgG fragments) from the plasma of large animals immunized with venoms. The synthetic compounds, small molecule inhibitors, and natural medicinal plants, obtainable nearby and used widely by traditional healers, villagers, therefore need attention [15]. However, all these alternate compounds to synthesized nanoparticles have to antivenom potential against different snake venom as an antidote.

Nanobiotechnology is a fast-developing research area and has the unique probable potential to develop new therapeutic agents. The nanoparticles are well-known metals like silver, gold, iron copper, zinc, mercury, and platinum to have significant applications against the venom. Silver nanoparticles (SNP) and gold nanoparticles (GNP) synthesized by engineered approaches and green synthesis utilize medicinal plants or extracts and have little or no toxicity. Medicinal plants and their secondary metabolite play a significant role during the green synthesis of nanoparticles by acting as reducing or capping agents.

Green synthesized silver nanoparticles produced by using the aqueous extract of *Zanthoxylum chalybeum* roots and evaluate anti-proteolytic activity against *Bitis arietans* snake venom [16]. Silver nanoparticles synthesized by the green synthesis Indian male fern, *Dryopteris Cochleata*, show *Naja naja* venom inhibition activity through in vitro phospholipases A_2 assay and tissue damage activity [17]. The gold nanoparticle is conjugated with andrographolide, a diterpenoid compound found in the aerial parts of *Andrographis paniculata,* and this nanoparticle has efficacy against *Daboia russellii russellii* venom (DRRV)-induced local damage, organ toxicity, and inflammatory response in Swiss albino male mice [18]. Gold nanoparticles (GNP) synthesized by 2-hydroxy 4 methoxy benzoic acid (HMBA) extracted from *H. indicus* root and revealed neutralization efficacy against viper venom [19]. Silver nanoparticles conjugated venom toxins of venomous animals are studies against the neuromuscular blockade, myotoxic effects induced by *Bothrops jararacussu* venom [20]. Engineered synthesized nanoparticles such as titanium dioxide nanoparticles have exhibit neutralized the *Daboia russelii* venom (DRV) and *Naja kaouthia* venom (NKV)-induced lethal activity [21] (Table 2).

6 Conclusion

In conclusion, nanotechnology in modern science is novel and has improved value for size, distribution, and morphology determination of various biomolecules. To fabricate nanoparticles, the source of the used metals are gold, silver, mercury, and platinum, which have conjugated with natural materials for significant applications in biomedical research. Since nanotechnology is a speedily developing fields with the noteworthy potential to develop a new generation of therapeutic agents. Venom research is a unique area of medical science that is combined with nanotechnology

Table 2 Anti-snake venom (antidote) made from nanoparticles in treat snakebites

S. No	Nanoparticles	Source	Activity	References
Green synthesized nanoparticles (Natural) as antidote				
1	Silver nanoparticles from rhizomes	*Dryopteris cochleata* (Indian male fern)	In vitro phospholipases A_2 assay and tissue damage activity	[17]
2	Silver nanoparticles from aqueous extract of roots	*Zanthoxylum chalybeum*	Anti-proteolytic activity against *Bitis arietans* snake venom and antimicrobial activity	[16]
3	Gold nanoparticles (AuNPs) conjugated with andrographolide	*Andrographis paniculata*	Efficacy against *Daboia russellii russellii* venom (DRRV) induced local damage, organ toxicity and inflammatory response in animal models	[18]
4	Silver nanoparticles from whole plants	*Trichodesma indicum*	Neutralization parameters of Naja naja (cobra) snake venom	[22]
5	Gold nanoparticles (GNP) from HMBA	*Hemidesmus indicus*	In vivo and in vitro neutralizing studies againstsnake (viper) venom	[19]
Engineered synthesized nanoparticles (Synthetic) as antidote				
6	Titanium dioxide nanoparticles	Titanium dioxide	Neutralized the *Daboia russelii* venom (DRV) and *Naja kaouthia* venom (NKV)-induced lethal activity	[21]
7	Engineered nanoparticles	Chemicals	Inhibited dose-dependently the dermonecrotic activity of *Naja nigricollis* venom	[23]
8	Gold nanoparticle-conjugated 2-hydroxy-4-methoxy benzoic acid (GNP-HMBA)3	Commercial herbal compound 2-hydroxy-4-methoxy benzoic acid	Neutralized RVV-induced toxicities (nephrotoxicity, myotoxicity and hepatotoxicity) in male albino mice	[24]

(continued)

Table 2 (continued)

S. No	Nanoparticles	Source	Activity	References
9	Silver nanoparticles synthesized by using chemical reduction method	Silver nitrate, and tri-sodium citrate	Inhibition of crude viper snake venom (*Doboia russellie*) action through protease and whole blood clotting test	[25]

and formed a novel research area "Venom Nanotechnology." Synthesized nanoparticles made from natural and metals sources that similar nanomaterials molecules and has a better chance to treat different diseases. Producing conventional antivenom (ASV) is not easy and have several problems. Alternative and recent approaches, nanoparticles capable of binding a prevailing one toxin but many.

References

1. Kumar S, Bhushan P, Bhattacharya S (2018) Fabrication of nanostructures with bottom-up approach and their utility in diagnostics, therapeutics, and others. In: Environmental, chemical and medical sensors. pp 167–198. https://doi.org/10.1007/978-981-10-7751-7_8
2. Jeevanandam J, Barhoum A, Chan YS, Dufresne A, Danquah MK (2018) Review on nanoparticles and nanostructured materials: history, sources, toxicity and regulations. Beilstein J Nanotechnol 9:1050–1074. https://doi.org/10.3762/bjnano.9.98
3. Mohamed Abd El-Aziz T, Garcia Soares A, Stockand JD (2019) Snake venoms in drug discovery: valuable therapeutic tools for life saving. Toxins (Basel). 11:564. https://doi.org/10.3390/toxins11100564
4. Singh P, Yasir M, Hazarika R, Sugunan S, Shrivastava R (2017) A review on venom enzymes neutralizing ability of secondary metabolites from medicinal plants. J Pharmacopuncture 20:173–178. https://doi.org/10.3831/KPI.2017.20.020
5. Saha P, Bhowmik T, Dasgupta A, Gomes A (2014) Nano gold conjugation, anti-arthritic potential and toxicity studies of snake Naja kaouthia (Lesson, 1831) venom protein toxin NKCT1 in male albino rats and mice. Indian J Exp Biol 52:763–772
6. Bhowmik T, Gomes A (2016) NKCT1 (purified Naja kaouthia protein toxin) conjugated gold nanoparticles induced Akt/mTOR inactivation mediated autophagic and caspase 3 activated apoptotic cell death in leukemic cell. Toxicon 121:86–97. https://doi.org/10.1016/j.toxicon.2016.08.004
7. Gomes A, Saha PP, Bhowmik T, Dasgupta AK, Dasgupta SC (2016a) Protection against osteoarthritis in experimental animals by nanogold conjugated snake venom protein toxin gold nanoparticle-Naja kaouthia cytotoxin 1. Indian J Med Res. https://doi.org/10.4103/ijmr.IJMR_1078_14
8. Bhowmik T, Saha PP, Sarkar A, Gomes A (2017) Evaluation of cytotoxicity of a purified venom protein from Naja kaouthia (NKCT1) using gold nanoparticles for targeted delivery to cancer cell. Chem Biol Interact 261:35–49. https://doi.org/10.1016/j.cbi.2016.11.007
9. Attarde SS, Pandit SV (2020) Anticancer potential of nanogold conjugated toxin GNP-NN-32 from Naja naja venom. J Venom Anim Toxins Incl Trop Dis 26:e20190047–e20190047. https://doi.org/10.1590/1678-9199-JVATITD-2019-0047

10. Fu Y, An N, Li K, Zheng Y, Liang A (2012) Chlorotoxin-conjugated nanoparticles as potential glioma-targeted drugs. J Neurooncol 107:457–462. https://doi.org/10.1007/s11060-011-0763-6
11. Jalaei J, Layeghi-Ghalehsoukhteh S, Hosseini A, Fazeli M (2018) Antibacterial effects of gold nanoparticles functionalized with the extracted peptide from Vespa orientalis wasp venom. J Pept Sci 24:e3124. https://doi.org/10.1002/psc.3124
12. Al-Sadoon MK, Rabah DM, Badr G (2013) Enhanced anticancer efficacy of snake venom combined with silica nanoparticles in a murine model of human multiple myeloma: molecular targets for cell cycle arrest and apoptosis induction. Cell Immunol 284:129–138. https://doi.org/10.1016/j.cellimm.2013.07.016
13. Badr G, Al-Sadoon MK, Rabah DM (2013) Therapeutic efficacy and molecular mechanisms of snake (Walterinnesia aegyptia) venom-loaded silica nanoparticles in the treatment of breast cancer- and prostate cancer-bearing experimental mouse models. Free Radic Biol Med 65:175–189. https://doi.org/10.1016/j.freeradbiomed.2013.06.018
14. Bala E, Hazarika R, Singh P, Yasir M, Shrivastava R (2018) A biological overview of Hyaluronidase: a venom enzyme and its inhibition with plants materials. In: Materials today: proceedings. https://doi.org/10.1016/j.matpr.2017.12.252
15. Singh P, Yasir M, Shrivastava R (2019) Screening and identification of plant metabolites against snake venom enzymes using in vitro to in silico approach. Indian J Pharm Educ Res 53:s607–s618. https://doi.org/10.5530/ijper.53.4s.156
16. Mahamadi C, Wunganayi T (2018) Green synthesis of silver nanoparticles using Zanthoxylum chalybeum and their antiprolytic and antibiotic properties. Cogent Chem 4:1–12. https://doi.org/10.1080/23312009.2018.1538547
17. Singh P, Yasir M, Khare R, Shrivastava R (2020) Green synthesis of silver nanoparticles using Indian male fern (Dryopteris Cochleata), operational parameters, characterization and bioactivity on Naja naja venom neutralization. Toxicol Res (Camb) 9:706–713. https://doi.org/10.1093/toxres/tfaa070
18. Ghosh S, Dasgupta SC, Dasgupta AK, Gomes A, Gomes A (2020) Gold Nanoparticles (AuNPs) conjugated with andrographolide ameliorated viper (Daboia russellii russellii) venom-induced toxicities in animal model. J Nanosci Nanotechnol 20:3404–3414. https://doi.org/10.1166/jnn.2020.17421
19. Gomes A, Sengupta J, Ghosh S, Gomes A (2016b) Application of gold nanoparticle conjugation with 2-hydroxy-4-methoxy benzoic acid (HMBA) from hemidesmus indicus root enhancing neutralization of snake (Viper) venom activity. J Nanosci Nanotechnol. https://doi.org/10.1166/jnn.2016.11777
20. Oliveira ICF, de Paula MO, Lastra HCB, de Alves BB, Moreno DAN, Yoshida EH, Amaral Filho J, Cogo JC, Varanda EA, Rai M, dos Santos CA, Oshima-Franco Y (2019) Activity of silver nanoparticles on prokaryotic cells and Bothrops jararacussu snake venom. Drug Chem Toxicol 42:60–64. https://doi.org/10.1080/01480545.2018.1478850
21. Chakrabartty S, Alam MI, Bhagat S, Alam A, Dhyani N, Khan GA, Alam MS (2019) Inhibition of snake venom induced sterile inflammation and PLA2 activity by Titanium dioxide Nanoparticles in experimental animals. Sci Rep 9:11175. https://doi.org/10.1038/s41598-019-47557-y
22. Bukke S, Beeram E, Divya B, Pallavi C, Venkataswamy M, Kemparaj KKT (2018) Effect of silver nano particles synthesized of Trichodesma indicum against Naja Naja (Cobra) Venom. Int J Pharm Sci Res. https://doi.org/10.13040/ijpsr.0975-8232.9(8).3291-96
23. O'Brien J, Lee S-H, Gutiérrez JM, Shea KJ (2018) Engineered nanoparticles bind elapid snake venom toxins and inhibit venom-induced dermonecrosis. PLoS Negl Trop Dis 12:e0006736–e0006736. https://doi.org/10.1371/journal.pntd.0006736
24. Saha K, Gomes A (2017) Russell's viper venom induced nephrotoxicity, myotoxicity, and hepatotoxicity—Neutralization with gold nanoparticle conjugated 2-hydroxy-4-methoxy benzoic acid in vivo. Indian J Exp Biol 55:7–14
25. Hingane VC, Pangam D, Dongre PM (2018) Inhibition of crude viper venom action by silver nanoparticles: a biophysical and biochemical study. Biophys Physicobiol 15:204–213. https://doi.org/10.2142/biophysico.15.0_204

Chapter 16
Ionogels for Biomedical Applications

Bojan Kopilovic, Francisca A. e Silva, Augusto Q. Pedro, João A. P. Coutinho, and Mara G. Freire

1 Introduction

With the aging of the global population and growing prevalence of diseases such as cancer, cardiovascular diseases, and diabetes [1, 2], the development of innovative approaches to improve diagnosis and treatment is in high demand. To address these needs, engineered materials based on basic molecular building blocks have been developed, with polymeric materials being one of the principals. Most of these have been inspired by the simplicity and efficiency with which nature builds complex self-assembled structures to carry out different biological processes [3, 4]. Along the years, a variety of functional polymers has been studied for biomedical applications [3, 5], with their use in micelles, gels, composites, fibers, particles, brush polymers, form memory polymers, among others [6–8]. The use of these novel materials in a biomedical perspective is quite recent. However, according to the findings reported so far, some outlined in this chapter, these materials have high potential due to their advanced features, including stimulus sensitivity, swelling ability, antimicrobial activity, high biocompatibility, resistance to rheological stress, and even self-healing capability [3, 9–11].

The definition of a gel varies in the literature, and there is still a terminological confusion [12–14]. Herein, a gel is considered as a rigid viscoelastic substance created by a main component, a solvent, and a minor component, a gelator [15]. The gelator forms a solid-like microfibrillar network that percolates the solvent and inhibits its flow due to capillary forces [16]. Many of the gel characteristics, such as thermal and chemical stability, shape, dimensions, surface chemistry, mechanical properties, and

B. Kopilovic · F. A. e Silva · A. Q. Pedro · J. A. P. Coutinho · M. G. Freire (✉)
Chemistry Department, CICECO—Aveiro Institute of Materials, University of Aveiro, Campus Universitário de Santiago, 3810-193 Aveiro, Portugal
e-mail: maragfreire@ua.pt

S. Gopi et al. (eds.), *Nanotechnology for Biomedical Applications*,
Materials Horizons: From Nature to Nanomaterials,
https://doi.org/10.1007/978-981-16-7483-9_18

porosity can be tailored to targeted functionalities to satisfy specifications of a particular biomedical application [17]. Among the various gels that can be prepared, those with different characteristics, such as softness, highly porous structure, relatively hydrophilic nature, enhanced biodegradability, and good biocompatibility, may be suitable for biomedical applications [3, 13]. In this field, novel stimuli-responsive gels and ionic liquid (IL)-containing gels (ionogels) have gained a growing interest [18–20].

Ionogels are an innovative class of soft materials that gained significant popularity in recent years [9, 21]. They are comprised of ILs, which are incorporated as a supramolecular, polymeric, or confined network. Some of these gels, with a melting point above 140 °C [17, 22], were shown to be smart polymeric conducting materials, merging the chemical versatility of an IL with the morphological flexibility of a biopolymer [23]. If properly designed, ionogels may exhibit numerous advantages, including high ionic conductivity [17, 22], high thermal and electrochemical stability [17, 22], enhanced biocompatibility and low (cyto)toxicity [24]. Ion conductivity values tend to be close to the conductivity values of neat ILs, suggesting that the ion mobility of the ions contained in the IL is not impaired by the presence of the gelator or confinement substrate [25]. In addition, cyclic voltammetry of these gels revealed that they are electrochemically stable, underscoring their usefulness for electrochemical applications [22]. For instance, bactericidal activity of imidazolium-based ionogels against gram-positive bacteria has also been reported in a study with polymeric ILs, allowing eradication of 100% of selected microorganisms [21]. Other biological activities can be designed by the incorporation of proper ILs [26]. All the described characteristics unlock the potential of using ionogels in many domains [27], namely as sensors [28], membranes [29], actuators and solar cells [17], but also in biomedicine for drug delivery [30], wound dressing [27], sensing [28], and as tissue regeneration matrices [24].

Figure 1 shows the distribution of works reporting on the application of ionogels in electronics and biomedicine (a), together with the incidence on different classes of ILs in the preparation of ionogels (b). Based on the previously mentioned properties, it is not surprising that ionogels remain mostly applied in the electrochemical field for sensing [28], batteries [31], and as solar cells [27], although they are being increasingly investigated within a biomedicine perspective (cf. Fig. 1a) [11, 30, 32]. The majority of studies on ionogels resorts to imidazolium-based ILs (cf. Fig. 1b), which are among the most studied and employed ILs across the related scientific and industrial communities [33–35]. For instance, imidazolium-based ILs bearing Cl^-, Br^-, $[BF_4]^-$, and $[NTf_2]^-$ anions were shown to have higher thermal stability, superior electrochemical performance, and good durability, justifying their interest in electronics [36–38]. However, various imidazolium-based ILs are challenged by their poor biocompatibility and high toxicity, making the transition toward bio-based ILs urgent to leverage the potential of ionogels within biomedical applications [39–41].

This chapter aims to highlight the physical and chemical properties of ionogels that turn these materials relevant for biomedical applications and to overview their most relevant applications in the biomedical field (Fig. 2). Herein, the terms ionogel and gel are used interchangeably, resorting to ionogels or polymers comprising ILs.

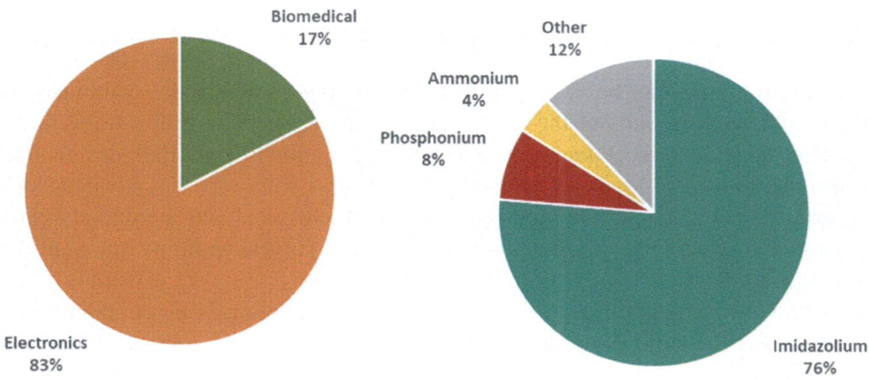

Fig. 1 Distribution of works reporting on the use of ionogels for both electronics and biomedicine (**a**) and usage incidence of distinct ILs families for their preparation (**b**) in the past ten years (starting from 2010). Literature search was done in Web of Knowledge on February 11, 2021, using the following keywords: ionogels, biomedicine, drug, cells, sensing, and electronics

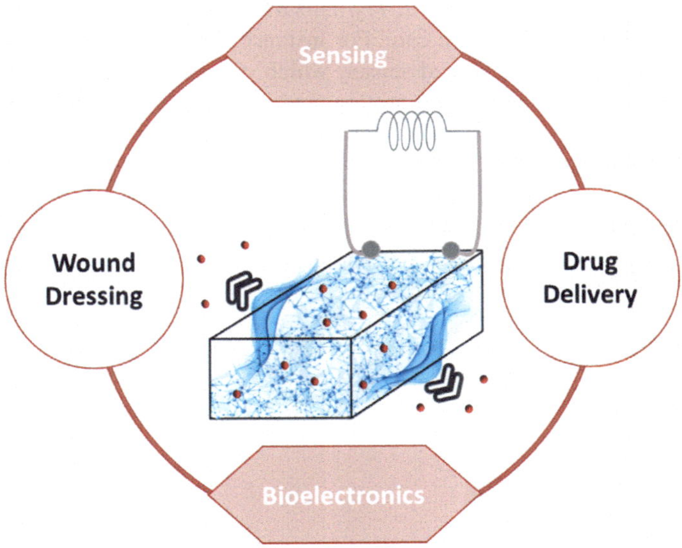

Fig. 2 Outline of the information reviewed in this chapter related to the application of ionogels in biomedical applications

2 Advances in Biomedical Applications by Ionogels

ILs are a subject of high scientific interest since the mid-1990s, many being nowadays recognized as sustainable solvents and functional materials in many areas of science [42]. ILs are comprised of a large organic cation and an organic or inorganic anion. Due to the asymmetry and low charge density of their ions that prevents the formation of an ordered crystalline structure, ILs usually display melting points lower than common electrolytes, and by general definition below 100 °C. ILs' intrinsic properties can be easily modified by simply changing the combinations and/or substituent groups of the cation and the anion, such an ability to be finely tuned and adapted to meet the needs of a given application justifies their designation of *designer solvents* [43, 44]. If correctly designed, ILs may present advantageous properties, such as negligible vapor pressure, high chemical and thermal stability, high ionic conductivity, non-flammability and high solvency/affinity for multiple compounds [45, 46].

As previously highlighted, a synergy is achieved in ionogels by combining the IL phase and the solid polymer phase: the liquid scattered or trapped within the solid retains the function and properties of the IL, yielding a gel with the desirable properties of both ILs (e.g., a range of tailored properties) and the solid (e.g., structural rigidity and practicality) [17]. Despite the intensive use of ionogels in electronics, the number of articles concerning ionogels for biomedical applications has been increasing throughout the last decade (Fig. 3). It should be additionally noticed that articles proposing new electronic platforms based on ionogels for biomedical application are also available [24, 32, 47]. This trend further highlights the potential of this kind of ionogels in health care. For instance, Yang et al. [48] have recently developed a wearable IL-activated sensor, which provides effective monitoring of breathing, pulse wave, human joint motion, and plantar pressure [48]. This breathable, waterproof, washable sensor is envisaged for clothing purposes, and it also has antibacterial properties [48].

Figure 4 showcases the incidence of several ILs' cations and anions in the preparation of ionogels for biomedical applications, together with their chemical structures. Table 1 provides these ILs' full names and corresponding abbreviations. In

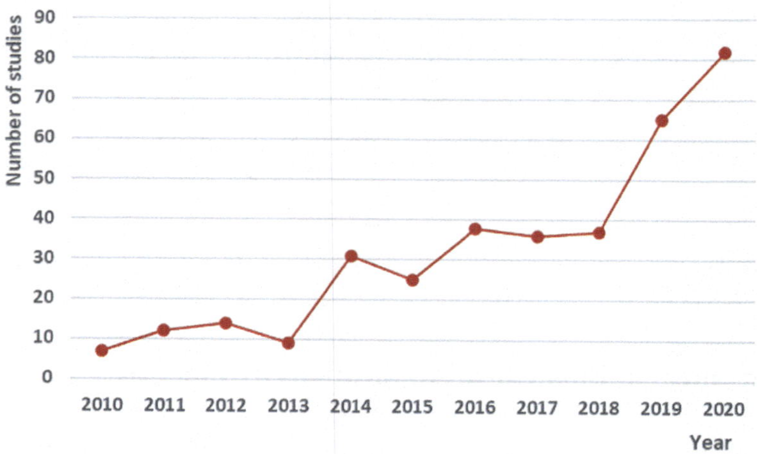

Fig. 3 Number of studies reporting the use of ionogels in the last decade, starting from 2010. Data was obtained from the Web of Knowledge on February 11, 2021, using the following keywords: ionogels, biomedicine, drug, cells, sensing, and electronics

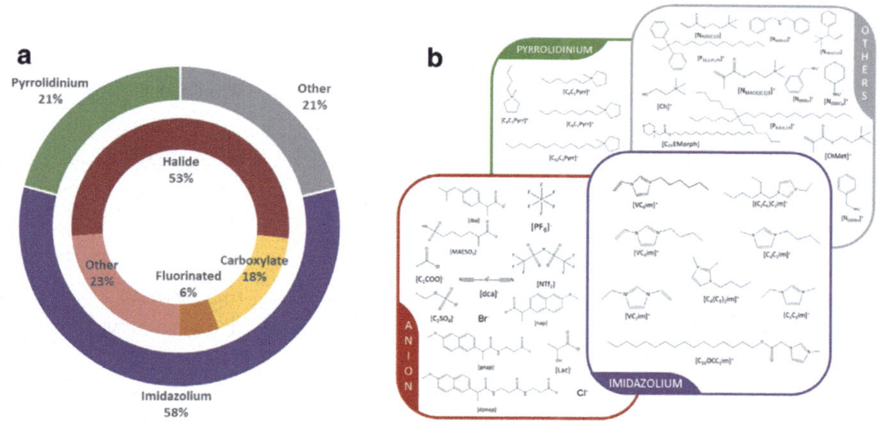

Fig. 4 Distribution of ILs used for ionogels preparation discussed in this book chapter (**a**), Chemical structures of cations and anions mentioned herein (**b**)

general, the ILs used are mostly comprised of imidazolium, but also of pyrrolidinium, quaternary ammonium, quaternary phosphonium, and morpholinium cations, with several substituents/functionalizations. Regarding anions, halides (e.g., Cl^-, Br^-) are the most used, followed by carboxylates (e.g., $[C_1COO]^-$, $[Lac]^-$), and fluorinated (e.g., $[PF_6]^-$, $[NTf_2]^-$) ones, among others. These incidence patterns also consider electronic-related applications, notwithstanding a 15% decrease in the adoption of imidazolium-based ILs is noted in the biomedical field. Moreover, regarding the preparation of ionogels, those used for biomedical applications are mostly obtained by polymerization of either ILs containing a polymerizable group, or by building a polymer network within an IL.

2.1 Preparation and Properties of Ionogels

Advances during the past decades in biomedicine had a significant influence on the development of products derived from natural and synthetic sources, with biomimetic products as an evolving field [23, 49]. This prospect has led to the development of novel materials, including ionogels. An overview of the distribution of ionogel preparation routes for biomedical applications is shown with Fig. 5.

Depending on the nature of the solid-like network, the synthetic pathways of ionogels fall into three groups, which may be organic (using low molecular weight gelators or polymers), inorganic (frequently using oxide nanoparticles, carbon nanomaterials or sol–gel oxide networks), or hybrid organic–inorganic (typically polymers reinforced with inorganic fillers) as depicted in Fig. 6 [17]. Herein, focus will be on every type of ionogel, except of oxide-based ionogels and without going into deeper

Table 1 Name and abbreviations of the ILs considered in this book chapter

IL	Abbreviation
(4-(2-(hexadecyloxy)-2-oxoethyl)-4-methylmorpholin-4-ium bromide	$[C_{16}EMorph]Br$
(vinylbenzyl)trimethylammonium chloride	$[N_{VBn(C1)3}]Cl$
[2-(acryloyloxy)ethyl]trimethylammonium chloride	$[N_{AOE(C1)3}]Cl$
[2-(methacryloyloxy)ethyl]trimethylammonium chloride	$[N_{MAOE(C1)3}]Cl$
1-(2-ethylhexyl)-3-ethylimidazolium hexafluorophosphate	$[(C_2C_6)C_2im][PF_6]$
1-butyl-1-methylpyrrolidinium bromide	$[C_4C_1Pyrr]Br$
1-butyl-2,3-dimethylimidazolium bis(trifluoromethanesulfonyl)imidate	$[C_4(C_1)_2im][NTf_2]$
1-butyl-3-methylimidazolium bis(trifluoromethanesulfonyl)imidate	$[C_4C_1im][NTf_2]$
1-butyl-3-methylimidazolium chloride	$[C_4C_1im]Cl$
1-butyl-3-methylimidazolium hexafluorophosphate	$[C_4C_1im][PF_6]$
1-dodecyl-1-methylpyrrolidinium bromide	$[C_{12}C_1Pyrr]Br$
1-ethyl-3-methylimidazolium acetate	$[C_2C_1im][C_1COO]$
1-ethyl-3-methylimidazolium dicyanamide	$[C_2C_1im][dca]$
1-ethyl-3-methylimidazolium ethylsulfate	$[C_2C_1im][C_2SO_4]$
1-hexyl-1-methylpyrrolidinium bromide	$[C_6C_1Pyrr]Br$
1-methyl-3-butylimidazolium ibuprofenate	$[C_4C_1im][ibu]$
1-octyl-1-methylpyrrolidinium bromide	$[C_8C_1Pyrr]Br$
1-vinyl-3-butyl-imidazolium acetate	$[VC_4im][C_1COO]$
1-vinyl-3-butyl-imidazolium bromide	$[VC_4im]Br$
1-vinyl-3-butyl-imidazolium chloride	$[VC_4im]Cl$
1-vinyl-3-ethyl-imidazolium acetate	$[VC_2im][C_1COO]$
1-vinyl-3-ethyl-imidazolium bromide	$[VC_2im]Br$
1-vinyl-3-ethylimidazolium dicynamide	$[VC_2im][dca]$
1-vinyl-3-hexylimidazolium bromide	$[VC_6im]Br$
2-cholinium methacrylate lactate	[ChMet][Lac]
3-methyl-1-(hexadecyloxycarbonylmethyl)imidazolium bromide	$[C_{16}OCC_1im]Br$
benzylammonium 2-(6-methoxynaphthalen-2-yl)propanoate	$[N_{000Bn}][nap]$
cholinium geranate	[Ch][Ge]
cyclohexylammonium 2-(6-methoxynaphthalen-2-yl)propanoate	$[N_{000Cy}][nap]$
dibenzylammonium 3-(2-(6-methoxynaphthalen-2-yl)propanamido)propanoate	$[N_{00(Bn)2}][pnap]$
benzylammonium 3-(3-(2-(6-methoxynaphthalen-2-yl)dipropanamido)propanoate	$[N_{000Bn}][dpnap]$
dibenzylammonium 3-(3-(2-(6-methoxynaphthalen-2-yl)dipropanamido)propanoate	$[N_{00(Bn)2}][dpnap]$
dodecylethyldiphenylphosphonium bis(trifluoromethanesulfonyl)imidate	$[P_{12,2,Ph,Ph}][NTf_2]$

(continued)

Table 1 (continued)

IL	Abbreviation
trihexyl(tetradecyl)phosphonium dicyanamide	$[P_{6,6,6,14}][dca]$

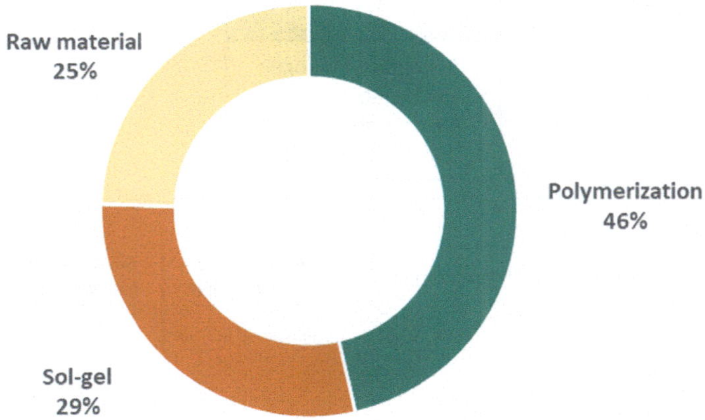

Fig. 5 Distribution of the preparation routes of ionogels used in biomedicine that are discussed in this book chapter. Data search was done in Web of Knowledge on February 11, 2021, using the following keywords: ionogels, biomedicine, drug, cells, sensing, and electronics

detail about silica-based ionogels, since they were recently overviewed elsewhere [50, 51].

It should be however highlighted that before using ionogels (or ILs) in medical purposes, their cyto/genotoxicities must be addressed. In fact, this represents a crucial step in the approval phase for medical and pharmaceutical products resorting to ILs and, hence, acts as an early exclusion factor in biomedical-related applications. As mentioned before, rather than having an organic solvent or water as the solvent phase, ionogels are comprised of ILs within a polymeric structure mutually assembling a gel. Most studies in the field of polymeric ionogels have only focused on 1-alkyl-3-methylimidazolium-based ILs, rather than on ammonium, pyrrolidinium, and other IL-based gels (Fig. 4) [17, 25]. Significant progress has been made on cytotoxicity assays for ILs. Among the various cell lines that have been tested are human spontaneously immortalized human keratinocyte cell line (HaCaT) [52, 53], cervical carcinoma [54, 55], colon carcinoma cells (Caco-2) [53, 56], human hepatocellular carcinoma cells (HepG2) [57], human umbilical vein cell line (EA.hy926) [53]. Results reveal that the head groups are of moderate concern about their effect on lipophilicity and hence on IL cytotoxicity. In contrast to this, the alkyl side chain is the primary effect influencing cytotoxicity considering the structural nature of IL cations [54, 58]. It has been shown that most of the studied commercially available IL anions had no or minor cytotoxic effects when using these cell lines. In particular, cholinium-based ILs have been recognized as ILs that exhibiting the lowest cytotoxicity toward Caco-2, HepG2 and Ea.hy926 cell lines [53]. Overall, one of

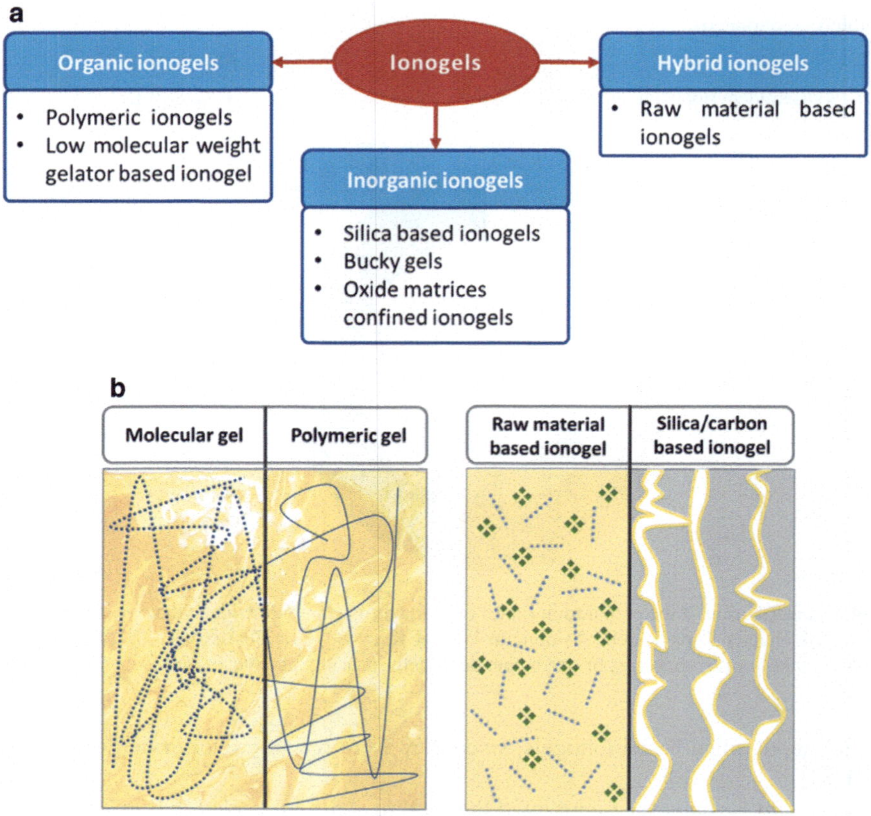

Fig. 6 (**a**) General classification of ionogels, adapted from [59]. (**b**) Categorization of gels based on synthesis route, adapted from [60]; gels can either be comprised of polymeric (right of the vertical line) or molecular fibers (left of the vertical line) depicted on the left image; the illustration on the right shows to synthesis routes, the raw material based and the silica/carbon confinement-based route; the yellow color represents the IL

the most critical issues of using ILs is related to their cytotoxicity and biocompatibility. However, researchers thrive to develop ionogels using more biocompatible ILs, such as cholinium-based ones [24, 61]. In particular, cholinium-based ILs have been recognized as ILs that exhibiting the lowest cytotoxicity toward Caco-2, HepG2, and Ea.hy926 cell line [53]. Overall, a deeper understanding of ionogels/ILs cytotoxicity is key to uncover their potential for broader implementation in biomedical applications.

2.1.1 Polymer-Based Ionogels

A common, efficient, and straightforward route of synthesizing ionogels is by radical polymerization of ILs monomers in another solvent [25]. One common feature for polymerizable ILs and ionogels formed using low molecular gelators, regardless of the type of solvent or gelator used, is the establishment of interconnected networked structures covering the entire volume of solvent during gelation [62, 63]. At this moment it is imperative to clarify that ILs can be polymerized (if they possess a polymerizable group in their chemical structure) and they can be solvents that fill in the cross-linked network of another polymer [25]. Polymerizable ILs are formed by dissolving the monomer and the corresponding amount of the cross-linker (usually a small quantity), adjusting the total concentration of the monomer [30]. Other compounds can subsequently be added to enhance the gels mechanical or self-healing properties, such as pharmaceuticals, metals, dextrose, and glycerol [19, 64]. The reaction mixture occurs at a constant temperature and sometimes at controlled humidity. As an example [65], the radical polymerization of ILs comprising a vinyl group, e.g., $[VC_2im]Br$, with a cross-linker, results in a polyionic liquid gel, namely poly$[VC_2im]Br$. Commonly used cross-linkers include poly(vinyl alcohol) (PVA) [66], N, N, N′, N′-tetramethylethylendiamine [9], glutaraldehyde [67], N, N0-methylenebis(acrylamide) [30], and 3-sulfopropylmethacrylate potassium (MAE-SO_3) [30]. The polymerization of several vinyl-alkyl-imidazolium ILs was already evaluated by disclosing the effect of the alkyl chain length and selected anion on the gelation process [18]. Overall, it was observed that both anion and alkyl chain length influenced the gelation process [18]. As an example, imidazolium-based ILs comprising I^- and $[Lac]^-$ did not form gels; on the other hand, similar ILs comprising $[PF_6]^-$ anion are not soluble in water, and therefore, no polymerization took place also in this case [68]. Furthermore, the polymerization of $[VC_4im][C_1COO]$ in aqueous media could not be accomplished unlike that of $[VC_2im][C_1COO]$, proving that longer alkyl chain lengths have a retarding effect on the polymerization process [18, 58]. The hydrophobicity also has an influence on polymerization, which was demonstrated by the unsuccessful polymerization of IL an with an octyl group in the side chain [18].

Supramolecular gels, or gels derived from small molecules, have been intensively studied. Small molecular weight gelators are known for being intrinsically diverse in chemical structures [63], and they can form hydrogels and organogels in solvents such as water and organic solvents, respectively [69, 70]. As a result of gelator self-assembly and aggregation by weak noncovalent interactions, such as H-bonding [71], hydrophobic interaction [72], π–π interaction [73], these gels are supramolecular in nature (i.e., supramolecular gels). The two pioneering works in this field were published nearly simultaneously in 2001, describing a glycolipid-derivative [68] and a cholesterol-derivative [62] as small-molecule IL gelators. The number of IL gelators has since then grown steadily, but not as significantly as the other two gelator types, hydrogelators and organogelators. Up to date the number of IL gelators does not exceed one hundred, while the corresponding number for organogelators must be beyond several hundreds [25]. Thereby, for researchers working in this area, relevant potential falls in the development of better IL gelators.

Ionogels have a significant advantage over hydrogels due to their negligible vapor pressure, which is a limiting factor in the case of water-based gels. Hydrogels gradually lose water during time, causing the material to become progressively brittle [74]. In addition to this, ionogels are more flexible and transparent than the equivalent hydrogels [74]. Given their suitable properties, polymeric ionogels are being used as drug loading devices [19], in disease monitoring and detection [75] and as dressing materials for wound healing [76], as it will be overviewed in Sect. 16.2.

2.1.2 Sol–gel Produced Ionogels

The synthesis of ionogels was recently reported by sol–gel processing in ILs [77, 78], with two approaches found for the synthesis route: (i) the confinement of ILs within a matrix, and (ii) the ionogel formation through the conditioning of raw materials.

Dai et al. [79] reported the pioneering work on sol–gel synthesis in an IL solvent, in 2000, where mesoporous silica with high surface area was prepared, from a mixture of tetramethoxysilane and formic acid in $[C_4C_1im][NTf_2]$. The IL, which had been removed at the end of the synthesis, was used as a template. Numerous sol–gel syntheses involving IL models have then been reported [17, 80, 81]. Confined ILs by the traditional sol–gel route are obtained after the removal of the liquid phase from the reaction mixture [82]. The solid matrix is commonly silica-based, although carbon-based matrices have also been reported [83]. After drying, the remaining porous solid loaded with a targeted compound is obtained. Traditionally, volatile solvents are being used for such purposes [78, 84]. However, the use of liquids without an apparent vapor pressure, such as ILs, delivers a stable solid–liquid system called ionogel [17]. It is also interesting to note that ILs are able to induce porosity in silica matrices [79, 85]. Within this solid matrix, the confined IL fills a 3D-interconnected mesoporous structure creating a percolating network of ionic conduction. Although confined, the IL keeps the liquid-like dynamics [86]. This approach is, however, out of the scope of this chapter due to the number or relevant previous studies [51, 59, 87].

"Bucky" gels, or carbon nanotube (CNT)-based gels, are comprised of hexagonally organized sp2-hybridized carbon atoms of graphene layers, which have strong electronic and optical properties [17]. CNTs have excellent mechanical, electrical, and thermal properties, but one major drawback is that they are difficult to process. A possible technique to improve solubility and processability is to absorb organic molecules on their surface [17]. Similarly as in the silica-based route, for the preparation "bucky" gels, Chernyshov et al. [83] weighed graphite and one of three imidazolium- or the phosphonium-based IL in a precise ratio and further mixed intensively in tetrahydrofurane. After the evaporation of the solvent, viscous materials were obtained. Previously to this work, Fukushima [88] noticed that single-walled carbon nanotubes when mixed with an imidazolium-based ionic IL formed gels after being ground. Phase transition and rheological properties indicated that the gels are formed by physical cross-linking of the bundles of nanotubes, mediated by local molecular IL ordering rather than by nanotubes' entanglement. The reported gels were thermally stable and did not show signs of shriveling [88, 89].

The breakthrough of using renewable sources led to an urge in developing sustainable materials. Accordingly, the recent field of ionogel synthesis via the conditioning of raw materials will be here briefly described. Renewable materials, such as biopolymers, can be utilized for many purposes and also for the synthesis of ionogels. Arvind Kumar and Srinivasa Rao [27] reported the use of agarose and chitosan for ionogels formation. Agarose is an algal polysaccharide comprising repeatied D-galactose and 3,6-anhydro-L-galactose units. It has excellent ability to form thermo-reversible gels in hot water [90]. The other component of the ionogel synthesis was chitosan, an extensively studied biodegradable, biocompatible, and antimicrobial compound. Both of these biopolymers have applications in food industry and biochemical fields and have been extensively studied for pharmaceutical and biomedical purposes, namely for controlled release of drug, wound dressing, and space filling implants [91–93]. Based on the knowledge that ILs have a high solvating ability for polymers derived from biomass, including agarose and chitosan [94, 95], powdered biopolymers were added to a preheated IL until their complete dissolution under continuous stirring and heating conditions. Afterward, silver nitrate was added to the mixture, resulting in the formation of silver oxide nanoparticles [27]. The inclusion of silver oxide nanoparticles is useful for improving the mechanical strength, conductivity, and antimicrobial properties of the composites, due to very strong optical/sensing, catalytic, and antimicrobial properties [27]. Precipitation of the nanocomposite materials can be achieved by adding an antisolvent, resulting in the nanocomposite ionogels being produced after cooling.

Sol–gel ionogels have high potential to be used in a variety of fields. In addition to the targeted biomedical applications, sol–gel ionogels are promising materials for solid-state electrolytes, fuel cells, lithium batteries, quasi-solid dye sensitized solar cells, actuators, sensor or electronic devices [26, 27, 78, 83, 96]. Although ionogels obtained through polymerization have been the most studied, and therefore applied in more fields, sol–gel ionogels tend to be doped with other and new compounds targeting a specific application.

2.2 *Ionogels in Biomedical Applications*

The use of ionogels in the biomedical context is quite recent; nevertheless, this type of material seems to hold a great potential in this field [17, 26]. As aforementioned, one key property of these gels is their intrinsic task-specific nature associated to ILs, enabling to tailor their properties and, hence, the nature of their interactions and application performance [25]. Therefore, it can be assumed that the development of ionogels could have a significant impact on biomedicine by providing innovative solutions that might outperform current state-of-the-art technologies.

One of the shortcomings of synthetic polymers constituents of hydrogels, organogels or ionogels is their inherent nonbiodegradability, which in turn raises concerns when foreseeing biomedical applications. If, on the one hand, this drawback is not a main obstacle if oral administration is envisaged, it is a critical issue for implantable biosensors or implantable drug delivery systems, further depending on the durability (temporary versus permanent) of the final application [97, 98]. Thus, current interest has been placed on developing materials with intrinsic biodegradability, namely based on polysaccharides, proteins, etc. [23]. Biopolymers such as cellulose, lignin, starch, and keratin are examples of renewable polymers, which due to their low price [27, 32, 96], are being increasingly applied to replace synthetic polymers [99, 100]. Nevertheless, even these biopolymers have drawbacks which limit their use in the development of new materials, e.g., the low solubility of lignin and cellulose. Remarkably, these limitations can be overcome with the use of ILs, in which lignin and cellulose dissolution are well-established [101, 102].

Ionogels for biomedical applications [26, 28] have focused on: (i) drug delivery and wound dressing and (ii) bioelectronics and sensing. Impressive advances have been accomplished in this field, disclosing the versatility and utility of ionogels. A summary comprising the ionogel, preparation method, performance, and target application is provided in Table 2. Their drug loading and drug delivery development is a well-researched area, where studies involving the loading and release of diclofenac, ibuprofen, insulin, and doxorubicin by encapsulation or patches are included [19, 30, 103]. Since imidazolium- and pyrrolidinium-based ILs have been reported to have antibacterial properties, it has been investigated whether ionogels comprised of these ILs are of the same nature [25]. As verified with ILs, the corresponding ionogels also display antimicrobial activity, further supporting their use for antimicrobial purposes, including in wound dressing, scaffolding, and dressing materials [19, 27, 74]. Moreover, ionogels are used in real-time monitoring and detection of diseases [75]. These robust smart materials have multifunctional sensing properties, including pressure, bending, and twisting. These were demonstrated for designed sensors, which could be used to detect human breathing characteristics and in wearable equipment, as they can sense even slight changes in the surrounding environment [48].

2.2.1 Drug Delivery and Wound Dressing

Throughout the past decades, drug delivery has been a target of research to improve therapeutic efficiency, aiming both a controlled and target-specific release. Developed materials for drug delivery include liposomes, microspheres, nanomaterials, gels, among others [4, 11, 104]. Drug carriers must have excellent biocompatibility and biodegradability. Recently, stimuli-responsive hydrogels have been developed since they can respond to external stimuli (e.g., heat and pH), enabling a controlled drug release [11, 104]. Meanwhile, major progress has been done for improving healthcare worldwide. In addition to drug delivery, there is a high interest in improving healing, with more than 3000 products designed to treat various forms of wounds [105]. Nonetheless, wound healing is a burdensome and sometimes painful process, which requires a suitable environment to promote healing process [106].

Table 2 Summary of ionogel biomedical applications included in this chapter

Ionogel	Preparation method	Performance (including main highlights, e.g., drug loading, performance indicators of therapeutic efficiency)	Application	References
$[N_{VBn(C1)3}]$ Cl; N, N′-methylenebis(acrylamide); N, N, N', N'-tetramethylethylendiamine; ammoniumpersulfate	Polymerization	$[VC_2im]Br$ achieved the best result in quantitative cell viability assay with a relative cell viability of 98.4%. The other gels are considered to have low toxicity, since they did not reduce the cell vitality by more than 20%	Drug delivery, implant coating	[9]
$[N_{AOE(C1)3}]$Cl; N, N′-methylenebis(acrylamide); N, N, N', N'-tetramethylethylendiamine; ammoniumpersulfate				
$[N_{MAOE(C1)3}]$Cl; N, N′-methylenebis(acrylamide); N, N, N', N'-tetramethylethylendiamine; ammoniumpersulfate				
$[VC_2im]Br$; N, N′-methylenebis(acrylamide); N, N, N', N'-tetramethylethylendiamine; ammoniumpersulfate				
$[VC_2im]Br$; tetramethylethylenediamine;	Polymerization	The gels were compressible more than 60%, pressing half of their initial length compared to Ca-alginate hydrogels	Drug delivery, scaffolds	[18]
$[VC_4im]Br$; tetramethylethylenediamine;				

(continued)

Table 2 (continued)

Ionogel	Preparation method	Performance (including main highlights, e.g., drug loading, performance indicators of therapeutic efficiency)	Application	References
[VC_4im]Cl; tetramethylethylenediamine;				
[VC_2im][C_1COO]; tetramethylethylenediamine;				
[C_4C_1im][ibu]; tetramethoxysilane	Sol–gel	The loading of [C_4C_1im][ibu] in the ionogels was about 0.80 g of ibuprofen/g of dried silica	Drug delivery	[26]
[C_2C_1im][C_2SO_4]; Poly(N-isopropylacrylamide); N, N′-methylenebisacrylamide; 2,2-Dimethoxy-2-phenylacetophenone	Polymerization	The ionogel is more flexible and transparent than the equivalent hydrogel. The rate of water uptake for the ionogel was found to be $\gtrapprox$30 times larger	Drug delivery	[74]
[$C_{16}OCC_1$im] Br	Sol–gel	The opaque gel encapsulated 40.4% of diclofenac, while 72.3% of imatinib mesylate was encapsulated. The transparent gel encapsulated 1.4% of diclofenac, whereas 53.59% of imatinib mesylate was encapsulated	Drug delivery and adsorbent	[103]
[C_4C_1im]Cl; Chitosan; agarose	Sol–gel	$T_{gelation}$ (°C): 38; $T_{melting}$ (°C): 80; gel strength (g cm^{-2}) at 30 °C: 590; conductivity κ ionogel (mS cm^{-1}) at 25 °C: 0.675	Drug delivery, antimicrobial agent	[27]

(continued)

Table 2 (continued)

Ionogel	Preparation method	Performance (including main highlights, e.g., drug loading, performance indicators of therapeutic efficiency)	Application	References
$[C_{16}EMorph]Br$; poly(vinyl alcohol); dextrose; $B(OH)^{4-}$	Polymerization	A ≊82.3% cumulative release of doxorubicin at 37 °C, pH 5.0 after 50 h was noticed (~53% release at 25 °C). Acidic pH accelerates the release of doxorubicin	Drug delivery	[19]
$[N_{MAOE(C1)3}]Cl$; N, N0-methylenebis(acrylamide); ammoniumpersulfate	Polymerization	Release times could be increased by holding the drug back through developed electrostatic interactions	Drug delivery	[30]
$K[MAESO_3]$; N, N0-methylenebis(acrylamide); ammoniumpersulfate				
[Ch][Ge]; Poly(vinyl alcohol)	Polymerization	In vitro transport experiments show that patches enhance insulin transport by more than 30% in a coculture model when compared to equivalent concentrations of [Ch][Ge] solution. A reduction in the long-term effect on the epithelium, while also limiting the amount of drug lost by epithelial cell uptake in oral delivery, can be achieved by the localization of patches	Insulin oral path delivery	[66]

(continued)

Table 2 (continued)

Ionogel	Preparation method	Performance (including main highlights, e.g., drug loading, performance indicators of therapeutic efficiency)	Application	References
$[VC_2im]Br$; N, N0-methylenebis(acrylamide);	Polymerization	Antibacterial effect. Outstanding results were achieved by using imidazolium-based hydrogels having a killing efficiency of at least 95% of *Staphylococcus aureus Xen 30* and *Pseudomonas aeruginosa Xen 5*	Implants, drug delivery systems, stent coatings	[21]
$[VC_4im]Cl$; N, N0-methylenebis(acrylamide);				
$[VC_4im]Br$; N, N0-methylenebis(acrylamide);				
$[Nvbn(C_1)_3]$ Cl; N, N0-methylenebis(acrylamide);				
$[NAOE(C_1)_3]Cl$; N, N0-methylenebis(acrylamide);				
$[NMAOE(C_1)_3]Cl$; N, N0-methylenebis(acrylamide);				
$[C_4C_1Pyrr]Br$; poly(vinyl alcohol); $B(OH)^{4-}$	Polymerization	ILs with longer alkyl chain exhibited better antibacterial activities against *Escherichia coli* and *Staphylococcus aureus*	Wound healing	[76]
$[C_6C_1Pyrr]Br$; poly(vinyl alcohol); $B(OH)^{4-}$				
$[C_8C_1Pyrr]Br$; poly(vinyl alcohol); $B(OH)^{4-}$				
$[C_{12}C_1Pyrr]Br$; poly(vinyl alcohol); $B(OH)^{4-}$				

(continued)

Table 2 (continued)

Ionogel	Preparation method	Performance (including main highlights, e.g., drug loading, performance indicators of therapeutic efficiency)	Application	References
$[C_4C_1im]Cl$; cellulose; chitin	Sol–gel	Ionogel reinforced with chitosan DD 84% showed the greatest viscoelastic properties (G': 10.6 kPa, G": 20.6–1.7 kPa, and η*: 200–0.05 kPa s)	Scaffolds for wound management	[96]
$[N_{000Cy}]$[nap]; methylsalicylate	Sol–gel	All the mice from two the groups were found to be healthy with fur grown on their back skin after two weeks of receiving the last dose of the topical gel	Skin patch	[20]
$[N_{000Bn}]$[nap]; methylsalicylate				
$[N_{00(Bn)2}]$[pnap]; methylsalicylate				
$[N_{000Bn}]$[dpnap]; methylsalicylate				
$[N_{00(Bn)2}]$[dpnap]; methylsalicylate				
[ChMet][Lac]; cellulose; 2-dimethylaminoethylmethacrylate; dimethoxy-2-phenylacetophenone	Polymerization	These ionogels were comprised of 60 wt% free ionic liquid and 5% cellulose. The ionogels were transparent, indicating good cellulose processability	Bioelectronics, sensor for health monitoring	[24]
$[C_4C_1im]$[dca]; poly(vinyl alcohol); gallic acid	Sol–gel	The supramolecular iongels are flexible and can withstand large elastic deformations with full recovery	No application, yet useful as scaffold	[107]
$[C_2C_1im][C_1COO]$; glutaraldehyde; chitosan	Sol–gel	High thermal stability was shown by the ionogel. The confined $[C_2C_1im][C_1COO]$ and the chitosan matrix start to thermally decompose at 185 °C and 259 °C, respectively	Sensor	[67]

(continued)

Table 2 (continued)

Ionogel	Preparation method	Performance (including main highlights, e.g., drug loading, performance indicators of therapeutic efficiency)	Application	References
[VC_2im][dca]; ammonium persulfate	Polymerization	Electrical properties recovery after 4 s. The repaired ionogels bear stretching to about 150%	Sensor for breathing monitoring	[75]
[C_4C_1im][PF_6]; graphite; ascorbic acid; nafion	Sol–gel	Small quantities of catecholamines were detected, e.g., 5.03 ± 0.03 μg mL^{-1} of adrenaline	Sensing for biomolecules	[83]
[C_4C_1im][NTf_2]; graphite; ascorbic acid; nafion				
[$C_4(C_1)_2$im][NTf_2]; graphite; ascorbic acid; nafion				
[(C_2C_6)C_2im][PF_6]; graphite; ascorbic acid; nafion				
[$P_{12,2,Ph,Ph}$][NTf_2]; graphite; ascorbic acid; nafion				
[$P_{6,6,6,14}$][dca]; poly(N-isopropylacrylamide); N, N-methylenebis(acrylamide); dimethoxy-phenylacetophenone	Polymerization	Ionogels were used for a visual monitoring the pH of sweat generated during exercise. Designed ionogel bars were comprised of the following pH indicators: bromocresol green (pH 3.8–5.6), bromocresol purple (pH 5.2–6.8), and bromothymol blue (pH 6.0–7.6)	Sensing for sweat pH	[47]

(continued)

Table 2 (continued)

Ionogel	Preparation method	Performance (including main highlights, e.g., drug loading, performance indicators of therapeutic efficiency)	Application	References
$[C_4C_1im][PF_6]$; cotton textile and CA film; silver nanowires	Sol–gel	The sensitivities for the sensor are 4.46 kPa^{-1} below the pressure of 0.5 kPa, 0.312 kPa^{-1} in the pressure range 0.5–10 kPa, and 0.0143 kPa^{-1} in the pressure range 10–120 kPa	Motion sensing	[48]
$[VC_2im][dca]$; N, N′-methylene bis(acrylamide), ammoniumpersulfate	Sol–gel	The ionogel-based sensor showed high sensitivity of 15.4 kPa^{-1}, a small hysteresis of 5.36%, and a low detection limit of 5 Pa	Motion sensing	[108]

The ionogels, preparation methods, performance, and target drug delivery of wound dressing application are summarized in Table 2. Kuddushi et al. [19] studied the loading ability of polymeric ionogels comprising IL-based surfactants polymerized with poly(vinyl) alcohol for doxorubicin release. ILs based on morpholinium cation with distinct alkyl side chain lengths ([C_nEMorph]Br, with $n = 12$, 14 and 16) were used. The authors observed that [C_{16}EMorph]Br-based ionogels loaded with doxorubicin, an anthracycline chemotherapy drug, accelerated doxorubicin release at acidic pH, with cumulative release of ≈82.3% after 50 h at 37 °C and pH 5.0. However, at 25 °C, doxorubicin release decreases to ≈53% for the same remaining conditions. The same authors [103] further found that the critical concentration of gelation was 25.20, 22.60, and 8.80% (w/v) for [C_{12}EMorph]Br, [C_{14}EMorph]Br, and [C_{16}EMorph]Br, respectively. The authors found that the inclusion of ester-functionalized groups in the cation side chain of IL-based surfactants exhibited better surface-active properties than the non-functionalized and vinyl-functionalized IL-based surfactants. Moreover, by substituting a methyl group in the IL by a methylcarboxylate ($-CH_2COOH$) group, it was found an increase in the surface activity [109] and biodegradability [110, 111]. As a result, ILs act as low molecular weight gelators (LMWGs), enabling the tailoring of ionogels properties. During the development of ionogels, the authors [103] observed a gel-to-gel transition (opaque to transparent) upon increasing the temperature above 48.5 °C, revealing that the solution undergoes a phase transition from isotropic to turbid due to the increased size of the aggregates. In this later work [103], the encapsulation efficiency of the opaque and transparent gels was investigated for two pharmaceuticals, sodium salt forms of diclofenac and imatinibmesylate. The opaque gel was able to encapsulate 40.4% of diclofenac and 72.3% of imatinibmesylate. On the other hand, the transparent gel encapsulates only 1.4% of diclofenac and 53.59% of imatinibmesylate. Overall, by controlling ionogels preparation conditions and gel-to-gel transition, the encapsulation capacity can therefore be improved.

Owing to the large and adjustable pore size of the polymer matrix, drugs are often rapidly washed out from polymer-based ionogels. Claus et al. [30] however, attempted to develop ionogels able to induce a decrease in the initial burst release and an eventual elongation of the total release. The authors decided to integrate an ionic drug into a counterionic gel matrix, in which sulfonic acid groups were selected to establish ion–ion interactions with the drug. Accordingly, the electrostatic nature of the interactions affected the release rate, demonstrating a strong decrease in the release rate. Polymerized IL gels comprised of poly[$N_{MAOE(C1)3}$]Cl and 3-sulfoprorylmethacrylate potassium (MAE-SO_3) were loaded with ibuprofen and timolol maleate, respectively. The loaded drugs concentrations were almost 2000 μg/mL. The material showed promising applications for the subconjunctival space, middle or inner ear [30]. Other drugs have confined in ionogels, such as ibuprofen that was encapsulated by a one-step procedure in novel porous functionalized silica [26]. The authors synthesized [C_4C_1im][ibu] by ion-exchange reaction and further encapsulated the IL within functionalized silica, forming an ionogel. Loading of [C_4C_1im][ibu] in ionogels was evaluated by thermogravimetric analysis and determined to be about 0.80 g of ibuprofen/g of dried silica. These novel ionogel materials had a release kinetics slower than with crystalline ibuprofen and pure ibuprofenate IL. At 10 min of drug release studies, around 35% ibuprofen was released from the ionogel, whereas 80% of ibuprofen was released from [C_4C_1im][ibu] in the same time period. These results demonstrate that ionogels act as drug depot for controlled delivery.

Due to the increasing number of patients with diabetes there is an urgent need for the development of a widely available, efficient and non-invasive, oral insulin drug delivery platform [112]. Peng et al. [66] developed tunable ionogels and investigated their potential and applicability in insulin drug delivery. The authors described a method of encapsulating [Ch][Ge] within a poly(vinyl alcohol solid network. The authors described two preparation methods of the patches—by drying and free thawing. The ionogels demonstrated the potential to improve intestinal absorption of insulin. Due to the ability of [Ch][Ge] to disrupt epithelial tight junctions, enabling drugs to pass through the paracellular route, they designed the application of the materials for oral delivery. It has also been shown that [Ch][Ge] has the potential to extend the storage of insulin and reduce the activity of enzymes such as trypsin, which usually pose an additional threat to biologics administered orally. These mucoadhesive ionogel patches are designed to adhere to the intestine. Insulin was released rapidly in the dried patch mentioned, completing 100% release within 2 h. In comparison, the freeze thawed type, with 40% release after 5 h, released insulin in a more regulated fashion. The tested cells showed a significantly lower uptake of drug, twofold lower when using the ionogel in comparison with the saline control system. Localizing the developed patches allows a reduction in the long-term effects on the epithelium and quite beneficially limits the amount of drug lost by epithelial cell uptake.

Among other applications, ionogels presenting antibacterial activity can be used as scaffolds and as wound dressing materials. Villar-Chavero et al. [96] developed new ionogels using [C_4C_1im]Cl reinforced with chitosan (deacetylation degree: 54, 62, 69, 77, and 84%) as scaffolds for wound management. The authors provided rheological and thermal properties of the materials, with the rheological properties revealing a strong dependence on the deacetylation degree (DD) of chitosan. Elasticity, as well as the gel strength and the complex viscosity, generally increased as the DD increased because of the physical interactions between chitosan and cellulose in the ionogel. The thermal properties of these thermoreversible materials were shown to be independent of the chitosan DD. Lastly, authors highlighted the ionogel reinforced with chitosan DD 84% for having the greatest viscoelastic properties (G': $\approx$10.6 kPa, G'': 20.6–1.7 kPa, and η^*: 200–0.05 kPa s), although no application of the developed ionogels was performed. Moreover, Arvind Kumar and Srinivasa Rao [27] designed a similar ionogel also based on [C_4C_1im]Cl and chitosan and agarose. These ionogels were coated with silver oxide nanoparticles, which along with chitosan also display antimicrobial activity.

Given the potential applications of ionogels in antibacterial dressings and topical administration, Yu et al. [76] designed multifunctional gels that are expected to overcome multiple limitations of currently available antibiotics. Authors subjected tetrahydroxyborate anions to interact with two distinct cis-diol groups on PVA to yield the formation of a gel in the presence of 1-alkyl-1-methylpyrrolidinium-based ILs ([C_nC_1Pyrr]Br, with $n = 4, 6, 8, 12$). Considering that [C_nC_1Pyrr]Br displays antibacterial activity, the corresponding ionogels were evaluated and proven to have a bactericidal effect on *Escherichia coli* and *Staphylococcus aureus*, which intensifies with the length of the alkyl chain of the IL cation. Also, the injectable gel has high plasticity and good adhesive properties, which can be adapted to the shape of various wounds, such as on human knuckle skin. A synergistic effect of multihydroxyl structures and dynamic covalent bonds results in a self-healing process occurring in approximately 2 min. These materials are also susceptible to the presence of saccharide molecules, such as glucose, meaning that they can be further employed for glucose sensing or self-regulated insulin release.

In summary, polymeric ionogels are predominantly used in drug delivery and wound dressing application. If properly designed, they can display better performance (e.g., drug loading capacity, mechanical characteristics, and thermal stability) than conventional materials. Contrarily to drug delivery and wound dressing, mainly obtained by polymerization, sol–gel produced ionogels are mainly used in bioelectronics and sensing. Recently, raw material-based ionogels are being used since the use of potentially toxic cross-linkers is avoided and ionogels properties can be further improved by incorporating additional compounds.

2.2.2 Bioelectronics and Sensing

The integration of ILs into gels is beneficial since it can lead to materials with the inherent advantages of ILs, ready for bioelectronics and sensing applications within a solid or semi-solid gel-type structure [28]. Especially interesting is the potential of wearable devices [113]. This field is growing quickly, as the healthcare economy gradually leads to the need for online control of patient progress rather than the existing hospital-focused model [28, 48, 75]. The ionogels, preparation methods, performance and target bioelectronics and sensing applications are summarized in Table 2.

An example of ionogels applied in bioelectronics was reported by [67] They have produced thermally stable ionogels based on chitosan chemically crosslinked with glutaraldehyde in [C_2C_1im][Ac]. This IL was already reported for biopolymer dissolution due to its desirable properties, such as low toxicity ($LD_{50} > 2000$ mg kg^{-1}), low corrosiveness, low melting point, low viscosity, and favorable biodegradability [110, 114]. This is due to its short alkyl side chain and benign acetate anion. In comparison with most used imidazolium-based ILs, [C_2C_1im][Ac] is predicted to be less toxic and already reached the REACH licensing [40]. Despite their relevance, the authors have not tested these biopolymer-based ionogels for any application. Isik et al. [24] produced a biocompatible cellulose/cholinium-based ionogel comprising [ChMet][Lac], which was afterwards photopolymerized in the presence of crosslinkers (2-dimethylaminoethylmethacrylate, dimethoxy-2-phenylacetophenone. Up to 5 wt% of cellulose, the obtained ionogel was a soft, jelly-like transparent material, but above this value the ionogel decreased in transparency.

Compared with ionogels without the crosslinker, ionogels with 4 wt% crosslinker displayed an enhanced integrity. These materials are important for emerging technologies comprising low toxicity cholinium-based ILs, widening their application in fields such as bioelectronics discussed below.

Adrenalin plays an important role in the central nervous and cardiovascular systems, and it is also active in a broad variety of metabolic processes, as well as other catecholamine neurotransmitters such as dopamine and dobutamine (the latter a cardiovascular pharmaceutical). Awareness of the effect of these biomolecules on the human body and their monitoring is highly important in clinical medicine [115]. Chernyshov et al. [83] designed printable sensors based on carbon paste electrodes modified with ILs ($[P_{12,2,Ph,Ph}][NTf_2]$, $[(C_2C_6)C_2im][PF_6]$, $[C_4(C_1)_2im][NTf_2]$, $[C_4C_1im][NTf_2]$, $[C_4C_1im][PF_6]$) for the detection of these biomolecules. They reported that catecholamine selectivity is improved in the presence of ascorbic acid (Nafion additive), and that analyte pre-concentration was achieved with different ILs. These ionogels were used as sensors for the electrochemical determination of aqueous solutions of adrenaline, dobutamine and dopamine with detection limits of $(5.3 \pm 0.1) \times 10^{-8}$ M, $(1.3 \pm 0.1) \times 10^{-7}$ M, and $(1.2 \pm 0.1) \times 10^{-6}$ M, respectively, as determined by cyclic voltammetry. Moreover, the authors reported that the detection limit can be lowered with the introduction of additional components (Co(III)tetrakis-(tert-butyl)-phthalocyanine additive) into the paste [83]. This type of sensor can be useful within biomedical applications for the monitoring of catecholamines, as they can be clinically relevant biomarkers and are associated with several diseases, such as Alzheimer's and Parkinson [116].

Monitoring human body signs comes along with illness treatment and recovery. Therefore, wearable sensors are relevant for personalized health monitoring and assessment, as well as disease diagnosis. They must be convenient, portable and available at any location to all, including the elderly people with serious diseases [117, 118]. Wang et al. [75] developed temperature-sensitive polymeric ionogels comprising $[VC_2im][dca]$ that could detect characteristic signs in human breathing and could be used for disease detection. The designed ionogels exhibited high transparency, high conductivity, exceptional stretchability and self-healing characteristics. Surprisingly, just 4 s after cutting the ionogel into two pieces, it recovers its electrical properties and the ionogels could endure stretching to about 150%. This temperature-sensitive sensor could detect various states of human respiration under different conditions, suggesting a potential opportunity for the use in health monitoring and illness recovery. These materials can be used as wearable devices; however, current wearable electronic devices often serve as an external part of clothing and demonstrate minimal flexibility with different textile materials. Remarkably, Yang et al. [48] overcame this limitation and reported a straightforward, low-cost, and user-friendly approach to the manufacturing of wearable sensors based on commercial textiles, represented in Fig. 7. Accordingly, the ionogel is an active sensing material and this is achieved by loading the IL onto the textile fabric. The authors selected $[C_4C_1im][PF_6]$ and loaded it onto the fabric skeleton, and then coated the material with silver nanowires as electrodes. The textile can detect different human

Fig. 7 Illustration of an IL-loaded textile material, acting as a wearable ionogel-based sensor for human health monitoring

movements from athletic or natural movements, and it can also track biomechanical signals such as breathing, heartbeat, and pulse waveform.

On the other hand, Zhang et al. [108] have fabricated ionogels by in situ polymerization of $[VC_2im][dca]$. These highly sensitive ionogels were able to detect slight ambient changes, including a gentle touch, a small leaf or flower petal and even weak gas flow. Other than exhibiting force-sensing characteristics and high conductivity, the developed ionogels showed high sensitivity of 15.4 kPa^{-1}, a hysteresis of 5.36%, and a low detection limit of 5 Pa. Having such features allow these materials to be used for the detection of human or robot interactions with the environment—e.g. as artificial skin scaffolds [119]. Benito-Lopez et al. [47] reported the design of an outstanding ionogel-based wearable, robust, flexible and disposable barcode system that can be used for the real-time monitoring of sweat pH. It is an entirely non-invasive method of assessing the bodily status of the wearer, guaranteeing well-being. Moreover, the authors could accurately obtain pH values by simply observing the $[P_{6,6,6,14}][dca]$ comprised barcode color and comparing them with a standard color chart. In addition, by using built-in Light Emitting Diodes (LEDs), the response can be tracked online, providing continuous feedback of the sweat chemistry with a minimal power demand.

Luque et al. [107] prepared ionogels by supramolecular crosslinking between polyphenol molecules such as gallic acid, pyrogallol, and tannic acid with a biocompatible polymer—poly(vinyl alcohol), in the presence of $[C_2C_1im]Br$ and $[C_2C_1im][DCA]$. The supramolecular interactions, which take place between poly(vinyl alcohol) and polyphenol molecules via hydrogen bonding while crosslinking the IL, affords the gel-like nature of these materials. The authors have also noted that these ionogels show a thermoreversible behavior with transition temperatures between 87 and 110 °C. Moreover, the formed gels are flexible and can withstand large elastic deformations (40% under compression) with full recovery. Authors have not reported any specific application. However, these polyphenol-based thermoreversible ionogels demonstrate ample justification for their future use as printable electrolytes for bioelectronics and as dressing materials due to their flexibility and ability to withstand large elastic deformations with full recovery.

In conclusion, sol–gel derived ionogels have been the main type investigated in the field of sensing and bioelectronics. Overall, ionogels have shown great potential for application in wearable electronics due to their excellent conductivity, stability, and biocompatibility.

3 Critical Assessment of Ionogels Application in the Biomedical Field

Ionogels, either polymer-based or prepared by sol–gel procedures, have been extensively investigated as advanced materials for wound dressing and drug delivery, as well as in bioelectronics and biosensing, as overviewed above. Based on the current state-of-the-art technologies, a SWOT analysis on the ionogels application in the biomedical field was carried out, being shown in Fig. 8.

Strengths. Unlike other products, ionogels represent a promising class of advanced materials in which the properties of ILs are integrated into polymers or organic/inorganic materials, being responsible for their enhanced performance in a wide range of applications. In general, as ILs exhibit a negligible vapor pressure

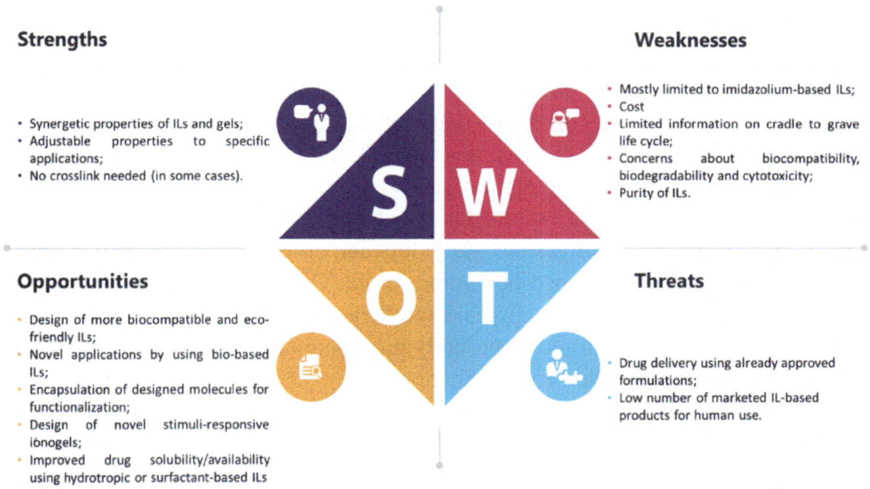

Fig. 8 SWOT analysis of the application of ionogels in biomedicine

at ambient temperature, solvent losses are minimized throughout time, which ultimately prevent brittleness while extending shelf-lives of ionogels. Furthermore, the tunable character displayed by ILs, achieved by the manipulation of the cation–anion design, allows to tailor the ionogels properties to meet the requirements of specific applications, namely: (i) ionic conductivity is used in biosensing; [43, 83, 108] (ii) the high ability displayed by ILs to bind biomolecules [66] has been investigated in drug delivery for the controlled release of bioactive compounds, and (iii) ILs can act themselves as active pharmaceutical ingredients with potential application in wound dressing and drug delivery [76]. Also, the ability to prepare ionogels without a crosslinker can be advantageous since a potentially harmful chemical is avoided. Ionogels often display improved rheological properties, namely elasticity [96], integrity [24], stretchability [75], which account for their improved performance in most applications. Moreover, ionogels show high transparency, which varies in color from white to yellowish, allowing the examination of the skin for potential infection and track healing progress. Ionogels that demonstrated good stretchability and exceptionally fast self-healing properties are pre-ordained as a wound dressing material, demonstrating up to fivefold faster self-healing than other products, such as hydrogels [75, 120].

Weaknesses. Ionogels fate inevitably follows that of ILs, which are sometimes costly and with limited information on cradle to grave life cycle. Limited data are available on ionogels comprising ILs other than imidazolium-based ILs. Concerns raised from the possible lack of biodegradability and biocompatibility as well as possible cytotoxic, occupational safety and health issues of some imidazolium-based ILs and, consequently, ionogels, limits their application for human use. Indeed, information on the risk of ionogels to health are still scarce [17, 80]. Up to date, the diversity of ILs chemical structures investigated is almost limited to imidazolium, and relatively few works dealing with ionogels for biomedical applications (if compared to electronics) are available, justifying the need for deeper and more extensive investigations on this field [28, 81, 116]. Moreover, ILs are viscous solvents with a high affinity toward moisture, resulting in additional laborious steps such as drying, which in turn dictate the cost of the final product. Also, related to their synthetic procedure, is the presence of impurities in the IL that, depending on their nature, may significantly change the properties/effectiveness of the IL itself and the prepared ionogel. If the development of (bio)sensing platforms is envisaged, impurities may lead to non-reliable and non-reproductible results. On the other hand, if human use is envisaged (e.g., drug delivery, wound dressing), the safety and efficiency of the ionogel-based biomedical approaches may also be compromised. Due to these reasons, not only the starting materials/solvents quality and synthetic procedure of the IL, but also the ionogel preparation route should be properly programmed to minimize impurities to levels acceptable/safe for the final application.

Opportunities. Given the wide range of advantages that ILs bring forth, the possibilities for ionogels application seem to be expanding steadily. By taking full advantage of their tunable nature, ILs may be engineered to present enhanced bio- and eco-friendliness, as well as potential recyclability. Another key opportunity is the design and use of bio-based ILs in ionogels [23, 61, 121–123]. It will unlock doors of opportunity for ionogels to be applied in relevant areas of the biomedical field. Moreover, the possibility of encapsulating designed molecules in the immobilized IL phase opens infinite possibilities of functionalization with potential application, for instance in biosensing [80]. The hydrotropic or surfactant character of some ILs can be explored to improve solubility and bioavailability of active pharmaceutical ingredients insoluble or sparingly soluble in water [124]. Additionally, the application of magnetic ILs can open new perspectives in drug delivery owing to their responsiveness to magnetic fields. On the other hand, the designer solvent ability of ILs, not yet fully explored, will enable tailored and improved performances of ionogels for distinct biomedical applications.

Threats. As an alternative to ionogels, drug delivery systems using formulations already approved by regulatory agencies will continue to be predominant due to the time and cost involved to reach a new approval. Even though the path for commercialization of ILs is being paved [35] and the suitability of ILs for biomedical uses has been shown in literature, commercial applications of ionogels and/or related to biomedicine still faltered. To be highlighted is the case of the pharmaceutical company MEDRX, which uses ILs to develop innovative transdermal medicines. Beyond other medicines undergoing preclinical or clinical studies, MEDRX has reached a NDA acceptance from FDA in 2020 with their IL-based technology for the transdermal delivery of lidocaine [125]. Considering the types of application overviewed in the present chapter, there is the need to assure the quality and safety of the proposed applications for human use. For instance, in bioelectronic and sensing applications, the developed ionogel-based devices should be validated using human samples. In turn, wound dressing and drug delivery related products may need to undergo clinical trials. These tasks involve multiple resources (e.g., time, money, volunteers, etc.), so that many proposed applications will not move toward validation/approval or take a long time to reach the final consumer. It is thus important to underline that to foster commercial implementation of ionogel-based biomedical products, a multistakeholder perspective should be adopted. Collaborations involving academia, industry, governments, healthcare professionals, and patients are essential to boost competition with currently available and/or approved options.

4 Concluding Remarks and Future Perspectives

Ionogels represent an innovative class of materials joining the chemical versatility of ILs with favorable mechanical properties (elasticity, stretchability, integrity, self-healing, among others) of (bio)polymers. If properly designed, ionogels may exhibit high ionic conductivity, high thermal and electrochemical stability, enhanced biocompatibility and low (cyto)toxicity, and be stimuli-responsive, and thus represent smart materials for advanced applications in biomedical applications, ranging from drug delivery and wound dressing to bioelectronics and sensing.

Depending on the solid-like network, synthetic pathways of ionogels fall within three groups, namely organic (using low molecular weight gelators or polymers), inorganic (*e.g.*, oxide nanoparticles or carbon nanomaterials), or hybrid organic–inorganic (polymers reinforced with inorganic fillers). Polymer-based ionogels can either be obtained from polymerizable ILs or rely on the application of ILs as solvents that fill the cross-linked network of a polymer, being the last one the most common method for preparing ionogels for drug delivery. Furthermore, an underdeveloped aspect that holds high promise is the development of better IL gelators once their diversity is comparatively lower than organogelators. Distinctly, ionogels preparation by sol–gel processing in ILs can be obtained by the confinement of ILs within a matrix, or by ionogel formation through the conditioning of raw materials, which have found wide application in sensing. Based on the high solvating ability of ILs for polymers derived from biomass, ionogels comprising, *e.g.*, chitosan, chitin, and cellulose have been a hot topic of research. In addition to the sustainable character of this approach for ionogels formation, biopolymers bring additional advantages to these ionogels such as biodegradability and biocompatibility.

The versatility and utility of ionogels have been demonstrated in drug delivery and wound dressing, as well as in bioelectronics and sensing. Ionogels allow improvements on the solubility and availability of pharmaceuticals (*e.g.*, doxorubicin) and maintain the stability of biomolecules (*e.g.*, insulin). By tailoring their composition, controlled release of therapeutic products can be achieved, with potential application in cancer therapies, diabetes mellitus, among others. On the other hand, ionogels have shown to present antimicrobial activity, derived either from pharmaceutically active ILs or from the polymeric material (*e.g.*, chitosan), which coupled to the improved rheological properties of these ionogels make them promising materials for wound dressing applications. The high ionic conductivity displayed by ionogels make them good candidates for bioelectronics and sensing applications. Thermally stable and high integrity ionogels have shown an improved performance for electrochemical detection of catecholamines. Moreover, wearable, flexible and stimuli-responsive ionogels have been applied in health monitoring and illness recovery, with enhanced ability to sense movements, vital signs, and biological characteristics of fluids, in a faster way and requiring minimal energy inputs.

Up to date, most studies on ionogels for biomedical applications resorts to imidazolium-based ILs, which however are often challenged by their poor biocompatibility and high toxicity. Moreover, some ionogels make use of cross-linkers which raise additional concerns with respect to the safety of these biodevices. Although the use of crosslinkers can be avoided, the transition toward bio-based ILs (*e.g.*, based on cholinium cation and glycine-betaine analogs) of a more biocompatible character remains an urgent need in the biomedical field. The development of tailored ionogels to fulfil unmet needs in drug delivery, while demonstrating an improved performance over the current state-of-the-art technologies based on compounds already approved for drug delivery is critical to meet the requirements of regulatory agencies when considering their application in humans.

Further investigation of ionogels is still required, namely: (i) to tailor ILs properties to specific applications by manipulation of the cation–anion design; (ii) to develop novel functionalized ionogels targeting unmet needs in biomedical research; and (iii) to further explore the hydrotropic or surfactant character of ILs in ionogels to improve solubilization of hydrophobic active pharmaceutical ingredients. It is also important to pinpoint that efficient recycling procedures should be designed to decrease the ILs carbon footprint. Research intensification on the field will be crucial to upgrade the performance of ionogels over cost-effective alternatives and devices already approved for human use.

An additional potential field in which ionogels could evolve substantially is in artificial muscle mimicking. Their properties, namely outstanding conductivity, rigidness, and self-healing, make them promising candidates in this field. Moreover, magnetic ILs can provide an additional and promising route toward the development of magnetic-responsive ionogels. On the other hand, the apparent disadvantage related to the low miscibility of polymers in ILs can in some cases be explored toward the development of novel and improved stimuli-responsive ionogels with potential application in drug delivery.

Overall, it was already demonstrated that ionogels exhibit an improved performance for distinct biomedical applications, spanning from wound dressing and drug delivery to bioelectronics and sensing. However, given the plethora of ILs chemical structures, types of inorganic and organic matrices that can be modified, and novel envisaged applications, this topic is still in its infancy. Accordingly, more related investigations are expected in coming years, ultimately resulting in materials with commercial potential in biomedical applications.

Acknowledgements This work was developed within the scope of the project CICECO Aveiro Institute of Materials, UIDB/50011/2020 & UIDP/50011/2020, financed by national funds through the Portuguese Foundation for Science and Technology/MCTES, and the project POCI-01-0145-FEDER-031106 (IonCytDevice) funded by FEDER, through COMPETE2020—Programa Operacional Competitividade e Internacionalização (POCI), and by national funds (OE), through FCT/MCTES. Francisca A. e Silva acknowledges FCT—Fundação para a Ciência e a Tecnologia, I.P. for the researcher contract CEECIND/03076/2018 under the Scientific Employment Stimulus—Individual Call 2018. Augusto Q. Pedro acknowledges the research contract under the project PTDC/BII-BBF/30840/2017. Bojan Kopilovic acknowledges the University of Aveiro for the PhD grant Contrato de Bolsa (RH) 85281.

References

1. Pilleron S et al (2019) Global cancer incidence in older adults, 2012 and 2035: a population-based study: global cancer incidence in older adults. Int J Cancer 144(1):49–58. https://doi.org/10.1002/ijc.31664
2. Sinclair A et al (2020) Diabetes and global ageing among 65–99-year-old adults: findings from the international diabetes federation diabetes atlas, 9th edition. Diabetes Res Clin Pract 162:108078. https://doi.org/10.1016/j.diabres.2020.108078
3. Ahmed EM (2015) Hydrogel: preparation, characterization, and applications: a review. J Adv Res 6(2):105–121. https://doi.org/10.1016/j.jare.2013.07.006
4. de las Alarcón H, Pennadam S, Alexander C (2005) Stimuli responsive polymers for biomedical applications. Chem Soc Rev 34(3):276–285. https://doi.org/10.1039/B406727D
5. Sun Z et al (2020) Hydrogel-based controlled drug delivery for cancer treatment: a review. Mol Pharm 17(2):373–391. https://doi.org/10.1021/acs.molpharmaceut.9b01020

6. Delaey J, Dubruel P, Van Vlierberghe S (2020) Shape-Memory Polymers for Biomedical Applications. Adv Func Mater 30(44):1909047. https://doi.org/10.1002/adfm.201909047
7. Talelli M et al (2015) Core-crosslinked polymeric micelles: Principles, preparation, biomedical applications and clinical translation. Nano Today 10(1):93–117. https://doi.org/10.1016/j.nantod.2015.01.005
8. Truong WT et al (2011) Self-assembled gels for biomedical applications. Chem Asian J 6(1):30–42. https://doi.org/10.1002/asia.201000592
9. Claus J, Brietzke A, et al (2020) Swelling characteristics and biocompatibility of ionic liquid based hydrogels for biomedical applications. PLOS ONE. 15(4):e0231421. https://doi.org/10.1371/journal.pone.0231421
10. Claus J, Sommer FO, Kragl U (2018) Ionic liquids in biotechnology and beyond. Solid State Ionics 314:119–128. https://doi.org/10.1016/j.ssi.2017.11.012
11. Sood N et al (2016) Stimuli-responsive hydrogels in drug delivery and tissue engineering. Drug Deliv 23(3):748–770. https://doi.org/10.3109/10717544.2014.940091
12. Douglas J (2018) Weak and Strong Gels and the Emergence of the Amorphous Solid State. Gels 4(1):19. https://doi.org/10.3390/gels4010019
13. Nayak AK, Das B (2018) Introduction to polymeric gels. In: Polymeric gels. Elsevier, pp 3–27. https://doi.org/10.1016/B978-0-08-102179-8.00001-6
14. Sakai T (2020) Definition of polymer gels and rubber elasticity. In Sakai T (ed) Physics of polymer gels, 1st edn. Wiley, pp 45–75. https://doi.org/10.1002/9783527346547.ch3
15. Qin L, Lv K, Shen Z, Liu M (2015) Self-assembly of organic molecules into nanostructures. In Chen X, Fuchs H (eds) (2015) Soft matter nanotechnology. Wiley-VCH Verlag GmbH & Co. KGaA, Weinheim, Germany, pp 21–80. https://doi.org/10.1002/9783527682157
16. Miravet JF, Escuder B (2013) Supramolecular gels for pharmaceutical and biomedical applications. In: Schneider H-J (ed) Monographs in supramolecular chemistry. Royal Society of Chemistry, Cambridge, pp 331–354. https://doi.org/10.1039/9781849737821-00331
17. Le Bideau J, Viau L, Vioux A (2011) Ionogels, ionic liquid based hybrid materials. Chem Soc Rev 40(2):907–925. https://doi.org/10.1039/C0CS00059K
18. Bandomir J et al (2014) Synthesis and characterization of polymerized ionic liquids: mechanical and thermal properties of a novel type of hydrogels. Macromol Chem Phys 215(8):716–724. https://doi.org/10.1002/macp.201400009
19. Kuddushi M et al (2020) Poly(vinyl alcohol) and functionalized ionic liquid-based smart hydrogels for doxorubicin release. ACS Appl Bio Mater 3(8):4883–4894. https://doi.org/10.1021/acsabm.0c00393
20. Majumder J et al (2014) Designing a simple organic salt-based supramolecular topical gel capable of displaying in vivo self-delivery application. Chem Commun 50(14):1671. https://doi.org/10.1039/c3cc48513g
21. Claus J, Jastram A, et al (2020) Polymerized ionic liquids-based hydrogels with intrinsic antibacterial activity: modern weapons against antibiotic-resistant infections. J Appl Polym Sci 138(16):50222. https://doi.org/10.1002/app.50222
22. Banerjee S, Das RK, Maitra U (2009) Supramolecular gels "in action." J Mater Chem 19(37):6649. https://doi.org/10.1039/b819218a
23. Secchi E et al (2013) Biopolymer gels with "physical" cross-links: gelation kinetics, aging, heterogeneous dynamics, and macroscopic mechanical properties. Soft Matter 9(15):3931. https://doi.org/10.1039/c3sm27153f
24. Isik M et al (2014) Cholinium lactate methacrylate: ionic liquid monomer for cellulose composites and biocompatible ion gels. Macromol Symp 342(1):21–24. https://doi.org/10.1002/masy.201300176
25. Singh WP, Koch U, Singh RS (2020) Gelation of ionic liquids by small-molecule gelators and their applications. Soft Mater 18(4):386–410. https://doi.org/10.1080/1539445X.2019.1697706
26. Viau L et al (2010) Ionogels as drug delivery system: one-step sol–gel synthesis using imidazolium ibuprofenate ionic liquid. Chem Commun 46(2):228–230. https://doi.org/10.1039/B913879J

27. Trivedi TJ, Srinivasa Rao K, Kumar A (2014) Facile preparation of agarose-chitosan hybrid materials and nanocomposite ionogels using an ionic liquid via dissolution, regeneration and sol-gel transition. Green Chem 16:320–330. https://doi.org/10.1039/C3GC41317A
28. Kavanagh A et al (2012) Stimuli responsive Ionogels for sensing applications—an overview. Membranes 2(1):16–39. https://doi.org/10.3390/membranes2010016
29. Di Noto V et al (2010) Structure-Relaxation Interplay of a New Nanostructured Membrane Based on Tetraethylammonium Trifluoromethanesulfonate Ionic Liquid and Neutralized Nafion 117 for High-Temperature Fuel Cells. J Am Chem Soc 132(7):2183–2195. https://doi.org/10.1021/ja906975z
30. Claus J, Eickner T et al (2020) Ion exchange controlled drug release from polymerized ionic liquids. Macromol Biosci 20(9):2000152. https://doi.org/10.1002/mabi.202000152
31. Tan L et al (2009) Gels of ionic liquid [C4mim]PF6 formed by self-assembly of gelators and their electrochemical properties. Electrochem Commun 11(5):933–936. https://doi.org/10.1016/j.elecom.2009.02.025
32. Liu X et al (2020) Silicone-ionic liquid elastomer composite with keratin as reinforcing agent utilized as pressure sensor. Macromol Rapid Commun 42(5):e2000602. https://doi.org/10.1002/marc.202000602
33. Greer AJ, Jacquemin J, Hardacre C (2020) Industrial applications of ionic liquids. Molecules 25(21):5207. https://doi.org/10.3390/molecules25215207
34. Park S, Oh KK, Lee SH (2018) Biopolymer-based composite materials prepared using ionic liquids. In Itoh T, Koo YM (eds) Adv Biochem Eng Biotechnol 168:133–176. https://doi.org/10.1007/10_2018_78
35. Shiflett MB (ed) (2020) Commercial applications of ionic liquids. Springer International Publishing (Green Chemistry and Sustainable Technology), Cham. https://doi.org/10.1007/978-3-030-35245-5
36. Chen X et al (2012) Bis-imidazolium based poly(ionic liquid) electrolytes for quasi-solid-state dye-sensitized solar cells. J Mater Chem 22(34):18018. https://doi.org/10.1039/c2jm33273f
37. Lv Z et al (2020) Electrochemical and physical properties of imidazolium chloride ionic liquids with pyrrolidinium or piperidinium cation addition and their application in dual-ion batteries. Energ Technol 8(9):2000432. https://doi.org/10.1002/ente.202000432
38. Muginova SV et al (2017) Applications of ionic liquids for the development of optical chemical sensors and biosensors. Anal Sci 33(3):261–274. https://doi.org/10.2116/analsci.33.261
39. Jastorff B et al (2003) How hazardous are ionic liquids? Structure–activity relationships and biological testing as important elements for sustainability evaluation. Green Chem 5:136–142. https://doi.org/10.1039/B211971D
40. Magina S et al (2021) Evaluating the hazardous impact of ionic liquids—challenges and opportunities. J Hazard Mater 412:125215. https://doi.org/10.1016/j.jhazmat.2021.125215
41. e Silva FA, Coutinho JAP, Ventura SPM (2019) Aquatic toxicology of ionic liquids (ILs). In Zhang S (ed) Encyclopedia of ionic liquids. Springer Singapore, Singapore, pp 1–18. https://doi.org/10.1007/978-981-10-6739-6_52-1
42. Md Moshikur R et al (2020) Biocompatible ionic liquids and their applications in pharmaceutics. Green Chem 22(23):8116–8139. https://doi.org/10.1039/D0GC02387F
43. Li Z et al (2012) Design of environmentally friendly ionic liquid aqueous two-phase systems for the efficient and high activity extraction of proteins. Green Chem 14(10):2941–2950. https://doi.org/10.1039/c2gc35890e
44. Shahriari S et al (2013) Aqueous biphasic systems: a benign route using cholinium-based ionic liquids. RSC Adv 3(6):1835–1843. https://doi.org/10.1039/C2RA22972B
45. Domínguez-Pérez M et al (2010) (Extraction of biomolecules using) aqueous biphasic systems formed by ionic liquids and aminoacids. Sep Purif Technol 72(1):85–91. https://doi.org/10.1016/j.seppur.2010.01.008
46. Freire, M. G. (ed.) (2016) *Ionic-Liquid-Based Aqueous Biphasic Systems: Fundamentals and Applications*. Berlin, Heidelberg: Springer Berlin Heidelberg (Green Chemistry and Sustainable Technology). doi: https://doi.org/10.1007/978-3-662-52875-4.

47. Benito-Lopez F et al (2010) Simple barcode system based on Inonogels for real time pH-sweat monitoring. In: 2010 International conference on body sensor networks. 2010 International conference on body sensor networks (BSN), IEEE, Singapore, Singapore, pp 291–296. https://doi.org/10.1109/BSN.2010.9
48. Yang J et al (2019) Ionic liquid–activated wearable electronics. Mater Today Phys 8:78–85. https://doi.org/10.1016/j.mtphys.2019.02.002
49. Naik RR, Singamaneni S (2017) Introduction: bioinspired and biomimetic materials. Chem Rev 117(20):12581–12583. https://doi.org/10.1021/acs.chemrev.7b00552
50. Ma Z, Yu J, Dai S (2010) Preparation of inorganic materials using ionic liquids. Adv Mater 22(2):261–285. https://doi.org/10.1002/adma.200900603
51. Pedro AQ, Coutinho JAP, Freire MG (2019) Immobilization of ionic liquids, types of materials, and applications. In Zhang S (ed) Encyclopedia of ionic liquids. Springer Singapore, Singapore, pp 1–12. https://doi.org/10.1007/978-981-10-6739-6_86-1
52. McLaughlin M et al (2021) Characterization of ionic liquid cytotoxicity mechanisms in human keratinocytes compared with conventional biocides. Chemosphere 270:129432. https://doi.org/10.1016/j.chemosphere.2020.129432
53. Vieira NSM et al (2019) Human cytotoxicity and octanol/water partition coefficients of fluorinated ionic liquids. Chemosphere 216:576–586. https://doi.org/10.1016/j.chemosphere.2018.10.159
54. Stepnowski P et al (2004) Evaluating the cytotoxicity of ionic liquids using human cell line HeLa. Hum Exp Toxicol 23(11):513–517. https://doi.org/10.1191/0960327104ht480oa
55. Xia X et al (2018) Toxicity of imidazoles ionic liquid [C16mim]Cl to Hela cells. Ecotoxicol Environ Saf 162:408–414. https://doi.org/10.1016/j.ecoenv.2018.07.022
56. Cvjetko M et al (2012) Cytotoxic effects of imidazolium ionic liquids on fish and human cell lines. Arch Ind Hyg Toxicol 63(1):15–20. https://doi.org/10.2478/10004-1254-63-2012-2132
57. Li X, Ma J, Wang J (2015) Cytotoxicity, oxidative stress, and apoptosis in HepG2 cells induced by ionic liquid 1-methyl-3-octylimidazolium bromide. Ecotoxicol Environ Saf 120:342–348. https://doi.org/10.1016/j.ecoenv.2015.06.018
58. Stolte S et al (2007) Effects of different head groups and functionalised side chains on the cytotoxicity of ionic liquids. Green Chem 9(7):760–767. https://doi.org/10.1039/B615326G
59. Sahrash R et al (2018) PVDF based ionogels: applications towards electrochemical devices and membrane separation processes. Heliyon 4(11):e00847. https://doi.org/10.1016/j.heliyon.2018.e00847
60. Truong WT, Lewis L, Thordarson P (2013) Biomedical applications of molecular gels. In Escuder B, Miravet JF. (eds) Soft matter series. Royal Society of Chemistry, Cambridge, pp 157–194. https://doi.org/10.1039/9781849737371-00157
61. Isik M et al (2013) Cholinium-based poly(ionic liquid)s: synthesis, characterization, and application as biocompatible ion gels and cellulose coatings. ACS Macro Lett 2(11):975–979. https://doi.org/10.1021/mz400451g
62. Ikeda A et al (2001) Gelation of ionic liquids with a low molecular-weight Gelator showing T_{gel} above 100 °C. Chem Lett 30(11):1154–1155. https://doi.org/10.1246/cl.2001.1154
63. Raeburn J, Adams DJ (2015) Multicomponent low molecular weight gelators. Chem Commun 51(25):5170–5180. https://doi.org/10.1039/C4CC08626K
64. Kavanagh A et al (2011) A two-component polymeric optode membrane based on a multifunctional ionic liquid. Analyst 136(2):348–353. https://doi.org/10.1039/C0AN00770F
65. Sommer FO et al (2020) UV-polymerized vinylimidazolium ionic liquids for permselective membranes. Membranes 10(11):308. https://doi.org/10.3390/membranes10110308
66. Peng K et al (2020) Mucoadhesive ionic liquid gel patches for oral delivery. ACS Biomater Sci Eng. In Press. https://doi.org/10.1021/acsbiomaterials.0c01024
67. Guyomard-Lack A et al (2015) Ion segregation in an ionic liquid confined within chitosan based chemical ionogels. Phys Chem Chem Phys 17(37):23947–23951. https://doi.org/10.1039/C5CP04198H
68. Kimizuka N, Nakashima T (2001) Spontaneous self-assembly of glycolipid bilayer membranes in sugar-philic ionic liquids and formation of Ionogels. Langmuir 17(22):6759–6761. https://doi.org/10.1021/la015523e

69. Estroff LA, Hamilton AD (2004) Water Gelation by Small Organic Molecules. Chem Rev 104(3):1201–1218. https://doi.org/10.1021/cr0302049
70. Terech P, Weiss RG (1997) Low molecular mass gelators of organic liquids and the properties of their gels. Chem Rev 97(8):3133–3160. https://doi.org/10.1021/cr9700282
71. van Gorp JJ, Vekemans JAJM, Meijer EW (2002) C_3—symmetrical supramolecular architectures: fibers and organic gels from discotic trisamides and trisureas. J Am Chem Soc 124(49):14759–14769. https://doi.org/10.1021/ja020984n
72. Singh WP, Singh RS (2017) A new class of organogelators based on triphenylmethyl derivatives of primary alcohols: hydrophobic interactions alone can mediate gelation. Beilstein J Org Chem 13:138–149. https://doi.org/10.3762/bjoc.13.17
73. Ghosh S, Praveen VK, Ajayaghosh A (2016) The chemistry and applications of π-gels. Annu Rev Mater Res 46(1):235–262. https://doi.org/10.1146/annurev-matsci-070115-031557
74. Gallagher S et al (2013) Temperature and pH triggered release characteristics of water/fluorescein from 1-ethyl-3-methylimidazolium ethylsulfate based ionogels. Chem Commun 49(41):4613. https://doi.org/10.1039/c3cc41272e
75. Wang F et al (2019) Facile fabrication of a self-healing temperature-sensitive sensor based on Ionogels and its application in detection human breath. Nanomaterials 9(3):343. https://doi.org/10.3390/nano9030343
76. Yu Y et al (2020) Multifunctional hydrogel based on ionic liquid with antibacterial performance. J Mol Liq 299:112185. https://doi.org/10.1016/j.molliq.2019.112185
77. Néouze M-A et al (2005) A route to heat resistant solid membranes with performances of liquid electrolytes. Chem Commun 8:1082–1084. https://doi.org/10.1039/B416267F
78. Néouze M-A et al (2006) Ionogels, new materials arising from the confinement of ionic liquids within silica-derived networks. Chem Mater 18(17):3931–3936. https://doi.org/10.1021/cm060656c
79. Dai S et al (2000) Preparation of silica aerogel using ionic liquids as solvents. Chem Commun 3:243–244. https://doi.org/10.1039/a907147d
80. Vioux A et al (2010) Use of ionic liquids in sol-gel; ionogels and applications. C R Chim 13(1–2):242–255. https://doi.org/10.1016/j.crci.2009.07.002
81. Zhang L et al (2020) Overview of Ionogels in flexible electronics. Chem Rec 20(9):948–967. https://doi.org/10.1002/tcr.202000041
82. Zhang J et al (2009) Room temperature ionic liquids as templates in the synthesis of mesoporous silica via a sol–gel method. Microporous Mesoporous Mater 119(1–3):97–103. https://doi.org/10.1016/j.micromeso.2008.10.003
83. Chernyshov DV et al (2008) Ionic liquid-based miniature electrochemical sensors for the voltammetric determination of catecholamines. Anal Chim Acta 621(2):178–184. https://doi.org/10.1016/j.aca.2008.05.042
84. Li Z, Zhang Y, Feng N (2019) Mesoporous silica nanoparticles: synthesis, classification, drug loading, pharmacokinetics, biocompatibility, and application in drug delivery. Expert Opin Drug Deliv 16(3):219–237. https://doi.org/10.1080/17425247.2019.1575806
85. Zhou Y (2005) Recent advances in ionic liquids for synthesis of inorganic nanomaterials. Curr Nanosci 1(1):35–42. https://doi.org/10.2174/1573413052953174
86. Le Bideau J et al (2007) Effect of confinement on ionic liquids dynamics in monolithic silica ionogels: 1H NMR study. Phys Chem Chem Phys 9(40):5419. https://doi.org/10.1039/b711539c
87. Borghi F, Podestà A (2020) Ionic liquids under nanoscale confinement. Adv Phys X 5(1):1736949. https://doi.org/10.1080/23746149.2020.1736949
88. Fukushima T (2003) Molecular ordering of organic molten salts triggered by single-walled carbon nanotubes. Science 300(5628):2072–2074. https://doi.org/10.1126/science.1082289
89. Mei X, Ouyang J (2010) Gels of carbon nanotubes and a nonionic surfactant prepared by mechanical grinding. Carbon 48(1):293–299. https://doi.org/10.1016/j.carbon.2009.09.032
90. Fernández E et al (2008) Rheological and thermal properties of agarose aqueous solutions and hydrogels. J Polym Sci, Part B: Polym Phys 46(3):322–328. https://doi.org/10.1002/polb.21370

91. Hubbell JA (1995) Biomaterials in Tissue Engineering. Nat Biotechnol 13(6):565–576. https://doi.org/10.1038/nbt0695-565
92. Muzzarelli R et al (1988) Biological activity of chitosan: ultrastructural study. Biomaterials 9(3):247–252. https://doi.org/10.1016/0142-9612(88)90092-0
93. Muzzarelli R et al (1989) Reconstruction of parodontal tissue with chitosan. Biomaterials 10(9):598–603. https://doi.org/10.1016/0142-9612(89)90113-0
94. Tan SSY, MacFarlane DR (2009) Ionic liquids in biomass processing. In Kirchner B (ed) Ionic liquids. Springer Berlin Heidelberg (Topics in Current Chemistry), Berlin, Heidelberg, pp 311–339. https://doi.org/10.1007/128_2008_35
95. Zakrzewska ME, Bogel-Łukasik E, Bogel-Łukasik R (2010) Solubility of carbohydrates in ionic liquids. Energy Fuels 24(2):737–745. https://doi.org/10.1021/ef901215m
96. Villar-Chavero MM et al (2019) Tuning the rheological properties of cellulosic ionogels reinforced with chitosan: the role of the deacetylation degree. Carbohyd Polym 207:775–781. https://doi.org/10.1016/j.carbpol.2018.12.041
97. Huang J et al (2018) Novel in situ forming hydrogel based on xanthan and chitosan re-gelifying in liquids for local drug delivery. Carbohydr Polym 186:54–63. https://doi.org/10.1016/j.carbpol.2018.01.025
98. Kumarasamy D, Ghosh MK, Giri TK (2018) Polymer-based responsive hydrogel for drug delivery. In Thakur VK, Thakur MK, Voicu SI (eds) Polymer gels. Springer Singapore (Gels Horizons: From Science to Smart Materials), Singapore, pp 1–25. https://doi.org/10.1007/978-981-10-6080-9_1
99. Thakur VK, Thakur MK (2015) Recent advances in green hydrogels from lignin: a review. Int J Biol Macromol 72:834–847. https://doi.org/10.1016/j.ijbiomac.2014.09.044
100. Wu R-L et al (2009) Green composite films prepared from cellulose, starch and lignin in room-temperature ionic liquid. Biores Technol 100(9):2569–2574. https://doi.org/10.1016/j.biortech.2008.11.044
101. Dias RM et al (2020) Investigation of Kraft Lignin Solubility in Protic Ionic Liquids and Their Aqueous Solutions. Ind Eng Chem Res 59(40):18193–18202. https://doi.org/10.1021/acs.iecr.0c02605
102. Freire MG et al (2011) Electrospun nanosized cellulose fibers using ionic liquids at room temperature. Green Chem 13(11):3173. https://doi.org/10.1039/c1gc15930e
103. Kuddushi M et al (2018) Thermo-switchable de novo ionic liquid-based gelators with dye-absorbing and drug-encapsulating characteristics. ACS Omega 3(9):12068–12078. https://doi.org/10.1021/acsomega.8b01984
104. Bawa P et al (2009) Stimuli-responsive polymers and their applications in drug delivery. Biomed Mater 4(2):022001. https://doi.org/10.1088/1748-6041/4/2/022001
105. Dhivya S, Padma VV, Santhini E (2015) Wound dressings – a review. Biomedicine 5(4):22. https://doi.org/10.7603/s40681-015-0022-9
106. Moeini A et al (2020) Wound healing and antimicrobial effect of active secondary metabolites in chitosan-based wound dressings: a review. Carbohydr Polym 233:115839. https://doi.org/10.1016/j.carbpol.2020.115839
107. Luque GC et al (2020) Elastic and thermoreversible Iongels by supramolecular PVA/phenol interactions. Macromol Biosci 20(11):2000119. https://doi.org/10.1002/mabi.202000119
108. Zhang S et al (2018) Flexible highly sensitive pressure sensor based on ionic liquid gel film. ACS Omega 3(3):3014–3021. https://doi.org/10.1021/acsomega.7b01575
109. Wang X et al (2012) Aggregation behavior of COOH-functionalized imidazolium-based surface active ionic liquids in aqueous solution. J Mol Liq 173:103–107. https://doi.org/10.1016/j.molliq.2012.06.023
110. Jordan A, Gathergood N (2015) Biodegradation of ionic liquids—a critical review. Chem Soc Rev 44(22):8200–8237. https://doi.org/10.1039/C5CS00444F
111. Petkovic M et al (2010) Novel biocompatible cholinium-based ionic liquids—toxicity and biodegradability. Green Chem 12(4):643. https://doi.org/10.1039/b922247b
112. Banerjee A et al (2018) Ionic liquids for oral insulin delivery. Proc Natl Acad Sci 115(28):7296–7301. https://doi.org/10.1073/pnas.1722338115

113. Mahmud MS, Fang H, Wang H (2019) An integrated wearable sensor for unobtrusive continuous measurement of autonomic nervous system. IEEE Internet Things J 6(1):1104–1113. https://doi.org/10.1109/JIOT.2018.2868235
114. Sun N et al (2009) Complete dissolution and partial delignification of wood in the ionic liquid 1-ethyl-3-methylimidazolium acetate. Green Chem 11(5):646–655. https://doi.org/10.1039/b822702k
115. Baluta S, Lesiak A, Cabaj J (2018) Graphene quantum dots-based electrochemical biosensor for catecholamine neurotransmitters detection. Electroanalysis 30(8):1781–1790. https://doi.org/10.1002/elan.201700825
116. Ribeiro JA et al (2016) Electrochemical sensors and biosensors for determination of catecholamine neurotransmitters: a review. Talanta 160:653–679. https://doi.org/10.1016/j.talanta.2016.06.066
117. Lin Z et al (2017) Triboelectric nanogenerator enabled body sensor network for self-powered human heart-rate monitoring. ACS Nano 11(9):8830–8837. https://doi.org/10.1021/acsnano.7b02975
118. Xue H et al (2017) A wearable pyroelectric nanogenerator and self-powered breathing sensor. Nano Energy 38:147–154. https://doi.org/10.1016/j.nanoen.2017.05.056
119. Debels H et al (2015) Dermal matrices and bioengineered skin substitutes: a critical review of current options. Plast Reconstr Surg Glob Open 3(1):e284. https://doi.org/10.1097/GOX.0000000000000219
120. Peng R et al (2014) Conductive nanocomposite hydrogels with self-healing property. RSC Adv 4(66):35149–35155. https://doi.org/10.1039/C4RA05381H
121. Fukumoto K, Yoshizawa M, Ohno H (2005) Room Temperature Ionic Liquids from 20 Natural Amino Acids. J Am Chem Soc 127(8):2398–2399. https://doi.org/10.1021/ja043451i
122. Hou X-D et al (2013) Evaluation of toxicity and biodegradability of cholinium amino acids ionic liquids. PLoS ONE 8(3):e59145. https://doi.org/10.1371/journal.pone.0059145
123. Pedro AQ et al (2018) Cholinium-based good's buffers ionic liquids as remarkable stabilizers and recyclable preservation media for recombinant small RNAs. ACS Sustain Chem Eng 6(12):16645–16656. https://doi.org/10.1021/acssuschemeng.8b03900
124. Pedro SN et al (2020) The role of ionic liquids in the pharmaceutical field: an overview of relevant applications. Int J Mol Sci 21(21):8298. https://doi.org/10.3390/ijms21218298
125. MEDRX (2021, accessed 12 March 2021) http://www.medrx.co.jp/english/business_5.html

Chapter 17
Composites in Hydrogel State with Nanostructured Components for Biomedical Applications

Denis A. Cabrera-Munguía, Martín Caldera-Villalobos, Tirso E. Flores-Guía, Lucía F. Cano-Salazar, and Jesús A. Claudio-Rizo

1 Introduction

Recently, the use of nanotechnology in biomedical applications has become a promising area of research. The nanoscale components offer the advantage of being internalized at the cellular level, optimizing biomedical performance; this can be associated with the regulation of cellular metabolism, detection of specific biochemical processes, and/or adaptation of the cellular phenotype [1, 2]. On the other hand, this type of technology also allows the coupling with molecules of therapeutic interest, offering the capacity for controlled release depending on the conditions of the application environment [3, 4]. To achieve these approaches, it is necessary to design devices that meet structural requirements and possess tailored properties for a successful biomedical application. In this sense, hydrogels have been characterized by having a wide field of application in said sectors of biomedicine [5–7].

Hydrogels are polymeric networks with characteristic crosslinking indices that show the capacity to absorb high masses of water, in a relation dependent on their structure and properties. Hydrogels can be synthesized through the physical or chemical coupling of natural and/or synthetic polymers, generating the physicochemical crosslinking of the polymeric chains to reach the hydrogel state [1, 7–9]. Hydrogels mainly exhibit an excellent biocompatibility associated with the high presence of water, allowing the biological environments of application to be kept moist, the diffusion of nutrients and oxygen linked to their structure with characteristic porosity, the absorption of exudates from wounds related to the presence of hydrophilic groups in the polymeric chains, and the regulation of tissue growth due to the relation of the degradation rate that the polymeric matrix presents [10–14]. Hydrogels can

D. A. Cabrera-Munguía · M. Caldera-Villalobos · T. E. Flores-Guía · L. F. Cano-Salazar · J. A. Claudio-Rizo (✉)
Faculty of Chemical Sciences, Autonomous University of Coahuila, Ing J. Cardenas Valdez S/N, República, 25280 Saltillo, Coahuila, México
e-mail: jclaudio@uadec.edu.mx

S. Gopi et al. (eds.), *Nanotechnology for Biomedical Applications*, Materials Horizons: From Nature to Nanomaterials,
https://doi.org/10.1007/978-981-16-7483-9_19

be coupled with molecules of therapeutic interest such as antimicrobial agents or modulators of the cellular phenotype, allowing optimal platforms to study cell–matrix interactions [5, 14–16]. Among the main hydrogel systems that have been designed, the hydrogels based on semi-interpenetrated (semi-IPN) or interpenetrated (IPN) networks can be highlighted. In semi-IPN systems, a (natural/synthetic) linear polymeric chain is superficially coupled on another polymer chain, generating a hybrid matrix with an improved physicochemical crosslinking that allows adapting the mechanical properties and degradation rate [17, 18]; while in IPN systems, polymeric chains are crosslinked with the formation of specific chemical bonds with another polymer chain of interest, resulting in matrices with a high crosslinking index, having structure and properties dependent on polymer crosslinking. With this in mind, these systems in the hydrogel state represent, from the chemical point of view, composite materials, since specific polymer mixtures are used to generate matrices with structure and properties dependent on the ratio of the polymeric phases [18–21].

The incorporation of nanostructured components in matrices in the hydrogel state represents an innovative approach for the development of composites with improved performance in biomedical applications. Different nanostructured components have been studied for this type of application (Fig. 1). Nanocellulose has been coupled in various matrices in the hydrogel state, providing composites with improved biocompatibility and regulation of the swelling and degradation process of the composite matrix [22–24]. Inorganic components such as hydroxyapatite have made it possible

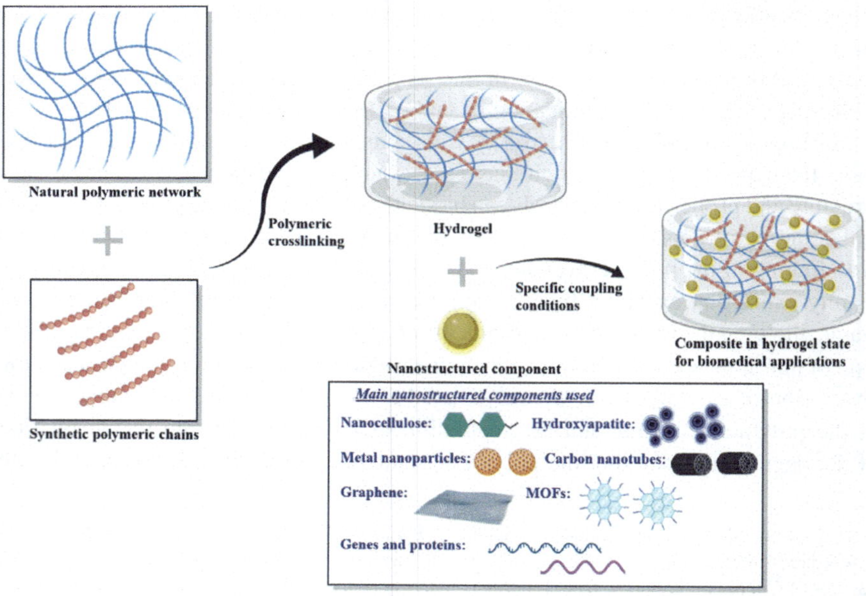

Fig. 1 General scheme for the preparation of novel composites in the hydrogel state with nanostructured components for biomedical applications (*Own authorship*)

to modulate the mechanical properties of hydrogels that have been coupled with this phase [25–27], also allowing the regulation of bone tissue growth and controlling the process of bone mineralization; also, the use of nanoparticles based on Si, Au, Ag, Ti, Zn, Cu, and Fe has been shown to influence the structure and properties of the designed hydrogel-state composite matrices, allowing the generation of detection systems for cancer therapies, tissue growth modulators, antibacterial character, and detection of specific biochemical processes such as inflammation or adapted fibrillogenesis [9, 20, 28–33]. The use of carbon nanotubes and graphene makes it possible to form composites in the hydrogel state for selective application in cancer therapies, specific biomedical imaging systems, and the ability to adsorb therapeutic molecules with controlled release [2, 34–36]. The metal–organic frameworks (MOFs) also allow the design of composite hydrogels with improved application in areas of biomedicine, since they allow to control of the structure and properties of the composite matrices, providing the capacity to inhibit bacterial growth, encapsulation, and controlled release of drugs with biomedical interest, as well as representing disease detection systems [37–39]. Finally, selective and specific modulation of the biological response can be efficiently achieved by developing composite hydrogels that have selective genes and proteins coupled for biomedical modulation [40–42]. The following sections address the topics of synthesis and application of various composite hydrogels that have nanometric components as agents to modulate biomedical performance.

2 Composites in Hydrogel State

Hydrogels are tridimensional materials constituted of hydrophilic crosslinked polymers to form entangled and insoluble polymeric matrices. Hydrogels are materials characterized to show excellent absorption capacity, swelling behavior, permeability, stability, degradation, bioadhesion, and bioactivity. The research on hydrogels synthesis has become of great interest because they are able to mimic the natural extracellular matrix (ECM), and thus offering support for cell migration, proliferation, and adhesion which are process involved in tissue healing [43, 44].

Typically, hydrogels are synthesized through chemical (covalent bonding) or physical crosslinking (hydrogen bonding, ionic or hydrophobic interactions). However, chemical crosslinking is preferred since hydrogels show higher mechanical properties than physical crosslinking hydrogels. Also, the tissue and cell adhesion ability of hydrogels can be enhanced by the use of linker molecules that enable non-covalently or covalently molecular interactions among the hydrogel network and the surroundings [43].

Hydrogels have been synthesized using synthetic polymers such as poly(2-hydroxyethyl methacrylate) (PHEMA), poly(ethylene glycol) (PEG), poly(vinyl alcohol) (PVA), poly(acrylic acid) (PAA), poly(vinylpyrrolidone) (PVP), polyurethanes (PU) as well as polypeptides and polyesters. The main advantages of synthetic polymers are their degradation rates and mechanical strength, but in general they tend to be bioaccumulated into the organism and do not possess binding sites limiting the cell proliferation and tissue regeneration. Whereas natural polymers such as chitosan, hyaluronic acid, gelatin, collagen, alginate, elastin, heparin, and chondroitin sulfate are biopolymers that not only allow cell adhesion, cell proliferation, and cell signaling, but also they show high biodegradability. Nonetheless, their application is limited due to their rapid structure degradation and poor mechanical properties. Nowadays, the trend is the synthesis of semisynthetic hydrogels for biomedical applications where natural and synthetic polymers are crosslinked to design novel hydrogels based on semi-IPN or IPN with improved properties. The molecular structure of hydrogel can be tuned by the appropriate selection of the polymer, crosslinking agent, crosslinking method, and crosslinking density [43].

In addition, some hydrogels are called "smart" since they show response under external stimuli like pH, temperature, ionic concentration, magnetic or electric fields. This characteristic can be applied in the development of materials for biomedical application such as site-specific drug delivery systems. The diffusion and release of therapeutic molecules can be regulated by manipulation of the crosslinking density of the hydrogel matrix and thus its porosity. Also, the response of hydrogels under different stimuli can use as sensors and actuators, diagnostic imaging and smart devices [43].

Nonetheless, current hydrogels applied in tissue engineering are still unable to replicate the complexities of the native biological environment. In general, hydrogels are inert under magnetic and electrical conditions, which limits the use of hydrogels in clinic applications. Moreover, hydrogels have shown a premature drug release, also some hydrophobic drug tends to precipitate causing poor formulation stability and low drug loading capacity [43]. To overcome these inconveniences of hydrogels for biomedical applications, the composite hydrogels have been developed by the incorporation of organic and inorganic fillers such as nanoparticles or nanostructures into the hydrogel network. The filler can be classified as carbon-based nanomaterials (carbon nanotubes, graphene), nanocrystals of cellulose, inorganic/ceramic nanoparticles (clays, silicates, hydroxyapatite), metal/metal oxide nanoparticles (gold, silver, and iron oxides), and recently metal–organic frameworks (Fig. 2).

The incorporation of these nanoparticles or nanostructures leads to unique properties such a "smart" materials that show response under stimulus to radiation, temperature, electric or magnetic fields, in this process the swelling behavior is induced leading to absorption or release of water or delivery of specific drugs [44, 45]. Another important characteristic is mechanical toughness, which is originated from the increment in the entanglements of polymeric networks caused by the incorporation of nanostructures that act as bridges among the polymer chains [45].

The introduction of nanostructures into the hydrogel network results in the formation of new materials called composites, getting advantages on their physical, chemical, and biological properties with promising applications in the biomedical field, such as drug delivery, biosensors, tissue engineering, wound healing, and antibacterial applications. According to the literature, it is observed that composites in state hydrogel with polymeric nanostructured components are used for hydrophobic and/or hydrophilic drug delivery applications. Whereas nanosilica-based composite

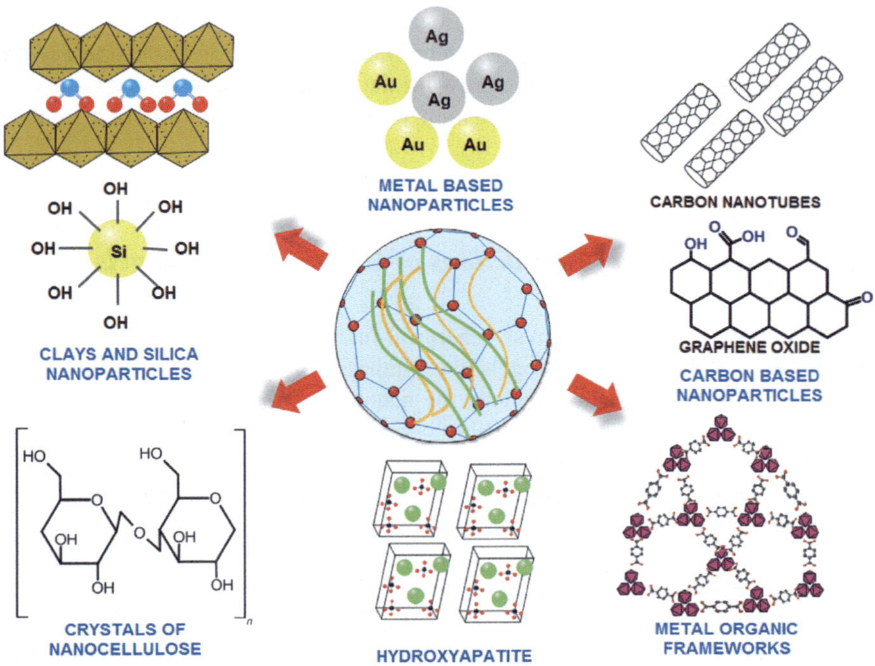

Fig. 2 Nanostructured components applied as fillers in composites in hydrogel state (*Own authorship*)

hydrogels are strong and mainly investigated for bone tissue engineering. Moreover, composite hydrogels based on metal nanoparticles are stimuli-responsive and have been investigated for remotely controlled drug delivery, modulation of cellular behavior, and antibacterial applications [21].

2.1 *Classification of Composite Hydrogels*

According to their dimension, composites in hydrogel state can be classified into nanogel, microgel, and macroscopic hydrogel. In the case of nanogels and microgels, which size is inherent in their name, the hydrogel matrix serves as coating of the nanostructures, obtaining a core–shell-type structure. The coating is chemically composed of natural polymers such as collagen, fibrin, chitosan, alginate, and hyaluronic acid. Thus, this matrix is able to load inorganic nanoparticles up to biomolecules such as enzymes and nucleic acids, making them desirable for applications such as drug delivery, sensors, imaging, and diagnostics. The macroscopic composite hydrogels involve entities from micrometers to centimeters where nanostructures are incorporated at various concentrations, depending on the particular enhancement of functionality or application. Similar to the microgel/nanogel class, the macroscopic composite hydrogels can be classified according to the dominating type of interaction of the nanostructure with the network chains, e.g., covalent or non-covalent, as well as their combination with polymeric networks being either chemical, physical, or topological [13].

2.2 *Physicochemical Interactions of Composites Hydrogels*

Similarly, to the bonding and physicochemical interaction involved in the physical or chemical crosslinking of hydrogels, the synthesis of composites in the hydrogel state comprises the interactions among the nanoparticle or nanostructure with the chemistry of the hydrogel network, in this case, the spaces at nanometric scale of the hydrogel matrix act as a "host" to be filled with nanostructures that act as a "guest" [21]. However, only a few systems have been deeply studied, and many composites in hydrogel state are still poorly understood [13].

In general, it is known that the entanglement of the composite in hydrogel state depends on the chemical affinity among the nanostructure surface and the inherent functional groups of the polymeric network. Thus, the formation of composite hydrogel is promoted by the adsorption of polymers on the nanostructure surface, and this adsorption depends on the polymer ability to adopt different structural conformations [13]. To increase the adhesion forces among the nanostructures and the hydrogel network is necessary as general rule to increase the specific surface area of nanostructure. Also, it is preferable to use nanostructured components with large pore diameter to absorb a higher amount of polymer [13]. In addition, it seems useful the employing of high concentration of crosslinking agents with the objective to get more reticulated polymer chains, leading to weaker adsorption energies, and thus favoring a better entanglement among the nanostructures and the hydrogel matrix [13]. To enhance the chemical affinity among the nanostructure surface and the hydrogel network (adhesion and stronger interactions), it is possible to include moieties supporting electrostatic interactions among the components of the 3D network. The most important key is that these electrostatic interactions can be increased by variation in ionic concentration and pH [13].

In spite of the poor knowledge about the physicochemical interactions involved in composites in hydrogel state, it has been observed that the formation of composites in hydrogel state based on metal nanoparticles involves the binding among metal nanoparticles, previously coated with organic ligands, and the host hydrogel matrix by Van der Waals forces and other non-covalent interactions that give rise to composites in hydrogel state with enhanced mechanical properties and bioactivity [21]. Composite hydrogels based on clay experiment electrostatic forces, for example, the positive amine groups of polymers guarantee a strong interaction with negatively charged clay surfaces [21].

Composite hydrogels based on nanosilica have been reported that hydrogen bonding from silanol groups (Si–OH) on the surface of silica particles is responsible for the adsorption of neutral polymers to their surface. Also, it is important to note that the hydrophilic or hydrophobic nature of polymer, and in the case of poly(ethylene oxide) (PEO), a hydrophilic polymer, its adsorption in silica particles is favored in regions with hydrophilic silanol sites instead of hydrated or ionizable surfaces. Thus, it is clear that polymers with carboxyl or ether groups will form hydrogen bonds with hydroxylated particle surfaces, and the strength of these interactions will depend on the chemistry of these groups, promoting the formation of composite hydrogel. Moreover, it is of great importance to take into account the hydrophilic or hydrophobic nature of the nanostructure surface and the polymer [13].

In the case of carbon-based nanoparticles such as graphene and carbon nanotubes, they naturally present π-π conjugated surfaces. Thus, π-π stacking interactions are predominantly involved in the development of composite hydrogels with outstanding conductivity and high mechanical properties, which are useful characteristics for tissue engineering and biomedical sensors [21]. However, in graphene oxide (GO) the presence of oxygenated groups such as hydroxyl and carboxyl groups promotes that the main forces are hydrogen bonds formed between GO nanoparticles and the hydrogel network [13]. At the end, the kind of interaction will depend on the surface chemistry of nanostructure and the click chemistry of the inherent functional groups found in the polymeric network [21].

2.3 Methods for Incorporation of Nanostructures in Composite Hydrogels

The synthesis of composites can be classified depending on the way with that the nanostructures are incorporated in the hydrogel matrix. The most common procedures are listed below:

1. *In-situ conversion of nanostructure into a pre-formed hydrogel*: In this case, the polymer network acts as scaffold, since the nanoscopic spaces of the hydrogel matrix are used for the nanostructure growing (Fig. 3a). Thus, in principle the nanostructured size can be controlled by the crosslinking density, where a high crosslinking density leads to thinner structures and spherical colloids. In addition, drugs or small proteins can be loaded in the hydrogels, where nanostructure function is to avoid the burst release effect. This synthesis is based on electrostatic interactions among components that comprise the gel matrix [13].
2. *Microemulsion of nanoparticles suspended in a polymer solution*: This is the simplest way to obtain composites in hydrogel state. The general idea is to homogeneously disperse the nanostructures over the polymer solution before gelling (Fig. 3b). This method has been widely applied for the synthesis of composite

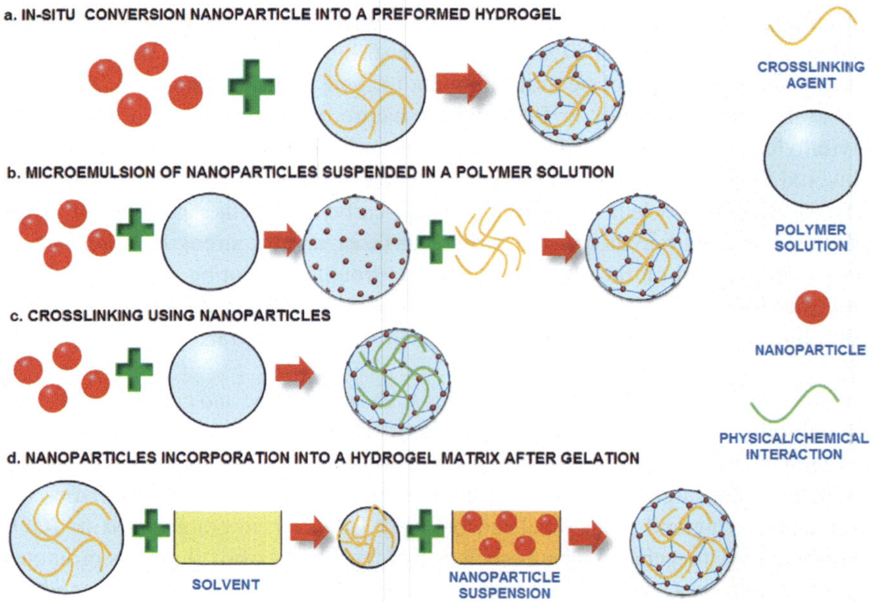

Fig. 3 Methods for the synthesis of composites in hydrogel state reinforced with nanocomponents (*Own authorship*)

hydrogels based on iron oxide nanoparticles. The main drawbacks of this method are that often heterogeneous composites are obtained, and it is possible the leaching of nanoparticles if the crosslinking density is low. This method should be avoided if it is possible the formation of covalent bonds among nanostructures and the polymeric network, since covalent bonding restricts the extensibility and self-healing potential of the hybrid material. Thus, it is suggested the mix of metal salts, instead of preformed metal nanoparticles, with polymer solutions for the in situ formation of nanoparticles and crosslinking of the polymer network with or without a crosslinking agent, with the aim to obtain homogeneously dispersed nanostructures on the hydrogel network, with good colloidal stability in just one step [13].

3. *Crosslinking using nanostructures*: Nanostructures can be used as crosslinking agents especially if nanostructures are functionalized with crosslinking groups (Fig. 3c). Nanostructures are able to form multiple bonds within the hydrogel networks, which is preferable that the electrostatic interactions involved in a traditional hydrogel synthesis. However, this characteristic relies on the ability of nanostructures to be adsorbed onto the polymeric network, and on the capacity of the polymer chains to reorganize them and dissipate energy under stress, when they are adsorbed onto the nanostructures. Some nanostructures used for this purpose are clays, silica nanoparticles, functionalized carbon nanotubes, and cellulose nanocrystals [13, 20].

4. *Nanostructures incorporation into a hydrogel matrix after gelation*: Nanostructures can be introduced after gelation of the hydrogel network by the hydrogel breathing ability (Fig. 3d). In this case, it is necessary to know if the hydrogel matrix is collapsed or expanded in a solvent of hydrophilic or hydrophobic nature. Then, the solvent where the hydrogel matrix is expanded will be chosen, to form the nanostructure solutions in the presence of a stabilizer, to avoid nanostructure aggregation. The method involves several cycles, alternating first the immersion of the hydrogel matrix into a solvent to collapse the structure with the expulsion of water, and then the hydrogel matrix is placed in a suspension of nanostructures to be swelled into the hydrogel network. This procedure is repeated several cycles until is obtained the desired nanostructure density [13, 20]. It is believed that the nanostructures are attached inside the hydrogel network by physical crosslinking or hydrogen bonds with the polymer chains. Another approach is the incorporation of nanoparticles into preformed hydrogels by cycles of heating, centrifugation, freezing, and re-dispersion followed by annealing [20].

3 Composite Hydrogels Reinforced with Nanostructured Cellulose

Cellulose and its derivatives are promising materials for biological and medical applications because they demonstrate excellent biocompatibility. Cellulose is the most abundant polymer occurring in nature, and it is a renewable and raw material. Cellulose-based nanomaterials are incorporated in composite hydrogels as a part of the polymeric matrix, or they are used as the reinforcement phase or filler. The most common cellulose-based nanomaterials are nanofibers and nanocrystals. Cellulose nanofibers are hydrophilic, and they interact by hydrogen bonding with other hydrophilic polymers. Cellulose nanofibers act as a crosslinking agent, and they form porous structures with a regular and uniform distribution of pores. Also, cellulose nanofibers reinforce the structure of the hydrogel and increase the mechanical properties [46].

The chemical properties of cellulose nanofibers are modified by oxidation. The oxidation of cellulose is typically catalyzed by (2,2,6,6-tetramethylpiperidin-1-yl)oxyl also known as TEMPO. Oxidized cellulose nanofibers by TEMPO (TOCNFs) are useful for the formulation of an injectable thermosensitive hydrogel. TOCNFs contribute to decreasing the gelation time significantly at physiologic temperature [47]. TOCNFs have a dual function in the formulation of composite hydrogels. They act as a reinforcement material for the matrix, and they are a physical crosslinker due to their ability for hydrogen bonding [48].

Bio-based hydrogels are interesting materials to be used as scaffolds for tissue engineering. However, typically hydrogels exhibit insufficient mechanical strength for bone tissue engineering. Composite hydrogels reinforced with cellulose-based nanomaterials have the best mechanical strength to be used as scaffolds in tissue engineering. A research work involves a chitosan-based hydrogel with cellulose nanofibers. Cellulose nanofibers increased the mechanical properties, improving the uniformity of porosity, and decreasing the size of the pore. The characteristics of the composite hydrogel are suitable to enhance the pre-osteoblast cell (MC3T3-E1) viability, attachment, and proliferation [49].

Composite hydrogels reinforced with nanostructured cellulose can be used for wound healing. This approach describes the preparation of a composite hydrogel with a gelatin/carboxylated cellulose nanofiber matrix and aminated silver nanoparticles. The hydrogen bonding interactions among cellulose and gelatin promote the formation of a three-dimensional (3D) porously crosslinked scaffold, and the Ag nanoparticles provide the antimicrobial activity. The composite hydrogel possesses low cell toxicity and excellent antibacterial activity. Further, it improves the wound healing ability with a high wound healing rate of 92.6% after two weeks of testing on in vivo rat model [50]. Due to the physically dynamic crosslinking forming the interpenetrating network of gelatin/carboxylated cellulose nanofibers, the hydrogel has the self-recovery capacity and a good mechanical and hemostatic performance for the wound treatment [50, 51]. Tavakoli and co-workers prepared a starch/cellulose nanofibers-based composite hydrogel coated with κ-carrageenan as a hemorrhage control material. In this composite, the cellulose nanofibers increase the mechanical strength of the polymer matrix. The κ-carrageenan coating enhances the swelling ratio and the degradation rate of the composite. In vitro test showed that the composite hydrogel is blood compatible and has an appropriate blood clotting potential. Further, the composite is not toxic toward human skin fibroblast promoting their viability and proliferation [52].

Another highlighted property of composite hydrogels containing cellulose nanofibers is their self-healing capacity. Taking advantage of this feature, a healable hydrogel with polyacrylamide, cellulose nanofibers, and Fe^{3+} ions has been reported. The authors attributed the healable capacity to cellulose nanofibers and their capacity to interact with the polyacrylamide chains and Fe^{3+} ions, promoting the fusion of surface components of two surfaces of the previously cut hydrogel [19].

Finally, 3D printing is a promising technology to build biomedical scaffolds and implants and it can be applied with composite hydrogels. For example, composite hydrogels of high/low methoxylated pectin reinforced with cellulose nanocrystals exhibit favorable rheological properties for the 3D printing. The shear-induced orientation enhances the alignment of cellulose nanocrystals along the print direction obtaining a degree of orientation (72–73%) [53]. This novel formulation of composite hydrogel can be used to design implants and/or prostheses with a specific shape, improving their biomedical performance.

4 Composite Hydrogel Reinforced with Hydroxyapatite Nanoparticles

Ceramic/polymer biocomposites emerged as potential biomaterials to fill, repair, replace, or regenerate injured or diseased bone, due to their excellent properties of biocompatibility, bioactivity, injectability, and biodegradability. Hydroxyapatite is a bioceramic and the major constituent of hard tissues such as bone and teeth. Hydroxyapatite is therefore of great interest as a transplant material to replace these tissues. Composites in hydrogel state containing hydroxyapatite nanoparticles have been used for a variety of biomedical applications, such as matrices for controlled drug release and bone tissue engineering materials. Up to now, several biopolymers have been studied as matrices in the formulation of composites hydrogels reinforced with hydroxyapatite nanoparticles. Some examples are chitosan [25], hyaluronic acid, glycol chitosan [54], collagen [55], and silk fibroin [56]. However, some synthetic polymers such as polyacrylamide [26] and poly(N-isopropylacrylamide) [57] have been used too. The resulting composites have great potential as biomaterials.

The properties of polymer-nanohydroxyapatite hydrogels such as biocompatibility, bioactivity, injectability, and biodegradability can be dependent on the amount of ceramic presents in the composite. It has been studied the effect of the nanohydroxyapatite content on alginate-based hydrogels to evaluate the best formulation for maximizing bone tissue regeneration. The results indicate that the main morphological characteristics of the alginate network are not affected by the addition of nanostructured hydroxyapatite. However, physicochemical features, like swelling capacity, stability at extreme pH values, apatite formation, and Ca^{2+} release depend on the concentration of hydroxyapatite nanoparticles. This work also established an optimal content of nanohydroxyapatite in the composites (30 wt.%) to enhance the osteoblastic cell proliferation and expression of osteogenic transcription factors. Further, it was also found that higher concentrations (50 and 70%) decreased the osteogenic cell response [58].

Other research work is focused on the preparation of a composite hydrogel with silk fibroin (SF) and hydroxyapatite nanoparticles suitable for bone regeneration. The results demonstrate that the compression modulus of composite hydrogels increased with the addition of hydroxyapatite nanoparticles (10–15 wt.%). Further, the interconnected porosity of the silk fibroin matrix and the uniform dispersion of hydroxyapatite nanoparticles improve the activity of phosphatase of osteoblastic cells [56].

Another biomedical application of these composite hydrogels is matrices for controlled drug release. Xu et al. prepared an injectable composite hydrogel with gellan gum-containing chlorhexidine and hydroxyapatite nanoparticles. The composite hydrogel promotes osteogenesis, and it has an antibacterial effect against *E. faecalis*. Additionally, loading the hydrogel with hydroxyapatite nanoparticles improved the mechanical, biodegradable, and osteogenic properties, also demonstrating a controlled release of chlorhexidine. With these results, this composite could be used for treating infectious bone defects caused by refractory periradicular periodontitis [59].

Zhang and co-workers prepared a composite hydrogel by in situ precipitation of nanostructured hydroxyapatite after photopolymerization of methacryloyl gelatin, quaternized chitosan, and polyhedral oligomeric silsesquioxane. Polyhedral oligomeric silsesquioxane was added to the composite to reinforce the structure of the hydrogel and for promoting mineralization. Test performed on rat skulls showed that the composite hydrogel accelerates the in situ bone regeneration. This novel material could be used as a promising substitute in bone tissue engineering [60].

Also, other work contemplates the development of a bone-mimetic nanohydroxyapatite/collagen hybrid scaffold in hydrogel state with high potential for bone tissue regeneration. The composite scaffold was prepared from the surface of a silanized mesoporous nanobioglass/collagen hybrid scaffold by biomimetic mineralization. The presence of mesoporous nanobioglass in the collagen fibrillar network enabled the progressive growth of hydroxyapatite nanocrystals and the generation of novel bone-mimetic hybrid structures while preserving the highly porous structure of the collagen scaffold [61]. This type of system demonstrates the effect of the structure and properties of these novel nanotechnology-based composites on the modulation of cell growth.

5 Composite Hydrogels Including Metallic Nanoparticles

Transition metal nanoparticles (NPs) such as Au, Ag, Fe, Ti, Zn, Cu and metalloids such as Si are characterized by having typical structures in a dimensional range of 1–100 nm. They are widely used in biomedical strategies due to their null reactivity, improved resistance to corrosion and oxidation in the environment and under physiological conditions [2, 11, 22]. This type of nanoparticles can be prepared by easily synthetic routes, either by direct precipitation, Stober's method, Turkevich's method, sol–gel process, and reduction, generating structures that have a high area/volume ratio, tailored optical properties, and high functionalization capacity chemistry [2, 11, 22, 62, 63]. Due to this, these inorganic components have been extensively studied for the generation of composite hydrogels with potential application in cancer therapies, physiological process detectors, radiotherapy, control of tissue growth (Fig. 4), loading and controlled release of therapeutics with biological interest, as well as the inhibition of the growth of pathogenic microorganisms [2, 11, 22, 62, 63]. Composites in hydrogel state based on metal nanoparticles integrate their properties on an hybrid material, according to the expression of those properties, composites in hydrogel state can be classified into: (i) composites where metal NPs express their optical, magnetic, and physical properties, (ii) the environmental changes alter the optical, electrical, or chemical properties of the incorporated NPs, and (iii) the NPs enhance the properties of the hydrogel matrix, but passively express their own properties [64].

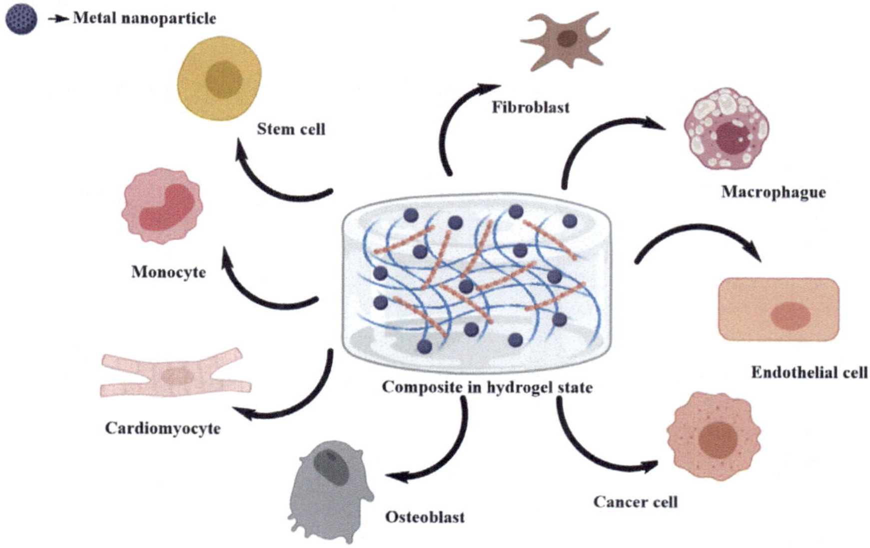

Fig. 4 Representative diagram for the modulation of the biological response in different cells by the composition of the composites in the hydrogel state based on metal (*Own authorship*)

The metal NPs are for general distributed homogeneously into the hydrogel matrix, but also they can be partitioned into the core, or the shell or in the core–shell interface, or act as the core. Metal NPs–hydrogel composite can be synthesized using the following approaches: (i) metal NPs are integrated by adsorbing precursor ions onto the polymer matrix, followed by in situ nucleation to generate matrix linked NPs, (ii) colloidal suspensions of metal NPs are suspended within the polymer sol before the composition is gelled, trapping the NPs within the matrix upon gelation (Fig. 5a), (iii) the polymer sol is gelled to produce a virgin hydrogel, and metal NPs are immobilized via physical or chemical methods post-gelation, (Fig. 5b) and iv) metal NPs are functionalized to act as crosslinking centers during gelation (Fig. 5c) [64].

All four types of composites are enabling to be applied as a sensor to detect and image biomolecules, with stimuli-responsive behavior for theranostic therapy, drug delivery nanocarrier, to form scaffolds for cell differentiation and proliferation (Fig. 4), and as antimicrobial biomaterials [64]. The synthesis of composite hydrogels with typical crosslinking agents results in irreversible covalent bonds, and these composites present cell toxicity. However, the use of metal NPs as crosslinking agent leads to a reversible interactions and composites for body's biological functions. Single metal ions from metal–ligand coordination bonds or complexes with the hydrogel matrix. In the case of bimetallic ions, they are used to crosslink the same polymeric chains at different positions with other functional groups. The stable metal–ligand coordination bonds are formed due to the high interaction and attraction among the metal ions having strong electron affinity and the polymer ligands

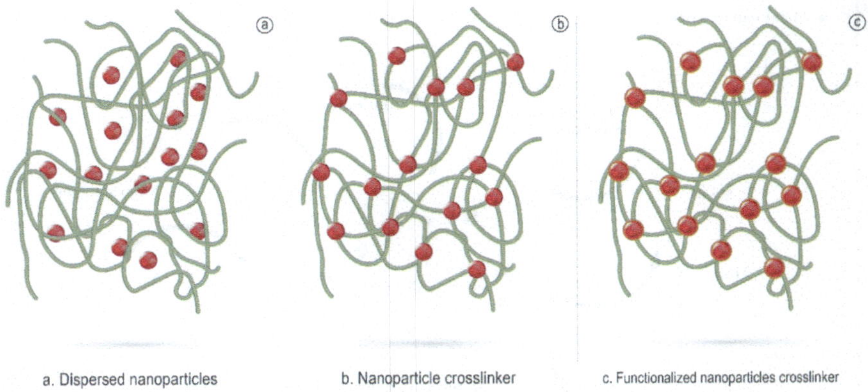

Fig. 5 Interactions of NPs with polymeric chains to generate composites in the hydrogel state (*Own authorship*)

consisting of more electronegative donor atoms with high polarizability. The strength of these interactions can be tuned by variation of the metal oxidation state, indeed a metal with multiple oxidation states is able to form a composite with unique shapes with different coordination site numbers in comparison with the organic ions or molecules applied as crosslinking agents [65]. The following sections detail the properties and biomedical applications of various composites in the hydrogel state that includes metal nanoparticles as modulators of the biomedical response.

5.1 Gold Nanoparticles

Recently, the use of Au nanoparticles has been proposed as a nanometric component that can modulate the growth of dermal, bone, and cardiac tissues, by being encapsulated in matrices in the hydrogel state. In strategies to control the growth of soft tissue, the generation of a composite hydrogel based on alginate and Au nanoparticles with an average diameter among 27–30 nm has been reported [66]; this composite hydrogel is formed by the crosslinking interactions favored by electrostatic interactions of the carboxylate groups of the alginate with the surface of the Au particles, generating a matrix in the hydrogel state that has the capacity to modulate the growth of endothelial cells derived from human umbilical veins, the presence of Au nanoparticles benefits the adhesion of this type of cells, as well as their proliferation and migration within the composite matrix, representing hydrogels with potential application for the regeneration of endothelial tissue, in addition demonstrating a high capacity to form new blood vessels (improved angiogenesis) in a rat-based in vivo model, allowing excellent vascularization after 21 days of implant exposure [66]. Another strategy involves the generation of a collagen-based composite matrix with 100 nm gold nanoparticles modified with 2-mercaptoethylamine groups [67]. The thiol groups of the modified nanoparticles can be condensed with the amino groups of collagen to generate a conjugated matrix in the hydrogel state. The typical fibrillar structure of collagen is unchanged and shows excellent durability for up to 6 months under an in vivo pig model; furthermore, these composite hydrogels do not exhibit irritability at the implant site, also presenting high biocompatibility; therefore, they are excellent candidates for the generation of composite hydrogels with in situ gelling capacity to fill defects in dermal tissues [67].

In the modulation related to bone tissues, composite matrices in hydrogel state based on chitosan and pectin have been studied by coupling Au nanoparticles of 20 nm in diameter [68]. These composites were prepared by adding the Au nanoparticles to pectin and chitosan solutions, generating the hydrogel state by physicochemical crosslinking at alkaline pH. The results indicate that the gelling temperature of the matrix in the hydrogel state depends inversely proportional to the concentration of gold nanoparticles, also increasing the pore diameter of the matrix based on polysaccharides. In vitro results using mouse preosteoblastic cells show that this type of composite does not have a cytotoxic character on these cells, not altering their metabolism when growing in these materials, and thus, this type of hydrogel represents a potential scaffold for the regeneration of bone tissues [68]. Another work proposes the inclusion of Au nanoparticles with an average diameter of 35.86 nm modified with *N*-acetylcysteine within a gelatin-tyramine matrix [69]. Tyramine can crosslink gelatin by the action of the horseradish peroxidase enzyme, generating a matrix in the hydrogel state that can withstand the loading of modified Au nanoparticles. The storage modulus of composite hydrogels presents a dependent relation with the nanoparticle loading, significantly improving it and generating an in situ temperature-dependent gelling system. In vitro experiments carried out with stem cells derived from human adipose tissue show that this type of composite hydrogels have high biocompatibility for this type of cells and allow their proliferation for up to 3 days, promoting osteodifferentiation, associated with the inclusion of the N-acetylcysteine group in the nanoparticles of gold; therefore, this type of hydrogels can be used to induce the regeneration of bone tissue [69].

In the modulation of cardiac tissue growth, composite hydrogels have been developed based on decellularized matrices with the inclusion of 4 and 10 nm diameter Au nanoparticles. A deposition–evaporation process was used for the generation of the composite matrices [70]. These novel materials exhibited superior mechanical behavior to pistrin patches, and degradation and conductivity controlled by the size of the nanoparticle, in addition they show excellent in vitro biocompatibility measured in neonatal rat cardiomyocytes and cardiac fibroblasts; these types of cells demonstrate a homogeneous alignment on the scaffold in the hydrogel state, allowing the growth of cardiac tissue to be modulated, representing devices with potential application in the treatment of infarcts in myocardial tissue [70]. In another approach addresses chitosan-based composite hydrogels with gold nanoparticles [71]. These composite matrices were prepared by gelling a chitosan solution induced by the crosslinking of Au nanoparticles formed in situ by reduction of $HAuCl_4{\cdot}3H_2O$ with citrate; the dispersed particles in the composite matrix exhibited a size of 7 nm in diameter. The gelling rate, as well as the electrical conductivity, depend directly on the concentration of the nanoparticles. The biocompatibility of these composites was evaluated using mesenchymal stem cells for up to 14 days under contact, indicating that the surfaces do not present cytotoxic effects on these cells, allowing their migration, growth and proliferation, generating homogeneous cell agglomerates. This type of association allows the cardiomyogenic differentiation of this type of cells measured by immunohistochemical tests. Cellular control by electrical stimuli can be tailored by this type of composites in the hydrogel state, making them excellent candidates for the modulation and growth of electroactive tissues [71]. Other researchers formulated gelatin/methacrylate-based composites by coupling Au nanorods [72]; these matrices in the hydrogel state were generated by the crosslinking of gelatin with methacrylate by the action of UV light, coupling the Au nanorods once the viscosity of the crosslinked matrix increased. The nanorods were characterized by having a size of 4 nm in diameter and being homogeneously dispersed in the composite matrix. The electrical conductivity and stiffness of the matrix depend on the concentration of nanorods. The in vitro results with cardiomyocytes indicate that these matrices in the hydrogel state exhibit controlled viability, allowing migration and organization of these cells to generate homogeneous layers of cardiac tissue. Furthermore, the immunocytochemistry results indicate that cardiomyocytes secrete specific proteins that regulate their ability to beat; indicating that this newly formed tissue acquires a synchronized rhythmic contraction; therefore, these novel composites in the hydrogel state can be used for potential application in repair strategies and modulation of cardiac tissue modulated by nanotechnology represented in Au nanorods [72]. Table 1 summarizes other composites in the hydrogel state designed with Au nanoparticles for biomedical applications, emphasizing the type of polymeric matrix, characteristics of the nanoparticles, and improved properties for biomedical applications.

5.2 *Silver Nanoparticles*

Silver nanoparticles of various sizes and structures have been used as nanostructured components to regulate cell behavior, benefiting biomedical applications such as cell growth modulation and tissue regeneration, characterizing composite hydrogels that are coupled with these components due that they have the ability to inhibit the growth of pathogenic bacteria [1, 75, 76]. In dermal tissue healing strategies, the formulation of a composition based on a matrix of chitosan and hydroxyapatite has been reported [31]. This material was prepared by crosslinking the components simultaneously by mixing at a temperature of 90 °C in the presence of calcium and phosphate ions. The silver nanoparticles determine a physicochemical crosslinking of the chitosan chains to improve the thermal degradation of the composite matrix, and in addition, the in vitro biocompatibility results show that these materials present high biocompatibility for dermal fibroblasts, not representing a cytotoxic character for their proliferation, also characterized by inhibiting the growth of pathogenic bacteria involved in chronic wounds such as gram-positive Staphylococcus aureus

Table 1 Composites in hydrogel state based on Au nanoparticles

Polymeric matrix	Characteristics of nanoparticles	Exhibited properties	Biomedical application	References
Methoxy poly(ethylene glycol)-b-poly(εcaprolactone-co-1,4,8-trioxa[4.6]spiro-9-undecanone) (mPECT) diblock copolymer	Nanorods of 40 nm diameter	Injectable hydrogel with controlled released of paclitaxel	Inhibits the growth of breast cancer in mice	[73]
Gelatin crosslinked with methacrylate	Gold nanoparticles of 27 nm of diameter	Controlled degradation	Human adipose stem cells are differentiated to osteoblastic tissue, allowing for improved bone regeneration	[28]
Collagen-CGS-LL37 peptide	Gold nanoparticles of 13 nm of diameter	Increment in stiffness and electrical conductivity	It benefits the activation of macrophages for tissue repair and allows the proliferation of neonatal cardiomyocytes, regulating their function by electrical stimuli	[74]
Gelatin/methacryloyl	Gold nanorods of 25 nm of diameter	Use as bioink with enhanced bioprinting due to its low viscosity	It allows the dispersion and organization of cardiac cells; exhibiting contraction synchronized to electrical stimuli	[72]

and gram-negative Pseudomonas aeruginosa; indicating that this type of matrices in the hydrogel state with temperature-dependent gelling capacity can be successfully used in the modulation of the wound healing process [31]. Another work that deals with an agarose-based composite matrix impregnated with chitosan spheres coupled with Ag nanoparticles also shows that this nanotechnology system represents the successful application in strategies for healing soft tissues; this type of composite matrix was prepared by dispersing the Ag-chitosan nanoparticles on an agarose solution that undergoes crosslinking with glutaraldehyde at basic pH [77]. The degradation rate, as well as the mechanical properties of the matrix, is adjusted by the concentration of Ag-chitosan nanoparticles; besides, these materials exhibit cytotoxic character for human cervical cancer cells (HeLa) and pancreatic epithelial cancer cells, representing materials for cancer therapies; on the other hand, these matrices in the hydrogel state also present high biocompatibility for human embryonic kidney cells and high hemocompatibility, representing hydrogels that have selectivity to modulate cellular behavior [77]. The results of bacterial inhibition for this type of matrices show to limit the growth of Escherichia coli2345 (*E. coli*), Bacillus subtilis, Klebsiella pneumoniae, and Staphylococcus aureus737 (*S. aureus*), which avoids the use of exogenous antibiotics, benefiting the success in the healing of soft tissues [77].

Various composites in the hydrogel state with Ag nanoparticles have been designed to control the regeneration and growth of bone tissue. A hybrid matrix based on chitin and bioactive bioglass coupled with Ag nanoparticles has been studied [78]. This type of composite was initially synthesized generating the chitin hydrogel matrix, inducing the crosslinking of this biopolymer in the presence of $CaCl_2$, later the hydrogels were dialyzed to add different concentrations of bioglass and Ag nanoparticles. The composite matrices were characterized by having interconnected porosity, improved swelling, and degradation rate controlled by the concentration of inorganic phases. The presence of silver nanoparticle in the matrix inhibits the growth of *S. aureus* and *E. coli*, and the biocompatibility results show that the surfaces of these composite hydrogels do not have cytotoxic effects on human osteoblasts, benefiting their proliferation and spreading throughout the composite, so these novel materials can be used successfully in the regeneration of bone tissue such as in remodeling of periodontal defects [78]. Hydrogels based on synthetic polymers have also been coupled with Ag nanoparticles to generate novel nanostructured composites; in this strategy nanohydrogels based on polyethylene glycol were prepared [79]; for this, the silver nanoparticles were dispersed for their adsorption in a matrix in the hydrogel state formed by the acid-catalyzed polymerization of poly(ethyleneglycol) diacrylate (PEGDA). These composites have the ability to inhibit the growth of *S. aureus*, and also, they have excellent biocompatibility for osteoblasts, regulating their metabolism for the generation of tissue remodeling genes; with these results, this type of composite matrix can be applied in the remodeling and/or reconstruction of bone defects [79].

For cardiac tissue strategies, the preparation of a composite matrix based on collagen and CGS-LL37 peptide has been reported by coupling silver nanoparticles [74]; the hydrogel was formed by the adsorption of the peptide on the surface of the silver nanoparticles of 7.3 nm in diameter; with the subsequent incorporation of these on a gelified collagen matrix by physicochemical crosslinking of phosphates. The presence of Ag nanoparticles provides the composite matrix with specific conductive properties, regulating the growth of cardiomyocytes under the effect of electrical stimuli. They also have a high capacity to attract macrophages and benefit their metabolism for tissue reconstruction; these results show composite matrices in hydrogel state with potential application for modulating the growth of cardiac tissue [74]. Table 2 presents various formulations of composite matrices in the hydrogel state coupled with Ag nanoparticles as nanostructured agents.

5.3 *Iron and Iron Oxide Nanoparticles*

Iron in comparison with other metals presents different oxidation states, and it possesses diverse coordination chemistry properties appropriate to be applied in biological systems. Fe^{+3} ions act as "hard" acids favoring "hard" oxygen ligands, whereas Fe^{+2} prefers nitrogen atoms than the oxygen-based ligands [83]. The Fe^{+3} has been crosslinked solely, coordinated to catechol or in combination with other metal with hydrogel matrices such poly(acrylic acid) (PAA), poly (acrylamide) (PAAM), xanthan gum, gelatin, polygalacturonate, sodium carboxymethyl cellulose (CMC), chitosan, and hyaluronic acid (HA). The main problem when Fe^{+3} is embedded in the hydrogel matrix is its partial reduction to Fe^{+2}, and thus, it is preferable the introduction of Fe^{+2} that with further oxidation becomes into Fe^{+3}. Furthermore, the combination of catechol and iron chelation offers enhanced hardness and extensibility to the hydrogels, where mono-, bis-, and tris-catechol forms containing Fe^{+3} ions give rise to pH-dependent crosslinking of hydrogels [83].

On the other hand, iron oxide (Fe_xO_y) NPs composite hydrogels have been developed for their ferromagnetic and semiconducting properties. Ferrogels have been produced using poly(acrylamide-co-maleic acid), 4-vinylpyridine, chitosan blends with PAA and methacrylate copolymers as hydrogel matrixes [20]. Several methods such as in situ precipitation, grafting on hydrogels, and physical blending have been used to prepare iron oxide NPs-based magnetic hydrogels [7, 21].

Iron oxide NPs containing composite hydrogels are fascinating due to that their response to the fast changes in the surrounding conditions such as pH, ionic strength, temperature, and magnetic and electric fields [7, 21]. Iron oxide nanoparticles have been utilized in biomedical fields (Table 3) such as biosensing, magnetic resonance imaging (MRI), tissue engineering, targeted drug delivery (cancer hyperthermia), and biomolecules separation [7, 21].

Magnetic nanoparticles such as magnetite (Fe_3O_4) and hematite (Fe_2O_3) are known to respond to alternating magnetic field (AMF). Composites in hydrogel state based on iron oxides have been applied in tissue engineering as magnetically driven actuators to mimic tissue movement (e.g., cardiac tissue, muscle) under magnetic field [20]. In addition, magnetic NPs embedded in thermo-responsive hydrogels can be employed for drug delivery by alternating magnetic fields. However, in cancer

Table 2 Composite hydrogels including Ag nanoparticles for biomedical applications

Polymeric matrix	Characteristics of nanoparticles	Exhibited properties	Biomedical application	References
Collagen	Ag nanoparticles with 3.5 nm of diameter	Concentration-dependent mechanical properties of Ag nanoparticles; and ability to inhibit S. aureus, *S. epidermidis*, *E. coli*, and *P. aeruginosa*	Ability to modulate anti-inflammatory activity, generating wound healing in mice with the accumulation of silver in the surrounding tissue	[80]
Hydroxyethyl cellulose	No reported	Regulation of the degradation rate and excellent swelling capacity	Excellent biocompatibility for human fibroblasts, allowing their proliferation and dispersion in the composite matrix	[81]
Poly(hydroxyethyl methacrylate)	No reported	Structure with interconnected porosity of 40 mm in diameter. Bacterial inhibition capacity of S. aureus and E. coli	They exhibit the ability to regulate dermal growth preventing scar formation in rat models	[82]
Alginate/polyacrylamide	Spherical particles with a diameter of 20–30 nm	Ciprofloxacin encapsulation capacity and controlled release, as well as pH-regulated swelling	Hydrogel with high capacity for absorption of wound exudates avoiding the formation of infections	[17]
Methacrylate	Ag nanoparticles with diameter of 30–80 nm	Improved mechanical properties, ability to inhibit the growth of staphylococcus epidermidis, and Methicillin-resistant Staphylococcus aureus	High biocompatibility for human osteoblasts, also allowing rapid mineralization of bone tissue	[76]

therapy this ability depends on the distance between the tumor and the magnet [20]. Also, iron oxide NPs-based hydrogels have the potential to be used in hyperthermia therapy (heating of cancerous tissue at 41–45 °C) due to that their capability to heat when exposed to AMF [7, 21]. Nonetheless, a common complication with iron oxide NPs is their aggregation and oxidation, which result in the diminishment or loss of the magnetic properties [44].

Table 3 Biomedical applications of composites in hydrogel stated based on Fe and Si nanoparticles

Hydrogel matrix	Nanoparticle	Application	References
Gelatin	Silicon particles	Controlled release of Tramadol	[84]
Alginate	Silica supported silver nanoparticles	Wound dressing with antimicrobial activity	[85]
Carboxyethyl chitosan/oxidized pullulan	Colchicine mesoporous silica nanoparticles	Transdermal drug delivery for osteoarthritis treatment	[86]
Agarose	Ferric tannate nanoparticles	Phothermal sterilization	[87]
Chitosan	Halloysite clay nanotubes, $Al_2(Si_2O_5)(OH)_4$	Drug delivery of Doxorubicin	[88]
Poly(acrylic acid) grafted onto κ-carrageenan polysaccharide	Entrapped magnetic iron oxide nanoparticles as crosslinker	Drug delivery of levodopa (L-DOPA)	[89]
Xanthan gum/chitosan	Iron oxide magnetic nanoparticles (Fe_3O_4 MNPs)	Proliferation and adhesion of dermal fibroblasts under the presence of magnetic fields	[90]
Alginate/guar gum	Mesoporous silica nanoparticles	Self-healing hydrogel for tissue engineering	[91]
Gelatin	Bentonite, (SiO_3 * H_2O)	High viability of encapsulated fibroblasts improves healing processes	[92]

5.4 *Silica and Silicate Nanoparticles*

Synthetic silica and silicate nanoparticles (nanoclays) are the essential minerals for normal tissue functions in the body. These inorganic particles are extensively used in implantable materials, especially for bone-related applications [45]. Silica nanoparticles offer high surface area, functionalizable and tunable particle size, biodegradability, and biocompatibility. A large number of bioactive molecules can be loaded into the porous structure of silica NPs, while their surfaces can be easily modified for improved biomedical performances. Silica NPs can be firmly bound with the polymeric networks and improve the mechanical strength of the composite hydrogels; this enhancement is attributed to an increment in the crosslinking density of hydrogel [7, 21].

Silica NPs also show a sustained drug release, and it is classified as a safe material for oral drug delivery in the concentration up to 1500 mg/day. Then, the incorporation of these nanoparticles into hydrogels, resulting in composites mainly investigated for tissue engineering, cell culture, drug delivery, and hemostasis [21].

On the other hand, composites in hydrogel state based on nanoclays lead to tissue-adhesive composite hydrogels with improved mechanical properties. The morphology of nanoclays is anisotropic and plate-like with a high aspect ratio, resulting in high surface reversible interactions (non-covalent) with polymer chains [45].

Depending on the clay type, the individual layers could be composed of two, three, or four sheets of either SiO_4^{4-} tetrahedral or $[AlO_3(OH)_3]^{6-}$ octahedral. These layers can organize themselves over one another, with a regular Van der Waals gap among them, called "interlayer" which possesses a net negative charge that is neutralized by cations occupying the inter-lamellar gap [45]. The studies of hydrogel composites based on nanoclays indicate that can be useful for the design of wound dressings, since they do not show cytotoxicity against Fibroblast cells, high swelling properties, and induce blood coagulation (Table 3). Also, recent studies have shown that nanoclay can induce osteogenic differentiation of human mesenchymal stem cells (hMSCs) without the use of exogenous growth factors [45].

5.5 Other Types of Metal Nanoparticles

Copper nanoparticles are used due to their low cost, compared to those contemplated in the previous subsections, and they have good thermal and electrical conductivity, associated with the fact that Cu (II) is a fundamental element used by our body for various functions. In addition to NPs of individual and/or functionalized metals, nanoparticles of metal oxides (MO–NPs) such as titanium, zinc, copper, and aluminum have been also used. The composite hydrogels formulated with these nanomaterials are multifunctional because they present photocatalytic, antibacterial activity, high stability, low cytotoxicity, to mention some of their properties [21]. Table 4 summarizes some examples of the use of these metallic nanoparticles in different composites in hydrogel state for biomedical applications.

6 Composite Hydrogels Based on Carbon Nanotubes and Graphene

Among the allotropes of carbon, carbon nanotubes are classified into fullerenes family (C_{60}), where a sheet of graphene is rolled into cylindrical tube-like structure. This configuration makes that sp^2 hybridized carbon atoms are arranged in a specific pattern to form hexagonal structural units, which imparts higher C–C bond stiffness leading to a tensile strength of 150 GPa [34]. According to their structure, carbon nanotubes are categorized in single-walled carbon nanotubes (SWCNTs) and multi-walled carbon nanotubes (MWCNTs) (Fig. 6). In SWCNTs, only a sheet of graphene is rolled leading to cylindrical single-walled tubes, which are one atom thick having

Table 4 Other metallic nanoparticles used in composite hydrogels for biomedicine

Metal	Biomedical application	Improved characteristics	References
Cupper	Drug release	CuO–NPs have demonstrated an increase in swelling capacity of composite hydrogel, achieving good loading and release behaviors of different drugs with therapeutic interest	[93, 94]
	Antibacterial activity	Cu and CuO–NPs are commonly used as nanofiller in hydrogels of natural polymers due to their antimicrobial effects	[24, 95, 96]
	Wound healing	Nanoparticles of Cu, CuO, and CuS used in composite hydrogels enhance the wound healing mechanism by increasing cell proliferation, antimicrobial property, and rapid initiation of inflammatory response	[97, 98]
	Nanocarriers	The modification and distortion of the skin architectural integrity is achieved by CuO–NPs composite hydrogels affecting hematological functions	[99]
Titanium	Tissue engineering	Multiple nanostructured TiO_2–NPs have been used as crosslinker and filler to obtain composite hydrogels with good mechanical properties and as scaffolds for cartilage regeneration	[9, 100–102]
	Wound healing	Composite hydrogels reinforced with TiO_2–NPs present potential use for this application due to antibacterial activity and good biocompatibility	[10]
	Cell implantation	Functionalized TiO_2–NPs have presented the ability to retain more and complex cell aggregates, making them a suitable substrate to analyze their potential use for in-vivo cell implantation experiments	[103]
Zinc	Wound healing	Natural and synthetic polymer-based composite hydrogels loaded with ZnO–NPs have been synthesized from different methods to obtain antibacterial and biocompatible materials	[30, 104–107]
	Contact lens	In the case of ophthalmologic applications, ZnO–NPs and ZnS–NPs provide good biocompatibility and antimicrobial properties	[75, 108, 109]

(continued)

Table 4 (continued)

Metal	Biomedical application	Improved characteristics	References
	Drug delivery	Functionalized ZnO–NPs obtained from different methods with specific structures and shapes have been used to control the release kinetics, swelling ratio, and loading drug capacity	[15, 110–112]

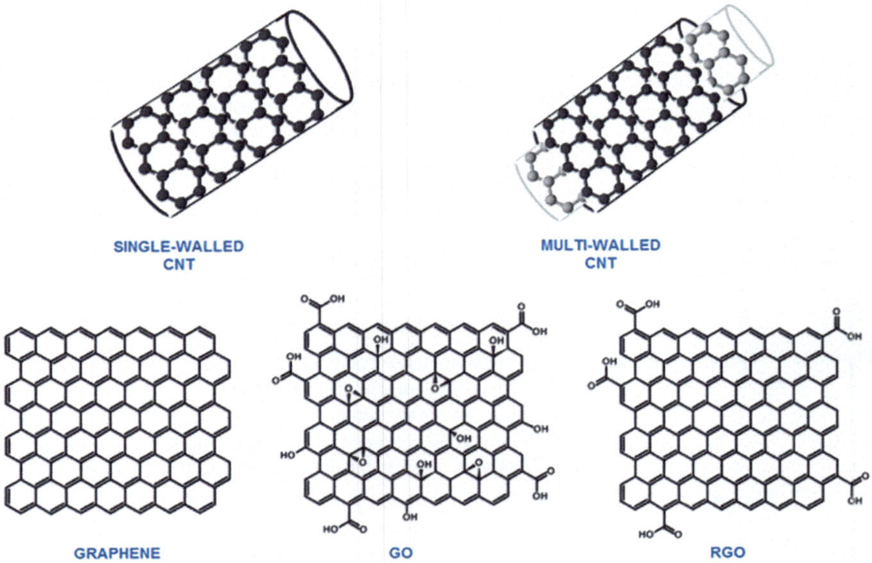

Fig. 6 Chemical structure of CNTs and graphene family (*Own authorship*)

a radius of up to 1 nm. Meanwhile, MWCNTs are composed of several layers of graphene sheets (2–10), obtaining an external diameter higher than 10 nm [34, 113].

These structural differences make that SWCNTs show better features for biomedical applications than MWCNTs, such as good electrical conductivity and good dispersability in aqueous media [113].

On the other hand, graphene is constituted by a single carbon layer of graphite with a two-dimensional planar and hexagonal array of carbon atoms forming a honeycomb crystal lattice. Graphene possesses a large specific surface area (2630 m^2g^{-1}), high intrinsic mobility, mechanical strength, high thermal (5000 W mK^{-1}), and electronic conductivity (104 S cm^{-1}) [114–118].

The graphene family includes graphene, graphene oxide (GO), and reduced graphene oxide (RGO) (Fig. 2). RGO is considered as an intermediate structure among the ideal graphene sheet and highly oxidized GO. While GO is the oxidized form of graphene with several functional groups like oxygen, epoxide groups, carbonyl, hydroxyl, and phenol groups that are present on the surface of GO, leading to a hydrophilic surface instead of the hydrophobic character of graphene [35, 118].

Carbon nanotubes are typically synthesized by the exposure of synthetic graphite to an electric arc discharge technique (EAD), a laser ablation method (LA), chemical vapor deposition (CVD), and high-pressure carbon monoxide (HiPCO). EAD can obtain only MWCNTs with uniform diameter, whereas LA can produce both CNTs types, but with impurities. Among these techniques, CVD and HiPCO are methods available for bulk production, but HiPCO is a costly technique that only obtains SWCNTs, on the contrary CVD is a promising technique for the industrial scale due to its low-cost yielding SWCNTs and MWCNTs [34, 119, 120].

The most common methods to synthesize graphene from graphite are via mechanical cleavage, liquid-phase exfoliation, graphite oxide/fluoride reduction, intercalation, and compound exfoliation. In the case of non-graphite sources, graphene is produced via epitaxial silicon carbide decomposition, chemical vapor deposition (CVD) growth, and bottom-up chemical synthesis [118, 121, 122]. In the case of graphene derivatives, GO can be synthesized via Hummer's method through oxidative exfoliation of graphite using $H_2SO_4/KMnO_4$. Also, RGO has been produced with the aid of reducing agents such as hydrazine, hydrazine hydrate, L-ascorbic acid, and sodium borohydride [118, 121, 122].

For biomedical applications, the toxicity of carbon nanotubes and graphene structures is related to their internalization into cellular membrane via diffusion, membrane fusion, and direct pore transport [123]. Another pathway is endocytosis (phagocytosis and pinocytosis) conducted by macrophages, monocytes, and neutrophils [123]. Also, carbon nanotubes cause cytotoxicity by blocking the ion channels that regulate the intracellular calcium level [36, 123].

The mechanisms of toxicity of carbon nanotubes and graphene structures include oxidative stress, cell membrane injury, and genotoxicity [34, 36, 114]. The oxidative stress represents the primary mechanism for carbon nanotubes toxicity, which is attribute to reactive oxygen species (ROS) and production of free radical which causes oxidative stress, inflammation, damage to proteins, and DNA of cells [34, 118, 124]. The mechanism of cell membrane injury for graphene included initial cell deposition on graphene-based materials, membrane stress caused by direct contact with sharp nanosheets, and ensuing superoxide anion-independent oxidation [36, 115].

The size and shape of CNTs and graphene structures must be regulated (<1 μm), to avoid cell membrane injury [36, 122]. Then, too large carbon nanotubes are unable to be phagocytosed by mononuclear cells that also provoke the macrophage activation, granulomatous inflammation, and progressive fibrosis [36, 115, 122]. Furthermore, the type of carbon nanotubes plays a key role in their cytotoxicity, since SWCNT contained hydroxyl (-OH) and carboxyl (-COOH) groups that made their surface more hydrophilic compared to MWCNTs. Then, SWCNTs with hydrophilic character are less toxic to macrophages than CNTs with hydrophobic surface (MWNT) [36, 115, 122]. This result is also related to the size, as MWCNTs form larger aggregates than 300 μm that cannot be phagocytosed in contrast with the small aggregates of SWCNTs ranging from 5 to 30 μm that facilitate their uptake by macrophages [36]. It is also important to avoid metals and other impurities of carbon in CNTs and graphene structures that come from the synthesis technique to avoid the inhibition of cell proliferation [34, 36, 114].

On the opposite, smaller flakes of graphene structures have been reported to be more cytotoxic and have shown the greatest hemolytic activity, whereas aggregated graphene sheets exhibited the lowest hemolytic activity, and this result is related to their higher cellular internalization [122]. Also, the number of oxygen functional groups that are attached to the surface also plays an important role, as larger C/O levels correlated to RGO structures are less cytotoxic [117].

The genotoxicity of carbon nanotubes and graphene structures can take place in a direct way when nanomaterials interact with DNA or mitotic apparatus, and the indirect way could result from oxidative stress and inflammatory responses [34]. Moreover, it has been reported than carbon nanotubes are able to cause toxic effects in lungs, liver, kidneys, the reproduction, and nervous system [120]. For instance, pristine graphene was shown to cause high oxidative stress by accumulation on the cell membrane, whereas GO is non-toxic for human lung cells even after internalized by cells, and it can be digested by neutrophils. On the other hand, nanographene sheets accumulated initially in the liver, and spleen, and later, they were cleared and induced no toxicity at a dose of 20 mg/kg [36, 117].

Strategies for the functionalization of carbon nanotubes and graphene structures can be used to decrease their cytotoxicity. The purpose of the functionalization is to enhance the solubility, dispersability, reactivity, selectivity, and biocompatibility for biomedical applications [34, 36]. The most common chemical agents employed for graphene functionalization and their derivatives are polyethylene glycol (PEG), polyvinylpyrrolidone (PVP), chitosan, deoxyribonucleic acid (DNA), enzymes, and proteins [125]. The oxygenated functional groups (–COOH, –OH) on the GO surface generate hydrogen bonds with water molecules obtaining a stable GO colloidal suspension for potential biomedical applications [118, 125]. The surface-functionalization of carbon nanotubes and graphene structures can be divided into covalent functionalization and non-covalent adsorption of required molecules.

In the case of graphene, the covalent functionalization is available by oxidation of graphene nanosheets to be easily functionalized with typical organic reactions (e.g., amidation, silanization, esterification, and cycloaddition reactions) and then reduced to graphene (RGO). However, several oxygen groups and structural defects remain which mainly affect the electrical and mechanical properties [126].

Non-covalent functionalization/wrapping and absorption are governed by the size of nanoparticles (<100 nm), which allows the direct attachment or supramolecular adsorption to the graphene sheet of polysaccharides, DNA, proteins, polypeptides, and synthetic polymers through different adsorption forces such as hydrogen bonds, Van der Waals forces, electrostatic forces, and p-stacking interactions, generating nanostructures with enhanced biocompatibility [34, 36, 118, 125], also causing less structural damage to the carbon nanotubes and the wrapped species [123]. However, these weak interactions limit the use for drug delivery, but can be exploited in imaging [124, 127]. It is reported that non-covalently functionalized carbon nanotubes predominately accumulate in liver and spleen tissues [120]. Similarly to carbon nanotubes, graphene surface can physically adsorb via π-π interaction surfactants, small aromatic molecules, interaction with porphyrins, or biomolecules such as deoxyribonucleic acid (DNA) and peptides without affecting the sp^2 carbon hexagonal lattice of graphene, enhancing its biocompatibility [126].

Carbon nanotubes and graphene structures have been considered as ideal nanofillers to produce composites in hydrogel state due to their high mechanical strength, effective surface area, and high electrical conductivity [34]. Diverse applications of these include controlled drug delivery (gene and cancer therapy), biomedical imaging, thermal phototherapy, antibacterial agents, and in the identification (biosensor) of nucleic acids, DNA, and RNA [34, 36, 117, 125, 126]. In addition, the high elasticity, flexibility, and adaptability to flat and irregular surfaces of graphene and their derivatives make them suitable for tissue engineering improving the adhesion, differentiation, and cell function, whose nanocomposites in hydrogel state are electro-conductive are promising for future applications in cardiac tissue engineering and regeneration [35].

The methods for the synthesis of nanocomposites based on carbon nanotubes include covalent and non-covalent techniques (physical crosslinking). Covalent techniques are classified into: (i) in situ and ex situ polymerization and (ii) grafting technique. In situ polymerization technique, the nanofiller is introduced during the reaction, resulting in composite hydrogels with excellent mechanical strength and stability, in a variety of shapes and surface forms. Also, the polymerization method is carried out under mild conditions with a high product yield [34]. In ex situ polymerization technique, it is involved the dispersion of premade nanoparticles into the polymeric matrix. This method is well known for large-scale production in comparison with in situ methods [34]. In grafting technique, it is based on shielding the surface of the carbon nanostructure, resulting in good dispersion of the nanofiller and better interaction with the polymer matrix. Since there is no stearic hindrance involved, high molecular weight polymers such as proteins can be grafted [34]. The physical crosslinking process involves that the nanoparticles are synthesized in situ during the formation of the crosslinked networks. Physically crosslinked hydrogels are known to show high biocompatibility and to be responsive to the various external stimuli, but they do not show high mechanical strength. This technique requires a high affinity of nanofillers for the polymeric matrix and thus limits the non-homogeneous distribution and agglomeration of the nanofillers [34]. Table 5 enlists the most recent research of nanocomposites in hydrogel state based on carbon nanotubes and graphene structures for biomedical applications.

7 Composite Hydrogels Based on Metal–Organic Frameworks (MOFs)

Metal–organic frameworks (MOFs) are porous coordination networks (PCNs) or porous coordination polymers (PCPs), and they are made of metal ions connected to

Table 5 Composites in hydrogel state based on carbon nanostructures for biomedical applications

Polymeric matrix	Carbon nanostructure	Biomedical application	References
Polyvinyl alcohol (PVA)	Graphene oxide	Antibacterial activity for wound healing treatment	[128]
Poly(lactic-co-glycolic acid) (PLGA)	Carbon nanotubes	Neural tissue engineering	[129]
Silk fibroin	Graphene oxide	Bone tissue engineering	[130]
Cyclodextrin	Carboxylated single-walled carbon nanotubes	Anticancer drug delivery	[42]
Poly(acrylic acid) (PAA)	Graphene oxide	Artificial muscle and tissue engineering scaffold	[131]
Gelatin	Multi-walled carbon nanotubes decorated with Transglutaminase (MTGase) enzyme	Tissue engineering	[132]
Chitosan	Graphene oxide	Cardiac tissue engineering	[133]
Acrylamide-co-acrylic acid	Carbon nanotubes decorated with chitosan	Antimicrobial properties against staphylococcus aureus	[134]
Poly(ethyleneglycol)-Caprolactone copolymer (PEG-PCL)	Carbon nanotubes	Fluorescence Imaging	[135]
Guanosine diboronic (GD)	Single-walled carbon nanotubes	Cell growing support	[136]
Poly(ethylene glycol) methyl ether methacrylate (PEGMA)-poly(acrylic acid) (PAA)	Carbon nanotubes functionalized with hydroxyl groups	Anticancer drug delivery: Cis-platinum	[137]
Poly(ethyleneglycol)-Caprolactone (PEG-PCL)	Multi-walled carbon nanotubes functionalized with chitosan	Thermosensitive hydrogel for drug delivery	[138]

rigid organic molecules obtaining open one, two, or three tridimensional frameworks without the wall but only with connected columns, resulting in a high surface area and pore volume at nanometric level [39, 139]. MOFs are built from the connexion of inorganic subunits to each other through polytopic organic ligands that define cavities of various shapes and sizes [110]. The methods reported for MOFs and nanoMOFs synthesis included solvothermal synthesis, microwave-assisted synthesis, sonochemical-assisted synthesis, electrochemical synthesis, mechanochemical synthesis, and ionothermal synthesis [141–144]. Table 6 exhibits the characteristics of these synthesis procedures, their conveniences, and drawbacks.

The future of MOFs synthesis involves: (i) the development of a cleaner and less expensive synthesis method, (ii) an experimental procedure scalable at the industrial level, and (iii) the development of MOF at nanoscale for applications in nanotechnology [140], with the finality to improve the success in biomedical applications. There are infinite possibilities to synthesize MOFs structures. However, not all of them have shown chemical stability in water. The most common ligands for MOFs synthesis include carboxylates, phosphonates, sulfonates, imidazolates, amines, and phenolates, while the metal ions mainly include transition elements and lanthanide metals [4]. The lack of stability of MOFs in water relies on a competition between water molecules and ligands to interact with metal ions or metal clusters, leading to the collapse of the framework. The MOF stability factors are their crystallinity, hydrophobicity, and the extent of the defective sites [4, 39].

Table 6 Characteristics of principal methods for nanoMOFs preparation [141, 142]

Method	Operation conditions	Advantages	Disadvantages
Solvothermal synthesis	High temperature and pressure, use of polar solvents,	Accelerate the discovery of new MOFs, optimize synthesis protocols	Long reaction time, harsh synthesis conditions: Avoiding scale-up production
Microwave-assisted synthesis	Mixture reaction put into a Teflon vessel and application of an oscillating electric field	Homogeneous size distribution, short-time reaction, fast crystallization, morphology control	Difficult scale-up production
Sonochemical-assisted synthesis	Mixture reaction put into a horn-type Pyrex reactor fitted to a sonicator bar	Homogeneous size distribution, short-time reaction, fast crystallization	Difficult scale-up production
Electrochemical synthesis	Metal ions are supplied through anodic dissolution of a metal source, which reacts with the dissolved ligand molecules and a conducting salt in the reaction medium	Scale-up production through a continuous process, fast synthesis at low temperature, avoid the use of metal salts, total use of the organic ligand	Use of protic solvents, generation of H_2
Mechanochemical synthesis	Metal oxides and the ligand are mixed at room temperature and put into a mill ball	Room temperature reaction, solvent-free conditions, small MOF particles, short reaction times	Limited to specific MOF types, low yield, use of expensive metal oxides instead of metal salts

As a hand rule, a combination of hard acidic metal ions such as Zr (IV), Cr (III), Al (III), and Fe (III) combined with hard bases as carboxylate-based ligands result in frameworks with a significant water resistivity/stability. Also, the coordination is in between weak acidic metal ions like Cu (II), Mn (II), Zn (II), Ag (I), and Ni (II) and weak basic ligands such as pyrazolates, triazolates, and imidazolates. Whereas the combination of strong acidic metal ions with weak basic ligands, and vice versa, results in frameworks vulnerability to water [4, 39].

Nanoparticles and salts represent the first-generation antibacterial agents, the metal substances and metal oxides represent the second-generation antibacterial agents, and MOFs are considered the third-generation antibacterial agents with topology, high surface area, adjustable pore structure, and controlled metal ion release rate compared with the traditional bactericidal materials. Therefore, MOFs are promising alternatives as antibacterial agents for medical diagnosis, antibacterial coating, and wound healing [143].

The antimicrobial properties of MOFs rely on their ability to act a reservoir of metal ions and to release those ions gradually and also as a reservoir of antibacterial agents due to their adsorptive properties [143, 145]. Then, MOFs composed of antibacterial metal ions such as Ag (I), Zn (II), Cu (II), and Ni (II) release gradually metal ions offering durable antibacterial activity. Also, they can encapsulate antibacterial molecules as a host–guest interactions or act as a ligand in the MOF. Moreover, the antibacterial properties of metal ions and the organic ligands (porphyrin and carboxyl based) can be combined to form a synergistic effect [142, 143].

According to the cell wall structure, bacteria can be classified into Gram-positive and Gram-negative. The peptidoglycan layer of the cell wall of Gram-positive bacteria is thicker than in the Gram-negative bacteria. Many study research indicate that Gram-negative bacteria are more sensitive to the antibacterial effects of MOFs, and their composites than Gram-positive bacteria [143].

Due to the lack of further investigation of MOF antimicrobial mechanisms, the antibacterial effects of MOF are still believed to be similar to those of metal ions such as (i) production of reactive oxygen species (ROS), (ii) depolarization of membrane (change of ion balance), (iii) chelation effects, (iv) diffusion-directed lipid oxidation, (v) interfering of cation transport, vi) interaction with thiol groups, (vii) inactivation of enzymes and (vi) cell wall destruction by direct damage by the crystal size and shape. However, MOFs had better antibacterial properties than metal ions, which may attribute to the controlled release of metal ions, crystal structures, components, and frameworks of coordination materials [142, 143].

The chemical composition of MOFs defines the biocompatibility and cytotoxicity of MOFs. The application of Ca (II), Mg (II), Zn (II), Fe (II), Ti (IV) and Zr (IV) are metal cations preferred for MOFs synthesis with low cytotoxicity [144]. Also, the use of carboxylates, imidazolates, or phosphonates as organic ligand provide biocompatibility [146]. In addition, the applied solvents in the synthesis of MOFs have shown toxic properties which limit their potential in biomedical applications [144].

Therefore, new MOFs called BioMOFs have emerged to solve these issues. BioMOFs are built with biomolecules that include nucleobases, aminoacids, peptides, proteins, porphyrins and saccharides which are applied as ligands for MOF construction [147].

Nucleobases such as purines (adenine and guanine) are better bioligands than pyrimidines due to their higher amount of oxygen and nitrogen sites with the ability to strongly coordinate with metal centers through hydrogen bonding, where typically nucleobases are deprotonated prior to coordination bonding. The convenience of nucleobases as organic ligand relies on its rigid structure which promotes the growth of porous and stable bioMOFs. In addition, polycarboxylate ligands can be applied as auxiliary ligands for BioMOFs preparation [147].

In the case of aminoacids, their inherent content of amino and carboxyl groups is ideal for BioMOFs synthesis. Also, side chains of aminoacids can tune the hydrophilic and hydrophobic character of its surface where guest molecules can be attached. Nonetheless, aminoacid base bioMOFs have shown low porosity probably because of their closed groups of –COOH and $-NH_2$, thus, modification of aminoacids and the use of auxiliary ligands is necessary to expand the porosity of bioMOFs [147]. Since peptides are constituted by aminoacids which provide an inherent chirality recommended for molecular recognition ability. Peptides structure can be adjusted changing the type and order of aminoacids obtaining a diversity of peptides and thus multiple bioMOFs [147]. Furthermore, proteins are built from peptides; hence, proteins have a flexible and complex structure that hinders the coordination among metal ions and protein interface. But, it is still considered due to the high diversity of protein molecules [147].

On the other hand, porphyrins are macromolecular compounds containing of four pyrrole subunits connected with methine bridge (=CH), with good biocompatibility and applied for specific targeting for tumor tissues that have not shown toxic effects in normal cells. Tetra (4-carboxyphenyl) porphyrin (TCPP) is one of the most commonly used porphyrin ligands for the construction of porphyrin metal–organic frameworks, and TCPP has been coordinated to Zr (IV) and Pt (IV) from a series of MOFs named PCN [147]. Another molecule is cyclodextrin (CD) a cyclic oligosaccharide composed of 6, 7, and 8 D (+)-glucopyranose, its molecule has a cylindrical shape, wide at the top and narrow at the bottom, opened at both ends; it is relatively hydrophobic inside the cavity, while all hydroxyl groups are outside the molecule [147]. In addition, small molecules with carboxyl functional groups such as formic acid, oxalic acid, and fumaric acid have been used as organic ligands [147]. Nonetheless, the numerous application areas for BioMOFs, there are many opportunities areas in this subject to be systematically studied as: (i) interaction between metal ions and biological ligands, (ii) synthesis procedure: effect of solvent, temperature, and pH on final MOF physicochemical characteristics, (iii) thermal and chemical stabilities, and water stability, and (iv) interaction between bioMOFs and cells for biomedical applications [147].

According to their structure flexibility, MOFs can be classified as rigid or flexible. Rigid MOFs have permanent porosity and robust porous frameworks similar to inorganic porous materials. The flexible MOFs have shown flexible structure able to modulate their pore size; according to the guest molecules adsorbed into the pores, this phenomenon is called "breathing," and then flexible MOFs respond also to guest molecules, temperature, and pressure. The breathing effect (pore opening) of MIL-53 family constituted by Al (III) and Cr (III) is explained by hydrogen bonding between water molecules and the inorganic hydrophilic parts of the pore [146].

These host–guest interactions can be modulated by their pose size, and also, by post-synthetic functionalization to vary the surface chemical groups of MOFs from the initial building block. The functionalization of MOF surface is used to encapsulate therapeutic molecules via anion–cation electrostatic interactions, π-π interactions, Van der Waals interactions, and hydrogen bonding [139]. Hence, host–guest interactions are attractive for sieving and sensing applications and in biomedical applications such as controlled drug delivery and molecular imaging; but in this case, MOFs need to be synthesized at nanoscale to avoid bioaccumulation [4].

Another strategy to enhance the biocompatibility of MOFs, and also, to prevent the agglomeration of their particles is the decoration of MOFs with different substances like folic acid, coating silica substrates, polymers (polyethylene glycol(PEG), polyvinylpyrrolidone (PVP), lipid bilayers, and iron oxides. Where MOFs that contain amino, hydroxyl, and carboxyl functional groups are easily functionalized since these functional groups serve as binding sites for covalent and non-covalent interactions [4].

The main techniques for MOFs functionalization are: (i) surface adsorption, (ii) pore encapsulation, (iii) covalent binding, and (iv) functional molecules as the building block. In the following text, each of these methods is described.

Surface adsorption: Consisting on the stirring of MOF particles in a solution of functional molecules, where Van der Waals interaction, π-π interaction, and hydrogen bonding are the dominant forces involved. There is no restriction about the pore size of MOF or type of functional groups, but leaching of functional molecules is a problem due to the weak interaction forces [4, 139–141].

Pore encapsulation: This method is based on the structural flexibility of MOFs (breathing) to encapsulate molecules into their framework. The encapsulation occurs at the same time that MOF is synthesized; thus, there is no restriction about the use of large molecules, but it is mandatory that the host molecule be stable under synthetic conditions [4, 141, 142].

Covalent binding: This procedure represents a solution to the leaching problems presented in the above methods. In this case amino, carboxyl, and hydroxyl groups are binding sites for covalent bond with molecules of interest. Also, inorganic metal clusters provide another type of reactive sites in MOFs to covalently bind functional molecules [4, 39, 139–145].

Functional molecules as the building block: This method involves the use of aminoacids, peptides, nucleobases, and saccharides to be coordinated to metal ions and synthesize BioMOFs with high biocompatibility and special biological functionality. However, biomolecules tend to be flexible and with low symmetry which hinders the obtention of the MOF, then symmetric and rigid carboxylate ligands can be used to assist the synthesis [148, 149].

First of all, MOFs have a high surface area and pore volume in a flexible structure where its porosity is adapted to the drug molecule shape and size, characteristics that indicated a high drug loading capacity and efficient release [144]. Moreover, MOFs can be synthesized to obtain hydrophilic pores able to carry positive or negative charges and also to encapsulate those drugs with opposite charges to MOFs [139].

MOFs also demonstrate the following advantages: (i) a library of thousands of MOFs for various bio-related applications, (ii) the stable but degradable structures guarantee their reuse and on-demand degradation, (iii) the thermal stability for human body conditions, and (iv) the good dispersability and biocompatibility (functionalization) of bioMOFs ensure the biosecurity for in vivo applications [142–144]. To improve the physiological and colloidal stabilities and to reduce the immune response, MOFs can be coated with polymers like PEG and liposomes [142].

Drug loading in MOFs can be mainly investigated with two different mechanisms: (1) covalent and (2) non-covalent binding, based on the drug-MOF interactions. In the encapsulation strategy (non-covalent binding), drug molecules are located within the pores of MOFs via π-π stacking, van der Waals forces, and electrostatic interactions. In the direct self-assembly method (covalent binding), cargo molecules are used as ligands when the framework is constructed via coordination bonds. In the post-synthesis strategy, drug molecules are located on the surface and/or metal sites, as well as the ligands of pre-synthesized MOFs via coordination and covalent bonds based on "click chemistry" [144].

The MOF structures applied in drug delivery applications include: chiral MOFs, Cr (III), Zn (II), Mn (II), Ca (II) and Cu (II)-based MOFs, MIL and HKUST family, isoreticular MOFs, dual metal–organic frameworks nanoparticles, zeolite-based MOFs, and nanoscale coordination polymers [139]. Where the zinc-based MOFs (ZIF-8), UiO-66, and MIL family are the most widely applied in drug delivery due to their hydrothermal stability, high surface area (3100–5900 m^2 g^{-1}), large pore sizes (25–34 Å), and possible functionalization [139, 144, 146] (Fig. 7).

The encapsulated molecules in MOFs include anticancer drugs (e.g., 5-fluoracil, doxorubicin), bioactive gas (NO and I2), analgesic drug (ibuprofen), and also biomolecules such as proteins, enzymes, and DNA [139]. Where anticancer drugs are the most widely combined with MOFs for drug delivery application. It is known that the pH of physiological environment in cancerous tissue is lower than in normal tissue. Then, it is convenient the use of MOFs as a drug delivery system as MOFs are degraded in a pH range from 4 to 6, and release of drug molecules only in the tumor region but not in the general circulation, and eliminate the unnecessary death of normal cells [4, 149].

The drawbacks of MOFs in drug delivery are the excessive amount of drugs and the longer period of time to load into their matrices. Also, the metal ions that constituted MOFs tend to leach out in the biological environment and could cause in high amounts cytotoxic effects. In addition, it is necessary that MOFs are coated with polymers or post-functionalized to enhance its chemical stability to prevent the leaching of metal ions and MOFs particles aggregation [139].

Molecular imaging is a medical technique for visualization and quantification of biological processes at the molecular levels in a noninvasive manner through the assistance of contrast agents [4]. NanoMOFs have been applied in bioimaging field because of (i) their easy cellular internalization, (ii) good dispersive stability, and (iii) possible functionalization by decoration with silica, organic polymers, or lipid

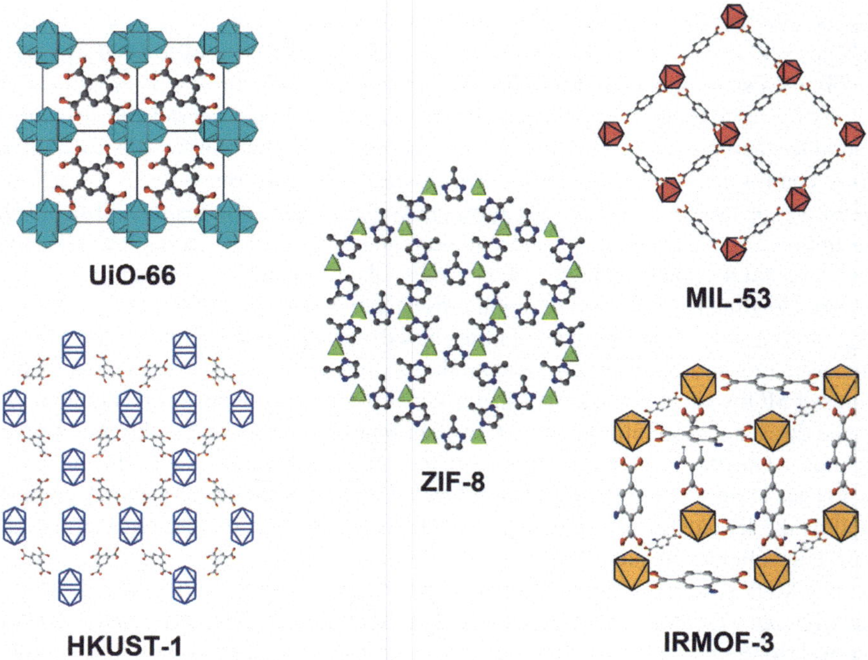

Fig. 7 MOFs applied in biomedical applications (*Own authorship*)

bilayers [4]. The molecular imaging involves magnetic resonance imaging (MRI), optical imaging (OI), X-ray computed tomography (CT), or their combinations.

The MOFs utilized in MRI contrasts agents are metallic coordination nanoparticles, Mn (II), Ga (III), and Fe(III)-based MOFs, isoreticular MOFs (IRMOF-3), and ZIF-90 nanoparticles. To prevent MOF aggregation and metal ions leaching out, MOFs frameworks have been functionalized with polymers, bioligands, or bioactive compounds [139]. However, Gd (III) ions had shown toxicity and bioaccumulation, then, Mn (II)-based MOFs present an attractive option due to the lower toxicity of Mn (II) ions [4].

The MOFs with luminescent building blocks (ligand or metal ion) or incorporating fluorescein could be used for OI [4]. Whereas, MOFs with iodinated ligands or iodide in payloads could be applied in CT as contrast agent [4].

The desired MOF for diagnostic imaging should be stable to complete the imaging tasks, biodegradable and not to be accumulated into the organism. Even though MOFs are modified for as MRI contrast agents, it is necessary for the suitable selection of a biocompatible metal ion that when leach out in the biological environment does not cause toxicity [139].

Introducing MOFs as nanofillers into hydrogels has been demonstrated as an effective method to strengthen the hydrophilic network, resulting in improved mechanical performance for composite hydrogels, and it also brings new functionalities into composite hydrogels for biomedical applications [150]. Where the polarized surface of MOFs nanoparticles with carboxyl or imidazole groups and metal ions acts as binding sites for hydrophilic polymers in hydrogels [150]. Also, composites in hydrogel state based on MOFs mitigate some MOFs drawbacks like biodegradability, cytotoxicity, and particle aggregation [140].

According to their physical state, MOF/polymer nanocomposite can be classified as (i) bulk material: MOF in polymer matrix or polymer in a MOF matrix, (ii) MOF/polymer membranes, and (iii) dispersed materials: polymer grafted MOF or MOF-coated polymer particles; this classification is also related to the intended application [151].

The typical physical state for biomedical applications is dispersed materials that provide good particles dispersability, protection against the physiological and biological environment, and stimuli response [151, 152]. Dispersed MOF/polymer composite can be formed by the blending of polymers and MOF particles, growth of MOFs on polymer particles, or coating of MOFs with polymers [151]. For the generation of composite hydrogels based on MOFs, the following routes have been proposed:

Microemulsion or directing mixing method: Integration of MOF particles into polymer substrate through blending has then been demonstrated as an alternative to prepare robust and flexible MOF–polymer composites [153]. The compatibility of MOF and polymer strongly depends on polymer rigidity. Less stiff polymers with Young's modulus less than 1 GPa can interact with the MOF surface more thoroughly, that is, cling to the rough surface and even penetrate the pores partially [151]. However, MOF–polymer composites prepared from simple blending often suffer from particle aggregation and poor particle–polymer matrix interaction, which can harm the composite performance [153]. Where is crucial the dispersion of MOF nanoparticles within the polymer matrix to the enhancement of its thermal and mechanical stability (Fig. 8a) [140].

Insitu MOF growth: MOFs are synthesized inside the pores of hydrogels by sequential addition of the metal precursor and the organic ligand. The MOF crystals are grown in situ inside the pores of the hydrogels to obtain MOF-based hydrogels [153]. It is a promising approach to overcome these drawbacks of microemulsion method; however, it has been much less studied (Fig. 8b) [153].

Coating of MOFs: In biomedical applications, MOF nanoparticles are coated with a polymer layer to form core–shell-like architectures. The polymer coating is generally set up by post-synthetic modification [39]. The ideal coating should be (i) selectively attached on the external surface: to avoid intrusion inside the porous structure, (ii) not interfere with the entrapped drugs, and (iii) be obtained in few steps under mild conditions; at present, the biggest problem is that the encapsulation of MOFs in polymers could cause agglomeration of the nanoparticles (Fig. 8c) [152].

Table 7 enlists the most recent research works about composites in hydrogel state based on MOFs for drug delivery, antimicrobial properties, molecular imaging, and tissue engineering.

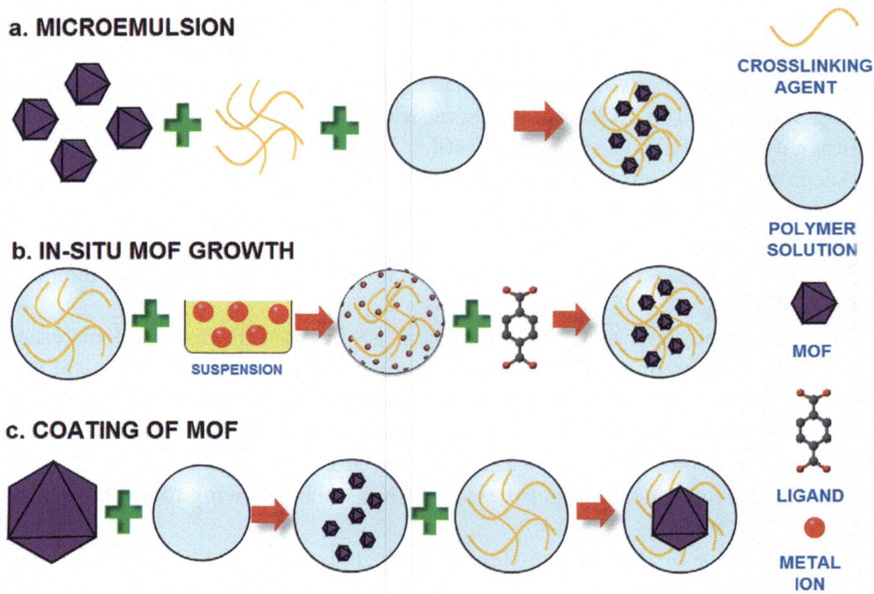

Fig. 8 Preparation methods for based MOFs composites in hydrogel state (*Own authorship*)

8 Composite Hydrogels Loaded with Proteins and Genes

The construction of composite hydrogels based on proteins and genes with specific biological activity represents a strategy that involves components at the nanometric level to guarantee biomedical performance. Proteins and genes can be released from the hydrogel matrix to regulate cell behavior in an efficient way (Fig. 9).

Globular proteins are macromolecules forming colloidal suspensions, and they can be dispersed in a hydrogel matrix forming a composite. In this type of composite, the disperse phase does not act as a reinforcement to improve the mechanical behavior of the hydrogel. The composite hydrogels act as an encapsulation system avoiding the premature decomposition of proteins. For example, the clinical efficacy of a therapeutic protein, the human growth hormone, is limited by its short plasma half-life and premature degradation. To overcome this issue, Singh et al. designed a protein delivery system by the self-assembly of human growth hormone negatively charged onto positively charged hydroxide nanoparticles. These hybrid nanoparticles were dispersed in a thermosensitive poly(amido ester urethane) matrix. The ionic interactions enhance the human growth hormone release maintaining a sustained profile for several days [41]. Kim and co-workers showed that the hydrolytic scission of the composite hydrogel resulted in a sustained release profile of human growth hormone which followed a mass erosion pattern [40].

Other work is based on the preparation of chitosan/sodium alginate hydrogels to deliver and protect lysozyme while eliminating food-borne microorganisms. The

Table 7 Biomedical applications of based MOFs composites in hydrogel state

Polymer	MOF	Metal	Organic ligand	Application	References
Polyurethane	UiO-67	Zr(IV)	4,4′-Biphenyldicarboxylic acid (BDPC)	Delivery of brimonidine tartrate, a drug in ocular therapeutics	[38]
2-hydroxyethyl methacrylate (HEMA), 2,3-dihydroxypropyl methacrylate (DHPMA), N-vinyl-2-pyrolidinone (VP) and ethylene glycol dimethacrylate (EGDMA)	Rho-ZMOF	In(IV)	4,5-imidazoledicarboxylic acid (H_3ImDC)	Delivery of Procainamide hydrochloride (PHCl), an antiarrhythmia drug	[37]
Carboxymethylcellulose (CMC)	MOF-5	Zr(IV)	1,4-benzenedicarboxylate (BDC)	Delivery of 5-fluorouracil (5-FU), colon cancer therapy	[23]
Polyacrylamide functionalized with nucleic acids	UiO-68	Zr(IV)	Amino-triphenyl dicarboxylic acid	Delivery of Doxorubicin, breast cancer therapy	[154]
Hyaluronic acid	ZIF-8	Zn(II)	2-methyl imidazole (2-mIM)	Delivery of 5-FU and titanocene in lactoferrin matrix, brain cancer therapy	[155]
Hyaluronic acid	PCN-224	Zr(IV)	Meso-tetra(4-carboxyphenyl)porphyrin (TCPP)	Delivery of Doxorubicin, breast cancer therapy	[156]
Polyethylene glycol (PEG)	UiO-66	Zr(IV)	Amino-terephthalic acid (NH_2-BDC)	Delivery of Docetaxel, breast cancer therapy	[157]

(continued)

Table 7 (continued)

Polymer	MOF	Metal	Organic ligand	Application	References
Poly(lactic-co-glycolic acid)(PLGA)-Polyethyleneglycol (PEG)	IRMOF-3	Zn(II)	Amino-terephthalic acid (NH_2-BDC)	Delivery of Doxorubicin (Dox) and celecoxib (Cel), oral cancer therapy	[158]
Functionalized chitosan with $-NH_2$ groups	PBNPs	Fe(II)	Polyvinylpyrrolidone (PVP)	Antibacterial properties against: *Staphylococcus aureus and Escherichia coli*	[159]
Diacrylated PEG 4-arm-thiolated PEG	Cu-MOF 1, Co-MOF 2, and Zn-MOF 3	Cu(II), Co(II), and Zn(II)	Glutarate and 1,2-bis(4-pyridyl)ethylene	Antibacterial properties against: *Staphylococcus aureus and Escherichia coli*	[160]
Hydroxyethyl-methacrylate (HEMA)	AGMNA	Ag(II)	2-thio-nicotinic acid (H_2MNA)	Antibacterial properties against: *Pseudomonas aeruginosa, Staphylococcus epidermidis, and Staphylococcus aureus*	[161]
Chitosan	MIL-53	Fe(II)	1,4-benzenedicarboxylate (BDC)	Delivery of vancomycin Antibacterial properties against: *Staphylococcus aureus*	[16]

(continued)

Table 7 (continued)

Polymer	MOF	Metal	Organic ligand	Application	References
Polylactic acid (PLA)	Ag_2[HBTC](im)	Ag(I)	1,3,5-benzenetricarboxylate (H_3BTC) and imidazole (im) ligand	Antibacterial properties against: Escherichia coli, Pseudomonas aeruginosa, Staphylococcus aureus, and Mycobacterium smegmatis	[162]
Polylactic acid (PLA) and gelatin	ZIF-67	Co(II)	2-methylimidazole (2-mIM)	Delivery of dimethyloxalylglycine (DMOG), diabetic wound healing	[3]
Polydopamine	Cu-BTC	Cu(II)	1,3,5-benzenetricarboxylate (H_3BTC)	Tissue engineering: cardiovascular stents	[163]
Chitosan-PEG	ZIF-8	Zn(II)	2-methylimidazole (2-mIM)	Delivery of cephalexin, wound dressing materials	[164]
Sodium alginate	MOF	Cu(II), Zn(II)	Nicotinic acid	Wound healing	[165]
Agarose	UiO-66-NH_2 a	Zr(III)	2-aminoterephthalic acid	Ratiometric fluorescence (rhodamine) detection of phosphate	[166]
Sodium alginate	Eu2(BPDC)(BDC)2(H_2O)$_2$]	Eu(II)	4,4′-Biphenyldicarboxylic acid (BDPC), 1,4-benzenedicarboxylate (BDC)	Detection of β-lactamase in serum via luminescence	[167]

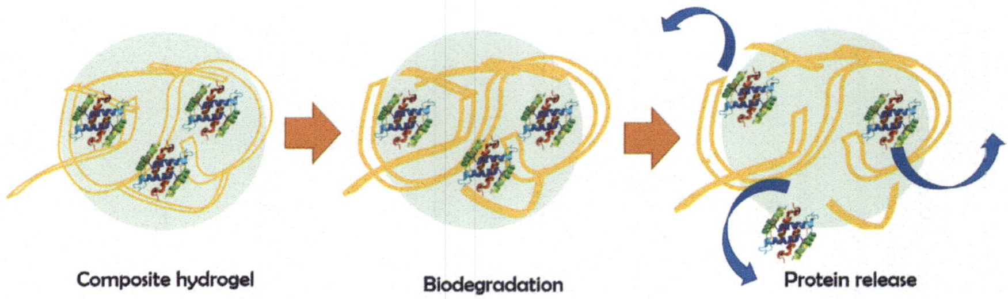

Fig. 9 Controlled protein release from composite hydrogels (*Own authorship*)

synergistic effect of chitosan and lysozyme improved the antibacterial activity against *E. coli* and *S. aureus*. This composite hydrogel has potential use in the food industry [168]; however, its lysozyme activation properties under physiological conditions could be used for the encapsulation and controlled release of drugs with therapeutic interest. Similarly, in this approach it was prepared composite materials with controllable degradation incorporating lysozyme-loaded Ca-alginate microparticles into a chitosan matrix. The encapsulation of lysozyme with Ca-alginate allows resisting the degradation rate variation caused by pH change. The controlled degradation of the composite leads to obtain a controlled release of lysozyme [169], representing an excellent composite hydrogel to inhibit bacterial growth that can be used in tissue healing strategies. A more controlled release of proteins can be obtained immobilizing temporally the protein on a bio-ceramic. For this, grafted human bone morphogenetic protein-2 (rhBMP-2) on the surface of mesoporous bioglass nanoparticles generating novel composite hydrogels. The hybrid nanoparticles were dispersed in a hydrogel matrix of crosslinked gelatin methacrylate. The composite hydrogel releases the rhBMP-2 during the first 4 weeks and then releases Ca (II) and Si (IV) ions during the 4 weeks after. The composite hydrogel promotes osteogenesis and accelerates bone repairing [170]. Another strategy involves the coupling of metalloproteinase-7 in a polymeric matrix of collagen crosslinked with acrylate; this novel composition induces the differentiation of mesenchymal stem cells to the chondrogenic phenotype, representing excellent candidates for lesions in cartilage tissue [171]. A highly bioactive composite hydrogel based on collagen, laminin-specific peptide, and CLG3 peptide has been designed; this formulation tends to generate a composite matrix in the hydrogel state due to the effect of ionic force; in vitro results demonstrate that neuronal stem cells can actively grow and proliferate for up to 7 days of evaluation, indicating that this bioactive material can be used to heal nervous tissue [172].

Finally, composite hydrogels are suitable systems for gene therapy delivering DNA or RNA that can up- or down-regulated genes. Yang and co-workers designed a composite hydrogel gene delivery system based on thermo-responsive hydrogel and a biodegradable poly(ester amine) modified with folic acid. The modified matrix condenses genes into particles and conduces the targeted delivery of genes because of the folate receptor's over-expression in tumors. The prepared composite hydrogel could slowly release genes into local tissue and selectively deliver genes into cancer cells rather than normal cells [173], representing selectively anticancer nanotechnology strategies and noninvasive for normal cells.

9 Perspectives and Conclusion

Composites in the hydrogel state that include nanostructured components have shown improved potential application in biomedical areas. The nanostructured component exhibits the capacity to internalize itself in cells promoting specific interactions that have been exploited for the development of biomedical sensors, biomedical imaging, tissue growth modulators, cancer therapies, and stimulation of cell phenotypes. In addition, these components also show a relation in the control of the structure and properties of the composite matrices in the hydrogel state, generating novel materials with antibacterial capacity, enhanced loading of drugs, proteins, and genes with therapeutic interest and controlled release of them. Although the current panorama of these nanotechnology systems continues to grow recently, it is important to find new strategies for the generation of these novel composites, representing viable alternatives for current problems in the biomedical sectors.

Declaration of Conflicts of Interest None.

References

1. Argenta DF, dos Santos TC, Campos AM, Caon T (2019) Hydrogel nanocomposite systems 81–131. https://doi.org/10.1016/b978-0-12-814033-8.00003-5
2. Amjad MS, Sadiq N, Qureshi H, Fareed G, Sabir S (2015) Nano particles: an emerging tool in biomedicine. Asian Pacific J Trop Dis 5(10):767–771. https://doi.org/10.1016/s2222-1808(15)60929-x
3. Li J, Lv F, Li J, Li Y, Gao J, Luo J, Xue F, Ke Q, Xu H (2020) Cobalt-based metal–organic framework as a dual cooperative controllable release system for accelerating diabetic wound healing. Nano Res 13(8):2268–2279. https://doi.org/10.1007/s12274-020-2846-1
4. Luo Z, Fan S, Gu C, Liu W, Chen J, Li B, Liu J (2019) Metal-Organic Framework (MOF)-based nanomaterials for biomedical applications. Curr Med Chem 26(18):3341–3369. https://doi.org/10.2174/0929867325666180214123500
5. Gaharwar AK, Peppas NA, Khademhosseini A (2014) Nanocomposite hydrogels for biomedical applications. Biotechnol Bioeng 111(3):441–453. https://doi.org/10.1002/bit.25160
6. Lee D, Heo DN, Nah HR, Lee SJ, Ko WK, Lee JS, Moon HJ, Bang JB, Hwang YS, Reis RL, Kwon IK (2018) Injectable hydrogel composite containing modified gold nanoparticles: implication in bone tissue regeneration. Int J Nanomed 13:7019–7031. https://doi.org/10.2147/IJN.S185715
7. Suhail S, Nair LS (2018) Hydrogel-nanoparticle composites for drug delivery 229–254. https://doi.org/10.1016/b978-0-323-50922-0.00011-0
8. Schexnailder P, Schmidt G (2008) Nanocomposite polymer hydrogels. Colloid Polym Sci 287(1):1–11. https://doi.org/10.1007/s00396-008-1949-0

9. Zazakowny K, Lewandowska-Lancucka J, Mastalska-Poplawska J, Kaminski K, Kusior A, Radecka M, Nowakowska M (2016) Biopolymeric hydrogels - nanostructured TiO_2 hybrid materials as potential injectable scaffolds for bone regeneration. Colloids Surf, B 148:607–614. https://doi.org/10.1016/j.colsurfb.2016.09.031
10. Ahmed A, Niazi MBK, Jahan Z, Ahmad T, Hussain A, Pervaiz E, Janjua HA, Hussain Z (2020) In-vitro and in-vivo study of superabsorbent PVA/Starch/g-C_3N_4/Ag@TiO_2 NPs hydrogel membranes for wound dressing. Eur Polymer J 130:109650. https://doi.org/10.1016/j.eurpolymj.2020.109650
11. Clasky AJ, Watchorn JD, Chen PZ, Gu FX (2020) From prevention to diagnosis and treatment: Biomedical applications of metal nanoparticle-hydrogel composites. Acta Biomater. https://doi.org/10.1016/j.actbio.2020.12.030
12. Conte R, De Luise A, Valentino A, Di Cristo F, Petillo O, Riccitiello F, Di Salle A, Calarco A, Peluso G (2019) Hydrogel nanocomposite systems 319–349. https://doi.org/10.1016/b978-0-12-814033-8.00010-2
13. Dannert C, Stokke BT, Dias RS (2019) Nanoparticle-hydrogel composites: from molecular interactions to macroscopic behavior. Polymers 11(2). https://doi.org/10.3390/polym11020275
14. Deo KA, Lokhande G, Gaharwar AK (2019). Nanostructured hydrogels for tissue engineering and regenerative medicine. https://doi.org/10.1016/b978-0-12-801238-3.11114-6
15. George D, Maheswari PU, Begum K (2020) Chitosan-cellulose hydrogel conjugated with L-histidine and zinc oxide nanoparticles for sustained drug delivery: Kinetics and in-vitro biological studies. Carbohyd Polym 236:116101. https://doi.org/10.1016/j.carbpol.2020.116101
16. Ghaffar I, Imran M, Perveen S, Kanwal T, Saifullah S, Bertino MF, Ehrhardt CJ, Yadavalli VK, Shah MR (2019) Synthesis of chitosan coated metal organic frameworks (MOFs) for increasing vancomycin bactericidal potentials against resistant *S. aureus* strain. Mater Sci Eng C, Mater Biol Appl 105:110111. https://doi.org/10.1016/j.msec.2019.110111
17. Basu S, Samanta HS, Ganguly J (2017) Green synthesis and swelling behavior of Ag-nanocomposite semi-IPN hydrogels and their drug delivery using Dolichos biflorus Linn. Soft Mater 16(1):7–19. https://doi.org/10.1080/1539445x.2017.1368559
18. Chen T, Hou K, Ren Q, Chen G, Wei P, Zhu M (2018) Nanoparticle-polymer synergies in nanocomposite hydrogels: from design to application. Macromol Rapid Commun 39(21):e1800337. https://doi.org/10.1002/marc.201800337
19. Niu J, Wang J, Dai X, Shao Z, Huang X (2018) Dual physically crosslinked healable polyacrylamide/cellulose nanofibers nanocomposite hydrogels with excellent mechanical properties. Carbohyd Polym 193:73–81. https://doi.org/10.1016/j.carbpol.2018.03.086
20. Thoniyot P, Tan MJ, Karim AA, Young DJ, Loh XJ (2015) Nanoparticle-hydrogel composites: concept, design, and applications of these promising, multi-functional materials. Adv Sci 2(1–2):1400010. https://doi.org/10.1002/advs.201400010
21. Wahid F, Zhao X-J, Jia S-R, Bai H, Zhong C (2020) Nanocomposite hydrogels as multifunctional systems for biomedical applications: current state and perspectives. Compos B Eng 200:108208. https://doi.org/10.1016/j.compositesb.2020.108208
22. Hasan A, Morshed M, Memic A, Hassan S, Webster TJ, Marei HE (2018) Nanoparticles in tissue engineering: applications, challenges and prospects. Int J Nanomed 13:5637–5655. https://doi.org/10.2147/IJN.S153758
23. Javanbakht S, Hemmati A, Namazi H, Heydari A (2020) Carboxymethylcellulose-coated 5-fluorouracil@MOF-5 nano-hybrid as a bio-nanocomposite carrier for the anticancer oral delivery. Int J Biol Macromol 155:876–882. https://doi.org/10.1016/j.ijbiomac.2019.12.007
24. Kumar H, Gehlaut AK, Gupta H, Gaur A, Park JW (2020) Development of copper loaded nanoparticles hydrogel made from waste biomass (sugarcane bagasse) for special medical application. Key Eng Mater 847:102–107. https://doi.org/10.4028/www.scientific.net/KEM.847.102
25. Kumar BYS, Isloor AM, Kumar GCM, Inamuddin AAM (2019) Nanohydroxyapatite reinforced chitosan composite hydrogel with tunable mechanical and biological properties for cartilage regeneration. Sci Rep 9(1):15957. https://doi.org/10.1038/s41598-019-52042-7

26. Li Z, Mi W, Wang H, Su Y, He C (2014) Nano-hydroxyapatite/polyacrylamide composite hydrogels with high mechanical strengths and cell adhesion properties. Colloids Surf, B 123:959–964. https://doi.org/10.1016/j.colsurfb.2014.10.050
27. Kim H, Che L, Ha Y, Ryu W (2014) Mechanically-reinforced electrospun composite silk fibroin nanofibers containing hydroxyapatite nanoparticles. Mater Sci Eng, C Mater Biol Appl 40:324–335. https://doi.org/10.1016/j.msec.2014.04.012
28. Tan HL, Teow SY, Pushpamalar J (2019) Application of metal nanoparticle(-)hydrogel composites in tissue regeneration. Bioengineering 6(1) https://doi.org/10.3390/bioengineeri ng6010017
29. Villamin ME, Kitamoto Y (2018) Synthesis of multifunctional clustered nano-Fe_3O_4 chitosan nanocomposite for biomedical applications. 1929:020014. https://doi.org/10.1063/1.5021927
30. Villanueva ME, Cuestas ML, Perez CJ, Campo Dall Orto V, Copello GJ (2019) Smart release of antimicrobial ZnO nanoplates from a pH-responsive keratin hydrogel. J Colloid Interface Sci 536:372–380. https://doi.org/10.1016/j.jcis.2018.10.067
31. Ali GW, El-Hotaby W, Hemdan B, Abdel-Fattah WI (2018) Thermosensitive chitosan/phosphate hydrogel-composites fortified with Ag versus Ag@Pd for biomedical applications. Life Sci 194:185–195. https://doi.org/10.1016/j.lfs.2017.12.021
32. Juby KA, Dwivedi C, Kumar M, Kota S, Misra HS, Bajaj PN (2012) Silver nanoparticle-loaded PVA/gum acacia hydrogel: synthesis, characterization and antibacterial study. Carbohyd Polym 89(3):906–913. https://doi.org/10.1016/j.carbpol.2012.04.033
33. Sharma G, Kumar A, Sharma S, Naushad M, Prakash Dwivedi R, Alothman ZA, Mola GT (2019) Novel development of nanoparticles to bimetallic nanoparticles and their composites: a review. J King Saud Univ- Sci 31(2):257–269. https://doi.org/10.1016/j.jksus.2017.06.012
34. Alshehri R, Ilyas AM, Hasan A, Arnaout A, Ahmed F, Memic A (2016) Carbon nanotubes in biomedical applications: factors, mechanisms, and remedies of toxicity. J Med Chem 59(18):8149–8167. https://doi.org/10.1021/acs.jmedchem.5b01770
35. Banerjee AN (2018) Graphene and its derivatives as biomedical materials: future prospects and challenges. Interface focus 8(3):20170056. https://doi.org/10.1098/rsfs.2017.0056
36. Gurunathan S, Kim JH (2016) Synthesis, toxicity, biocompatibility, and biomedical applications of graphene and graphene-related materials. Int J Nanomed 11:1927–1945. https://doi.org/10.2147/IJN.S105264
37. Ananthoji R, Eubank JF, Nouar F, Mouttaki H, Eddaoudi M, Harmon JP (2011) Symbiosis of zeolite-like metal–organic frameworks (rho-ZMOF) and hydrogels: Composites for controlled drug release. J Mater Chem 21(26):9587. https://doi.org/10.1039/c1jm11075f
38. Gandara-Loe J, Souza BE, Missyul A, Giraldo G, Tan JC, Silvestre-Albero J (2020) MOF-based polymeric nanocomposite films as potential materials for drug delivery devices in ocular therapeutics. ACS Appl Mater Interfaces 12(27):30189–30197. https://doi.org/10.1021/acsami.0c07517
39. Giliopoulos D, Zamboulis A, Giannakoudakis D, Bikiaris D, Triantafyllidis K (2020) Polymer/Metal Organic Framework (MOF) nanocomposites for biomedical applications. Molecules 25(1). https://doi.org/10.3390/molecules25010185
40. Mee Ryang Kim TGP (2002) Temperature-responsive and degradable hyaluronic acid/Pluronic composite hydrogels for controlled release of human growth hormone. J Control Release 80:69–77. https://doi.org/10.1016/s0168-3659(01)00557-0
41. Singh NK, Nguyen QV, Kim BS, Lee DS (2015) Nanostructure controlled sustained delivery of human growth hormone using injectable, biodegradable, pH/temperature responsive nanobiohybrid hydrogel. Nanoscale 7(7):3043–3054. https://doi.org/10.1039/c4nr05897f
42. Liu X, Xu D, Liao C, Fang Y, Guo B (2018) Development of a promising drug delivery for formononetin: cyclodextrin-modified single-walled carbon nanotubes. J Drug Del Sci Technol 43:461–468. https://doi.org/10.1016/j.jddst.2017.11.018
43. Zhao F, Yao D, Guo R, Deng L, Dong A, Zhang J (2015) Composites of polymer hydrogels and nanoparticulate systems for biomedical and pharmaceutical applications. Nanomaterials 5(4):2054–2130. https://doi.org/10.3390/nano5042054

44. Utech S, Boccaccini AR (2015) A review of hydrogel-based composites for biomedical applications: enhancement of hydrogel properties by addition of rigid inorganic fillers. J Mater Sci 51(1):271–310. https://doi.org/10.1007/s10853-015-9382-5
45. Rafieian S, Mirzadeh H, Mahdavi H, Masoumi ME (2019) A review on nanocomposite hydrogels and their biomedical applications. Sci Eng Compos Mater 26(1):154–174. https://doi.org/10.1515/secm-2017-0161
46. Qiao K, Zheng Y, Guo S, Tan J, Chen X, Li J, Xu D, Wang J (2015) Hydrophilic nanofiber of bacterial cellulose guided the changes in the micro-structure and mechanical properties of nf-BC/PVA composites hydrogels. Compos Sci Technol 118:47–54. https://doi.org/10.1016/j.compscitech.2015.08.004
47. Nguyen THM, Abueva C, Ho HV, Lee SY, Lee BT (2018) In vitro and in vivo acute response towards injectable thermosensitive chitosan/TEMPO-oxidized cellulose nanofiber hydrogel. Carbohyd Polym 180:246–255. https://doi.org/10.1016/j.carbpol.2017.10.032
48. Li N, Chen W, Chen G, Tian J (2017) Rapid shape memory TEMPO-oxidized cellulose nanofibers/polyacrylamide/gelatin hydrogels with enhanced mechanical strength. Carbohyd Polym 171:77–84. https://doi.org/10.1016/j.carbpol.2017.04.035
49. Maharjan B, Park J, Kaliannagounder VK, Awasthi GP, Joshi MK, Park CH, Kim CS (2021) Regenerated cellulose nanofiber reinforced chitosan hydrogel scaffolds for bone tissue engineering. Carbohyd Polym 251:117023. https://doi.org/10.1016/j.carbpol.2020.117023
50. Gou L, Xiang M, Ni X (2020) Development of wound therapy in nursing care of infants by using injectable gelatin-cellulose composite hydrogel incorporated with silver nanoparticles. Mater Lett 277:128340. https://doi.org/10.1016/j.matlet.2020.128340
51. Liu R, Dai L, Si C, Zeng Z (2018) Antibacterial and hemostatic hydrogel via nanocomposite from cellulose nanofibers. Carbohyd Polym 195:63–70. https://doi.org/10.1016/j.carbpol.2018.04.085
52. Tavakoli S, Kharaziha M, Nemati S, Kalateh A (2021) Nanocomposite hydrogel based on carrageenan-coated starch/cellulose nanofibers as a hemorrhage control material. Carbohyd Polym 251:117013. https://doi.org/10.1016/j.carbpol.2020.117013
53. Ma T, Lv L, Ouyang C, Hu X, Liao X, Song Y, Hu X (2021) Rheological behavior and particle alignment of cellulose nanocrystal and its composite hydrogels during 3D printing. Carbohyd Polym 253:117217. https://doi.org/10.1016/j.carbpol.2020.117217
54. Huang Y, Zhang X, Wu A, Xu H (2016) An injectable nano-hydroxyapatite (n-HA)/glycol chitosan (G-CS)/hyaluronic acid (HyA) composite hydrogel for bone tissue engineering. RSC Adv 6(40):33529–33536. https://doi.org/10.1039/c5ra26160k
55. Zhang X, Zhou J, Ying H, Zhou Y, Lai J, Chen J (2020) Glycogen as a cross-linking agent of collagen and nanohydroxyapatite to form hydrogels for bMSC differentiation. ACS Sustain Chem Eng 8(4):2106–2114. https://doi.org/10.1021/acssuschemeng.9b07051
56. Ribeiro M, de Moraes MA, Beppu MM, Garcia MP, Fernandes MH, Monteiro FJ, Ferraz MP (2015) Development of silk fibroin/nanohydroxyapatite composite hydrogels for bone tissue engineering. Eur Polymer J 67:66–77. https://doi.org/10.1016/j.eurpolymj.2015.03.056
57. Wang S, Zhang Z, Zhang Q, Li L (2018) Physical crosslinked poly(N-isopropylacrylamide)/nano-hydroxyapatite thermosensitive composite hydrogels. J Inorg Organomet Polym Mater 28(5):2069–2079. https://doi.org/10.1007/s10904-018-0893-9
58. Barros J, Ferraz MP, Azeredo J, Fernandes MH, Gomes PS, Monteiro FJ (2019) Alginate-nanohydroxyapatite hydrogel system: optimizing the formulation for enhanced bone regeneration. Mater Sci Eng, C Mater Biol Appl 105:109985. https://doi.org/10.1016/j.msec.2019.109985
59. Xu L, Bai X, Yang J, Li J, Xing J, Yuan H, Xie J, Li J (2020) Preparation and characterisation of a gellan gum-based hydrogel enabling osteogenesis and inhibiting Enterococcus faecalis. Int J Biol Macromol 165(Pt B):2964–2973. https://doi.org/10.1016/j.ijbiomac.2020.10.083
60. Zhang X, He Y, Huang P, Jiang G, Zhang M, Yu F, Zhang W, Fu G, Wang Y, Li W, Zeng H (2020) A novel mineralized high strength hydrogel for enhancing cell adhesion and promoting skull bone regeneration in situ. Compos B Eng 197:108183. https://doi.org/10.1016/j.compositesb.2020.108183

61. El-Fiqi A, Kim JH, Kim HW (2020) Novel bone-mimetic nanohydroxyapatite/collagen porous scaffolds biomimetically mineralized from surface silanized mesoporous nanobioglass/collagen hybrid scaffold: Physicochemical, mechanical and in vivo evaluations. Mater Sci Eng, C Mater Biol Appl 110:110660. https://doi.org/10.1016/j.msec.2020.110660
62. Khan I, Saeed K, Khan I (2019) Nanoparticles: properties, applications and toxicities. Arab J Chem 12(7):908–931. https://doi.org/10.1016/j.arabjc.2017.05.011
63. Nasrollahzadeh M, Sajadi SM, Sajjadi M, Issaabadi Z (2019) An introduction to nanotechnology 28:1–27. https://doi.org/10.1016/b978-0-12-813586-0.00001-8
64. Mooney EASaDJ, (2004) Synthetic extracellular matrices for tissue engineering and regeneration. Curr Top Dev Biol 64:181–205
65. Lutolf MP, Hubbell JA (2005) Synthetic biomaterials as instructive extracellular microenvironments for morphogenesis in tissue engineering. Nat Biotechnol 23(1):47–55. https://doi.org/10.1038/nbt1055
66. Wang B, Wang W, Yu Y, Zhang Y, Zhang J, Yuan Z (2017) The study of angiogenesis stimulated by multivalent peptide ligand-modified alginate. Colloids Surf, B 154:383–390. https://doi.org/10.1016/j.colsurfb.2017.03.049
67. Grant SA, Zhu J, Gootee J, Snider CL, Bellrichard M, Grant DA (2018) Gold nanoparticle-collagen gels for soft tissue augmentation. Tissue Eng Part A 24(13–14):1091–1098. https://doi.org/10.1089/ten.TEA.2017.0385
68. Tentor FR, de Oliveira JH, Scariot DB, Lazarin-Bidoia D, Bonafe EG, Nakamura CV, Venter SAS, Monteiro JP, Muniz EC, Martins AF (2017) Scaffolds based on chitosan/pectin thermosensitive hydrogels containing gold nanoparticles. Int J Biol Macromol 102:1186–1194. https://doi.org/10.1016/j.ijbiomac.2017.04.106
69. Niu C, Yuan K, Ma R, Gao L, Jiang W, Hu X, Lin W, Zhang X, Huang Z (2017) Gold nanoparticles promote osteogenic differentiation of human periodontal ligament stem cells via the p38 MAPK signaling pathway. Mol Med Reports 16:4879–4886. https://doi.org/10.3892/mmr.2017.7170
70. Shevach M, Fleischer S, Shapira A, Dvir T (2014) Gold nanoparticle-decellularized matrix hybrids for cardiac tissue engineering. Nano Lett 14(10):5792–5796. https://doi.org/10.1021/nl502673m
71. Baei P, Jalili-Firoozinezhad S, Rajabi-Zeleti S, Tafazzoli-Shadpour M, Baharvand H, Aghdami N (2016) Electrically conductive gold nanoparticle-chitosan thermosensitive hydrogels for cardiac tissue engineering. Mater Sci Eng, C Mater Biol Appl 63:131–141. https://doi.org/10.1016/j.msec.2016.02.056
72. Zhu K, Shin SR, van Kempen T, Li YC, Ponraj V, Nasajpour A, Mandla S, Hu N, Liu X, Leijten J, Lin YD, Hussain MA, Zhang YS, Tamayol A, Khademhosseini A (2017) Gold nanocomposite bioink for printing 3D cardiac constructs. Adv Funct Mater 27(12). https://doi.org/10.1002/adfm.201605352
73. Meiyan Liu PH, Wang W, Feng Z, Zhang J, Deng L, Dong A (2019) Injectable Nanocomposite Hydrogel Co-Constructed by Gold Nanorods and Paclitaxel-Loaded Nanoparticles for Local Chemo-Photothermal Synergetic Cancer Therapy. Journal of Materials Chemistry B 7:2667–2677. https://doi.org/10.1039/C9TB00120D
74. Hosoyama K, Ahumada M, McTiernan CD, Bejjani J, Variola F, Ruel M, Xu B, Liang W, Suuronen EJ, Alarcon EI (2017) Multi-functional thermo-crosslinkable collagen-metal nanoparticle composites for tissue regeneration: nanosilver vs. nanogold. RSC Adv 7(75):47704–47708. https://doi.org/10.1039/c7ra08960k
75. Araby E, Elbastawisy H, El-Tablawy S (2018) Combating bacterial adhesion and protein deposition on cosmetic contact lenses using zinc oxide nanoparticles. Egypt J Radiat Sci Appl. https://doi.org/10.21608/ejrsa.2018.2092.1030
76. Gonzalez-Sanchez MI, Perni S, Tommasi G, Morris NG, Hawkins K, Lopez-Cabarcos E, Prokopovich P (2015) Silver nanoparticle based antibacterial methacrylate hydrogels potential for bone graft applications. Mater Sci Eng, C Mater Biol Appl 50:332–340. https://doi.org/10.1016/j.msec.2015.02.002

77. Kumar N, Desagani D, Chandran G, Ghosh NN, Karthikeyan G, Waigaonkar S, Ganguly A (2018) Biocompatible agarose-chitosan coated silver nanoparticle composite for soft tissue engineering applications. Artif Cells, Nanomed Biotechnol 46(3):637–649. https://doi.org/10.1080/21691401.2017.1337021
78. Srinivasan S, Kumar PT, Nair SV, Nair SV, Chennazhi KP, Jayakumar R (2013) Antibacterial and bioactive alpha- and beta-chitin hydrogel/nanobioactive glass ceramic/nano silver composite scaffolds for periodontal regeneration. J Biomed Nanotechnol 9(11):1803–1816. https://doi.org/10.1166/jbn.2013.1658
79. Xu H, Zhang G, Xu K, Wang L, Yu L, Xing MMQ, Qiu X (2018) Mussel-inspired dual-functional PEG hydrogel inducing mineralization and inhibiting infection in maxillary bone reconstruction. Mater Sci Eng, C Mater Biol Appl 90:379–386. https://doi.org/10.1016/j.msec.2018.04.066
80. Emilio I, Alarcon KU, Noel CW, Gagnon L-P, Taylor PK, Vulesevic B, Simpson MJ, Spyridon Gkotzis M, Islam M, Lee C-J, Richter-Dahlfors A, Mah T-F, Suuronen EJ, Scaiano JC, Griffith M (2015) Safety and efficacy of composite collagen-silver nanoparticle hydrogels as tissue engineering scaffolds. Nanoscale 7(44):18789–18798. https://doi.org/10.1039/C5NR03826J
81. Zulkifli FH, Hussain FSJ, Zeyohannes SS, Rasad M, Yusuff MM (2017) A facile synthesis method of hydroxyethyl cellulose-silver nanoparticle scaffolds for skin tissue engineering applications. Mater Sci Eng, C Mater Biol Appl 79:151–160. https://doi.org/10.1016/j.msec.2017.05.028
82. Xu T, Zhang J, Zhu Y, Zhao W, Pan C, Ma H, Zhang L (2018) A poly(hydroxyethyl methacrylate)-Ag nanoparticle porous hydrogel for simultaneous in vivo prevention of the foreign-body reaction and bacterial infection. Nanotechnology 29(39):395101. https://doi.org/10.1088/1361-6528/aad257
83. Janarthanan G, Noh I (2021) Recent trends in metal ion based hydrogel biomaterials for tissue engineering and other biomedical applications. J Mater Sci Technol 63:35–53. https://doi.org/10.1016/j.jmst.2020.02.052
84. Rocha-García D, Betancourt-Mendiola MdL, Wong-Arce A, Rosales-Mendoza S, Reyes-Hernández J, González-Ortega O, Palestino G (2018) Gelatin-based porous silicon hydrogel composites for the controlled release of tramadol. Eur Polymer J 108:485–497. https://doi.org/10.1016/j.eurpolymj.2018.09.033
85. Ambrogi V, Pietrella D, Donnadio A, Latterini L, Di Michele A, Luffarelli I, Ricci M (2020) Biocompatible alginate silica supported silver nanoparticles composite films for wound dressing with antibiofilm activity. Mater Sci Eng, C Mater Biol Appl 112:110863. https://doi.org/10.1016/j.msec.2020.110863
86. Mohamed AL, Elmotasem H, Salama AAA (2020) Colchicine mesoporous silica nanoparticles/hydrogel composite loaded cotton patches as a new encapsulator system for transdermal osteoarthritis management. Int J Biol Macromol 164:1149–1163. https://doi.org/10.1016/j.ijbiomac.2020.07.133
87. Deng H, Sun J, Yu Z, Guo Z, Xu C (2021) Low-intensity near-infrared light-triggered spatiotemporal antibiotics release and hyperthermia by natural polysaccharide-based hybrid hydrogel for synergistic wound disinfection. Mater Sci Eng, C Mater Biol Appl 118:111530. https://doi.org/10.1016/j.msec.2020.111530
88. Huang B, Liu M, Zhou C (2017) Chitosan composite hydrogels reinforced with natural clay nanotubes. Carbohyd Polym 175:689–698. https://doi.org/10.1016/j.carbpol.2017.08.039
89. Bardajee GR, Khamooshi N, Nasri S, Vancaeyzeele C (2020) Multi-stimuli responsive nanogel/hydrogel nanocomposites based on kappa-carrageenan for prolonged release of levodopa as model drug. Int J Biol Macromol 153:180–189. https://doi.org/10.1016/j.ijbiomac.2020.02.329
90. Rao KM, Kumar A, Han SS (2018) Polysaccharide-based magnetically responsive polyelectrolyte hydrogels for tissue engineering applications. J Mater Sci Technol 34(8):1371–1377. https://doi.org/10.1016/j.jmst.2017.10.003

91. Rao Z, Liu S, Wu R, Wang G, Sun Z, Bai L, Wang W, Chen H, Yang H, Wei D, Niu Y (2019) Fabrication of dual network self-healing alginate/guar gum hydrogels based on polydopamine-type microcapsules from mesoporous silica nanoparticles. Int J Biol Macromol 129:916–926. https://doi.org/10.1016/j.ijbiomac.2019.02.089
92. Sakr MA, Mohamed MGA, Wu R, Shin SR, Kim D, Kim K, Siddiqua S (2020) Development of bentonite-gelatin nanocomposite hybrid hydrogels for tissue engineering. Appl Clay Sci 199:105860. https://doi.org/10.1016/j.clay.2020.105860
93. Ahmadian Y, Bakravi A, Hashemi H, Namazi H (2018) Synthesis of polyvinyl alcohol/CuO nanocomposite hydrogel and its application as drug delivery agent. Polym Bull 76(4):1967–1983. https://doi.org/10.1007/s00289-018-2477-9
94. Namazi H, Pooresmaeil M, Hasani M (2020) Oxidized starch/CuO bio-nanocomposite hydrogels as an antibacterial and stimuli-responsive agent with potential colon-specific naproxen delivery. Int J Polym Mater Polym Biomater 1–10. https://doi.org/10.1080/00914037.2020.1798431
95. Yadollahi M, Gholamali I, Namazi H, Aghazadeh M (2015) Synthesis and characterization of antibacterial carboxymethylcellulose/CuO bio-nanocomposite hydrogels. Int J Biol Macromol 73:109–114. https://doi.org/10.1016/j.ijbiomac.2014.10.063
96. Jayaramudu T, Varaprasad K, Reddy KK, Pyarasani RD, Akbari-Fakhrabadi A, Amalraj J (2020) Chitosan-pluronic based Cu nanocomposite hydrogels for prototype antimicrobial applications. Int J Biol Macromol 143:825–832. https://doi.org/10.1016/j.ijbiomac.2019.09.143
97. Zhou W, Zi L, Cen Y, You C, Tian M (2020) Copper sulfide nanoparticles-incorporated hyaluronic acid injectable hydrogel with enhanced angiogenesis to promote wound healing. Front Bioeng Biotechnol 8:417. https://doi.org/10.3389/fbioe.2020.00417
98. Tao B, Lin C, Deng Y, Yuan Z, Shen X, Chen M, He Y, Peng Z, Hu Y, Cai K (2019) Copper nanoparticles-embedded hydrogel for killing bacteria and promoting wound healing with photothermal therapy. J Mater Chem B 7:2534–2548. https://doi.org/10.1039/C8TB03272F
99. Ezealisiji KM, Siwe-Noundou KR (2019) Copper oxide nano-hydrogel composite and their toxicology studies: a green chemistry approach. J Mater Sci Nanotechnol 7(1):1–5
100. Mostakhdemin M, Nand A, Arjmandi M, Ramezani M (2020) Mechanical and microscopical characterisation of bilayer hydrogels strengthened by TiO_2 nanoparticles as a cartilage replacement candidate. Mater Today Commun 25:101279. https://doi.org/10.1016/j.mtcomm.2020.101279
101. Cross LM, Shah K, Palani S, Peak CW, Gaharwar AK (2018) Gradient nanocomposite hydrogels for interface tissue engineering. Nanomed Nanotechnol Biol Med 14(7):2465–2474. https://doi.org/10.1016/j.nano.2017.02.022
102. Ulu A, Birhanli E, Koytepe S, Ates B (2020) Chitosan/polypropylene glycol hydrogel composite film designed with TiO_2 nanoparticles: a promising scaffold of biomedical applications. Int J Biol Macromol 163:529–540. https://doi.org/10.1016/j.ijbiomac.2020.07.015
103. Toledo L, Racine L, Perez V, Henriquez JP, Auzely-Velty R, Urbano BF (2018) Physical nanocomposite hydrogels filled with low concentrations of TiO_2 nanoparticles: swelling, networks parameters and cell retention studies. Mater Sci Eng, C Mater Biol Appl 92:769–778. https://doi.org/10.1016/j.msec.2018.07.024
104. Raguvaran R, Manuja BK, Chopra M, Thakur R, Anand T, Kalia A, Manuja A (2017) Sodium alginate and gum acacia hydrogels of ZnO nanoparticles show wound healing effect on fibroblast cells. Int J Biol Macromol 96:185–191. https://doi.org/10.1016/j.ijbiomac.2016.12.009
105. Zhai M, Xu Y, Zhou B, Jing W (2018) Keratin-chitosan/n-ZnO nanocomposite hydrogel for antimicrobial treatment of burn wound healing: characterization and biomedical application. J Photochem Photobiol, B 180:253–258. https://doi.org/10.1016/j.jphotobiol.2018.02.018
106. Khorasani MT, Joorabloo A, Moghaddam A, Shamsi H, MansooriMoghadam Z (2018) Incorporation of ZnO nanoparticles into heparinised polyvinyl alcohol/chitosan hydrogels for wound dressing application. Int J Biol Macromol 114:1203–1215. https://doi.org/10.1016/j.ijbiomac.2018.04.010

107. Swaroop K, Somashekarappa HM (2018) In vitro biocompatibility and antibacterial activity of gamma ray crosslinked ZnO/PVA hydrogel nanocomposites. Mater Today: Proc 5(10):21314–21321. https://doi.org/10.1016/j.matpr.2018.06.534
108. Hoyo J, Ivanova K, Guaus E, Tzanov T (2019) Multifunctional ZnO NPs-chitosan-gallic acid hybrid nanocoating to overcome contact lenses associated conditions and discomfort. J Colloid Interface Sci 543:114–121. https://doi.org/10.1016/j.jcis.2019.02.043
109. Zhao P, Xu J, Zhang Y, Zhu W, Cui Y (2018) Polymerizable-group capped ZnS nanoparticle for high refractive index inorganic-organic hydrogel contact lens. Mater Sci Eng, C Mater Biol Appl 90:485–493. https://doi.org/10.1016/j.msec.2018.04.086
110. Radwan-Praglowska J, Janus L, Piatkowski M, Sierakowska A, Matysek D (2020) ZnO nanorods functionalized with chitosan hydrogels crosslinked with azelaic acid for transdermal drug delivery. Colloids Surf, B 194:111170. https://doi.org/10.1016/j.colsurfb.2020.111170
111. Swaroop Kumaraswamy GB, Badalamoole V, Mallaiah SH (2017) Gamma irradiation synthesis and in vitro drug release studies of ZnO/PVA hydrogel nanocomposites. Adv Mater Lett 8(1):2–7. https://doi.org/10.5185/amlett.2016.6819
112. Agban Y, Mugisho OO, Thakur SS, Rupenthal ID (2020) Characterization of zinc oxide nanoparticle cross-linked collagen hydrogels. Gels 6(4). https://doi.org/10.3390/gels6040037
113. Vashist A, Kaushik A, Vashist A, Sagar V, Ghosal A, Gupta YK, Ahmad S, Nair M (2018) Advances in carbon nanotubes-hydrogel hybrids in nanomedicine for therapeutics. Adv Healthcare Mater 7(9):e1701213. https://doi.org/10.1002/adhm.201701213
114. Shen H, Zhang L, Liu M, Zhang Z (2012) Biomedical applications of graphene. Theranostics 2(3):283–294. https://doi.org/10.7150/thno.3642
115. Foo ME, Gopinath SCB (2017) Feasibility of graphene in biomedical applications. Biomed Pharmacotherapy 94:354–361. https://doi.org/10.1016/j.biopha.2017.07.122
116. Singh DP, Herrera CE, Singh B, Singh S, Singh RK, Kumar R (2018) Graphene oxide: an efficient material and recent approach for biotechnological and biomedical applications. Mater Sci Eng, C Mater Biol Appl 86:173–197. https://doi.org/10.1016/j.msec.2018.01.004
117. Tadyszak K, Wychowaniec JK, Litowczenko J (2018) Biomedical applications of graphene-based structures. Nanomaterials 8(11). https://doi.org/10.3390/nano8110944
118. Dasari Shareena TP, McShan D, Dasmahapatra AK, Tchounwou PB (2018) A review on graphene-based nanomaterials in biomedical applications and risks in environment and health. Nanomicro Lett 10(3). https://doi.org/10.1007/s40820-018-0206-4
119. Beg S, Rizwan M, Sheikh AM, Hasnain MS, Anwer K, Kohli K (2011) Advancement in carbon nanotubes: basics, biomedical applications and toxicity. J Pharm Pharmacol 63(2):141–163. https://doi.org/10.1111/j.2042-7158.2010.01167.x
120. Saliev T (2019) The advances in biomedical applications of carbon nanotubes C5(2):29. https://doi.org/10.3390/c5020029
121. Potts JR, Dreyer DR, Bielawski CW, Ruoff RS (2011) Graphene-based polymer nanocomposites. Polymer 52(1):5–25. https://doi.org/10.1016/j.polymer.2010.11.042
122. Pattnaik S, Swain K, Lin Z (2016) Graphene and graphene-based nanocomposites: biomedical applications and biosafety. J Mater Chem B 4(48):7813–7831. https://doi.org/10.1039/c6tb02086k
123. Fisher CE, Rider A, Jun Han Z, Kumar S, Levchenko I, Ostrikov K (2012) Applications and nanotoxicity of carbon nanotubes and graphene in biomedicine. J Nanomater 2012:1–19. https://doi.org/10.1155/2012/315185
124. Prajapati SK, Malaiya A, Kesharwani P, Soni D, Jain A (2020) Biomedical applications and toxicities of carbon nanotubes. Drug Chem Toxicol 1–16. https://doi.org/10.1080/01480545.2019.1709492
125. Nanda SS, Papaefthymiou GC, Yi DK (2015) Functionalization of graphene oxide and its biomedical applications. Crit Rev Solid State Mater Sci 40(5):291–315. https://doi.org/10.1080/10408436.2014.1002604
126. Silva M, Alves NM, Paiva MC (2018) Graphene-polymer nanocomposites for biomedical applications. Polym Adv Technol 29(2):687–700. https://doi.org/10.1002/pat.4164

127. Sajid MI, Jamshaid U, Jamshaid T, Zafar N, Fessi H, Elaissari A (2016) Carbon nanotubes from synthesis to in vivo biomedical applications. Int J Pharm 501(1–2):278–299. https://doi.org/10.1016/j.ijpharm.2016.01.064
128. Li Y, Wang J, Yang Y, Shi J, Zhang H, Yao X, Chen W, Zhang X (2021) A rose bengal/graphene oxide/PVA hybrid hydrogel with enhanced mechanical properties and light-triggered antibacterial activity for wound treatment. Mater Sci Eng, C Mater Biol Appl 118:111447. https://doi.org/10.1016/j.msec.2020.111447
129. Shafiee A, Kehtari M, Zarei Z, Soleimani M, Varshochian R, Ahmadi A, Atyabi F, Dinarvand R (2021) An in situ hydrogel-forming scaffold loaded by PLGA microspheres containing carbon nanotube as a suitable niche for neural differentiation. Mater Sci Eng, C 120:111739. https://doi.org/10.1016/j.msec.2020.111739
130. Wang L, Lu R, Hou J, Nan X, Xia Y, Guo Y, Meng K, Xu C, Wang X, Zhao B (2020) Application of injectable silk fibroin/graphene oxide hydrogel combined with bone marrow mesenchymal stem cells in bone tissue engineering. Colloids Surf, A 604:125318. https://doi.org/10.1016/j.colsurfa.2020.125318
131. Qiao K, Guo S, Zheng Y, Xu X, Meng H, Peng J, Fang Z, Xie Y (2018) Effects of graphene on the structure, properties, electro-response behaviors of GO/PAA composite hydrogels and influence of electro-mechanical coupling on BMSC differentiation. Mater Sci Eng, C Mater Biol Appl 93:853–863. https://doi.org/10.1016/j.msec.2018.08.047
132. Fatima SW, Barua S, Sardar M, Khare SK (2020) Immobilization of Transglutaminase on multi-walled carbon nanotubes and its application as bioinspired hydrogel scaffolds. Int J Biol Macromol 163:1747–1758. https://doi.org/10.1016/j.ijbiomac.2020.09.091
133. Jing X, Mi H-Y, Napiwocki BN, Peng X-F, Turng L-S (2017) Mussel-inspired electroactive chitosan/graphene oxide composite hydrogel with rapid self-healing and recovery behavior for tissue engineering. Carbon 125:557–570. https://doi.org/10.1016/j.carbon.2017.09.071
134. Bellingeri R, Mulko L, Molina M, Picco N, Alustiza F, Grosso C, Vivas A, Acevedo DF, Barbero CA (2018) Nanocomposites based on pH-sensitive hydrogels and chitosan decorated carbon nanotubes with antibacterial properties. Mater Sci Eng, C Mater Biol Appl 90:461–467. https://doi.org/10.1016/j.msec.2018.04.090
135. Wei C, Dong X, Zhang Y, Liang J, Yang A, Zhu D, Liu T, Kong D, Lv F (2018) Simultaneous fluorescence imaging monitoring of the programmed release of dual drugs from a hydrogel-carbon nanotube delivery system. Sens Actuators, B Chem 273:264–275. https://doi.org/10.1016/j.snb.2018.06.064
136. Ursu EL, Gavril G, Morariu S, Pinteala M, Barboiu M, Rotaru A (2020) Single-walled carbon nanotubes-G-quadruple hydrogel nanocomposite matrixes for cell support applications. Mater Sci Eng, C Mater Biol Appl 111:110800. https://doi.org/10.1016/j.msec.2020.110800
137. He Z, Jiang R, Long W, Huang H, Liu M, Chen J, Deng F, Zhou N, Zhang X, Wei Y (2020) The combination of Diels-Alder reaction and redox polymerization for preparation of functionalized CNTs for intracellular controlled drug delivery. Mater Sci Eng, C Mater Biol Appl 109:110442. https://doi.org/10.1016/j.msec.2019.110442
138. Dong X, Wei C, Liang J, Liu T, Kong D, Lv F (2017) Thermosensitive hydrogel loaded with chitosan-carbon nanotubes for near infrared light triggered drug delivery. Colloids Surf, B 154:253–262. https://doi.org/10.1016/j.colsurfb.2017.03.036
139. Chowdhury MA (2017) Metal-organic-frameworks for biomedical applications in drug delivery, and as MRI contrast agents. J Biomed Mater Res, Part A 105(4):1184–1194. https://doi.org/10.1002/jbm.a.35995
140. El Hankari S, Bousmina M, El Kadib A (2019) Biopolymer@Metal-organic framework hybrid materials: a critical survey. Prog Mater Sci 106:100579. https://doi.org/10.1016/j.pmatsci.2019.100579
141. Lee Y-R, Kim J, Ahn W-S (2013) Synthesis of metal-organic frameworks: a mini review. Korean J Chem Eng 30(9):1667–1680. https://doi.org/10.1007/s11814-013-0140-6
142. Yang J, Yang YW (2020) Metal-organic frameworks for biomedical applications. Small 16(10):e1906846. https://doi.org/10.1002/smll.201906846

143. Shen M, Forghani F, Kong X, Liu D, Ye X, Chen S, Ding T (2020) Antibacterial applications of metal-organic frameworks and their composites. Comprehensive reviews in food science and food safety 19(4):1397–1419. https://doi.org/10.1111/1541-4337.12515
144. Gulcay E, Erucar I (2020) Metal-organic frameworks for biomedical applications173–210. https://doi.org/10.1016/b978-0-12-817650-4.00006-1
145. Wyszogrodzka G, Marszalek B, Gil B, Dorozynski P (2016) Metal-organic frameworks: mechanisms of antibacterial action and potential applications. Drug Discovery Today 21(6):1009–1018. https://doi.org/10.1016/j.drudis.2016.04.009
146. Keskin S, Kızılel S (2011) Biomedical applications of metal organic frameworks. Ind Eng Chem Res 50(4):1799–1812. https://doi.org/10.1021/ie101312k
147. Wang H-S, Wang Y-H, Ding Y (2020) Development of biological metal–organic frameworks designed for biomedical applications: from bio-sensing/bio-imaging to disease treatment. Nanoscale Advances 2(9):3788–3797. https://doi.org/10.1039/d0na00557f
148. Sun Y, Zheng L, Yang Y, Qian X, Fu T, Li X, Yang Z, Yan H, Cui C, Tan W (2020) Metal–organic framework nanocarriers for drug delivery in biomedical applications. Nano-Micro Lett 12(1). https://doi.org/10.1007/s40820-020-00423-3
149. Chu C, Su M, Zhu J, Li D, Cheng H, Chen X, Liu G (2019) Metal-organic framework nanoparticle-based biomineralization: a new strategy toward cancer treatment. Theranostics 9(11):3134–3149. https://doi.org/10.7150/thno.33539
150. Huaizhi Liu HP, Yumeng X, Jiuyang Z, Metal-organic frameworks: a universal strategy towards super-elastic hydrogels. Polym Chem 10:2263–2272. https://doi.org/10.1039/C9PY00085B
151. Schmidt B (2020) Metal-organic frameworks in polymer science: polymerization catalysis, polymerization environment, and hybrid materials. Macromol Rapid Commun 41(1):e1900333. https://doi.org/10.1002/marc.201900333
152. Zhong J, Kankala RK, Wang SB, Chen AZ (2019) Recent advances in polymeric nanocomposites of metal-organic frameworks (MOFs). Polymers 11(10). https://doi.org/10.3390/polym11101627
153. Wang L, Xu H, Gao J, Yao J, Zhang Q (2019) Recent progress in metal-organic frameworks-based hydrogels and aerogels and their applications. Coord Chem Rev 398:213016. https://doi.org/10.1016/j.ccr.2019.213016
154. Chen W-H, Liao W-C, Sohn YS, Fadeev M, Cecconello A, Nechushtai R, Willner I (2018) Stimuli-responsive nucleic acid-based polyacrylamide hydrogel-coated metal-organic framework nanoparticles for controlled drug release. Adv Func Mater 28(8):1705137. https://doi.org/10.1002/adfm.201705137
155. Pandey A, Kulkarni S, Vincent AP, Nannuri SH, George SD, Mutalik S (2020) Hyaluronic acid-drug conjugate modified core-shell MOFs as pH responsive nanoplatform for multimodal therapy of glioblastoma. Int J Pharm 588:119735. https://doi.org/10.1016/j.ijpharm.2020.119735
156. Kim K, Lee S, Jin E, Palanikumar L, Lee JH, Kim JC, Nam JS, Jana B, Kwon TH, Kwak SK, Choe W, Ryu JH (2019) MOF x biopolymer: collaborative combination of metal-organic framework and biopolymer for advanced anticancer therapy. ACS Appl Mater Interfaces 11(31):27512–27520. https://doi.org/10.1021/acsami.9b05736
157. Gupta V, Mohiyuddin S, Sachdev A, Soni PK, Gopinath P, Tyagi S (2019) PEG functionalized zirconium dicarboxylate MOFs for docetaxel drug delivery in vitro. J Drug Deliv Sci Technol 52:846–855. https://doi.org/10.1016/j.jddst.2019.06.003
158. Tan G, Zhong Y, Yang L, Jiang Y, Liu J, Ren F (2020) A multifunctional MOF-based nanohybrid as injectable implant platform for drug synergistic oral cancer therapy. Chem Eng J 390:124446. https://doi.org/10.1016/j.cej.2020.124446
159. Han D, Li Y, Liu X, Li B, Han Y, Zheng Y, Yeung KWK, Li C, Cui Z, Liang Y, Li Z, Zhu S, Wang X, Wu S (2020) Rapid bacteria trapping and killing of metal-organic frameworks strengthened photo-responsive hydrogel for rapid tissue repair of bacterial infected wounds. Chem Eng J 396:125194. https://doi.org/10.1016/j.cej.2020.125194

160. Gwon K, Han I, Lee S, Kim Y, Lee DN (2020) Novel Metal-Organic framework-based photocrosslinked hydrogel system for efficient antibacterial applications. ACS Appl Mater Interfaces 12(18):20234–20242. https://doi.org/10.1021/acsami.0c03187
161. Rossos AK, Banti CN, Kalampounias AG, Papachristodoulou C, Kordatos K, Zoumpoulakis P, Mavromoustakos T, Kourkoumelis N, Hadjikakou SK (2020) pHEMA@AGMNA-1: a novel material for the development of antibacterial contact lens. Mater Sci Eng, C Mater Biol Appl 111:110770. https://doi.org/10.1016/j.msec.2020.110770
162. Zhang S, Ye J, Sun Y, Kang J, Liu J, Wang Y, Li Y, Zhang L, Ning G (2020) Electrospun fibrous mat based on silver (I) metal-organic frameworks-polylactic acid for bacterial killing and antibiotic-free wound dressing. Chem Eng J 390:124523. https://doi.org/10.1016/j.cej.2020.124523
163. Fan Y, Zhang Y, Zhao Q, Xie Y, Luo R, Yang P, Weng Y (2019) Immobilization of nano Cu-MOFs with polydopamine coating for adaptable gasotransmitter generation and copper ion delivery on cardiovascular stents. Biomaterials 204:36–45. https://doi.org/10.1016/j.biomaterials.2019.03.007
164. Mazloom-Jalali A, Shariatinia Z, Tamai IA, Pakzad SR, Malakootikhah J (2020) Fabrication of chitosan-polyethylene glycol nanocomposite films containing ZIF-8 nanoparticles for application as wound dressing materials. Int J Biol Macromol 153:421–432. https://doi.org/10.1016/j.ijbiomac.2020.03.033
165. Yu Y, Chen G, Guo J, Liu Y, Ren J, Kong T, Zhao Y (2018) Vitamin metal–organic framework-laden microfibers from microfluidics for wound healing. Mater Horiz 5(6):1137–1142. https://doi.org/10.1039/c8mh00647d
166. Gao N, Huang J, Wang L, Feng J, Huang P, Wu F (2018) Ratiometric fluorescence detection of phosphate in human serum with a metal-organic frameworks-based nanocomposite and its immobilized agarose hydrogels. Appl Surf Sci 459:686–692. https://doi.org/10.1016/j.apsusc.2018.08.092
167. Lian X, Yan B (2018) Diagnosis of penicillin allergy: a MOFs-based composite hydrogel for detecting beta-lactamase in serum. Chem Commun 55(2):241–244. https://doi.org/10.1039/c8cc08245f
168. Wu T, Huang J, Jiang Y, Hu Y, Ye X, Liu D, Chen J (2018) Formation of hydrogels based on chitosan/alginate for the delivery of lysozyme and their antibacterial activity. Food Chem 240:361–369. https://doi.org/10.1016/j.foodchem.2017.07.052
169. Wei L, Cai C, Lin J, Wang L, Zhang X (2011) Degradation controllable biomaterials constructed from lysozyme-loaded Ca-alginate microparticle/chitosan composites. Polymer 52(22):5139–5148. https://doi.org/10.1016/j.polymer.2011.09.006
170. Xin T, Mao J, Liu L, Tang J, Wu L, Yu X, Gu Y, Cui W, Chen L (2020) Programmed sustained release of recombinant human bone morphogenetic protein-2 and inorganic ion composite hydrogel as artificial periosteum. ACS Appl Mater Interfaces 12(6):6840–6851. https://doi.org/10.1021/acsami.9b18496
171. Parmar PA, Chow LW, St-Pierre JP, Horejs CM, Peng YY, Werkmeister JA, Ramshaw JA, Stevens MM (2015) Collagen-mimetic peptide-modifiable hydrogels for articular cartilage regeneration. Biomaterials 54:213–225. https://doi.org/10.1016/j.biomaterials.2015.02.079
172. Nakaji-Hirabayashi T, Kato K, Iwata H (2012) Improvement of neural stem cell survival in collagen hydrogels by incorporating laminin-derived cell adhesive polypeptides. Bioconjug Chem 23(2):212–221. https://doi.org/10.1021/bc200481v
173. Yang Y, Zhao H, Jia Y, Guo Q, Qu Y, Su J, Lu X, Zhao Y, Qian Z (2016) A novel gene delivery composite system based on biodegradable folate-poly (ester amine) polymer and thermosensitive hydrogel for sustained gene release. Sci Rep 6:21402. https://doi.org/10.1038/srep21402

Chapter 18
Nanotechnology and Its Applications: Insight into Bacteriological Interactions and Bacterial Gene Transfer

Saba Yousaf, Hitesh Chopra, Muhammad Arbaz Khan, Faheem Mustafa, Mohammad Amjad Kamal, and Atif Amin Baig

Nanotechnology combines science, technology, and engineering at a 1–100 nm scale. It is the study and application of tiny things, and it can be used in other science domains such as biology, chemistry, physics, engineering, and material science. The concept and idea of nanotechnology and nanoscience are not new; the idea started back in the 1950s. The idea was actually brought by physicist Richard Feynman at an American physical society meeting at California Institute of Technology with a talk on "**There's Plenty of Room at the Bottom**" on December 29, 1959. The process of manipulating and controlling atoms and molecules individually was explained by Feynman, and after the development of a scanning microscope that could see atoms at the individual level, the beginning of modern nanotechnology occurred.

The term nanotechnology was invented by Norio Tanguichi of Tokyo Science University in 1974; the definition stands as the introductory statement "*nanotechnology mainly consists of the processing of separation, consolidation, and deformation of materials by one atom or one molecule*." The term nanotechnology, also defined by The U.S National Nanotechnology Initiative (NNI), is as follows "the understanding and control of matter at dimensions between approximately 1

S. Yousaf · M. A. Khan · F. Mustafa
School of Health Sciences, University of Management and Technology, Lahore, Pakistan

H. Chopra
Chitkara College of Pharmacy, Chitkara University, Rajpura, India

M. A. Kamal
King Fahd Medical Research Center, King Abdulaziz University, P.O. Box 80216, Jeddah 21589, Saudi Arabia

Enzymoics, Novel Global Community Educational Foundation, 7 Peterlee Place, Hebersham, NSW 2770, Australia

F. Mustafa · A. A. Baig (✉)
Faculty of Medicine, Universiti Sultan Zainal Abidin, Kampung Gong Badak, Malaysia
e-mail: atifamin@unisza.edu.my

S. Gopi et al. (eds.), *Nanotechnology for Biomedical Applications*,
Materials Horizons: From Nature to Nanomaterials,
https://doi.org/10.1007/978-981-16-7483-9_20

and 100 nm, where unique phenomena enable novel nanotechnology applications. Nanotechnology involves imaging, measuring, modeling, and manipulating matter at this length scale, enabling nanoscale science, engineering, and technology. A nanometer is one-billionth of a meter. A paper sheet is about 100,000 nm thick; a single gold atom is about a third of a nanometer in diameter. Dimensions between approximately 1 and 100 nm are known as the nanoscale. Unusual physical, chemical, and biological properties can emerge in materials at the nanoscale. These properties may differ in essential ways from the properties of bulk materials and single atoms or molecules."

Nanotechnology has earned the status of one of the critical research ventures in the twenty-first century. Scientists utilize the distinctive characteristics of molecular and atomic accumulations at the nanometer scale in the 1–100 nm range. At this scale, the things such as "nanoparticles" that cash the novel characteristics and functions that are dramatically different from those seen in the bulk scale [1]. Nanoparticles are the key component of nanotechnology. These particles lie between 1 and 100 nm in size and are probably made up of organic matter, carbon, metal, or metal oxides [2]. These particles have distinct magnetic, electrical, optical, and thermal characteristics in a nano-range [3].

The surface adaptability, enhanced solubility, multifunctionality, and the small size of nanoparticles attract scientists for many new research types [1]. Due to their unique properties, nanoparticles are used in different scientific areas such as pharmaceutical, biomedical, cosmetic, energy, environmental, optoelectronic, electronic, magnetic, catalytic, and material applications. The investment in nanotechnology and development has increased widely in these decades due to their uses [4].

The nanoparticles have different structures, shapes, and sizes. The particles can be cylindrical, tubular, hollow core, conical, spiral, flat, or spherical in shape. They can be flat, irregular, and different in size from 1 to 100 nm. The particles can have a uniform or irregular surface. Nanoparticles can be 0D, 1D, 2D, or 3D; it depends on their shape. A zero-dimensional nanoparticle has fixed length, breadth, and height at a single point; for example, nanodots; one dimensional nanoparticle has one parameter, for example, grapheme; two dimensional nanoparticle processes breadth and length, for example, nanotubes; while three dimensional nanoparticle has length, breadth, and height, all these parameters, for example, gold nanoparticles [5, 6].

The nanoparticle structure is not that simple as we think; according to studies, the particles are composed of three layers—the surface layer, operationalized with different small molecules, metal ions polymers, and surfactants. The shell layer is different from the core layer chemically in all aspects. The core is the nanoparticle's central and main part; it is actually considered the nanoparticle itself [7].

Nanoparticles are divided into four main categories [8].

1 Carbon-Based Nanoparticles

The nanoparticles totally made up of carbon are categorized as carbon-based nanoparticles. These nanoparticles are present in different forms like carbon nanotubes (CNTs), carbon nanofibers, fullerenes, carbon black, graphene, and carbon onions [9].

2 Organic Nanoparticles

These nanoparticles are mostly prepared from organic matter, except carbon-based nanoparticles. These organic nanoparticles can be transformed into desired forms such as dendrimers, liposomes, micelles, and polymers nanoparticles due to their weak interactions for self-assembly and design.

3 Inorganic Nanoparticles

These nanoparticles include metal and metal oxide nanoparticles, metal nanoparticles such as Au or Ag nanoparticles, a metal oxide such as TiO_2 and ZnO NPs, and silicon and ceramics as semiconductors.

4 Composite-Based Nanoparticles

The composite nanoparticles are multiphase with one phase on the nanoscale dimension that can be combined with another nanoparticle, larger material like hybrid nanofibers, and these nanoparticles can be a complex structure like metal–organic framework [8].

5 Origin-Based Classification of Nanoparticles

The nanoparticles can be divided into two groups depending on their origin.

6 Natural

The nanoparticles are produced in nature by anthropogenic activities or biological species. The natural nanoparticles are found in the earth's sphere, such as in the atmosphere, hydrosphere, lithosphere, and biosphere, other than human activities. Nanoparticles are naturally present in oceans, lakes, rivers, microorganisms, and higher organisms, including humans and in the atmosphere [10, 11].

7 Synthetic Nanoparticles

Synthetic nanoparticles are the engineered nanoparticles produced by engine exhaust, smoke, and mechanical grinding, rarely synthesized by hybrid, biological or chemical methods. There may be a risk of risk assessment due to enhanced creation and delivery of engineered nanoparticles. Does the synthetic nanoparticle exhibit a distinct environment-related behavior, different forms of naturally occurring nanoparticles, is the major challenge among engineered nanoparticles? At present, there are various sources related to potential applications being used for producing synthetic nanoparticles [12].

8 Methods of Nanoparticles Preparation

There are several methods used for nanoparticle preparation and well will discuss some mostly used methods. Majorly, there are three divisions as chemical, physical, and biogenic methods [13].

9 Chemical Methods

9.1 Polyol Method

In this method, the nano-aqueous liquid (polyol) is used as a solvent and reducing agent. The polyol has an edge for minimizing the oxidation surface and accumulation. By this method, the shape, size, and texture of the nanoparticles can be controlled [14].

This can be used for synthesizing oxide nanoparticles at increased temperature and precise particle growth control [15].

Several studies suggest using oxide nanoparticles by this method, including Mn_3O_4, $CoTiO_3$, SnO_2, PbO, ZnO, and TiP_2 [16–22].

Ethylene glycol is the most commonly used polyol liquid to make metal oxide nanoparticles through polyol method [23]. The polyol method has also been used for making bimetallic alloys and core–shell nanoparticles [24–26].

9.2 Thermal Decomposition

This method is the chemical decomposition by heat, which is also called thermolysis. The chemical bond of compound breaks; when the decomposition is exothermic, it generates a positive feedback loop for thermal runaway. Many studies explain the thermal decomposition process of different metals such as copper, zinc, cadmium, nitrates, metal acetates, lead, and rare earth metals for producing nanoparticles [27–29].

9.3 Electrochemical Synthesis

The method of preparing chemical compounds in an electrochemical cell is called an electrochemical synthesis method. This method is better than other chemical processes as it rejects the potential wasteful alternative half-reaction and precisely tunes the preferred potential (Wikipedia). Several studies reported silver, colloidal silver, red fluorescent silicon, long-lived silver, and platinum nanoparticle through the electrochemical method [30–35].

9.4 Microemulsions

The liquid in liquid dispersion is called an emulsion. Depending upon the size of droplets, these are macro-, mini-, and microemulsions [36]. The method of microemulsion synthesis is used for the preparation of inorganic nanoparticles widely [37]. The method of making inorganic nanoparticles through microemulsion is comprised of mixing two microemulsions containing a reducing agent and metallic salt. Some studies report that iron oxide III, zinc oxide, and chitosan nanoparticles can be formed using this method by adding the right microemulsions [38–40].

10 Physical Methods

10.1 Plasma

In this method, the plasma is created by radio frequency (RF) heating coils. The metal is enclosed in a Muller, and that Muller is then held in a chamber that is evacuated. The metal is heated above its evaporation point by the coils around the evacuated chamber. Helium (He) gas is used to form a high-temperature plasma when it floats in the coils. The metal vapor nucleates are collected by a cold collector where the nanoparticles being passivized by oxygen gas are collected [41].

10.2 Chemical Vapor Deposition

This process involves a chemical reaction and most commonly used for the synthesis of semiconductors. The process requires volatile precursor to which the substrate is exposed to decompose on it and make deposits that are desired. The precursor in vaporized form is kept into chemical vapor deposition reactor and adsorbed at high temperature onto the substrate. The molecules are absorbed and react with other molecules. This method can produce particles of less than 1 μm by chemical reaction that takes place gaseous phase [41].

10.3 Gamma Radiation

This is the most used method for the production of metallic nanoparticles. This method is preferred for the metallic synthesis because it is cheap, produces less toxins, reproducible, reaction occurs on room temperature, and completes in less steps with less wastage [42]. The gamma photons with high energy exited and ionize the metal ions in solvent [43].

Gold and silver nanoparticles have been produced by this technique that were further investigated by UV–visible spectroscopy [44–46].

10.4 Pulsed Laser Method

This method is widely used for the synthesis of silver nanoparticles. In this process, the solution of silver nitrate and a reducing agent were added in a blender-like device. The solid disk of the device rotates with solution. Hotspots are created on the disk by irradiating it with pulses from laser beam. The reducing agent in the device reacts with silver nitrate and produces silver particles that can be separated later by the process of centrifugation. The size of the particle can be controlled by laser energy and angular velocity of disk [41].

10.5 Sonochemical Reduction

This method is mostly used for metal nanoparticles synthesis. The process involves the reduction of metals in absence or presence of organic additives. There are many studies that report the synthesis of MnO4−, Au+, Pd2+, and platinum nanoparticles [47, 48].

10.6 Microwave Irradiation

The method is used for the synthesis of organic, inorganic, and hybrid materials [49].

This method is mostly used for making organic nanoparticles because it provides higher yields, enhanced purity and works ease with improved reaction rates. The synthesis of benzimidazolyl chalcone was done using this method [50]. Past studies report the synthesis of silver (AgNPs) and zinc (ZnS) by microwave irradiation method [51].

10.7 Biogenic Methods

Nanoparticles can be prepared using biological systems. Metal and metal oxide ions can be prepared by this method. In the biogenic method, the nanoparticles can be prepared using different bacteria and other microorganisms [52, 53].

Unlike bacteria, fungi, yeast, plants, and waste materials are also used for nanoparticle synthesis, as reported by different studies [54–57].

10.8 Applications of Nanoparticles

Nanoparticles found uses in almost every field of life due to their unique properties and large surface area; they are used in science, biology, medicine, physics, chemistry, cosmetics, industries, genetics, drug delivery, wastewater treatment, food, agriculture, catalysis, and environmental sciences, etc.

Nanoparticles are used in biosensing due to their unique physicochemical properties. The sensing of diseases, toxic substances, and other biological agents is essential for forensic analysis, environmental monitoring, and biomedical diagnosis. Here is a summary of nanoparticles applications (Table 1).

11 Nanoparticle and Transformation in Bacteria

This era has evolved from molecular to atomic-level studies and cell to genetics studies. The nanoparticles have found their applications in genetics. The particles affect the transfer of DNA into the bacterial cells from its surrounding. This process is called transformation.

The process by which the bacteria import and insert exogenous DNA into their genome is natural transformation (NT). The process is essential for horizontal gene transfer (HGT), and by this, the bacteria acquire new functions such as nutrition utilization and antibiotic resistance [76].

Table 1 Types of nanoparticles and their functions

Type of particle	Function	References
Gold nanoparticles	Biosensing agent	[58]
Metallic nanoparticles also	They possess the ability of quenching and photoluminescence	[59–63]
Nanoparticles	Nanoparticles are used in several molecular imaging techniques like magnetic resonance imaging (MRI), optical imaging (OI), positron emission tomography (PET), and ultrasound imaging (USI) They possess high uptake efficacy in target cells and prevent normal cells from interacting with the biological environment; improve cellular uptake and decreasing toxicity	[64, 65]
Metallic nanoparticles	Metallic nanoparticles synthesized by plant extracts or biomass are widely used for the cytotoxic effects against cancer cells	[66–68]
The silver and gold	Nanoparticles synthesized by guava and clove plant extracts have anti-cancer properties against many types of cancer such as leukemia, bone marrow cancer, colorectal adenocarcinoma, and human cervix cancer	[69]
Gold nanoparticles	Stable and versatile source of drug delivery for their distinctive size, chemical, and physical properties	[70]
Most commonly, gold and silver nanoparticles	Antimicrobial agents against pathogens	[71–73]
The metal nanoparticles of nanosized as gold, silver, and platinum	In personal care products like shampoo, anti-aging creams, perfumes, and soaps	[74]
	Also used in agriculture and crops, they decrease the microbial activity in wastewater effluents	[75]

Environmental, physiological, and genetic factors affect the transfer of genes in bacteria; now, more studies have been conducted in the last two decades to understand nanoparticles' role in gene transfer. Different studies report the role of nanoparticles in transfer of genes in bacteria at different environmental conditions.

The nanoparticles perform their role in gene transfer into the bacterial cells by different mechanisms. The following table is the elaboration of how nanoparticles increase the gene transfer in other bacterial species (Table 2).

The above table portrays nanoparticles' usage in gene transfer in bacteria, but many studies need to be conducted for many types of bacteria getting noticed as they are concerned with public health, such as *Streptococcus pneumonia*. The exact

Table 2 Role of nanoparticles in gene transfer

Nanoparticles	Affect	Method	References
Electrosprayed chitosan nanoparticles	Enhance bacterial transformation in *E. coli* (K12) strain	Due to their molecular size and physicochemical properties, they enhance the process of transformation	[77]
Silver nanoparticles	Increase the transformation efficiency in *E. coli* DH5α	The transformation efficiency varies with the concentration of silver nanoparticles	[78]
TiO and nanosized alumina	Promote horizontal transfer of plasmid-borne ARGs in pure bacteria cultures	By harming the cell membrane and improving the mating pair expression formation and transfer and replication of genes	[79, 80]
Zno	Increase horizontal gene transfer in *E. coli*	Enhance bacterial cell membrane permeability	[81]
Copper and copper oxide nanoparticles	Promote the transfer of antibiotic-resistance genes across genera of bacterial species, i.e., from *Escherichia coli* to Pseudomonas putida	Increase the reactive oxygen species (ROS) production	[82]
Silver and silver oxide nanoparticles	Enhance the transfer of plasmid-borne antibiotic resistance genes(ARG) across genera, i.e., from *E. coli* K-12LE392 to *Pseudomonas putida* KT2440	Increase oxidative stress in cells and membrane permeability	[83]
Natural saphelerite (NS)	Increase the horizontal gene transfer of antibiotic-resistance genes in bacteria	Accelerate the production of reactive oxygen species and damages cell membrane	[83]
Zinc oxide nanoparticle	Increase the process of transformation in *Bacillus subtilis*	Enhance the competence in bacteria	[84]
Carbon nanotubes	Increase transformation efficacy in *E. coli* cells	The carbon nanotubes attack on the surface of bacteria; it makes nanochannels temporarily across the cell wall and delivers plasmids into the cell	[85]

(continued)

Table 2 (continued)

Nanoparticles	Affect	Method	References
Gold, chloride, chitosan, multiwalled carbon nanotubes, sepiolite	The nanoparticles increase the efficiency of transformation in *Lactococcus lactis* subsp., *Enterococcus faecalis*, *Bacciluc* species, *Ralstonia eutropha*, and *Methulomonas* species	These nanoparticles help in piercing the cell membrane and increase the permeability of the membrane	[85]

physiology of how the nanoparticles take to DNA or plasmid into the cell is unknown and needs to be well-studied.

Bacteria themselves are reported to mediate the synthesis of gold and silver nanoparticles. The field of nanoscience and nanotechnology is ever-expanding, such that it has attracted researchers around the globe from various areas such as energy, medicine, drug delivery and designing, electronics. Nanotechnology generally deals with material in the range of few nanometers to 100 nm. They presented high surface to volume area and showed the size and shape-dependent properties. The critical application of nanotechnology-based application is the synthesis of nanoparticles. Many studies currently focus on developing nanoparticles using eco-friendly methods such as enzymes, microbial enzymes, vitamins, polysaccharides. The approach for the synthesis of nanoparticles using microbes has shown promising potential in the past years. They have been shown to develop monodispersed and highly stable nanoparticles. Moreover, the researchers use biomass or cell extracts of bacteria to synthesize nanoparticles [86].

Nanoparticles prepared using biogenic enzymatic processes were far better than those produced by chemical or physical approaches. Though nanoparticles' chemical or biological formation is advantageous in respect to the number of nanoparticles synthesized with definite reproducibility of shape and size and consumes less time, they are more complicated and are dangerous to human and wildlife health. The metallic ions are generally toxic for bacteria, resulting in the formation of a defense mechanism known as drug resistance. The bacterial population resides in varying degrees of temperature, pH, and pressure, resulting in more catalytic activity, specific area, and improved contact between enzyme and metal salt. The nanoparticles are synthesized in bacteria when they derive the target ions from the environment itself and turn the metal ions into elemental metal through the various cell processes. The formation of elemental ions may be classified based on the location of the formation of nanoparticles [87]. In the intracellular approach, the microbial cell wall consisting mainly of negatively charged species attracts the positively charged surrounding ions. Whereas in the case of the extracellular method, the reductase enzymes secreted by the cell result in bioreduction of metallic ions into corresponding metal nanoparticles (MtNP) [88].

12 Biological Synthesis of Metallic Nanoparticles

Gold nanoparticles: Gold nanoparticles (AuNPs) have been in use since the Roman Empire where it was primarily used for the decorative purposes. Its usage in modern era started about 150 years ago where it was observed that properties of colloidal gold were different in comparison to the bulk gold [89]. Since then, it has found use in sensors, photo-imaging applications, drug delivery, diagnosis, phototherapy, etc. The synthesis of AuNPs, by chemical process, consists of multistep reactions, expensive chemicals, and reagents. Moreover, stabilizing agents are added to cause stabilization over the lifetime. Furthermore, the chemical processes produce end products that are hazardous in nature. Furthermore, they are not so biocompatible in nature. They needed to have an extra layer of coating to make them biocompatible. Recently, researchers produced AuNPs using the bacterial strain of *B. Indica* DC1 isolated from the *Panax ginseng* rhizospheric soil [90]. The microbial synthesis of AuNPs was carried using the tryptic soy broth, and flower-shaped nanoparticles were formed. Das et al. studied the synthesis of AuNPs using *Rhizopus oryzae* mycelia, maintained on potato culture [91]. Thus, the AuNPs were of various shapes such as star, triangle, urchin, depending on nanoplates used for fabrication. Mishra and team [92] reported the fabrication of AuNPs using the *Penicillium brevicompactum,* which was found to possess cytotoxic effects on the mouse mayo blast cancer C_2C_{12} cells. The pH range of 5–8 was chosen as these conditions increased the concentration of gold salt and resulting in bigger sized and aggregated nanoparticles.

Barabadi et al. [93] studied the eco-friendly and cost-effective method for the AuNPs synthesis from the *Penicillium crustosum* isolated from soil. The NPs thus formed showed well-defined dimensions and monodispersity [93]. The AuNPs are generally accounted to be formed by a mechanism involving the reduction of Au^{3+} to AuNPs. Various enzymes are produced by microbes and play an essential role in NP nucleation and growth. He and colleagues [94] biosynthesized AuNPs using *Rhodopseudomonas capsulata* bacteria. It was also found that pH played an important role as synthesis performed at pH 7 resulted in the formation of spherical-based NPs, while at pH 4, it resulted in the formation of nanoplates shaped. Singh et al. [90] reported biosynthesis of AuNPs using the *Sporosarcina koreensis* DC4 via an extracellular pathway. The nanoparticles thus produced were useful in catalytic reduction of 4-nitrophenol to aminophenol using the sodium borohydride [90]. Syed et al. studied the synthesis of AuNPs using the *Pseudomonas fluorescens* 417 inhabiting *Coffea arabica* L. with size ranging from 5 to 50 nm spherical shaped. The NPs were shown to possess bactericidal activity against the pathogens such as *Pseudomonas aeroginosa* followed by *Escherichia coli*, *Staphylococcus aureus*, *Bacillus subtilis*, and *Klebsiella pneumoniae.*

Zhang and team reported the synthesis of AuNPs from yeast such as *Magnusiomyces ingens* LH-F1, and the end product thus formed was able to have excellent catalytic activity against the reduction of nitrophenols [95]. Narayan et al. reported the facile method of biosynthesis of AuNPs from the cell-free filtrate of fungus, *Sclerotium rolfsii.* During the study, the *S. rolfsii,* when incubated with the chloroauric acid, can synthesize the NPs within 15 min with the ambient conditions using the NADPH-dependent enzyme. The size and shape of NPs can be manipulated by causing the variation in the ratio of the metal salt and cell-free filtrate of *S. rolfsii*; as the concentration of filtrate was varied, it was found spherical-shaped triangles, hexagonal, and rod-shaped NPs were produced.

Fayaz et al. reported the synthesis of AuNPs using the thermophilic bacterium *Geobacillus stearothermophilus.* The TEM results showed the formation of monodispersed particles. The bacteria secreted capping enzymes that helped stabilize the AuNPs during their lifetime (Fayaz et al. 2011).

Gurunathan et al. studied the AuNPs using the *Ganoderma* spp. by treating with the $HAuCl_4$ with a particle size of 20 nm. The color changes were observed from yellow to reddish pink, and robust surface plasmon resonance was monitored at 520 nm. The NPs were shown to be non-cytotoxic and biocompatible, as results of MDS-MB-231 breast cancer cells suggested. Prema et al. studied the formation of AuNPs from the *Klebsiella pneumoniae* (MTCC-4030). SEM images showed the spherical nature of NPs and were widely dispersed. And, microbial studies showed that the NPs were effective against the *E. coli* and *S. aureus*. Researchers developed a green synthesis route for the AuNPs biosynthesis using the *Gracilaria verrucosa.* The FTIR results showed the presence of proteins and other phenolic compounds that helped in the stabilization of NPs.

In the case of *Lactobacillus casei,* it has been found that the concentration of glycolipids present varies with the formation of AuNPs. The lipids such as diglycosyldiacylglycerol (DGDG) and triglycosyldiacylglycerol (TGDG) were found to get reduced after the formation of NPs. It concludes that they play some role in reducing Au^{3+} to Au [96].

13 Silver Nanoparticles

Silver nanoparticles (AgNPs) are increasingly used in various fields such as medical, food, health care, industrial properties mainly due to their unique physical and chemical properties. They are mostly used as antibacterial agents, medical device coating, optical sensors, anti-cancer agents, etc.

Balakuraman et al. described the synthesis of AgNPs using the soil fungi, *Aspergillus terreus,* which was capable of synthesizing the NPs within 12 h. Thus, the NPs were 8–20 nm sized, spherical-shaped and found to possess antimicrobial properties against the *S. aureus, B. subtilis*. Due to leakage of intracellular components such as proteins and nucleic acid, it has also been found responsible for cell death and, hence, its antibacterial action [97].

Ibrahim et al. reported the method of producing AgNPs using *Bacillus cereus* from the contaminated soil. Researchers also optimized various variables in conditions such as $AgNO_3$, inoculum size, temperature, time, and pH. It was found that conditions such as $AgNO_3$ (1 Mm) 10 ML, 8.7 ml of inoculum, 48.5 °C, 69 h, and pH of 9 were optimized for the formation of AgNPs. The NPs were 5–7.06 nm in size and spherical in shape. The NPs were effective against the *Staphylococcus epidermidis, Staphylococcus aureus, Escherichia coli, Salmonella enterica, Porteus mirabilis* by disk diffusion method. Some bacteria may produce extracellular AgNPs when exposed to silver salts. It may precede by two processes such as biomolecules released by bacteria into external medium, which reduces silver ions to AgNPs. And, NPs formed inside the cell are excreted outside. The extracellular AgNPs may get themselves attached to the cell membrane, detached by mild sonication for the recovery procedure [98].

Researchers isolated bacterial isolates from the Kashan textile industry effluent [99]. The silver strain (strain N) was selected for the synthesis. The 16SrDNA sequence was found to be quite similar to the *Marinospirillum alkaliphilum*. The strain was able to decolorize the Disperse Blue 183 using the immobilized AgNPs with 100% efficiency. Das et al. isolated CS11 belonging to the Bacillus sp. bacterial strain from the heavy metal contaminated soil with a size range of 42–92 nm. The strain was able to form the NPs within 24 h. Feroze et al. developed novel method of synthesis of AgNPs using the fungal metabolites of *Penicillium oxalicum*. The antibacterial activity was evaluated against the *S. aureus*, *S. dysenteriae*, and *S. typhi* and was found to be effective by the inhibition zone and UV spectroscopy results. Aziz et al. developed method of AgNPs synthesis routed through *Mucor hiemalis* [100]. The AgNPS thus developed showed antimicrobial properties against the *Klebsiella pneumoniae, Pseudomonas brassicacearum, Aeromonas hydrophila, Escherichia coli, Bacillus cereus*, and *Staphylococcus aureus* and three pathological fungal strains (*Candida albicans, Fusarium oxysporum*, and *Aspergillus flavus*). They were able to act synergistically for acting against the microbial growth. Paul et al. described method of synthesis of AgNPs using the *Pseudomonas aeruginosa* KUPSB12, phosphate solubilizing bacterial strain isolated from effluent of jute mill from India. The NPs synthesized were of the size range of 50-85 nm. The NPs were effective against the various pathogenic bacteria such as *Escherichia coli* MTCC 443, *Vibrio cholerae* MTCC 3904, *Shigella flexneri* MTCC 1457, *Bacillus subtilis* MTCC 441, *Staphylococcus aureus* MTCC 3160, and *Micrococcus luteus* MTCC 1538.

Researchers described facile method of synthesis of AgNPs using fruit extract from the *Gmelina arborea* of size range 8-32 nm. They were found to be effective in catalytic degradation of methylene blue dye. Arya et al. developed method of synthesis of silver nanoparticles from extract of green alga *Botryococcus braunii*. The FTIR studies showed that all functional groups of nanoparticles participate in reduction and stabilization of nanoparticles. The NPs were crystalline in nature with FCC geometry. They were found to be effective against the gram negative strains of *Pseudomonsa aeruginosa* (MTCC441), *E. coli* (MTCC442) and gram positive strains of *Klebsiella pneumoniae* (MTCC 109) and *Staphylococcus aureus* (MTCC 96), and a fungal strain *Fusarium oxysporum* (MTCC 2087).

14 Conclusion

The chapter has been written to get a better insight and understanding of nanotechnology, nanoparticles, their types, synthesis methods, applications of different kinds of nanoparticles in different areas, how nanoparticles are useful in gene transfer in bacteria. Although the chapters cover many things and bring better picture of nanoparticles understanding, there is still a need to get intensified knowledge of nanoparticles and transformation. We need to know different processes and exact physiology of how the particles actually work in gene transfer in bacteria.

There are only certain types of bacteria used for checking nanoparticles impact on DNA transformation, i.e., *E. coli*, and there is a need to work on many other bacteria species. We need to work on some bacteria that contain importance as a concern for global public health; if we work on them, we may find effective and unique methods to fight the bacteria.

The use of nanoparticles in DNA uptake and bacteria transformation is the key stimulator toward the future advances in gene transfer. From the above information mentioned in this chapter, few studies have been conducted on many bacteria important for public health, such as Streptococcus pneumonia and influenza. Nanotechnology can be utilized to transfer genes of interest in the bacteria through transformation, which can be fruitful for treating this bacterial infection.

References

1. McNeil SE (2005) Nanotechnology for the biologist. J Leukoc Biol 78(3):585–594
2. Hasan S (2015) A review on nanoparticles: their synthesis and types. Res J Recent Sci 2277:2502
3. Luo C-H et al (2015) Nanoparticle biosynthesis using unicellular and subcellular supports. NPG Asia Mater 7(8):e209–e209
4. Roco MC (2003) Broader societal issues of nanotechnology. J Nanopart Res 5(3):181–189
5. Ealia SAM, Saravanakumar M (2017) A review on the classification, characterisation, synthesis of nanoparticles and their application. In: IOP conference series: materials science and engineering. IOP Publishing
6. Machado S et al (2015) Characterization of green zero-valent iron nanoparticles produced with tree leaf extracts. Sci Total Environ 533:76–81
7. Shin W-K et al (2016) Cross-linked composite gel polymer electrolyte using mesoporous methacrylate-functionalized SiO_2 nanoparticles for lithium-ion polymer batteries. Sci Rep 6(1):1–10
8. Jeevanandam J et al (2018) Review on nanoparticles and nanostructured materials: history, sources, toxicity and regulations. Beilstein J Nanotechnol 9(1):1050–1074
9. Kumar N, Kumbhat S (2016) Essentials in nanoscience and nanotechnology. Wiley Online Library
10. Hochella MF et al (2015) Nanotechnology: nature's gift or scientists' brainchild? Environ Sci Nano 2(2):114–119
11. Sharma VK et al (2015) Natural inorganic nanoparticles–formation, fate, and toxicity in the environment. Chem Soc Rev 44(23):8410–8423
12. Wagner S et al (2014) Spot the difference: engineered and natural nanoparticles in the environment—release, behavior, and fate. Angew Chem Int Ed 53(46):12398–12419

13. Avramescu SM et al (2020) Engineered nanomaterials: Health and Safety, BoD–Books on Demand
14. Meshesha BT et al (2009) Polyol mediated synthesis & characterization of Cu nanoparticles: effect of 1-hexadecylamine as stabilizing agent. Nanotechnology 1:87–91
15. Quievryn C et al (2014) Polyol-based synthesis of praseodymium oxide nanoparticles. Nanomater Nanotechnol 4:7
16. Flores-Gonzalez M et al (2005) Preparing nanometer scaled Tb-doped Y_2O_3 luminescent powders by the polyol method. J Solid State Chem 178(4):989–997
17. Jiang X et al (2003) Monodispersed spherical colloids of titania: synthesis, characterization, and crystallization. Adv Mater 15(14):1205–1209
18. Jiang X et al (2004) Ethylene glycol-mediated synthesis of metal oxide nanowires. J Mater Chem 14(4):695–703
19. Lee S et al (2008) ZnO nanoparticles with controlled shapes and sizes prepared using a simple polyol synthesis. Superlattices Microstruct 43(4):330–339
20. Ragupathy P et al (2008) Preparation of nanostrip V2O5 by the polyol method and its electrochemical characterization as cathode material for rechargeable lithium batteries. J Phys Chem C 112(42):16700–16707
21. Sicard L et al (2010) Polyol synthesis and magnetic study of Mn_3O_4 nanocrystals of tunable size. J Magn Magn Mater 322(18):2634–2640
22. Siemons M, Simon U (2007) Gas sensing properties of volume-doped $CoTiO_3$ synthesized via polyol method. Sens Actuators, B Chem 126(2):595–603
23. Sun Y et al (2002) Uniform silver nanowires synthesis by reducing $AgNO_3$ with ethylene glycol in the presence of seeds and poly (vinyl pyrrolidone). Chem Mater 14(11):4736–4745
24. Bonet F et al (2003) Synthesis and characterization of bimetallic Ni–Cu particles. J Solid State Chem 172(1):111–115
25. Garcia-Gutierrez DI et al (2005) Temperature effect on the synthesis of Au–Pt bimetallic nanoparticles. J Phys Chem B 109(9):3813–3821
26. Tekaia-Elhsissen K et al (1999) Finely divided platinum–gold alloy powders prepared in ethylene glycol. J Alloy Compd 292(1–2):96–99
27. Arshad M et al (2008) Thermal decomposition of metal complexes of Type MLX_2 (M= Co (II), Cu (II), Zn (II), and Cd (II); L= DIE; X= NO_3^{1-}) by TG-DTA-DTG techniques in air atmosphere. Turk J Chem 32(5):593–604
28. Baykara S (2004) Hydrogen production by direct solar thermal decomposition of water, possibilities for improvement of process efficiency. Int J Hydrogen Energy 29(14):1451–1458
29. Patil K et al (1968) Infrared spectra and thermal decompositions of metal acetates and dicarboxylates. Can J Chem 46(2):257–265
30. Cele T (2020) Preparation of nanoparticles. Engineered Nanomaterials-Health and Safety, IntechOpen
31. Choi J et al (2014) Electrochemical synthesis of red fluorescent silicon nanoparticles. Bull Korean Chem Soc 35(1):35–38
32. Islam M, Islam MS (2013) Electro-deposition method for platinum nano-particles synthesis. Saidul, Electro-Deposition Method for Platinum Nano-Particles Synthesis
33. Khaydarov RA et al (2009) Electrochemical method for the synthesis of silver nanoparticles. J Nanopart Res 11(5):1193–1200
34. Rodriguez-Sanchez L et al (2000) Electrochemical synthesis of silver nanoparticles. J Phys Chem B 104(41):9683–9688
35. Sharma S et al (2014) Synthesis of Silver Nanoparticles by 'Electrochemical Route' through pure metallic Silver electrodes, and evaluation of their Antimicrobial Activities. Int J Pharm Sci Rev Res 28(2):272–277
36. Tauer K, Nazaran P (2010) Mechanism and modeling of emulsion polymerization: new ideas and concepts–1. Particle nucleation. In: Macromolecular symposia. Wiley Online Library
37. Dong Y et al (2010) Controllable synthesis of $CaCO_3$ micro/nanocrystals with different morphologies in microemulsion. Chem Res Chin Univ 26(5):678–682

38. Maitra A et al (1999) Process for the preparation of highly monodispersed polymeric hydrophilic nanoparticles. Google Patents
39. Sarkar D et al (2011) Formation of zinc oxide nanoparticles of different shapes in water-in-oil microemulsion. Colloids Surf, A 381(1–3):123–129
40. Wongwailikhit K, Horwongsakul S (2011) The preparation of iron (III) oxide nanoparticles using W/O microemulsion. Mater Lett 65(17–18):2820–2822
41. Cele T (2018) Synthesis of platinum group metals nanoparticles by gamma radiolysis
42. Rao Y et al (2010) Gamma irradiation route to synthesis of highly re-dispersible natural polymer capped silver nanoparticles. Radiat Phys Chem 79(12):1240–1246
43. Abidi W, Remita H (2010) Gold based nanoparticles generated by radiolytic and photolytic methods. Recent Patents Eng 4(3):170–188
44. Chen Q et al (2010) Radiolytic syntheses of nanoparticles and inorganic-polymer hybrid microgels
45. Hornebecq V et al (2003) Stable silver nanoparticles immobilized in mesoporous silica. Chem Mater 15(10):1993–1999
46. Krklješ A (2011) Radiolytic synthesis of nanocomposites based on noble metal nanoparticles and natural polymer, and their application as biomaterial
47. Obreja L et al (2008) Alcoholic reduction platinum nanoparticles synthesis by chemical method, biomaterials in biophysics. Med Phys Ecol
48. Okitsu K, Nishimura R (2010) Sonochemical reduction method for controlled synthesis of metal nanoparticles in aqueous solutions. In: Proceedings of the 20th International Congress on Acoustics, ICA, Sydney, Australia (Citeseer)
49. Horikoshi S, Serpone N (2013) Introduction to nanoparticles. In: Microwaves in nanoparticle synthesis: fundamentals and applications, pp 1–24
50. 정유진 (2015) The mechanism of catalyzing reduction of 4-nitrophenol by quasimetallic nanoparticles and the critical function of NaBH4, 서울대학교 대학원
51. Tiwary K et al (2013) Structural and optical properties of ZnS nanoparticles synthesized by microwave irradiation method. Chalcogenide Lett 10(9):319
52. Shivaji S et al (2011) Extracellular synthesis of antibacterial silver nanoparticles using psychrophilic bacteria. Process Biochem 46(9):1800–1807
53. Stephen JR, Macnaughtont SJ (1999) Developments in terrestrial bacterial remediation of metals. Curr Opin Biotechnol 10(3):230–233
54. Akhtar MS et al (2013) Biogenic synthesis of metallic nanoparticles by plant extracts. ACS Sustain Chem Eng 1(6):591–602
55. Arumugam P, Berchmans S (2011) Synthesis of gold nanoparticles: an ecofriendly approach using *Hansenula anomala*. ACS Appl Mater Interfaces 3(5):1418–1425
56. Kanchi S et al (2018) Exploitation of de-oiled jatropha waste for gold nanoparticles synthesis: a green approach. Arab J Chem 11(2):247–255
57. San Chan Y, Don MM (2013) Biosynthesis and structural characterization of Ag nanoparticles from white rot fungi. Mater Sci Eng, C 33(1):282–288
58. Jain PK et al (2006) Calculated absorption and scattering properties of gold nanoparticles of different size, shape, and composition: applications in biological imaging and biomedicine. J Phys Chem B 110(14):7238–7248
59. Lakowicz JR (2005) Radiative decay engineering 5: metal-enhanced fluorescence and plasmon emission. Anal Biochem 337(2):171–194
60. Sapsford KE et al (2006) Materials for fluorescence resonance energy transfer analysis: beyond traditional donor–acceptor combinations. Angew Chem Int Ed 45(28):4562–4589
61. Thomas KG, Kamat PV (2003) Chromophore-functionalized gold nanoparticles. Acc Chem Res 36(12):888–898
62. Van Dijk MA et al (2005) Far-field optical microscopy of single metal nanoparticles. Acc Chem Res 38(7):594–601
63. Zheng J et al (2004) Highly fluorescent, water-soluble, size-tunable gold quantum dots. Phys Rev Lett 93(7):077402

64. Margolis DJ et al (2007) Molecular imaging techniques in body imaging. Radiology 245(2):333–356
65. Wang S et al (2014) Application of nanotechnology in improving bioavailability and bioactivity of diet-derived phytochemicals. J Nutr Biochem 25(4):363–376
66. Anand K et al (2015) Agroforestry waste Moringa oleifera petals mediated green synthesis of gold nanoparticles and their anti-cancer and catalytic activity. J Ind Eng Chem 21:1105–1111
67. Kuppusamy P et al (2016) Biosynthesis of metallic nanoparticles using plant derivatives and their new avenues in pharmacological applications—an updated report. Saudi Pharma J 24(4):473–484
68. Sathishkumar M et al (2015) Biosynthesis of gold nanoparticles and related cytotoxicity evaluation using A549 cells. Ecotoxicol Environ Saf 114:232–240
69. Raghunandan D et al (2011) Anti-cancer studies of noble metal nanoparticles synthesized using different plant extracts. Cancer Nanotechnol 2(1):57–65
70. Han G et al (2007) Drug and gene delivery using gold nanoparticles. NanoBiotechnology 3(1):40–45
71. Islam NU et al (2019) Pistacia integerrima gall extract mediated green synthesis of gold nanoparticles and their biological activities. Arab J Chem 12(8):2310–2319
72. Khatami M et al (2015) Synthesis of silver nanoparticles using seed exudates of *Sinapis arvensis* as a novel bioresource, and evaluation of their antifungal activity. Bioresources Bioprocess 2(1):1–7
73. Sadeghi B (2015) *Zizyphus mauritiana* extract-mediated green and rapid synthesis of gold nanoparticles and its antibacterial activity. J Nanostruct Chem 5(3):265–273
74. Kumar V, Yadav SK (2009) Plant-mediated synthesis of silver and gold nanoparticles and their applications. J Chem Technol Biotechnol: Int Res Process Environ Clean Technol 84(2):151–157
75. Durán N et al (2007) Antibacterial effect of silver nanoparticles produced by fungal process on textile fabrics and their effluent treatment. J Biomed Nanotechnol 3(2):203–208
76. Ambur OH et al (2016) Steady at the wheel: conservative sex and the benefits of bacterial transformation. Philos Trans R Soc B: Biol Sci 371(1706):20150528
77. Abyadeh M et al (2017) Electrosprayed chitosan nanoparticles: facile and efficient approach for bacterial transformation. Int Nano Lett 7(4):291–295
78. Nagamani G et al (2019) A novel approach for increasing transformation efficiency in *E. coli* DH5α cells using silver nanoparticles. 3 Biotech 9(3):1–7
79. Qiu Z et al (2015) Effects of nano-TiO_2 on antibiotic resistance transfer mediated by RP4 plasmid. Nanotoxicology 9(7):895–904
80. Qiu Z et al (2012) Nanoalumina promotes the horizontal transfer of multiresistance genes mediated by plasmids across genera. Proc Natl Acad Sci 109(13):4944–4949
81. Wang X et al (2018) Bacterial exposure to ZnO nanoparticles facilitates horizontal transfer of antibiotic resistance genes. NanoImpact 10:61–67
82. Zhang S et al (2019) Copper nanoparticles and copper ions promote horizontal transfer of plasmid-mediated multi-antibiotic resistance genes across bacterial genera. Environ Int 129:478–487
83. Lu J et al (2020) Both silver ions and silver nanoparticles facilitate the horizontal transfer of plasmid-mediated antibiotic resistance genes. Water Res 169:115229
84. Eymard-Vernain E et al (2018) Impact of nanoparticles on the Bacillus subtilis (3610) competence. Sci Rep 8(1):1–8
85. Ren J et al (2018) Optimization of chemico-physical transformation methods for various bacterial species using diverse chemical compounds and nanomaterials. J Biotechnol 288:55–60
86. Korbekandi H, Iravani S, Abbasi S (2009) Production of nanoparticles using organisms. Crit Rev Biotechnol 29(4):279–306
87. Simkiss K, Wilbur KM (1989) Biomineralization. Academic, New York, NY
88. Hulkoti NI, Taranath T (2014) Biosynthesis of nanoparticles using microbes—a review. Colloids Surf B 121:474–483. https://doi.org/10.1016/j.colsurfb.2014.05.027

89. Hayat MA (1989) Colloidal gold: principles, methods, and applications. Academic Press, San Diego
90. Singh P, Singh H, Kim YJ, Mathiyalagan R, Wang C, Yang DC (2016) Extracellular synthesis of silver and gold nanoparticles by *Sporosarcina koreensis* DC4 and their biological applications. Enzyme Microb Technol 86:75–83
91. Das SK, Das AR, Guha AK (2010) Microbial synthesis of multishaped gold nanostructures. Small 6(9):1012–1021
92. Mishra A, Tripathy SK, Wahab R et al (2011) Microbial synthesis of gold nanoparticles using the fungus *Penicillium brevicompactum* and their cytotoxic effects against mouse mayo blast cancer C2C12 cells. Appl Microbiol Biotechnol 92:617–630. https://doi.org/10.1007/s00253-011-3556-0
93. Barabadi H, Honary S, Ebrahimi P, Mohammadi MA, Alizadeh A, Naghibi F (2014) Microbial mediated preparation, characterization and optimization of gold nanoparticles. Braz J Microbiol 45(4):1493–1501
94. He S, Guo Z, Zhang Y, Zhang S, Wang J, Gu N (2007) Biosynthesis of gold nanoparticles using the bacteria *Rhodopseudomonas capsulata*. Mater Lett 61(18):3984–3987
95. Zhang X, Qu Y, Shen W, Wang J, Li H, Zhang Z, Li S, Zhou J (2016) Biogenic synthesis of gold nanoparticles by yeast Magnusiomyces ingens LH-F1 for catalytic reduction of nitrophenols. Colloids Surfaces A: Physicochem Eng Aspects 497:280–285
96. Fumiya K et al (2016) Formation of gold nanoparticles by glycolipids of *Lactobacillus casei*. Sci Rep (Nature Publisher Group) 6(1)
97. Balakumaran MD et al (2016) Mycosynthesis of silver and gold nanoparticles: optimization, characterization and antimicrobial activity against human pathogens. Microbiol Res 182:8–20
98. Ibrahim S, Ahmad Z, Manzoor MZ et al (2021) Optimization for biogenic microbial synthesis of silver nanoparticles through response surface methodology, characterization, their antimicrobial, antioxidant, and catalytic potential. Sci Rep 11:770
99. Nazari N, Kashi FJ (2021) A novel microbial synthesis of silver nanoparticles: its bioactivity, Ag/Ca-Alg beads as an effective catalyst for decolorization Disperse Blue 183 from textile industry effluent. Separat Purificat Technol 259:118117
100. Aziz N et al (2016) Leveraging the attributes of Mucor Hiemalis-derived silver nanoparticles for a synergistic broad-spectrum antimicrobial platform. Front Microbiol 7:1984
101. Chellapandian C et al (2019) Gold nanoparticles using red seaweed *Gracilaria verrucosa*: green synthesis, characterization and biocompatibility studies. Process Biochem 80:58–63
102. Das VL et al (2014) Extracellular synthesis of silver nanoparticles by the Bacillus strain CS 11 isolated from industrialized area. 3 Biotech 4(2):121–126
103. Feroze N et al (2020) Fungal mediated synthesis of silver nanoparticles and evaluation of antibacterial activity. Microscopy Res Tech 83(1):72–80
104. Gurunathan S et al (2014) A green chemistry approach for synthesizing biocompatible gold nanoparticles. Nanoscale Res Lett 9(1):1–11
105. Mann S (2001) Biomineralization: principles and concepts in bioinorganic materials chemistry. Oxford University Press, Oxford
106. Li G et al (2020) Natural sphalerite nanoparticles can accelerate horizontal transfer of plasmid-mediated antibiotic-resistance genes. Environ Int 136:105497
107. Parikh RY et al (2011) Genus-wide physicochemical evidence of extracellular crystalline silver nanoparticles biosynthesis by Morganella spp. PLoS One 6(6):e21401
108. Saha J et al (2017) A novel green synthesis of silver nanoparticles and their catalytic action in reduction of Methylene Blue dye. Sustain Environ Res 27(5):245–250
109. Singh P et al (2016) Microbial synthesis of flower-shaped gold nanoparticles. Artif Cells Nanomed Biotechnol 44(6):1469–1474
110. Paul D, Sinha SN (2014) Extracellular synthesis of silver nanoparticles using *Pseudomonas aeruginosa* KUPSB12 and its antibacterial activity. Jordan J Biol Sci 7(4):245–250

111. Prema P, Iniya PA, Immanuel G (2016) Microbial mediated synthesis, characterization, antibacterial and synergistic effect of gold nanoparticles using *Klebsiella pneumoniae* (MTCC-4030). RSC Adv 6(6):4601–4607
112. Wikipedia (2016) Electrochemical_synthesis. From https://en.wikipedia.org/wiki/Electrochemical_synthesis

The manufacturer's authorised representative in the EU is Springer Nature Customer Service Centre GmbH, Europaplatz 3, 69115 Heidelberg, Germany. If you have any concerns regarding our products, please contact ProductSafety@springernature.com

Printed and bound by CPI Group (UK) Ltd, Croydon, CR0 4YY
01/10/2026
02226185-0009